PLANT BIOTECHNOLOGY

VOLUME 2

Transgenics, Stress Management, and Biosafety Issues

PLANT BIOTECHNOLOGY

VOLUME 2

Transgenics, Stress Management, and Biosafety Issues

Edited by

Sangita Sahni, PhD

Bishun Deo Prasad, PhD

Prasant Kumar, PhD

Apple Academic Press Inc.
3333 Mistwell Crescent
Oakville, ON L6L 0A2 Canada

Apple Academic Press Inc.
9 Spinnaker Way
Waretown, NJ 08758 USA

First issued in paperback 2021

Plant Biotechnology (2-volume set)
ISBN-13: 978-1-77463-111-9 (pbk)
ISBN-13: 978-1-77188-581-2 (hbk)
ISBN-13: 978-1-77188-582-9 (2-volume set)

Library and Archives Canada Cataloguing in Publication

Library and Archives Canada Cataloguing in Publication
Plant biotechnology (Oakville, Ont.)
Plant biotechnology / edited by Bishun Deo Prasad, PhD, Sangita Sahni, PhD, Prasant Kumar, PhD, Mohammed Wasim Siddiqui, PhD.
Volume 2 edited by Sangita Sahni, PhD, Bishun Deo Prasad, PhD, and Prasant Kumar, PhD.
Includes bibliographical references and indexes.
Contents: Volume 2. Transgenics, stress management, and biosafety issues.
Issued in print and electronic formats.
ISBN 978-1-77188-581-2 (v. 2 : hardcover).--ISBN 978-1-315-21373-6 (v. 2 : PDF)
1. Plant biotechnology. I. Prasad, Bishun Deo, editor II. Sahni, Sangita, editor III. Kumar, Prasant, editor IV. Siddiqui, Mohammed Wasim, editor V. Title.
TP248.27.P55P63 2017 660.6 C2017-905057-5 C2017-905058-3

Library of Congress Cataloging-in-Publication Data

Names: Sahni, Sangita, editor.
Title: Plant biotechnology. Volume 2, Transgenics, stress management, and biosafety issues / editors: Sangita Sahni, Bishun Deo Prasad, Prasant Kumar.
Other titles: Transgenics, stress management, and biosafety issues
Description: Waretown, NJ : Apple Academic Press, 2017. | Includes bibliographical references and index.
Identifiers: LCCN 2017037913 (print) | LCCN 2017046510 (ebook) | ISBN 9781315213736 (ebook) | ISBN 9781771885812 (hardcover : alk. paper)
Subjects: LCSH: Plant biotechnology.
Classification: LCC TP248.27.P55 (ebook) | LCC TP248.27.P55 P55452 2017 (print) | DDC 630--dc23
LC record available at https://lccn.loc.gov/2017037913

Apple Academic Press also publishes its books in a variety of electronic formats. Some content that appears in print may not be available in electronic format. For information about Apple Academic Press products, visit our website at **www.appleacademicpress.com** and the CRC Press website at **www.crcpress.com**

ABOUT THE EDITORS

Sangita Sahni, PhD

Dr. Sangita Sahni is a Junior Scientist and Assistant Professor in the Department of Plant Pathology, Tirhut College of Agriculture, Dholi, Rajendra Agricultural University, Pusa, Samastipur, Bihar, India. She has published several research papers in reputed peer-reviewed national and international journals. She has published two authored book and several book chapters. She has isolated several bacterial isolates from different sources and submitted their sequences to the National Center for Biotechnology Information (NCBI).

Dr. Sahni acquired a BSc (Agriculture) degree from A.N.G.R.A.U, Hyderabad, India, and an MSc (Agriculture) in Mycology and Plant Pathology from Banaras Hindu University, Varanasi, India. She received her PhD (Agriculture) in Plant Pathology from the B.H.U, Varanasi. Subsequently, she worked as a postdoctoral research fellow at the University of Western Ontario University, London, Ontario, Canada.

Dr. Sahni has been awarded with the Dr. Rajendra Prasad National Education Shikhar Award for outstanding contribution in the field of education, a Young Scientist Award in 2014 from the Society for Scientific Development in Agriculture and Technology (SSDAT), and an Innovative Scientist of the Year Award, 2015, from the Scientific Education Research Society for outstanding contribution in the field of Plant Pathology. She is a Principal Investigator in All India Co-ordinated Research Programme at MULLaRP and Chickpea Pathology at T.C.A., Dholi. She is an officer in-charge of ARIS cell, TCA, Dholi, and a member of different committees of RAU, Pusa. She has been an active member of the organizing committees of several national and international seminars.

Dr. Sahni has been associated with molecular host–pathogen interaction studies in *Arabidopsis* and *B. napus*. She is also associated with pathological aspect of chickpea and MULLaRP. She is actively involved in teaching graduate and post-graduate courses in Plant Pathology and Biotechnology. She has proved herself as an active scientist in the area of Molecular Plant Pathology.

Bishun Deo Prasad, PhD

Dr. Bishun Deo Prasad is an Assistant Professor and Scientist in the Department of Molecular Biology and Genetic Engineering, Bihar Agricultural University, Sabour, India. He has published several research papers in reputed peer-reviewed international journals which have been cited more than 100 times. He has also contributed to two authored book, has written several book chapters, and has submitted 10 sequences of different isolates to the National Center for Biotechnology Information (NCBI). He is a reviewer of the *International Journal of Agriculture Sciences* and *Journal of Environmental Biology.*

Dr. Prasad has received the DAE—Young Scientist Research award in 2013 and the Fast Track Scheme for Young Scientists award by the Department of Science and Technology (DST), India, in 2012. He has also been awarded with an Outstanding Achievement Award in 2014 from the Society for Scientific Development in Agriculture and Technology (SSDAT) and an Inventor of the Year Award, 2015 in the discipline of Molecular Biology and Genetic Engineering from the Society of Scientific and Applied Research Centre at an international conference (iCiAsT-2016) held at the Faculty of Science, Kasetsart University, Bangkok, Thailand in 2016.

Dr. Prasad acquired his BSc (Agriculture) degree from MPKV, Rahuri, Maharashtra, India, MSc (Agricultural Biotechnology) from Assam Agricultural University and his PhD from M. S. University from Baroda, Gujarat, India, with a thesis in the field of Plant Biotechnology. He also worked at the John Innes Centre (JIC), Norwich, UK, during his PhD. Subsequently, he worked as a postdoctoral research fellow at the University of Western Ontario University, London, Ontario, Canada. He also worked at V.M.S.R.F., Bangalore, as a Scientist and S. D. Agricultural University, Gujarat, as an Assistant Professor. He has received grants from various funding agencies to carry out his research projects. He is a member secretary of Biosafety Committee and member of different committees of Bihar Agricultural University, Sabour.

Dr. Prasad has been associated with biotechnological aspects of rice, *Brassica napus*, *Arabidopsis*, linseed, lentil, vegetable (bitter guard and pointed guard), and horticultural (mango, litchi, and banana) crops. He is also associated with host–pathogen interaction studies in rice, *B. napus*, and *Arabidopsis* as well as mutational breeding aspect in rice for abiotic stress tolerance. He is dynamically involved in teaching graduate and post-graduate courses of Biotechnology, Plant Breeding and Genetics, Vegetable Crops, and Horticultural Crops.

Prasant Kumar, PhD

Dr. Prasant Kumar is an Assistant Professor at the C. G. Bhakta Institute of Biotechnology, Department of Fundamental and Applied Science at Uka Tarsadia University, Surat, Gujarat, India, and is the author or co-author of 10 peer-reviewed journal articles and eight conference papers and a newsletter.

He is a reviewer and editorial board member of several peer-reviewed journals. He has been an active member of the organizing committees of several national and international seminars and conferences.

Dr. Kumar received a BSc (Agriculture) from Acharya N. G. Ranga Agriculture University through the all India combined entrance exam conducted by the Indian Council of Agriculture Research, India. After graduating from Acharya N. G. Ranga Agriculture University, he was selected for the MSc Biotechnology program of The Maharaha Sayajirao University of Baroda, Gujarat, through the all India combined biotechnology entrance exam conducted by Department of Biotechnology (Govt. of India) and Jawaharlal Nehru University, New Delhi. Along with completion of his postgraduation, with first class with distinction in Biochemistry, he qualified GATE, ICMR-JRF, UGC-NET exam of national repute. Later, he joined the PhD program in Biochemistry from The Maharaha Sayajirao University of Baroda. He was awarded an Indian Council of Medical Research Fellowship Award for the PhD from the Indian Council of Medical research, New Delhi, India. He worked as an Assistant Professor in Sardar Patel University, Anand, Gujarat, from August 2011 to June 2012.

CONTENTS

LIST OF CONTRIBUTORS

Anupam Adarsh
Department of Horticulture, Bihar Agricultural University, Sabour, Bihar 813210, India

Subhankar Roy-Barman
Department of Biotechnology, National Institute of Technology, Durgapur 713209, India. E-mail: subhankarroy.barman@bt.nitdgp.ac.in

Chandrashekar S. Y.
College of Horticulture, Mudigere, Andhra Pradesh 577132, India. E-mail: chandrashekar.sy@gmail.com

Sanjukta Chakraborty
Department of Biotechnology, National Institute of Technology, Durgapur 713209, India

Nandlal Choudhary
Amity institute of Virology and Immunology, Amity University, Noida, Uttar Pradesh 201313, India

Kishore Dey
Florida department of Agriculture and Consumer Sciences (FDACS), North Port, FL, USA

Shikha Dhatwalia
Amity Institute of Virology and Immunology, Amity University, Noida, Uttar Pradesh 201313, India

John Hu
Department of Plant and Environmental Protection Sciences, University of Hawaii, Honolulu, Hawaii, USA

Aruna Jangid
Amity institute of Virology and Immunology, Amity University, Noida, Uttar Pradesh 201313, India

Vijay Kumar Jha
Department of Botany, Patna University, Patna, Bihar, India

Chandan Kishore
Department of Plant Breeding and Genetics, BAU, Sabour, Bhagalpur 813210, Bihar, India

Deepak Kumar
Research and Development Division, Shri Ram Solvent Extractions Pvt. Ltd., Jaspur, Uttarakhand, India

Pankaj Kumar
Department of Plant Breeding and Genetics, Bihar Agricultural University, Sabour, Bihar 813210, India

Prasant Kumar
C.G. Bhakta Institute of Biotechnology, Department of Fundamental and Applied Science, Uka Tarsadia University, Bardoli, Surat 394350, Gujarat, India

Randhir Kumar
Department of Horticulture (Vegetable and Floriculture), Bihar Agricultural University, Sabour, Bihar 813210, India

Ravi Ranjan Kumar
Department of Molecular Biology and Genetic Engineering, Bihar Agricultural University, Sabour, Bihar 813210, India. E-mail: ravi1709@gmail.com

Vinod Kumar
Department of Molecular Biology and Genetic Engineering, Bihar Agricultural University, Sabour, Bihar 813210, India. E-mail: biotech.vinod@gmail.com

Mahesh Kumar
Department of Molecular Biology and Genetic Engineering, Bihar Agricultural University, Sabour, Bihar 813210, India. E-mail: maheshkumara2zbau@gmail.com

Michael Melzer
Department of Plant and Environmental Protection Sciences, University of Hawaii, Honolulu, HI, USA

Nimmy M. S.
National Research Centre on Plant Biotechnology, New Delhi 110012, India

Varsha C. Mohanan
Genome Research Centre, Faculty of Science, The M. S. University of Baroda, Vadodara, Gujarat 390002, India. E-mail: varshaschyk@gmail.com

Suhail Muzaffar
National Centre for Biological Sciences, GKVK Campus, Bellary Road, Bangalore 560065, India. E-mail: suhail.bt@gmail.com

Hemla Naik B.
College of Horticulture, Mudigere, Andhra Pradesh 577132, India

Nagateja Natra
Department of Plant Pathology, Irrigated Agriculture Research & Extension Center, Washington State University, 24106 N. Bunn Road, Prosser, WA 99350, United States

Ganesh Patil
Vidya Pratisthan's College of Agriculture Biotechnology, Vidyanagari, Baramati 413133, India

Ankush Prasad
Biomedical Engineering Research Center, Tohoku Institute of Technology, Sendai, Japan; Department of Biophysics, Centre of the Region Haná for Biotechnological and Agricultural Research, Faculty of Science, Palacký University, Olomouc, Czech Republic

Bishun Deo Prasad
Department of Plant Breeding and Genetics, Bihar Agricultural University, Sabour, Bihar-813210, India. E-mail: dev.bishnu@gmail.com

Kumari Rajani
Department of Seed technology, Bihar Agricultural University, Sabour, Bihar 813210, India

Ravi Ranjan
Department of Molecular Biology and Biotechnology, BAU, Sabour, Bhagalpur 813210, Bihar, India

Tushar Ranjan
Department of Basic Science and Humanities Genetics, Bihar Agricultural University, Sabour, Bihar 813210, India. E-mail: mail2tusharranjan@gmail.com

Rasmi R.
College of Horticulture, Mudigere, Andhra Pradesh 577132, India

Ravindra A. Raut
Department of Biotechnology, National Institute of Technology, Durgapur 713209, India

Nazmiara Sabnam
Department of Biotechnology, National Institute of Technology, Durgapur 713209, India

Pallabi Saha
Department of Biotechnology, National Institute of Technology, Durgapur 713209, India

Tamoghna Saha
Department of Entomology, Bihar Agricultural University, Sabour, Bihar 813210, India

Sangita Sahni
Department of Plant Pathology, T.C.A, Dholi, Muzaffarpur, Bihar, India

Atrayee Sarkar
Department of Biotechnology, National Institute of Technology, Durgapur 713209, India

Md. Shamim
Department of Molecular Biology and Genetic Engineering, Bihar Agricultural University, Sabour, Bihar 813210, India

Vaishali Sharma
DOS in Biotechnology, University of Mysore, Mysuru, Karnataka, India

Abhishek Mani Tripathi
Global Change Research Institute, Academy of Sciences of the Czech Republic, Brno, Czech Republic

P. Verma
Division of Crop Protection, ICAR—Central Institute of Cotton Research, Nagpur 440010, Maharashtra, India

Deepak Kumar Yadav
Department of Biophysics, Centre of the Region Haná for Biotechnological and Agricultural Research, Faculty of Science, Palacký University, Olomouc, Czech Republic. E-mail: deepak.yadav@ceitec.muni.cz

Shailesh Yadav
International Rice Research Institute (IRRI), ICRISAT, Hyderabad 502324, India

LIST OF ABBREVIATIONS

2,4-D	2,4-dichloro pheonoxyacetic acid
2D-GE	two-dimensional gel electrophoresis
ABA	abscisic acid
ACC	1-aminocyclopropane-1-carboxylic acid
AFLP	amplified fragment length polymorphism
AM	arbuscular mycorrhizal
AMP	adenosine monophosphate
AR	Amplex Red
ATMT	*Agrobacterium tumefaciens*-mediated transformation
AUR	Amplex Ultra Red
BAP	6-benzylaminopurine
BC	backcross population
Bt	*Bacillus thuringiensis*
CIM	callus induction medium
CRISPR	clustered regularly interspaced short palindromic repeats
DAB	3,3-diaminobenzidine
DAF	DNA amplification fingerprinting
DAGT	diacylglycerol acyltransferase
DHL	double haploid lines
DNA	deoxyribonucleic acid
ELISA	enzyme linked immunosorbent assay
EPR	electron paramagnetic resonance
ETS	expressed tagged sites
FACS	fluorescence activated cell sorting
GA3	gibberelic acid
GBS	genotyping-by-sequencing
GC–MS	gas chromatography–mass spectrometry
GC–TOF–MS	gas chromatography–time-of-flight–mass spectrometry
GM	genetic modification
GMOs	genetically modified organisms
IAA	indole-3-acetic acid
IBA	indole-3-butyric acid
ILs	introgression Lines
ISSR	inter-simple sequence repeats

KASPar	KBioscience competitive allele specific PCR
Kn	kinetin
LC–MS	liquid chromatography–mass spectrometry
LD	linkage disequilibrium
MAGIC	multiparent advanced generation intercrosses
MAS	marker-assisted selection
miRNA	microRNA
mRNA	messenger RNA
MS	mass spectrometry
MS	Murashige and Skoog
NAA	α-naphthalene acetic acid
NADP	nicotinamide adenine dinucleotide phosphate
NBT	nitroblue tetrazolium
NGS	next generation sequencing
NILs	near-isogenic lines
NMR	nuclear magnetic resonance
PAGE	polyacrylamide gels
PCD	programmed cell death
PCR	polymerase chain reaction
PEG	polyethylene glycol
PGPR	plant growth promoting rhizobacteria
PHA	polyhydroxyalkanoate
PHB	poly(3-hydroxybutyrate)
PMC	pollen mother cell
PTGS	posttranscriptional gene silencing
PVP	polyvinylpyrrolidone
QRT-PCR	quantitative real-time PCR
QTL	quantitative trait locus
RAD	restriction site-associated DNA
RAPD	rapid amplified polymorphic DNA
REMI	restriction enzyme-mediated integration
RFLP	restriction fragment length polymorphism
RILs	recombinant inbred lines
RISC	RNA-induced silencing complex
RNAi	RNA interference
ROS	reactive oxygen species
RT-PCR	reverse transcription PCR
SCARs	sequence characterized amplified regions
SDS	sodium dodecyl sulphate
SIM	simple interval mapping

siRNAs	small interfering RNAs
SNP	single nucleotide polymorphism
SSRs	simple sequence repeats
STMs	sequence-tagged microsatellites
STR	simple tandem repeats
STSs	sequence-tagged sites
TAIL-PCR	thermal asymmetric interlaced PCR
Ti	tumor inducing
TILLING	targeting induced local lesions IN genomes
T-RFLP	terminal restriction fragment length polymorphism
UV	ultra violet
VIGS	virus-induced gene silencing
vir	virulence

PREFACE

Recent advances in the development of transgenic plants revolutionized our concepts of sustainable food production, cost-effective alternative energy strategies, microbial biofertilizers and biopesticides, disease diagnostics, etc. through plant biotechnology. As a result, a number of transgenic plants have been developed with improved traits. With the advancement of plant biotechnology many of the customary approaches are out of date, and understandings on new updated approaches are needed. To this end, this book has been written to share the information related to recent methods of genetic transformation, gene silencing, development of transgenic crops, biosafety issues, microbial biotechnology, oxidative stress, and plant disease diagnostics and management. This book comprises 13 chapters dealing with various aspect of plant biotechnology. Chapters 1–3 provide an in-depth knowledge of various techniques of genetic transformation of plants, chloroplast, and fungus. Chapter 4 describes advances in gene silencing in plant. Recently, gene silencing has rapidly become the method of choice for functional genomics analysis in plants. Importantly, Chapter 4 discusses specifically on recent advances in virus-induced gene silencing (VIGS) in plants. VIGS technique has been extensively used in plant reverse genetics studies.

Over the last three decades, a large number of transgenic plants have been developed across different classes of plants with various traits of interest. In order to give the comprehensive discussion on transgenic plants, five chapters (Chapters 6–10) are incorporated in the book which elaborately discuss the transgenic plants for various traits. Chapter 6 describes on transgenic plants and their application in crop improvement. Chapter 7 discusses intensively on genetically modified foods and biodiesel production. Chapters 8 and 9 describe biotechnological approaches in horticultural and ornamental plants, respectively. Chapter 10 has extensively described about the recent advances in the development of transgenic crop plants. Transgenic plants have unprecedented potential in crop improvement; however, several cancers are still associated with release of transgenic crops. In Chapter 10 also discusses on biosafety aspect associated with transgenic crops. Chapter 11 deals with microbial biotechnology in which role of microbes in sustainable agriculture has been discussed in depth. Chapter 12 describes about the oxidative stress in plants. Finally, Chapter 13 discusses about biotechnological approaches for plant disease diagnosis and management.

This book is a blend of basics, new advances and their application so that students don't feel either it's very basic or too advanced. The aim of this book will be to nurture the graduate and postgraduate students in the field of biotechnology. Each chapter has been written by one or more leading experts in their field and then carefully edited to ensure thoroughness and consistency.

In the end, we would like to emphasize that though every possible care has been exercised in writing and proofreading the book, still we don't claim to be infallible. Thus, suggestions for further improvement from teachers, researchers, and students (our real strength) would be gratefully acknowledged.

—Sangita Sahni
—Bishun Deo Prasad
—Prashant Kumar

ACKNOWLEDGMENT

At the end of editing this book, I close my eyes and remember the day when the idea of writing this book was seeded in my mind, followed by discussion about this with my other colleagues, which led to the foundation of this project. From that initial day to now, when we are finally publishing our book, there have been several ups and downs. However, with blessings of "Almighty God," we were able to convert our ideas, teaching, and research experiences to a logical end in the form of this book. Therefore, first of all, we would like to thank "Almighty God" from whom all blessings come. Further, I would like to express my gratitude to the many people who saw us through this book; to all those who provided support, talked things over, read, wrote, offered comments, allowed me to quote their remarks, and assisted in the editing, proofreading, and design. I would like to thank Dr. Tusar Ranjan and Dr. Mitesh Dwivedi for helping us in the process of editing this book.

With a profound and unfading sense of gratitude, I wish to express our sincere thanks to the Bihar Agricultural University, India, for providing me with the opportunity and facilities to execute such an exciting project and for supporting me toward research and other intellectual activities around the globe.

We feel privileged to acknowledge our immense sense of devotion to our parents and family members for their infinitive love, cordial affection, and incessant inspiration. Last not least: we beg forgiveness of all those who have been with us during the course of writing this book and whose names we have failed to mention.

PART I
Genetic Transformation

CHAPTER 1

DIRECT AND INDIRECT METHODS OF GENE TRANSFER IN PLANTS

NANDLAL CHOUDHARY*, ARUNA JANGID, and SHIKHA DHATWALIA

Amity institute of Virology and Immunology, Amity University, Noida 201313, UP, India

**Corresponding author. E-mail: nandlalc@gmail.com*

CONTENTS

ABSTRACT

Transformation is a molecular biology method to alter the genetic material of cells by incorporation of desired foreign genetic material by direct and indirect methods. The Agrobacterium tumefaciens and A. rhizogenes mediated are the direct transformation methods whereas, the protoplast transformation, electroporation, particle bombardment, microinjection, sonoporation, lipofection, calcium phosphate, laser transfection, chloroplast transformation, and mediated transformation are the indirect methods. The plant transformation has become a versatile method for the incorporation of desired characteristics in the selected plant for the benefits of human society. Agrobacterium tumefaciens is the most successful and popular technique compared to other physical/mechanical techniques. Agrobacterium infection has been used for transfer of foreign DNA into a number of dicotyledonous species (utilizing its plasmids as vectors). Arabidopsis thaliana was stably transformed with high efficiency using T-DNA transfer by agrobacterium. Agrobacterium-mediated transformation using the floral dipping method is the most widely used method to transform Arabidopsis. Genetic transformation mediated by agrobacterium is a simple and comparatively less expensive than other methods of transformation. Transgenic crop obtained through agrobacterium-mediated genetic transformation have better fertility percentage.

1.1 INTRODUCTION

Plants are important sources of many important products, such as food, fibers, medicines, and energy, which fulfill the need of human beings. Humans have been cultivating these plants to meet desired products. These selective plants are being improved for better quality and quantity ofproduct by breeding to meet the needs of growing human population on Earth. The plant breeders are dependent on the existing gene pool and sexual compatibility of the plant species, which is a limitation.

In 1928, Griffith suspected the transfer of genetic material in an experiment when an nonpathogenic pneumococcus strain became pathogenic when mixed with heat-killed pathogenic pneumococcus strain. This was the first report of gene transfer but transforming substance was not identified (Griffith, 1928). After two decades, Hershey and Chase (1952) successfully demonstrated the transfer of genetic material DNA of bacteriophage into the *Escherichia coli* cells. This finding led the researcher to introducedesired

DNA into wide variety of organisms. The genetic material of some species were altered by the incorporation of the selective foreign DNA following the molecular biology methods are called transformation. These transformation techniques if performed directly or indirectly in the plant cells are called plant genetic transformation. In 1981, the first successful gene transformation was demonstrated in tobacco plant using the soil bacterium *Agrobacterium tumefaciens* (Otten et al., 1981). Until now, more than 100 different plant species have been transformed with the desired foreign genes using *A. tumefaciens* or other available methods.

The plant genetic transformation has become a versatile method for the production of agricultural and medicinal value product for the benefits of human society (Campbell, 1999; Lorence and Verpoorte, 2004; Uzogara, 2000). The transfers of gene into plants cells are difficult because plant cells are impermeable which acts as a barrier to diffuse through the cell membrane. Because of diversity of plant species and their diverse genotypes, various gene transformation methods have been developed to overcome this barrier in plants.

1.2 TISSUE CULTURE

The capability of growing the plants from cells after gene transformation depends on the kind of selected. This is one ofthe most important steps for a successful gene transformation technology. The identification of correct cell types is difficult in plants because plant cells are totipotent which can be regenerated to become whole plant in vitro by theorganogenesis or embryogenesis process. But these processes may force a degree of genome stress which might lead to a somaclonal variation, if the whole plant is regenerated via callus phase. If the gene is transferred into pollen or egg cells to produce the genetically transformed gametes, and used for fertilization (in vivo), then they will rise to transformed whole plants. Similarly, gene is inserted intozygote by in vivo or in vitro, and then rescued embryos can also be used to produce transgenic plants. The individual cells in embryos or meristem can also be grown in vitro to produce the transgenic plants.

To generate successful transgenic plants, some kind of tissue-culture step depending on the plant species is necessary. In tissue-culture process, the explants or small piece of living tissue are isolated from the plant, grown aseptically on artificial nutrient medium into an undifferentiated mass known as callus. The explants, such as buds, root tips, nodal stem segments, or germinating seeds, are most preferred because they are rich in undetermined cells and are capable for rapid proliferation. The selected explant should first be disinfected by washing with the sodium hypochlorite or hydrogen peroxide before placing them on the medium because the medium also supports the growth of microorganism. The nutrient medium should contain the appropriate quantity of phytohormones to maintain the cells in an undifferentiated condition and develop into callus. The correct proportion of phytohormones, auxin, and cytokinins depends on species and explant types are important for the growth of callus culture. Thecytokinin is required for shoot culture and auxin for root culture, therefore, low auxin:cytokinin ratio favors the shoot growth, whereas high ratio leads to root growth in callus. Gibberellins, GA3, are required by some explant for their continuousgrowth, whereas abscisic acid boosts specific development actions like somatic embryogenesis. In addition to phytohormones, the nutrient medium also contains the macroelements and microelements, essential vitamins, amino acids, and sucrose. Some plant medium also containscasein hydrolysate, coconut water, yeast extract, and gelling agent. The plant material that can be manipulated in culture provides excellent opportunities for gene transfer methods and generation of transgenic plants.

1.3 PLANT'S GENETIC TRANSFORMATION METHODS

Various available plant transformation methods is discussed in this chapter such as biological methods: *A. tumefaciens* and protoplast-mediated transformation; and chemical methods: calcium phosphate, co-precipitation and lipofection, physical methods, electroporation, biolistics, agitation with glass beads, vacuum infiltration, silicon-carbide whisker, laser microbeams, ultrasound and shock-wave-mediated method.

Plant genetic transformations are classified into direct and indirect gene transfer methods.

1.3.1 INDIRECT GENE TRANSFER METHODS

For indirect gene transformation methods, two bacterial strains, *A. tumefaciens*and *A. rhizogenes*, have been discovered to transfer the desired gene into plant cells; however *A. tumefaciens* is widely used.

1.3.1.1 AGROBACTERIUM TUMEFACIENS

Among the various available vectors for plant transformation, the Ti plasmid of *A. tumefaciens*has been widely accepted. *Agrobacterium*was first discovered in grape plant in 1897 by Fridiano Cavara. In 1907, crown gall diseasewas reported in plants caused by T-DNA of Ti-plasmid of *Agrobacterium* (Smith andTownsend, 1907). *Agrobacterium* has been characterized as Gram-negative bacteria or soil phytopathogen that belongs to the rhizobiaceae family. *Agrobacterium* prefer to infect mostly the dicotyledonous plants and at the infection or wound site produces an unorganized growth of cells that is known as crown gall tumors. This bacterium harbors the tumor-inducing plasmid, known as Ti plasmids, which are exploited to transfer desired gene into target plant tissue. *Agrobacterium*got the natural ability to transfer the T-DNA, part of Tiplasmid, into the plants' genome, and because of this unique ability, *Agrobacterium*is known as natural genetic engineer of plants (Binns and Thomashaw, 1988; Nester et al., 1984). The unwanted sequence of T-DNA region of Ti-plasmid is the crucial region to replace with foreign desire gene.

1.3.1.2 MOLECULAR BASIS OF AGROBACTERIUM-MEDIATED TRANSFORMATION

1.3.1.2.1 Vectors for Gene Transfer

Vectors usually contain the selectable markers to recognize the transformed cells from the untransformed cells, multiple rare restriction sites, and bacterial origins of replication (e.g., ColE1). However, these features in vectors do not help in transfer of gene and integration into plant nuclear genome. Ti-plasmid of *Agrobacterium* has wide host range and capable to transfer gene that makes them preferred vector over other available vectors.

The T-DNA nucleotide sequence end are flanked by 25-bp direct repeat sequences known as left border (LB) and right border (RB) and both border sequences collectively known as T-DNA border. Plasmid DNA comprising the T-DNA with border sequencesis called mini- or micro-Ti-plasmid (Waters et al., 1991).Nucleotide sequences of T-DNA borders are essential and play an important role to transfer the T-DNA into the plant cell upon infectionif present in *cis* orientationbut border sequence itself does not get transferred. Any DNA sequence flanked by repeat of 25 bp in the correct orientation can be transferred to plant cells and similar attribute were exploited with *Agrobacterium*-mediated gene transfer to produce transgenics of higher plants. It was shown in the experiment that onlyRB sequence has been used and observean enhance sequence or sometimes refer overdrive sequence located upstream to RB sequence is also necessary for high-efficiency transfer of T-DNA (Peralta et al., 1986; Shaw et al., 1984). However, the left-border sequencehas little activity alone (Jen and Chilton, 1986).

1.3.1.2.2 Structure and Functions of Ti Plasmids

Agrobacterium harbor a large Ti-plasmid of 200–800 kbp which contains four main regions: T-DNA, vir region, origin of replication, region enabling conjugative transfer, and *o*-cat region (Hooykaas and Schilperoort, 1992; Zupan and Zambrysky, 1995).The molecular understanding of crown gall disease caused created an opportunity to develop the gene transfer system in plant.

1.3.1.2.2.1 T-DNA

It has been demonstrated that T-DNA of Ti-plasmid of *Agrobacterium* transfers to plant nuclear genome that causes the crown gall disease. Even if the *Agrobacterium*is killed with antibiotics, then also undifferentiated callus can be cultivated in in vitro retaining the tumorous properties. This property represents the oncogenic transformation of crown gall tissues and has the ability to form tumor if grafted onto a healthy plant. It is clear that tumor-inducing agents has been transferred from the *Agrobacterium* to plants at wounded site, and while maintaining the plant in their transformed condition, the continued presence of *Agrobacterium* is not required. T-DNA, a small and specific element of Ti plasmid, is of ~24-kbp size whichcomprisesthe following important regions, (1) two tms genes responsible for indole acetic acid (an auxin) biosynthesis and tmr gene responsible for isopentyladenosine 5′-monophosphate (a cytokinin) synthesis. This is the reason when T-DNA sequence transferred to the plant nuclear genome leads to form crown gall because of the synthesis of two phytohormones, auxin and cytokinin; (2) os region responsible for synthesis of unusual amino acid or sugar derivatives, known as opines. Opines metabolism is the chief feature of crown gall disease formation. Two common opines, octopine and nopaline, synthesized in the plant cells from octopine and nopaline synthase, respectively. *Agrobacterium* strain determine the type of opine to be produced, not by host plant. On the basis ofthe type of opine produced, Ti-plasmid is further described as octopine type Ti-plasmid or nopaline type Ti-plasmid. Octopine type of Ti-plasmid is more closely related to each other and composes the TL and TR segment that carry genes for tumor formation and opine synthesis, respectively. TL and TR are transferred to the plant genome independently and may be available in multiple copies. Nopalinetype of Ti-plasmid isunrelated to each other and the structure and organization of DNA sequences are usually simple. Higher plant not capable of using opines but *Agrobacterium* utilize efficiently. Ti-plasmid contains the gene outside of T-DNA and their gene products catabolize opines which supply the carbon and nitrogen source to *Agrobacterium*.

1.3.1.2.2.2 Virulence Gene

Vir gene is also essential for T-DNA transfer and unlike border sequence of T-DNA vir gene can function even in *trans* orientation. T-DNA and *vir* gene, present on two different plasmids, do not affect the T-DNA transfer provided both are present in same *Agrobacterium* cells. The virulence region ofTi-plasmid are of approximately 35 kbp in size and organized in six operons known as Vir A, Vir B, Vir C, Vir D, Vir E, and Vir G. All operon except Vir A and Vir G are polycistronic in nature. Vir A, B, D, and G genes are required for virulence while Vir C and E genes required for tumor formation (Hooykaas and Schilperoort, 1992; Jeon et al., 1998; Zupan and Zambryski, 1995).The vir gene product probably functions as chemoreceptor, which can sense the phenolic compounds, such as acetosyringone and β-hydroxyaceae in exudates of wounded plant tissue.

1.3.1.2.2.3 Origin of Replication

In general, the *Agrobacterium* has broad host range of origin of replication for multiplication in different type of host plant.

1.3.1.2.2.4 Region Enabling Conjugative Transfer and o-Cat Region

Ti-plasmid encodes two functional andactive separate conjugal transfer systems. First system known as Vir-associated system, which has a role to transfer T-DNA region to the plant cell and another system, Tra-associated system plays the role in transferring the whole plasmid from bacterial donor to bacterial recipient (Rogowsky et al., 1990).

1.3.1.3 MECHANISM OF T-DNA TRANSFER TO THE PLANT GENOME

The foreign gene transfer using the *Agrobacterium* has been considered a highly improved form of bacterial conjugation. The complete Ti-plasmid does not transfer to plant instead only a small segment or T-DNA get transferred and integrated into the plant genome. Virulence (vir) genes responsible for T-DNA transfer are located in a separate part of the Ti-plasmid. T-DNA carries the genes for unregulated growth and also to synthesize opines in the transformed plant tissues. These genes are not necessary to transfer the T-DNA and therefore this region generates an opportunity to be replaced with the desired foreign genes.

1.3.1.3.1 Recognition and Induction of Vir genes

Plants secrets number of chemotactic signal or phenolic compounds but acetosyringone and β-hydroxyacetosyringone particularly and certain monosaccharides are recognized to induce the *vir* genes expression, processing, and transfer and may also for the integration of T-DNA into the plant genome (Pan et al., 1993). The bacteria respond well to the simple molecules like sugars and amino acids but not to the acetosyringone compounds to the injured plant cells. The *vir* genes induce only after attachment of agrobacterium to the plant cells (Loake et al., 1988; Parke et al., 1987).Because of synergy of most of sugars, the action of phenolic signals and the *vir* gene expression increases (Shimada et al., 1990). The chemotactic signal turns on Vir A and G genes component belonging to the bacterial regulatory system which controls the other *vir* genes (Pan et al., 1993). Vir A and G gene are expressed constitutively at low level. Vir A and G product encodes a membrane-bound sensor kinase and cytoplasmic regulator protein, respectively. It is well established that the signal transduction process involves Vir A autophosphorylation and then subsequently transfer phosphate to Vir G. Mainly, *vir* G gene along with additional gene on the *Agrobacterium* chromosome encodestranscriptional activator; they play an important role to regulate the other *vir* genes (Gelvin, 2000 & 2003; Huang et al., 1990; Jin et al., 1990; Kado, 1998). The list of *vir* genes of Ti-plasmid of *Agrobacterium* and their functions are summarized in Table 1.1.

TABLE 1.1 List of VirLocus of Ti-Plasmid of *Agrobacterium* and Their Function.

S. No.	Virulence Gene	Functions
1	*Vir* A	Encodes acetosyringone receptor protein which activates Vir G by phosphorylation leading to expression of other vir genes
2	*Vir* B	Encodes membrane protein which involved in conjugal tube formation for T-DNA transfer
3	*Vir* C	Encodes helicase enzyme for unwinding of T-DNA
4	*Vir* D	Topoisomerase activity, vir D2 is an endonuclease
5	*Vir* E	Single strand binding protein binds to T-DNA during transfer
6	*Vir* F	Activity not known
7	*Vir* G	Master controller DNA binding protein, vir A activates the vir G by phosphorylation, Vir G dimerise and activates expression other vir genes
8	*Vir* H	Activity not known

1.3.1.3.2 Transfer of T-DNA to Plant Cells

Transfer of T-DNA initiated by VirD1 and D2 products which act as anendonuclease enzymesarespecifically recognizesT-DNA border sequences. These enzymes create nick either onsingle strand or double strand at T-DNA border sequences which leads to release of ss-T-DNA from the Ti-plasmid. This process is enhanced by the Vir C2 and C12 proteins by recognizing and binding to the enhancer elements. VirD2 binds covalently to 5′ end of processed ss-T-DNA, forming an immature T-DNA complex (Dürrenberger et al., 1989). Single-stranded T-DNA intermediate favored by the octopine type Ti-plasmid, whereas double-stranded favored by nopaline type Ti-plasmid (Steck, 1997). VirD2 protein protects the T-DNA intermediate complex from nuclease degradation to target the DNA to cytoplasm and nucleus then integrate into plant genome (Christie, 1997). T-DNA intermediate are coated with Vir E2which encodes a single-stranded DNA-binding protein (SSBP).The induction of *vir* gene expression forms the conjugative pilus to transfer the T-DNA to plant cells. Vir B gene operon product also involved to make part of conjugative pilus (Lai and Kado, 1998, 2000).Vir B and Vir D products then transport the T-DNA complex through membrane channel a type IV secretion system to cytoplasm of plant cell (Zupan et al., 1995). Vir D4 protein acts as a linker facilitating the interaction of processed T-DNA complex with membrane channel. Vir B2–B11 and Vir D products are important for forming a membrane-associated export apparatus includes hydrophobicity, membrane-spanning domains, and/or N-terminal signal sequences in cytoplasm. The interaction between Vir B7 and VirB9 helps to form heterodimer that stabilizes the other Vir B proteins (Fernández et al., 1996). Membrane channel is composed of Vir D4 and Vir B11 protein which is necessary for transport of T-DNA complex (Christie, 1997). Vir B protein also serves as ATPases to provide energy for channel assembly or export process (Firth et al., 1996). VirB1 has transglycosidase activity which utilizes to assemble other Vir proteins (Berger and Christie, 1993, 1994). With the T-DNA complex, the other Vir protein H and F are also transported into plant cells which are necessary for efficient transport of T-DNA complex and nuclear transport (Hooykaas and Schilperoort, 1992). Next, Vir D2 and Vir E2 proteins play an important role for nuclear transport of T-DNA complex because they contain the nuclear localization signal (Hooykaas and Schilperoort, 1992). The nucleus of injured plant cells is often associated with the cytosolic membrane facilitating the rapid transfer of T-DNA into nucleus without much exposure to the cytosolic environment (Kahland Schell, 1982). After reaching the nucleus, T-DNA probably integratesto plant genome by illegitimate recombination process exploiting naturally occurring chromosome breaks (Tinland, 1996; Tzfira et al., 2004).

1.3.1.4 TI-PLASMID DERIVATIVES FORGENETRANSFER

T-DNA of Ti-plasmid transfers and expresses in to plant cells because T-DNA carry the promoter element and polyadenylation site similar to eukaryotic one. This sequence acquired by agrobacterium may be during the evolution of Ti-plasmid (Bevan et al., 1983; Depicker et al., 1982; De Greve et al., 1982).Ti-plasmid of *Agrobacterium* can transfer its T-DNA to the plant genome and due to this specific reason; Ti-plasmid is qualified as natural vector to engineer the plant cells. The wild-type Ti-plasmid is suitable due to presence of oncogenes on the T-DNA which causes the uncontrolled growth of plant cells. So the oncogenes region of T-DNA must disarmed to be qualified as successful natural vector for revival of plant well. T-DNA should be left with the LB and RB sequence and the *nos* gene in modified vector. While the plant cells transform with *Agrobacterium*-containing modified vector, no tumor should be produced and nopaline production will be evident for positive transformation. To make the screening easier to identify, the transformed plants cells, selectable marker like drug or herbicide resistance could be inserted to the T-DNA because the enzymatic assay for nopaline at every step of transformation is a cumbersome process. Nopaline positive cells could be cultured to callus tissueswhich provided the required phytohormones.

The modified vector of Ti-plasmid is also not convenient for plant transformation because of their large size that makes them difficult for manipulation. The absence of unique restriction enzymes sites in T-DNA sequence is another problem to manipulate this vector.

This problem was resolved by constructing anintermediate vector, in which T-DNA was subcloned into *E. coli* plasmid vector for easy manipulation (Matzke and Chilton, 1981). However, intermediate vector does not replicate in *Agrobacterium* and also lacks conjugation functions. To succeed the gene transfer process, the triparental mating was introduced mixing the three bacterial strains like: (1) *E. coli* strain carrying the recombinant intermediate vector; (2) *E. coli* strain containing the helper plasmid to mobilize intermediate vector in *trans*; and (3) *Agrobacterium* carrying the Ti-plasmid. Conjugation between *E. coli* strains—1 and 2, transferred them to the 3, recipient *Agrobacterium*. Homologous recombination occurs between the T-DNA and intermediate vector and forms a large cointegrate plasmid, from where recombinant T-DNA is transferred to plant genome. Intermediate vector has been widely used but large cointegrate vector are still not required for gene transformation in plant.

T-DNA sequence is not essential and the necessary *vir* gene region can function in *trans* during the transfer of the gene by Ti-plasmid to the plant. Therefore, *vir* gene and disarmed T-DNA sequence part of Ti-plasmids can supply on separate plasmids in *Agrobacterium* and this principle was termed T-DNA binary vector system (Hoekema et al., 1983).In binary vector system, maintaining the T-DNA on a shuttle vector is beneficial because the copy number is not determined by Ti-plasmid and is not dependent on recombination. This event makes the identification of transformants much easier. The gene of interest to transfer including origin of replication and antibiotic resistance genes will be maintained on T-DNA regionin binary vector system, whereas *vir* gene is maintained on separate replicon known as Vir helper plasmid. The *vir* gene products will help processing the T-DNA and export further to plant cells (Fig. 1.1). Ti-plasmid with *vir* gene region without the T-DNA sequence will be transformed into the *Agrobacterium*. T-DNA sequence will introduced in to *Agrobacterium* by triparental mating or methods like electroporation (Cangelosi et al., 1991). To achieve the efficient transformation, the binary vector should have some properties: (1) RB and LB sequence of T-DNA; (2) selectable marker gene compatible to plant usually antibiotic or herbicide resistance (Wang et al., 1984); (3) multiple rare-cutting restriction endonuclease site on T-DNA (Tzfira et al., 2004) and the *lacZ* and gene for blue-white screening (McBride and Summerfelt, 1990) and cos site for preparing cosmid libraries; (4) origin(s) of replication for *E. coli* and *Agrobacterium* facilitate the replication in broad host range; and (5) antibiotic-resistance genes in binary vector for selection in both *E. coli* and *Agrobacterium* (Ditta et al., 1980).

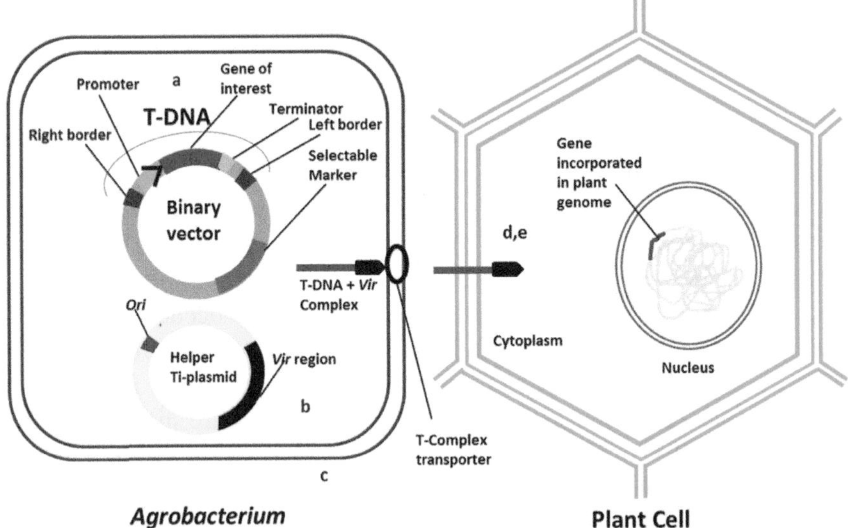

FIGURE 1.1 *Agrobacterium*-mediated gene transformation processes in host plant cells. (a) Isolated desired foreign gene along with promoter sequence inserted between right and left border sequences into the binary vector; (b) vir genes region inserted separately into helper Ti-plasmid; (c) both binary and helper plasmid vectors transferred into *Agrobacterium*; (d) leaf disc of host plant was inoculated in medium containing transformed *Agrobacterium*; (e) T-DNA strand excised from T-DNA complex and then transferred and integrated into plant genome.

The exact upper limit size of T-DNA has not been determined to transfer successfully in the *Agrobacterium*. Following the standard methods, inserting greater than 30 kbp is difficult because of instability in bacterial host. Now, the high-capacity binary vector has been developed to transfer multiple genes together. The first vector, named BIBAC2, is on the basis of the artificial chromosome type vector exercise in *E. coli* containing the F-plasmid origin of replication (Hamilton, 1997). This vector has the kanamycin and hygromycin-resistant gene for the selection in bacteria and transgenic plants, respectively. BIBAC2 vector has been used to transfer about

150-kbp human DNA into the tobacco plant genome. As mentioned above, the role of virulence helper plasmid provides *vir* G and *vir* E in *trans* is very important for successful gene transfer (Hamilton et al., 1996).

1.3.1.5 AGROBACTERIUM-MEDIATED GENE TRANSFORMATION IN PLANTS

For the transfer of gene in dicot plants, a few millimeters diameter of leaves were surface sterilized and inoculated in medium containing *Agrobacterium* cells transformed with recombinant disarmed binary or cointegrate vector (Horsch et al., 1985). The leaf disk was first grown for 2 days and then transferred to the medium containing the kanamycin and carbenicillin. In the medium, kanamycin was added because the foreign chimeric gene has kanamycin resistance gene for selection and carbenicillin to kill *Agrobacterium* cells. The shoots were usually developed in 2–4-week time from the leaf disk. The grown shoots were removed from callus and transferred to the medium containing auxin for root development. The roots were developed in 2–3-week time and then plantlets were transplanted to the soil. This is a superior, simple, and rapid method compared to methods where transformed plants were recovered from the protoplast-derived callus which transformed with agrobacterium by co-cultivation (De Block et al., 1984; Horsch et al., 1984).

In case of monocot plants, only few of the monocots plant such as rice (Chan et al., 1992, 1993; Raineri et al., 1990), wheat (Cheng et al., 1997), barley (Tingay et al., 1997), and sugarcane (Arencibiaet al., 1998) were reported susceptible to *Agrobacterium* infection with the modified culture condition and transformation procedures. The use of explant, embryo, and apical meristem and super virulent strain of *Agrobacterium* like AGL-1 was the key factor for successful transformation. AGL-1 has the ability with increased expression of *vir G* and *virE1* to turn other *vir* gene and to enhance the T-DNA transfer, respectively (Sheng and Citovsky, 1996). Transformation efficiency in rice was achieved by adding 100 mM of acetosyringone in co-cultivating medium of *Agrobacterium* and rice embryos (Hiei et al., 1994). Next, a superbinary vector was created using part of Ti-plasmid of supervirulent agrobacterium strain A281 by transferring to T-DNA carrying plasmid, and this superbinary vector can be used with any strain of agrobacterium (Komari et al., 1996).

1.3.1.6 AGROBACTERIUM RHIZOGENES

A. rhizogenes is another bacterial strain used to transfer the gene of interest to the plant cells. The molecular understanding of hairy root diseases helps to utilize this *Agrobacterium* for gene transfer system, analogous to *A. tumefaciens*. This *Agrobacterium* harbor the Ri-plasmid which are responsible to produce the characteristic hairy root disease symptoms upon infection to dicotyledonous plant. Ri-plasmid also has T-DNA region which transfer into the plant nuclear genome (Chilton et al., 1982).T-DNA integrates into the plant genome and in turn *iaaM* and *iaaH* gene are induced toproduce excess phytohormones tryptophan 2-monooxygenase and indoleacetamide hydrolase, respectively. There are no major differences observed between the Ri-plasmid and Ti-plasmid (Tepfer, 1984). However, it is not accepted commercially because of the problem involved in scale-up of transformed roots (Giri andNarassu, 2000).

1.3.2 DIRECT DNA TRANSFER TO PLANTS

Direct gene transfer methods can also be called as vector fewer methods because no vector is used to transfer gene to plant cells. The polar molecule such as foreign gene does not get transferred directly to the plant cells because of hydrophilic and hydrophobic nature of plasma membrane of plant cell. Many physical and mechanical methods were developed that facilitate the entry of this foreign DNA into plant cells. Both physical and mechanical methods use the selective chemical environment for uptake of foreign gene into recipient cells. The recipient cells were placed in chemical environment that enlarges the pore size of cell membrane for easy uptake of desired gene from the surrounding chemical solution to the inside of the cells. The selected chemical compounds must have some properties to qualify for gene transformation experiments in plant cells: (1) do not induce the nuclease for foreign gene degradation; (2) do not obstruct the foreign gene to pass to nucleus; and (3) facilitate the transport of foreign gene through the plasma membrane.

1.3.2.1 PROTOPLAST TRANSFORMATION

The protoplast cells are capable to take up the gene of interest from their surrounding liquid environment. After entry of gene of interest into protoplast, it gets integrated into the genome of transfected cells. The selectable marker can be also added with the gene of interest that is required for the selection of desired gene in the protoplast. The gene transfer process can be induced and accelerated under the influence of some chemicals like polyethylene glycol (PEG) (Negrutiu et al., 1987). After addition of PEG, the transformation efficiency goes upto 100% and results almost every protoplast to be transformed with the gene of interest. Alternatively, the electroporation methods can also be used for gene transfer to the protoplast (Shillito et al., 1985). The putative protoplasts containing gene of interest were grown on selective medium; where protoplast regenerate their cell wall, cell division begins and eventually produces the callus. Callus then produces the roots and shoots by inducing with phytohormones. Major problems are often observed for the regeneration of host plant from the protoplast; however, dicots found are more responsive than the monocots. For protoplast transformation experiment, the kanamycin tolerant gene, *npt*II marker is very successful in dicot and hygromycin or phosphinothricin for monocot because monocot is naturally tolerant to kanamycin. First transformation experiment in protoplast was conducted transforming the maize cDNA-encoding enzyme dihydroquercetin 4-reductase for anthocyanin biosynthesis with *npt*II gene in petunia (Meyer et al., 1987).

1.3.2.2 ELECTROPORATION

Electroporation method was first demonstrated studying gene transfer in mouse cells (Wong and Newmann, 1982) and can be applied also with bacterial, fungal, and plant cells. It is a simple and efficient method for integration of gene of interest into protoplast or intact plant cells. For electroporation, with high-voltage (1.5 kV) short duration and with low-voltage (350V) long duration of pulse was used for gene transfer. Electroporation pulse increases the permeability of membrane by disrupting the phospholipid bilayer of protoplast. This is in turn facilitating the entry of gene of interest into cells if present on protoplast membrane. The target cells can be pretreated with enzymes or wounded for ease of gene transfer process (D'Halluin et al., 1992; Laursen et al., 1994). Without any form of pretreatment also, gene transfer has been successfully achieved in immature rice, maize, and wheat embryosby electroporation method (Kloti et al., 1993; Sorokin et al., 2000; Xu and Li, 1994).

The efficient and successful gene transfer by electroporation methods depends on the following factor like applied electric field strength, electric pulse length, temperature, DNA conformation, DNA concentration, and ionic composition of transfection medium, etc. PEG can also be used to stimulate and enhance the uptake of liposome and also improve the transformation efficiency. As discussed above, regeneration from protoplast to plants is difficult, so for electroporation culture of cells or nodal meristem, explants are often selected for gene transfer which does not require tissue culture. Successful transfer of gene of interest using electroporation methods has been already achieved in maize, petunia, rice, sorghum, tobacco crops. Some new measure has been suggested to increase the transformation efficiency such as (1) uses of 1.25 kV/cm, (2) add first DNA followed PEG, (3) heat shock at 45°C for 5 min, and (4) use linear DNA in place of circular. The modified conditionsare suitable to transfer gene of interest in both monocot and dicot protoplast (Fromm et al., 1985).

1.3.2.3 PARTICLE BOMBARDMENT

Particle bombardment method of gene transfer is also known as biolistics, or particle gun, or gene gun, or short gun or microparticle gun, or projectile bombardment method. This method is especially useful when some of the live-plant tissues, like intercellular organelles, leaves, meristem, immature embryos, callus or suspension cultured cells, and live pollen, are impermeable to foreign DNA. The type of plant material used for DNA delivery is not a limitation in this method because the DNA delivery is governed by physical parameters (Altpeter et al., 2005). Plant cell wall is hard and it is not easy to deliver anything from outside, so the powerful particle bombardment method is very useful for efficient gene transfer in plant (Rasco-Gaunt et al., 2001). Particle bombardment method was developed at Cornell University (Kikkert et al., 2005; Klein et al., 1987). The gene of interest, DNA or RNA coated to tiny biologically inert high-density particles like gold or tungsten of 1pm–3-µm sizeare placed on the target tissue in vacuum condition. Then, gene coated high-density particles are accelerated for high velocity (1400 ft/s) by powerful shot using gene gunto enter inside the tissue membrane. The explosive charge like cordite explosion or shock waves initiated by high voltage can be used to get the high velocity acceleration in the gene gun. The success of particle bombardment method are governed by some factors such as particle size, acceleration (for penetration and determine the tissue damage), amount, and conformation of DNA. These four factors must optimize for each species and type of target tissue using for gene transfer for the success of gene transfer (Finer et al., 1999; Twyman and Christou, 2004).

Using this method, the first successful transgenic were produced in soybean transferring gene to meristem tissues which were isolated from immature seeds (McCabe et al., 1988). Gene gun method has been successfully used to transfer gene of interest in crops like barley, cotton, maize, oat, papaya, rice, soybean, sugarcane, tobacco, and wheat (Twyman and Christou, 2004). This method has also been used to transfer gene of interest for transient expression in onion, maize, rice, and wheat.

Over the years, the particle bombardment method has been modified for better control over particle delivery, efficient transformation of gene, and enhanced reproducibility of transformation states. For example, particle bombardment based on electric discharge has been designed for gene transfer in recalcitrant cereals and legume crops (McCabe and Christou, 1993). Other gene gun modification includes like pneumatic apparatus (Iida et al., 1990), particle inflow gun using flowing helium (Finer et al., 1992; Takeuchi et al., 1992), and device utilizing compressed helium

1.3.2.4 MICROINJECTION

Microinjection is widely used and efficient technique for transfer of desire gene into animal cells, tissues, or embryo cells nuclear genome. This technique is not efficient to direct transfer gene to the plant cells. In this method, gene transfer to the cytoplasm or nucleus of recipient protoplast or plant cells were performed with the glass micropipette of 0.5–10-μm diameter needle tip. The target recipient cells for gene transfer are many, such as immature embryos, meristems, immature pollen, germinating pollen, isolated ovules, embryogenic suspension cultured cells, etc. The recipient cells were immobilized on a solid support like depression slide under suction and then the cell membrane and nuclear envelope of plant cells were penetrated with the glass micropipette tipunder specialized micromanipulator microscope set up. The modified and improved method of microinjection is termed as holding pipette method, in which plant protoplasts utilize a holding pipette for immobilizing the protoplast while an injection pipette is utilized to inject the macromolecule. With the help of holding pipette hold, the protoplast and DNA were injected into protoplast nucleus by injection pipette. Many genetic manipulation experiments were widely performed using this technique for cell modification, silencing of gene, etc. This microinjection technique was also performed and demonstrated successful gene transformation and transient expression in green algae, *Acetabularia* (Neuhaus et al., 1984). The drawback of this microinjection technique in that process is very slow, expensive, requires highly trained technician and only a part of plantis transformed with the desired gene. However, the success rate of transfer of gene is very high. This technique has been employed successfully in oilseed rape (*Brassica napus*) and obtained the transgenic chimera.

1.3.2.5 SONOPORATION

In this method, the explant like leaves is chopped into pieces, and using the ultrasound waves creates the permeability function in cell wall. Through this permeable cell wall, the gene of interest can be uptake from the surrounding environment by cell wall. Sonoporation process uses the sound waves which help to form tiny bubbles that enhance the DNA entry into cell walls (Miller et al., 1999). The explants were further transferred to the culture nutrient medium for the growth of shoots and roots.

1.3.2.6 CALCIUM PHOSPHATE MEDIATED

Calcium-phosphate-mediated gene transformation method was also considered promising for plant cells. In this method, the desired gene with Ca^{2+} ions precipitated and forms calcium phosphate which coats the cells and is released inside the plant cells. Using this method, the desired gene can be transferred to study the molecular, biochemical, cellular, genomic, and proteomic aspects in in vitro and in vivo of plant cells (Feher et al., 2003; Yoo et al., 2007).

1.3.2.7 LIPOFECTION

Lipofection is a liposome-mediatedgene transfer method. This methodemploys a liposome-containing desired gene which induced by PEG to transfer of gene and then fuse into protoplasts. Liposome is cationic in nature and is made up of phospholipid layer similar to cell membrane. Liposome and target cells adhere and form aggregates easily because of similar phospholipid bilayer (Felgner et al., 1987). The aggregate of liposome and cell wall are positively charged that enhances the efficiency of negatively charged DNA uptake. The desired gene enters into protoplast by endocytosis process of liposome that includes adhesion of the liposomes to the protoplast surface, liposomes fusion at the site of adhesion, and then finally release of

DNA inside the protoplast cell. There are many advantages with this method over other gene-transfer method like desired gene not exposed to nuclease, stability due to encapsulation, low cell toxicity, high degree of reproducibility, and suitable for wide range of cell types. Lipofection method of gene transfer has been successfully used in a number of plant species like tobacco, petunia, and carrot.

1.3.2.8 LASER TRANSFECTION

Laserporation or laser-mediated method is a method in which laser beams are used to transfer desired gene into plant cells (Berns et al., 1983; Weber et al., 1988). The plant cell walls are punctured around 0.5 μm by laser beam which are sealed very fast in 5–6-s time. The foreign desired gene is transferred easily by laser pulses through this temporary opening in the cell wall membrane to the nuclear genome. The stringent laser beam system (nitrogen lasers, titanium–sapphire lasers) attached with appropriate microscope is necessary for accurate and efficient transfer of gene to plant cells (Greulich et al., 2000). With this method, a large number of plant cells can be transformed with the foreign DNA and transformed cells can be recovered upto 100%. However, this method has not become popular like other direct gene transfer methods because it needs very expensive equipment for laser beam (Lin et al., 1981).

1.3.2.9 CHLOROPLAST TRANSFORMATION

Until now, all the above methods have been discussed about how to transfer foreign gene in the nuclear genome of plant cells. Another promising choice is chloroplast which has the large number of important gene in photosynthetic system. The photosynthetic DNA can be manipulated with foreign gene to achieve the expression level up to 50 times higher compared to nuclear transformation. Moreover, gene integration in chloroplast genome does not create any problem of silencing or position effect that may affect foreign gene expression of nuclear genome. Another advantage with this method is that they provide a natural containment since the foreign gene cannot be transmitted through pollen (Maliga, 1993). The selectable marker gene, *aad* (aminoglycoside adenyltransferase) conferring resistance to streptomycin and spectinomycin was best choice which usesmostly in chloroplast gene-transfer method (Zoubenko et al., 1994).The application of marker gene (*aad*) with green fluorescent protein gene allows tracking of chlorophyll synthesis for rapid identification of transformed plant tissue (Khan and Maliga, 1999). With the advanced technology, now it has become possible to remove the selectable marker from desired gene from transformed chloroplast genome (Corneille et al., 2001; Iamthamand Day, 2000; Klaus et al., 2004). Using the chloroplast transformation method, foreign gene has been transferred into many crop-plant like tobacco (Svab et al., 1990a, 1990b), tomato (Ruf et al., 2001), potato (Sidorov et al., 1999), rapeseed (Hou et al., 2003; Skarjinskaia et al., 2003), and soybean (Dufourmant el et al., 2004).

1.4 FUTURE PROSPECTS

There are many direct and indirect ways of gene transfer methods that are available for plant species. These gene transfer methods has already delivered many transgenic plants that includes the characteristic such as insect pest resistance, high yielding, enhanced nutrition, virus resistant, herbicide resistance, etc. However, the successful and efficient transformation methods can be improved in future if we focus to include thoughtful controls for treatments and analysis, correlation between treatment and predicted result, correlation between physical and phenotypic data, southern analysis for hybrid fragments of host DNA and foreign DNA, absence of contaminating fragments, and no false positives or negatives in evaluation. The world population is increasing and the transgenic plant cultivation seems the only potential solution to meet food shortage. But, the production of transgenic crops is a great concern for environment safety and human health, that's why future of this technology is still uncertain. The population in India is increasing at a high rate, so India should urgently simplify and develop the mechanism to commercialize the transgenic crops. The seed industries should be given permission under strict laws to develop superior transgenic varieties and supply at an economic rate to the Indian farmers. Apart from this, agricultural scientist should be encouraged to develop transgenic crops with trait such as edible vaccine, deficient vitamins and minerals, antibodies, produce biofuels and bioenergy, etc.

In future, such an advanced technology should be also developed for foreign gene transfer to the plant cells that should provide efficient transformation, high rate of recovery of modified plants. Some promising technology has been tried to transfer foreign gene into target genome of germinating pollen tube, dry seed, and embryos tissue simply by simple incubation with foreign gene. Pollen tube pathway is another hope to integrate the foreign gene in to genome of either sperm nuclei or zygote. This approach applied and observed the phenotypic changes indicating the gene transfer; however, no proof for gene transfer has been obtained so far. The major problems exist with this natural gene transfer methods are the presence of cell wall, nuclease to degrade DNA, and callose plug in pollen tube which need to be resolved. Though this method has extreme potential and seems conclusive and highly acceptable, but, so far, no transgenic has been obtained.

KEYWORDS

- ***Agrobacterium tumefaciens***
- **Ti plasmids**
- **plant transformation**
- **protoplast**
- **electroporation**
- **particle bombardment**

REFERENCES

Altpeter, F.; Baisakh, N.; Beachy, R.; Bock, R.; Capell, T.; Christou, P.; Daniell, H.; Datta, K.; Datta, S.; Dix, P. J.; Fauquet, C.; Huang, N.; Kohli, A.; Mooibroek, H.; Nicholson, L.; Nguyen, T. T.; Nugent, G.; Raemakers, K.; Romano, A.; Stoger, E.; Taylor, N.; Visser, R. Particle Bombardment and the Genetic Enhancement of Crops: Myths and Realities. *Mol. Breed.* **2005,** *15*, 305–327.

Arencibia, A. D.; Carmona, E. R.; Tellez, P.; Chan, M.; Yu, S.; Trujillo, L. E.; Oramas, P. An Efficient Protocol for Sugarcane (*Saccharum* spp. L) Transformation Mediated by *Agrobacterium tumefaciens*. *Transgenic Res.* **1998,** *7*, 213–222.

Berger B. R; Christie P. J. The *Agrobacterium tumefaciens* virB4 Gene Product Is an Essential Virulence Protein Requiring an Intact Nucleoside Triphosphate-binding Domain. *J. Bacteriol.* **1993,** *175*, 1723–1734.

Berns M. W; Aist J, Edward J; Strahs K; Girton J; Mcneill P. Laser microsurgery in cell and developmental biology. *Science*, **1983,** *213*, 505–513.

Bevan, M. W.; Flavell, R. B.; Chilton, M. D. A Chimeric Antibiotic Resistance Gene as a Selectable Marker for Plant Cell Transformation. *Nature* **1983,** *304*, 184–187.

Binns, A. N.; Thomashow, M. F. Cell Biology of Agrobacterium Infection and Transformation of Plants. *Annu. Rev. Microbiol.* **1988,** *42*, 575–606.

Campbell, K. H. Nuclear Transfer in Farm Animal Species. *Semin. Cell Dev. Biol.* **1999,** *10*, 245–252.

Cangelosi, G. A.; Best, E. A.; Martinetti, C.; Nester, E. W. Genetic Analysis of *Agrobacterium tumefaciens*. *Methods Enzymol.* **1991,** *145*, 177–181.

Chan, M. T.; Chang, H. H.; Ho, S. L.; Tong, W. F.; Yu, S. M. Agrobacterium-Mediated Production of Transgenic Rice Plants Expressing a Chimeric α-Amylase Promoter/β-Glucuronidase Gene. *Plant Mol. Biol.* **1993,** *22*, 491–506.

Chan, M. T.; Lee, T. M.; Chang, H. H. Transformation of Indica Rice (*Oryza sativa* L.) Mediated by *Agrobacterium tumefaciens*. *Plant Cell Physiol.* **1992,** *33*, 577–583.

Cheng, M.; Fry, J. E.; Pang, S. Z. Genetic Transformation of Wheat Mediated by *Agrobacterium tumefaciens*. *Plant Cell Physiol.* **1997,** *115*, 971–980.

Chilton, M. D.; Tepfer, D. A.; Petit, A.; David, C.; Casse-delbart, F.; Tempe, J. *Agrobacterium rhizogenes* Insert T-DNA into the Genomes of the Host Plant Root Cells. *Nature* **1982,** *295*, 432–434.

Christie, P. J. *Agrobacterium tumefaciens* T-Complex Transport Apparatus: A Paradigm for a New Family of Multifunctional Transporters in Eubacteria. *J. Bacteriol.* **1997,** 179, 3085–3094.

Corneille, S.; Lutz, K.; Svab, Z.; Maliga, P. Efficient Elimination of Selectable Marker Genes from the Plastid Genome by the Cre–Lox Site-Specific Recombination System. *Plant J.* **2001,** *27*, 171–178.

D'Halluin, K.; Bossut, M.; Bonne, E. Transformation of Sugar Beet (*Beta vulgaris* L.) and Evaluation of Herbicide Resistance in Transgenic Plants. *Biotechnology* **1992,** *10*, 309–314.

De Block, M.; Herrera-Estrella, L.; Van Montagu, M.; Schell, J.; Zambryski, P. Expression of Foreign Genes in Regenerated Plants and their Progeny. *EMBO J.* **1984,** *3*, 1681–1689.

De Greve, H.; Leemans, J.; Hernalsteens, J. P. Regeneration of Normal Fertile Plants that Express Octopine Synthase from Tobacco Crown Galls after Deletion of Tumor-Controlling Functions. *Nature* **1982,** *300*, 752–755.

Depicker, A.; Stachel, S.; Dhaese, P.; Zambryski, P.; Goodman, H. M. Nopaline Synthase: Transcript Mapping and DNA Sequence. *J. Mol. Appl. Genet.* **1982,** *1*, 561–574.

Ditta, G.; Stanfield, S.; Corbin, D.; Helinski, D. R. Broad Host Range DNA Cloning System for Gram-Negative Bacteria: Construction of a Gene Bank of *Rhizobium meliloti*. *Proc. Nat. Acad. Sci. USA.* **1980,** *77*, 7347–7351.

Dufourmantel, N.; Pelissier, B.; Garc, F.; Peltier, G.; Ferullo, J. M.; Tissot, G. Generation of Fertile Transplastomic Soybean. *Plant Mol. Biol.* **2004,** *55*, 479–489.

Dürrenberger, F.; Crameri, A.; Hohn, B.; Koukolikova-Nicola, Z. Covalently Bound VirD2 Protein of *Agrobacterium tumefaciens* Protects the T-DNA from Exonucleolytic Degradation. *Proc. Natl. Acad. Sci. USA.* **1989,** *86*, 9154–9158.

Feher A, Pasternak T. P, Duttis D . Review of Plant Biotechnology and Applied Genetics. Transistion of Somatic Plant Cells to an Embryogenic State. *Plant Cell Tissue Organ Cult.* **2003,** *74*, 201–228.

Felgner P. L, Gadek T. R. Holm M, Roman R, Chan H W, Wenz M, Northrop J P, Ringold G M, Danielsen M. (1987). Lipofection: a highly efficient, lipid-mediated DNA-transfection procedure. Proc. Natl. Acad. Sci. U.S.A. 84: 7413–7417.

Fernández, D.; Spudich, G. M.; Zhou, X. R.; Berger, B. R.; Christie, P. J. *Agrobacterium tumefaciens* VirB7 Lipoprotein is Required for Stabilization of VirB Proteins During Assembly of the T-Complex Transport Apparatus. *J. Bacteriol.* **1996,** *178*, 3168–3176.

Finer, J. J.; Finer, K. R.; Ponappa, T. Particle Bombardment Mediated Transformation. *Curr. Top. Microbiol. Immunol.* **1999,** *240*, 59–80.

Finer, J. J.; Vain, P.; Jones, M. W.; McMullen, M. D. Development of the Particle Inflow Gun for DNA Delivery to Plant Cells. *Plant Cell Rep.* **1992,** *11*, 323–328.

Firth, N.; Ippen-Ihler, K.; Skurray, R. A. Structure and Function of the F-Factor and Mechanism of Conjugation. In: *Escherichia coli and Salmonella: Cellular and Molecular Biology*, second ed.; Neidhardt, F., et al., Eds.; ASM Press: Washington, DC, 1996; p 2377–2401.

Fromm, M.; Taylor, L. P.; Walbot, V. Expression of Genes Transferred into Monocot and Dicot Plant Cells by Electroporation. *Proc. Natl. Acad. Sci. USA* **1985,** *82*, 5824–5828.

Gelvin, S. B. Agrobacterium and Plant Genes Involved in T-DNA Transfer and Integration. *Ann. Rev. Plant Physiol. Plant Mol. Biol.* **2000,** *51*, 223–256.

Gelvin, S. B. Agrobacterium-Mediated Plant Transformation: The Biology Behind the Gene Jockeying Tool. *Mol. Biol. Rev*. **2003,** *67*, 16–37.

Giri, A.; Narassu, M. L. Transgenic Hairy Roots: Recent Trends and Applications. *Biotechnol. Adv*. **2000,** *18*, 1–22.

Greulich K. O.; Pilarczyk G.; Hoffmann A.; Meyer Z.; Horste G.; Scafer B. Micro Manipulation by Laser Microbeam and Optical Tweezers: from Plant Cells to Single molecules. *J. Microscopy Oxford, 2000,* 198, 182–187.

Griffith, F. The Significant of Pneumoccopal Types. *Hyg. J.* **1928,** *27*, 113.

Hamilton, C. M. A Binary–BAC System for Plant Transformation with High Molecular Weight DNA. *Gene* **1997,** *200*, 107–116.

Hamilton, C. M.; Frary, A.; Lewis, C.; Tanksley, S. D. Stable Transfer of High Molecular Weight DNA into Plant Chromosomes. *Proc. Nat. Acad. Sci. USA* **1996,** *93*, 9975–9979.

Hershey, A. D.; Chase, D. Independent Functions of Viral Protein and Nuclei Acid in Growth of Bacteriophage. *J. Gen Physiol.* **1952,** *36*, 39–56.

Hiei, Y.; Ohta, S.; Komari, T.; Kumashiro, T. Efficient Transformation of Rice (*Oryza sativa* L.) Mediated by Agrobacterium and Sequence Analysis of the Boundaries of the T-DNA. *Plant J.* **1994,** *6*, 241–282.

Hoekema, A.; Hirsch, P. R.; Hooykaas, P. J. J.; Schilperoort, R. A. A Binary Plant Vector Strategy Based on Separation of Vir- and T-Region of the *Agrobacterium tumefaciens*Ti-plasmid. *Nature* **1983,** *303*, 179–180.

Hooykaas, P. J. J.; Shilperoort, R. A. Agrobacterium and Plant Genetic Engineering. *Plant Mol. Biol.* **1992,** *19*, 15–38.

Horsch, R. B.; Fraley, R. T.; Rogers, S. G. Inheritance of Functional Genes in Plants. *Science* **1984,** *223*, 496–498.

Horsch, R. B.; Fry, J. E.; Hoffmann, N. L. A Simple and General Method for Transferring Genes into Plants. *Science* **1985,** *227*, 1229–1231.

Hou, B. K.; Zhou, Y. H.; Wan, L. H.; Zhang, Z. L.; Shen, G. F.; Chen, Z. H.; Hu, Z. M. Chloroplast Transformation in Oilseed Rape. *Transgenic Res*. **2003,** *12*, 111–114.

Huang, Y.; Morel, P.; Powell, B.; Kado, C. I. VirA, a Coregulator of Ti-Specified Virulence Genes, Is Phosphorylated In Vitro. *J. Bacteriol.* **1990,** *172*, 1142–1144.

Iamtham, S.; Day, A. Removal of Antibiotic Resistance Genes from Transgenic Tobacco Plastids. *Nat. Biotechnol.* **2000,** *18*, 1172–1176.

Iida, A.; Morikawa, H.; Yamada, Y. Stable Transformation of Cultured Tobacco Cells by DNA-Coated Gold Particles Accelerated by Gas Pressure Driven Particle Gun. *Appl. Microbiol. Biotechnol.* **1990,** *33*, 560–563.

Jen, G. C.; Chilton, M. D. The Right Border Region of pTiT37 T-DNA Is Intrinsically More Active than the Left Border Region in Promoting T-DNA Transformation. *Proc. Nat. Acad. Sci. USA.* **1986,** *83*, 3895–3899.

Jeon, G. A.; Eum, J. S.; Sim, W. S. The Role of Inverted Repeat (IR) Sequence of the Vir E Gene Expression in *Agrobacterium tumefaciens* pTiA6. *Mol. Cells* **1998,** *8*, 49–53.

Jin, S.; Prusti, R. K.; Roitsch, T.; Ankenbauer, R. G.; Nester, E. W. The VirG Protein of *Agrobacterium tumefaciens* is Phosphorylated by the AutophosphorylatedVirA Protein and This Is Essential for Its Biological Activity. *J. Bacteriol.* **1990,** *172*, 4945–4950.

Kado, C. I. Agrobacterium-Mediated Horizontal Gene Transfer. *Genet. Eng.* **1998,** *20*, 1–24.

Kahl, G.; Schell, J. S., Eds. *Molecular Biology of Plant Tumours*; Academic Press: New York, 1982; pp 211–267.

Khan, M. S.; Maliga, P. Fluorescent Antibiotic Resistance Marker for Tracking Plastid Transformation in Higher Plants. *Nat. Biotechnol.* **1999,** *17*, 910–915.

Kikkert, J. R.; Vidal, J. R.; Reisch, B. I. Stable Transformation of Plant Cells by Particle Bombardment/Biolistics. *Methods Mol Biol.* **2005,** *286*, 61–78.

Klaus, S. M. J.; Huang, F.-C.; Golds, T. J.; Koop, H. U. Generation of Marker-Free Plastid Transformants Using a Transiently Cointegrated Selection Gene. *Nat. Biotechnol.* **2004,** *22*, 225–229.

Klein, T. M.; Wolf, E. D.; Wu, R.; Sanford, J. C. High-Velocity Micro-projectiles for Delivering Nucleic Acids into Living Cells. *Nature* **1987,** *327*, 70–73.

Kloti, A.; Iglesias, V. A.; Wunn, J. Gene Transfer by Electroporation into Intact Scutellum Cells of Wheat Embryos. *Plant Cell Rep.* **1993,** *12*, 671–675.

Komari, T.; Hiei, Y.; Saito, Y.; Murai, N.; Kumashiro, T. Vectors Carrying Two Separate T-DNAs for Co-transformation of Higher Plants Mediated by *Agrobacterium tumefaciens* and Segregation of Transformants Free from Selection Markers. *Plant J.* **1996,** *10*, 165–1474.

Lai, E. M.; Kado, C. I. Processed VirB2 is the Major Subunit of the Promiscuous Pilus of *Agrobacterium tumefaciens*. *J Bacteriol.* **1998,** *180*, 2711–2717.

Lai, E. M.; Kado, C. I. The T-Pilus of*Agrobacterium tumefaciens*. *Trends Microbiol.* **2000,** *8*, 361–369.

Laursen, C. M.; Krzyzek, R. A.; Flick, C. E.; Anderson, P. C.; Spencer, T. M. Production of Fertile Transgenic Maize by Electroporation of Suspension Culture Cells. *Plant Mol. Biol.* **1994,** *24*, 51–61.

Lin, P-F; Ruddle F. Photo Engraving of Coverslips and Slides to Facilitate Monitoring of Micromanupulated Cells or Chromosome Spreads. *Exp Cell Res.*, **1981,** *134*, 485–488.

Loake, G. J.; Ashby, A. M.; Shaw, C. H. Attraction of *Agrobacterium tumefaciens* C58C1 towards Sugars Involves a Highly Sensitive Chemotaxis System. *J. Gen. Microbiol.* **1988,** *134*, 1427–1432.

Lorence, A.; Verpoorte, R. Gene Transfer and Expression in Plants. *Methods Mol. Biol.* **2004,** *267*, 329–350.

Maliga, P. Towards Plastid Transformation in Flowering Plants. *Trends Biotechnol.* **1993,** *11*, 101–107.

Matzke, A. J. M.; Chilton, M. D. Site-Specific Insertion of Genes into T-DNA of the *Agrobacterium*Tumour-Inducing Plasmid: An Approach to Genetic Engineering of Higher Plant Cells. *J. Mol. Appl. Genet.* **1981,** *1*, 39–49.

McBride, K. E.; Summerfelt, K. R. Improved Binary Vectors for *Agrobacterium*-Mediated Plant Transformation. *Plant Mol. Biol.* **1990,** *14*, 269–276.

McCabe, D. E.; Swain, W. F.; Martinell, B. J.; Christou, P. Stable Transformation of Soybean (*Glycine max*) by Particle Acceleration. *Biotechnology* **1988,** *6*, 923–926.

McCabe, D.; Christou, P. Direct DNA Transfer Using Electric Discharge Particle Acceleration (Accell® Technology).*Plant Cell Tissue Organ Cult.* **1993,** *33*, 227–236.

Meyer, P.; Heidmann, I.; Forkmann, G.; Saedler, H. A New Petunia Flower Colour Generated by Transformation of a Mutant with a Maize Gene. *Nature* **1987,** *330*, 677–678.

Miller, D.; Bao, S.; Morris, J. Sonoporation of Cultured Cells in the Rotating Tube Exposure System. *Ultrasound Med. Biol.* **1999,** *25*, 143–149.

Negrutiu, I.; Shillito, R.; Potrykus, I.; Biasini, G.; Sala, F. Hybrid Genes in the Analysis of Transformation Conditions. I. Setting Up a Simple Method for Direct Gene Transfer in Plant Protoplasts. *Plant Mol. Biol.* **1987,** *8*, 363–373.

Nester, E. W.; Gordon, M. P.; Amasino, R. M.; Yanofsky, M. F. Crown Gall: A Molecular and Physiological Analysis. *Annu. Rev. Plant Phys.* **1984,** *35*, 387–413.

Neuhaus, G.; Neuhaus-Url, G.; Gruss, P.; Schweiger, H. G. Enhancer-Controlled Expression of the Simian Virus 40 T-Antigen in the Green Alga Acetabularia. *EMBO J.* **1984,** *3*, 2169–2172.

Otten, L.; DeGreve, H.; Hernalsteens, J. P.; Van Montagu, M.; Schieder, O. Mendelian Transmission of Genes Introduced into Plants by the Ti Plasmids of *Agrobacterium tumefaciens*. *Mol. Gen. Genet.* **1981,** *183*, 209–213.

Pan, S. Q.; Charles, T.; Jin, S.; Wu, Z. L.; Nester, E. W. Preformed Dimeric State of the Sensor Protein VirA is Involved in Plant *Agrobacterium* Signal Transduction. *Proc. Nat. Acad. Sci. USA.* **1999,** *90*, 9939–9943.

Parke, D.; Ornston, L. N.; Nester, E. W. Chemotaxis to Plant Phenolic Inducers of Virulence Genes Is Constitutively Expressed in the Absence of the Ti Plasmid in *Agrobacterium tumefaciens*. *J. Bacteriol.* **1987,** *69*, 5336–5338.

Peralta, E. G.; Hellmiss, R.; Ream, W. Overdrive, a T-DNA Transmission Enhancer on the *A. tumefaciens*Tumour-Inducing Plasmid. *EMBO J.* **1986,** *5*, 1137–1142.

Raineri, D. M.; Bottino, P.; Gordon, M. P.; Nester, E. W. Agrobacterium-Mediated Transformation of Rice (*Oryza sativa* L.).*Biotechnology* **1990,** *9*, 33–38.

Rasco-Gaunt, S.; Riley, A.; Cannell, M.; Barcelo, P.; Lazzeri, P. A. Procedures Allowing the Transformation of a Range of European Elite Wheat (*Triticumaestivum* L.) Varieties via Particle Bombardment. *J. Exp. Bot.* **2001,** *52*, 865–874.

Rogowsky, P. M.; Close, T. J.; Chimera, J. A.; Shaw, J. J.; Kado, C. I. Regulation of the Vir Genes of *Agrobacterium tumefaciens* Plasmid pTiC58. *J. Bacteriol.* **1987,** *169*, 5101–5112.

Ruf, S.; Hermann, M.; Berger, I. J.; Carrer, H.; Bock, R. Stable Genetic Transformation of Tomato Plastids and Expression of a Foreign Protein in Fruit. *Nat. Biotechnol.* **2001,** *19*, 870–875.

Shaw, C. H.; Watson, M. D.; Carter, G. H.; Shaw, C. H. The Right Hand Copy of the NopalineTi Plasmid 25 bp Repeat is Required for Tumor Formation. *Nucl. Acids Res.* **1984,** *12*, 6031–6041.

Sheng, O. J.; Citovsky, V. Agrobacterium Plant Cell DNA Transport: Have Virulence Proteins Will Travel. *Plant Cell* **1996,** *8*, 1699–1710.

Shillito, R. D.; Saul, M. W.; Pazkowski, J.; Muller, M.; Potrykus, I. High Efficiency Direct Gene Transfer to Plants. *Biotechnology* **1985,** *3*, 1099–1103.

Shimada, N.; Toyoda-Yamamoto, A.; Nagamine, J. Control of Expression of *Agrobacterium*Vir Genes by Synergistic Actions of Phenolic Signal Molecules and Monosaccharides. *Proc. Nat. Acad. Sci. USA.* **1990,** *87*, 6684–6688.

Sidorov, V. A.; Kasten, D.; Pang, S. Z.; Hajdukiewicz, P. T.; Staub, J. M.; Nehra, N. S. Stable Chloroplast Transformation in Potato: Use of Green Fluorescent Protein as a Plastid Marker. *Plant J.* **1999,** *19*, 209–216.

Skarjinskaia, M.; Svab, Z.; Maliga, P. Plastid Transformation in *Lesquerellafendleri*, an Oilseed Brassicacea. *Transgenic Res.* **2003,** *12*, 115–122.

Smith, E. F.; Townsend, C. O. A Plant-Tumor of Bacterial Origin. *Science* **1907,** *25*, 671–673.

Sorokin, A. P.; Ke, X.; Chen, D.; Elliott, M. C. Production of Fertile Transgenic Wheat Plants via Tissue Electroporation. *Plant Sci.* **2000,** *156*, 227–233.

Steck, T. R. Ti Plasmid Type Affects T-DNA Processing in *Agrobacterium tumefaciens. FEMS Microbiol. Lett.* **1997,** *147*, 121–125.

Svab, Z.; Hajdukiewcz, P.; Maliga, P. Stable Transformation of Plastids in Higher Plants. *Proc. Nat. Acad. Sci. USA* **1990a,** *87*, 8526–8530.

Svab, Z.; Harper, E. C.; Jones, J. D. G.; Maliga, P. Aminoglycoside 3′-Adenyltransferase Confers Resistance to Spectinomycin and Streptomycin in *Nicotianatabacum* Plants. *Plant Mol. Biol.* **1990b,** *14*, 197–205.

Takeuchi, Y.; Dotson, M.; Keen, N. T. Plant Transformation: A Simple Particle Bombardment Device Based on Flowing Helium. *Plant Mol. Biol.* **1992,** *18*, 835–839.

Tepfer, D. Transformation of Several Species of Higher Plants by *Agrobacterium rhizogenes*: Sexual Transmission of the Transformed Genotype and Phenotype. *Cell* **1984,** *37*, 959–967.

Tingay, S.; McElroy, D.; Kalla, R.; Fieg, S.; Wang, M.; Thornton, S.; Brettell, R. *Agrobacterium tumefaciens*-Mediated Barley Transformation. *Plant J.* **1997,** *11*, 1369–1376.

Tinland, B. The Integration of T-DNA into Plant Genomes. *Trends Plant Sci.* **1996,** *1*, 179–184.

Twyman, R. M.; Christou, P. Plant TransformationTechnology–Particle Bombardment. In: *Handbook of Plant Biotechnology*; Christou, P., Klee, H., Eds.; John Wiley & Sons Inc.: New York, 2004; pp 263–289.

Tzfira, T.; Li, J.; Lacroix, B.; Citovsky, V. Agrobacterium T-DNA Integration: Molecules and Models. *Trends Genet.* **2004,** *20*, 375–383.

Uzogara, S. G. The Impact of Genetic Modification of Human Foods in the 21st Century: A Review. *Biotechnol Adv.* **2000,** *18*, 179–206.

Wang, K.; Herrera-Estrella, L.; Van Montagu, M.; Zambryski, P. Right 25 bp Terminus Sequence of the Nopaline T-DNA is Essential for and Determines Direction of DNA Transfer from *Agrobacterium* to the Plant Genome. *Cell* **1984,** *38*, 455–462.

Waters, V. L.; Hirata, K. H.; Pansegrau, W.; Lanka, E.; Guiney, D. G. Sequence Identity in the Nick Regions of IncP Plasmid Transfer Origins and T-DNA Borders of Agrobacterium Ti Plasmids. *Proc. Nat. Acad. Sci. USA.* **1991,** *88*, 1456–1460.

Weber, G.; Monajembashi, S.; Greulich K. D; Wolfrum J. Infection of DNA into Plant Cells with UV Laser Microbeam. NaturWissenschaften, **1988,** *75,* 35–36.

Wong, T. K.; Neumann, E. Electric Field Mediated Gene Transfer. *Biochem. Biophys. Res. Commun.* **1982,** *107*, 584–587.

Xu, X. P.; Li, B. J. Fertile Transgenic Indica Rice Plants Obtained by Electroporation of Seed Embryo Cells. *Plant Cell Rep.* **1994,** *13*, 237–242.

Yoo, S. D; Cho, Y. H; Sheen Jen. Arabidopsis Mesophyll Protoplasts: A Versatile Cell System for Transient Gene Expression Analysis. *Nat. Protoc.*, **2007,** *2*, 1565–1572.

Zoubenko, O. V.; Allison, L. A.; Svab, Z.; Maliga, P. Efficient Targeting of Foreign Genes into the Tobacco Plastid Genome. *Nucl. Acids Res.* **1994,** *22*, 3819–3824.

Zupan, J. R.; Zambryski, P. Transfer of T-DNA from Agrobacterium to the Plant Cell. *Plant Physiol.* **1995,** *107*, 1041–1047.

CHAPTER 2

CHLOROPLAST TRANSFORMATION

TUSHAR RANJAN[1], SANGITA SAHNI[2], VINOD KUMAR[3], PANKAJ KUMAR[3], NIMMY M. S.[4], VIJAY KUMAR JHA[5], VAISHALI SHARMA[6], PRASANT KUMAR[7], and BISHUN DEO PRASAD[3*]

[1]*Department of Basic Science and Humanities Genetics, BAU, Sabour, Bhagalpur, Bihar, India*

[2]*Department of Plant Pathology, T.C.A., Dholi, Muzaffarpur, Bihar, India*

[3]*Department of Molecular Biology & Biotechnology, BAU, Sabour, Bhagalpur, Bihar, India*

[4]*ICAR-N.R.C. on Plant Biotechnology, IARI, Pusa Campus, New Delhi 110012, India*

[5]*Department of Botany, Patna University, Patna, Bihar, India*

[6]*DOS in Biotechnology, University of Mysore, Mysore, Karnataka, India*

[7]*C.G. Bhakta Institute of Biotechnology, Department of Fundamental and Applied Science, Uka Tarsadia University, Bardoli, Surat 394350, Gujarat, India*

[*]*Corresponding author. E-mail: dev.bishnu@gmail.com*

CONTENTS

ABSTRACT

Chloroplasts are subcellular organelles (plastids) of plant cells generally considered to be derived from the prokaryotes, that is, probably from endosymbiotic cyanobacterium that was taken up by eukaryotic cells in symbiotic associations very early during the course of evolution. Chloroplasts have their own genome, and it resembles that of bacteria not that of the nuclear genome. There are no histones associated with the DNA. Chloroplasts have their own protein-synthesizing machinery, and it more closely resembles that of bacteria than that found in the cytoplasm of eukaryotes. Till date, genetic engineering has been experienced mostly in the nuclear genome. However, inserting transgene(s) into the nuclear genome has led to an increasing public concern of the possibility of escape of the transgene through pollen to wild or weedy relatives of the transgenic crops. Scientists suggested that since plastids are compared with prokaryotes, they can take up DNA as in bacterial transformation using naked DNA. Therefore, during the past few years, researchers have begun to evaluate application of plastid transformation in plant biotechnology as a viable alternative to conventional technologies for transformation of the nuclear DNA.

2.1 INTRODUCTION AND EVOLUTIONARY ORIGIN OF CHLOROPLAST

Chloroplasts are subcellular organelles (plastids) of plant cells generally considered to be derived from the prokaryotes, that is, probably from endosymbiotic cyanobacterium that was taken up by eukaryotic cells in symbiotic associations very early during the course of evolution (Fig. 2.1). They are no longer capable to exist independently in eukaryotic cell, but they have retained a small autonomous genome that contains few hundreds of genes, many of the genes from the organelle transferred to the nucleus. So many proteins that function in chloroplast are encoded by nuclear genes and are being transported to the organelle (Weber and Osteryoung, 2010). Plastid genomes resemble bacterial genomes in many aspects and also contain some features of multicellular organisms, such as RNA editing and split genes. Both chloroplasts and mitochondria can arise only from preexisting mitochondria and chloroplasts. They cannot be formed in a cell that lacks them because nuclear genes encode only some of the proteins of which they are made. Chloroplasts have their own genome, and it resembles that of bacteria not that of the nuclear genome (Bhattacharya et al., 2007). There are no

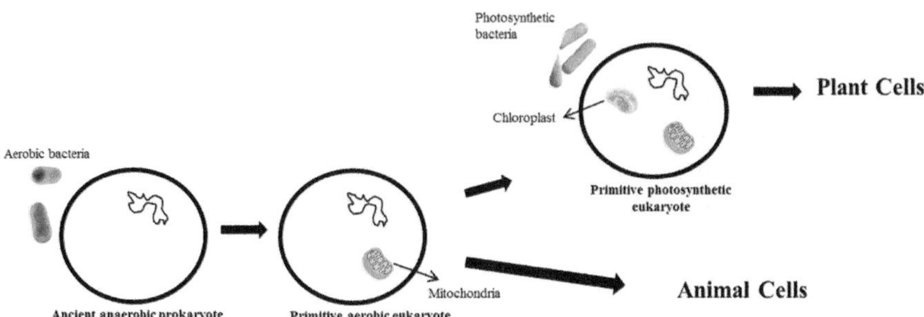

FIGURE 2.1 Endosymbiosis and the origin of eukaryotes.

histones associated with the DNA. Chloroplasts have their own protein-synthesizing machinery, and it more closely resembles that of bacteria than that found in the cytoplasm of eukaryotes. The first amino acid of their transcripts is always *fMet* as it is in bacteria (not methionine [Met] that is the first amino acid in eukaryotic proteins). A number of antibiotics (e.g., streptomycin) that act by blocking protein synthesis in bacteria also block protein synthesis within mitochondria and chloroplasts. They do not interfere with protein synthesis in the cytoplasm of the eukaryotes. Conversely, inhibitors (e.g., diphtheria toxin) of protein synthesis by eukaryotic ribosomes do not sensibly enough have any effect neither on bacterial protein synthesis nor on protein synthesis within mitochondria and chloroplasts. The antibiotic rifampicin, which inhibits the RNA polymerase of bacteria, also inhibits the RNA polymerase within mitochondria. It has no such effect on the RNA polymerase within the eukaryotic nucleus (Bhattacharya et al., 2007). Chloroplasts are organelles present in photosynthetic plant cells. Their principal function is to capture energy from light to fix atmospheric CO_2 and convert it into sugars. They possess their own genome with a variable size up to several hundred kilobases; each chloroplast can contain up to 100 copies of its own genome. The number of chloroplast varies between 1 and more than 100 in higher plants. They are present in shoots and leaves of green plants and contain pigment called chlorophyll. They are also present in several forms as colorless plastids (amyloplasts) in roots and as colored plastids (chromoplasts) in fruits. The chloroplast genome is circular in nature, double-stranded DNA located in stroma of the chloroplast. Genome is highly conserved amongst plant species. The exact copy number varies during development, but mesophyll cell in young leaves contain approximately 100 copies of genome. The size of the chloroplast genome in most of the plant species ranges between 120 and 160 kb and contain 120–140 genes (Bhattacharya et al., 2007; Wambugu et al., 2015). Most of the genes encode for either protein synthesis or photosynthesis and are located in nucleus. Most of the genes are aligned in clusters that allow expression in the form of large polycistronic primary transcripts which are processed to oligo or monocistronic messenger RNAs (mRNAs). Chloroplast transformation is becoming more popular and an alternative to nuclear gene transformation because of various advantages like high protein levels, the feasibility of expressing multiple proteins from polycistronic mRNAs, and gene containment through the lack of pollen transmission.

Till date, genetic engineering has been experienced mostly in the nuclear genome. However, inserting transgene(s) into the nuclear genome has led to an increasing public concern of the possibility of escape of the transgene through pollen to wild or weedy relatives of the transgenic crops. Scientists suggested that since plastids are compared with prokaryotes, they can take up DNA as in bacterial transformation using naked DNA. Therefore, during the past few years, researchers have begun to evaluate application of plastid transformation in plant biotechnology as a viable alternative to conventional technologies for transformation of the nuclear DNA. Recently, plastids have become attractive targets for genetic engineering efforts. Plants with transformed plastid genomes are termed transplastomic (Singh et al., 2010). Chloroplast transformation technologies were developed to transform these organelles into cell factories for production of molecules of high commercial value (therapeutic compounds, recombinant proteins, precursors for biofuel production). These technologies use the repetition of the chloroplast genetic material (up to 10,000 copies of the genome per cell) to produce high amounts of protein. Moreover, chloroplasts are not usually transmitted by pollen (maternal inheritance); therefore, the risk of the transfer of modified genes to other plants is minimal. Plastid transformation was first achieved in a unicellular green eukaryotic alga *Chlamydomonas reindhartii* in 1990, which helped to pave the way for the development of stable chloroplast transformation. In higher plants, *Nicotiana tabaccum* has been used, because of the relative ease of its tissue culture and regeneration. Gene delivery into plastid initially done using Agrobacterium-mediated transformation method. After discovery gene gun method, stable transformation was achieved by introduction of *Escherichia coli* plasmids contained marker, and gene of interest is introduced into plastids. The foreign genes are inserted into host plasmid DNA by homologous recombination (Fig. 2.3A). PEG-mediated transformation is also utilized for plastid transformation. However, biolistic method has been used for manipulation of wide variety of explants such as leaves, cotyledons, or cultured cells than PEG treatment. Many vectors have been developed specifically for chloroplast transformation. In addition, the aadA marker and methods for removal of marker were first demonstrated in higher plants, tobacco due to its ease of culture and regeneration, gained significant attention for chloroplast transformation (Boynton et al., 1988). Tobacco protoplasts were cocultivated with Agrobacterium but the resulted transgenic lines showed the unstable integration of foreign DNA into the chloroplast genome. The candidate genes were introduced in isolated intact chloroplasts and then into protoplasts resulting in transgenic plants. Recently, tobacco plastid has been engineered to express the E7 HPV type 16 protein, which is an attractive candidate for anticancer vaccine development. In addition to model plant tobacco, many transplastomic crop plants have been generated that possess higher resistance to biotic and abiotic stresses and molecular pharming. Gene gun, a transformation device, was developed by John Sanford to enable the transformation of plant chloroplasts without using isolated plastids (Svab and Maliga, 1993).

2.2 ADVANTAGE OF CHLOROPLAST TRANSFORMATION OVER NUCLEAR TRANSFORMATION

Introduction of foreign genes into nuclear genome of higher plant has become routine and a number of useful genes have been transformed into crop plants. Despite tremendous application, such transformed genes may get leaked into wild relatives due to cross pollination and pose big environmental problems. Also, the expression level is sometimes very low due to chromosomal position effects and trans-gene interacting. All such problems may be overcome in plastid transformation.

There are several advantages of chloroplast transformation over nuclear transformation (Table 2.1). Insertion of foreign gene into plastid genome may result in amplification of 50–100 copies of the gene per cell. In many particular species, all plastid types carry identical, multiple copies of the same genome (Palmer, 1991). There are 10–15 proplastids in meristematic cells and each containing 50 genome copies. Leaf cell contains as much as 100 chloroplasts and each chloroplast contains approximately 100 copies of the plastid genome giving in total 10,000 copies of the plastid genome per cell. It should be noted that there may be significant species-specific deviation from these mean values, with a total number of genomes per leaf cell in the range of 1900–50,000 (Maliga, 1993). In plastid transformation, there is no damage of the introduced gene getting leaked into wild relatives

TABLE 2.1 Advantages of Plastid Transformation Compare to Nuclear Transformation.

Plastid Transformation	Nuclear Transformation
Plastid genome is highly polyploid leading to high accumulation of proteins	Nuclear genome is not highly polyploidy, hence low level of protein expression
Plastid possesses prokaryotic expression system, which facilitates the expression of several genes simultaneously from the single operon	The nucleus does not possess prokaryotic expression system, hence can't express several genes simultaneously
Facilitate the expression of multi-subunit complex protein from polycistronic mRNA under a single promoter	Several promoters are needed for the expression of the individual gene encoding the respective subunit
Polycistronic multigene expression enables enhanced sequential metabolic reactions in a single transformation procedure	Multiple transformation procedure will be required to achieve multigene expression
The use of single operon to express several genes removes the burden of using several selection markers	Several selection markers will be used to independently select for integration events of these individual gene
The plastid genome is versatile in codon usage for recombinant protein production	Widespread codon usage bias exist, hence the codon optimization is common in order to optimize translation efficiency
Provide substantial degree of natural biocontainment of transgene flow by out-crossing, as plastids are inherited through maternal tissues in most species	There is always a risk of out-crossing through pollination
No positional effect and epigenetics inheritance because integration is guarded into the functional region of the genome through homologous targeting	There are position effect and epigenetics inheritance because integration is random
Absence of transgene instability and gene silencing	Presence of transgene instability and gene silencing

(Reprinted from Obembe, O,O.; Popoola, J. O.; Leelavathi, S.; Reddy, V.S. Recent advances in plastid transformation. Indian J.Sci.Technol, 2010, 3(2): 1229-1235. http://www.indjst.org/index.php/indjst/article/view/29871/25831. https://creativecommons.org/licenses/by/3.0/)

as plastid genes are inherited in (almost all) crop plant by the female parent only. Therefore, relocation of nuclear genes to the plastid genome will confine to the transferred genes to the crop. Relocation of genes, such as that encoding herbicide resistance, to the plastids would prevent the transfer of herbicide resistance to the other species by cross pollination (Maliga, 1993). The codon usages of chloroplast genes which are close to prokaryotic genes are therefore a suitable place to express useful bacterial genes. Chloroplast genes are preceded by the −35 and −10 elements typical of prokaryotic promoters. These genes transcribed by RNA polymerase containing plastid-encoded subunits homologous to the α, β, and β′ subunits of *E. coli* RNA polymerase (Igloi and Kossel, 1992). On the other hand, transformation of chloroplast genomes has some potential problems. First, it may be more difficult for DNA to cross the plastid double membrane than the nuclear membrane (Maliga, 1993). Second, chloroplast genomes are present in much higher copy number than nuclear genomes as mentioned above. For a transformed genome to replace all copies of the original genome, strong selection pressure must be applied. The transgenic plastid genomes are products of a multiple-step process, involving DNA recombination, copy correction, and sorting out of plastid DNA copies (Svab et al., 1990). Chloroplast genome can become somewhat unstable following transformation and that gene amplification represents highly specialized phenomena that are not easily manipulated (Suzuki et al., 1997). Unintegrated plasmid DNA has also been detected in chloroplast transformants (Turkec, 1999).

2.3 CHLOROPLAST TRANSFORMATION: A HIGH-THROUGHPUT CELLULAR PROTEIN FACTORIES

Chloroplast genome is maternally inherited and there is rare occurrence of pollen transmission. It provides a strong level of biological containment and thus reduces the escape of transgene from one cell to other. It exhibits higher level of transgene expression and thus higher level of protein production due to the presence of multiple copies of chloroplast transgenes per cell and remains unaffected by phenomenon such as pre- or posttranscriptional silencing (Daniell et al., 2004). Chloroplast transformation involves homologous recombination and is therefore precise and predictable. This minimizes the insertion of unnecessary DNA that accompanies in nuclear genome transformation. This also avoids the deletions and rearrangements of transgene DNA, and host genome DNA at the site of insertion. RNA interference does not occur in genetically engineered chloroplasts. Absence of position effect due to lack of a compact chromatin structure and efficient transgene integration by homologous recombination made this system very fascinating for scientific community. Genetic engineering in chloroplast avoids inadvertent inactivation of host gene by transgene integration. Ability to form disulfide bonds and folding human proteins results in high-level production of biopharmaceuticals in plants. High level of expression and engineering foreign genes without the use of antibiotic resistant genes makes this compartment ideal for the development of edible vaccines (Daniell et al., 2004). Chloroplast is originated from cyanobacteria through endosymbiosis.

It shows significant similarities with the bacterial genome. Thus, any bacterial genome can be inserted in chloroplast genome (Daniell, 2007).

2.3.1 BASIC PRINCIPLE OF CHLOROPLAST TRANSFORMATION

First, the gene or genes to be introduced into the plastid genome are coated onto microscopic gold particles (0.6–1 mm in diameter) (Fig. 2.2). These DNA-coated gold particles are then shot into tobacco/plant cells using a helium-driven biolistic gun. Following shooting, transformed plant cells (plant cells that contain a plastid or plastids with the gene of interest) are selected and a new transplastomic plant is regenerated from this plant cell (Fig. 2.2). Although simple in principle, the selection and regeneration of transplastomic plants is prone to errors using current conventional antibiotics-based selection methods (Wani et al., 2010). New selection and regeneration method has been developed which is more efficient and reliable compared to currently available technologies.

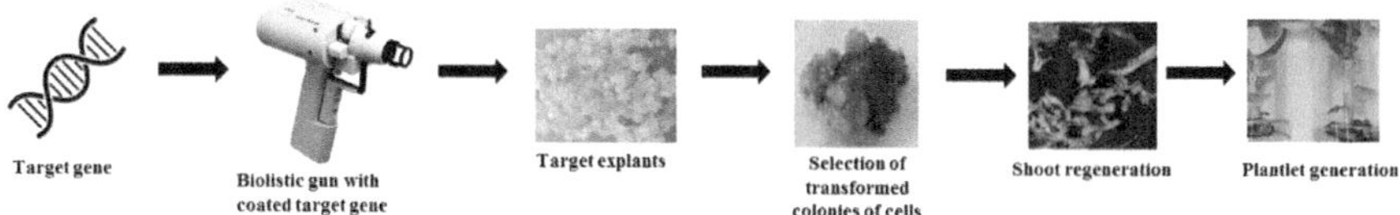

FIGURE 2.2 Basic principle lies behind the success of chloroplast transformation.

2.3.2 VECTOR FOR CHLOROPLAST TRANSFORMATION

Some special plastid transformation vectors, for example, pZS148, are widely used. Vector pZS148 contains the following modules: (1) 16S ribosomal DNA from the tobacco genotype SPC2 (it specifies resistance to both streptomycin and spectinomycin), (2) the origin (cp-ori) for replication in chloroplasts, and (3) a segment of a phagemid vector (pBluescript IKS+) for cloning in *E. coli*. This vector has been designed as a shuttle vector to replicate both in *E. coli* and chloroplasts. The vector pZS148 can be introduced into tobacco leaves by many methods which include particle gun delivery, and transformants can be recovered. These recombinants plants show maternal inheritance for resistance to the two antibiotics. It is estimated that gene integration in cpDNA was only 1% of that in nuclear DNA. Stable chloroplast transformation depends on the integration of the transgene into the chloroplast exclusively by homologous recombination (Maliga, 2003) (Fig. 2.3).

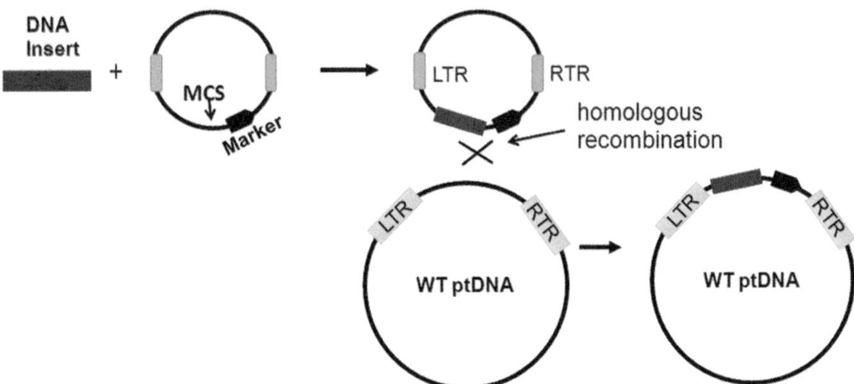

FIGURE 2.3 Transformation of plastid DNA (ptDNA) by homologous recombination.

2.3.3 SELECTABLE MARKERS

Some selectable markers for chloroplast transformation are now available. Of these, spectinomycin and kanamycin resistance serve as dominant markers, while recessive antibiotic resistance markers are provided by genes encoding antibiotic-insensitive alleles of ribosomal RNA genes and by genes that restore photoautotrophic growth by complementing nonphotosynthetic mutants (Maliga, 2003). Figure 2.4 represents development and selection of transgenic plants after chloroplast transformation.

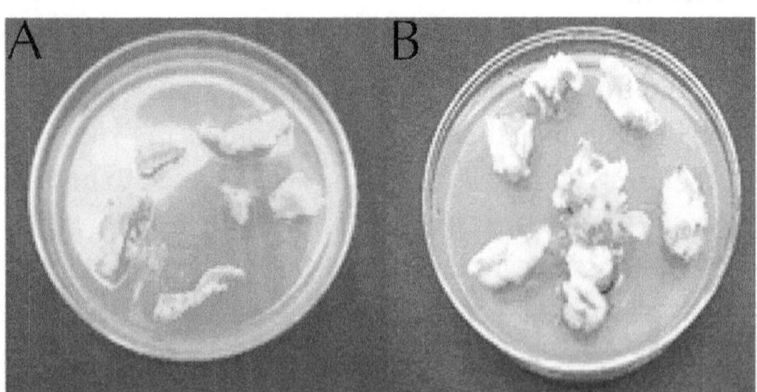

FIGURE 2.4 Transformation of chloroplast genome by bombarding tobacco leaves with microprojectiles coated with DNA. Following bombardment, leaf disc are placed on to medium containing antibiotics (Panel A). Transgenic plants are regenerated from the transformed tissue transformed tissue that is able to develop green chloroplast (Panel B). (Reprinted from https://www.slideshare.net/SACHINEKATPURE/chloroplast-transformation).

2.4 CHLOROPLAST TRANSFORMATION METHODS

Some of the methods used for gene transfer into nuclei also bring about chloroplast transformation; these methods are (1) particle gun or biolistic DNA delivery, (2) PEG treatment, and (3) microinjection. Table 2.2 represents the milestone of chloroplast transformation till date. Particle gun is the most widely used and effective method of transforming chloroplasts. This method yields a high efficiency rate and can be used to transform a variety of explants. The first successful chloroplast transformation in higher plants was achieved in 1989, when spectinomycin resistance was transferred into tobacco. Protoplasts take up DNA in the presence of PEG; this DNA is also transported by some yet unknown means into the chloroplast where it may become integrated into the chloroplast genome. PEG-mediated DNA uptake is less efficient than the biolistic approach. Biolistic transfection represents a direct physical gene-transfer approach in which nucleic acids are precipitated on biologically inert high-density microparticles (usually gold or tungsten) and delivered directly through cell walls and/or membranes into the nucleus of target cells by high velocity acceleration using a ballistic device such as the gene gun (Wani et al., 2010). This method uses He as a propellant (Fig. 2.5A and Table 2.2). In the Galistan expansion femtosyringe approach, DNA is microinjected directly into chloroplasts using a very small syringe (Fig. 2.5B). The heat-induced expansion of a liquid metal, Galistan, within a glass syringe forces the plasmid DNA through a capillary tip (diameter 0.1 nm). This method is not widely used. An efficient and inexpensive technique to transform chloroplasts in a wide range of crop species is yet to be devised. The currently used methods have only been optimized for the transformation of tobacco, and only a limited success has been obtained with tomato and a few other species. Agrobacterium-mediated gene delivery is not very much successful for the chloroplast transformation so far (Wani et al., 2010).

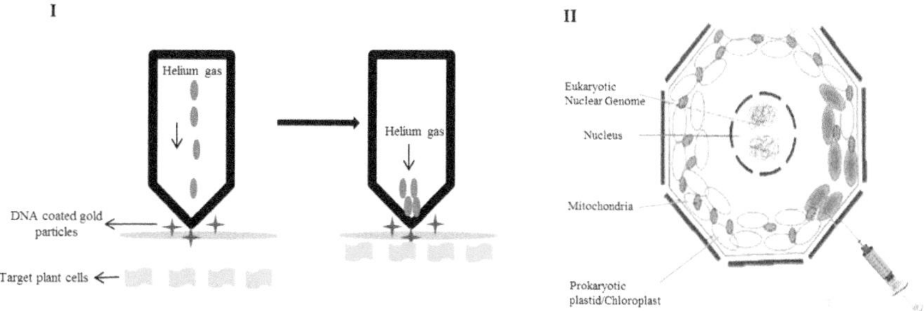

FIGURE 2.5 Biolistic DNA delivery (A) and Galistan expansion femtosyringe or microinjection approach (B).

Plastid transformation has become an attractive alternative to nuclear gene transformation due to several other advantages. The high ploidy number of the plastid genome allows high levels (up to 1–40% of total protein) of protein expression or expression of the transgene. It has been reported that while nuclear transgenes typically result in 0.5–3% of total proteins, concentration of proteins expressed by plastid transgenes is much higher, up to 18% (Daniell et al., 2004; Hou et al., 2003).

The greater production of the expressed protein is possible because plastid transgenes are present as multiple copies per plant cell, and they are little affected by phenomena like pre- or posttranscriptional silencing. Other advantages of plastid engineering are the capacity to express multiple genes from polycistronic mRNA, and the absence of epigenetic effects and gene silencing (Table 2.1). The major difficulty in engineering plastid genome for production of transplastomic plants is to generate homoplasmic plants in which all the plastids are uniformly transformed. This requires a long process of selection, thus, hampering the production of genetically stable transplastomic plants (e.g., rice). This is due to the presence of about 10–100 plastids, each of which has up to 100 copies of the plastid genome, in one cell, that does not allow achieving homoplastomic state. It was also stated that getting high level of protein expression, even though the gene copy number is high,

TABLE 2.2 Milestones of Chloroplast Transformation.

Crop	Transformation Method
Plant regeneration by embryogenesis	
Carrot	Fine cell suspension cultures derived from stem
Cotton	Biolistic using 0.6-mm gold particles
Rice	Biolistic using 0.6-mm gold particles
Soybean	Biolistic using 0.6-mm gold particles
Plant regeneration by organogenesis from protoplasts	
Cauliflower	PEG4000 mediated
Lettuce	PEG4000 mediated
Plant regeneration by organogenesis from leaf	
Cabbage	Biolistic using 1.0-mm gold particles
Lettuce	Biolistic using 0.6-mm gold particles
Oilseed rape	Biolistic using tungsten particles
Potato	Biolistic using 0.6-mm gold particles
Tomato	Biolistic using 0.6-mm gold particles

is another problem. In 2005, however, Nguyen et al. (2005) described the generation of homoplasmic plastid transformants of a commercial cultivar of potato (*Solanum tuberosum* L.) using two tobacco specific plastid transformation vectors, pZS197 (*Prrn/aadA/psbA3'*) and pMSK18 (*trc/gfp/Prrn/aadA/psbA3'*). Similarly, Liu et al. (2007) were able to develop homoplasmic fertile plants of *Brassica oleracea* L. var. *capitata* L. (cabbage). Among other higher plants of which fertile homoplasmic plants with genetically modified plastid genomes have been produced are *Nicotiana tabacum* (tobacco), *Nicotiana plumbaginifolia* (texmex tobacco), *Solanum lycopersicum* (tomato), *Glycine max* (soybean), *Lesquerella fendleri* (bladder pod), *Gossypium hirsutum* (cotton), *Petunia hybrid* (petunia), and *Lactuca sativa* (lettuce). The amino glycoside 3-adenylyl transferase *(aad*A) gene, which confers dual resistance to spectinomycin–streptomycin antibiotics, is still the selectable marker that is routinely used efficiently for plastid transformation. Since the antibiotic resistant genes used in transformation are not desirable in the final products, different strategies have been developed to eliminate the necessity of using such selectable markers.

2.5 IMPROVEMENT OF AGRONOMIC TRAITS BY PLASTID ENGINEERING

The transplastomic technology used to improve agronomic traits such as insect resistance, herbicide resistance, and tolerance/resistance of stresses such as disease, drought, insect pests, salinity, and freezing that limit plant growth. Some of the examples are given below (Table 2.3).

2.5.1 INSECT RESISTANCE

Introduction of Bt toxin into plastid genome showed high level of toxin accumulation in leaf protein (5%) as compared to total soluble protein through nuclear genome transformation. For example, the high level of cry1Ac expression in plastid with rbcS:tp system resulted in high levels of plant resistance to three kinds of pests such as leaf folder, green caterpillar, and skipper in rice plant. Similarly, several other *cry* proteins have also been expressed in plastids of tobacco (Gatehouse, 2008) (Table 2.3).

TABLE 2.3 Lists of Transplastomic Plants are Engineer for Agronomic Traits.

Plant species	Gene	References
Nicotiana tabacum	*rrn1*	Maliga and Tungsuchat-Huang (2014)
Nicotiana tabacum	*nptII*	Verma and Daniell (2007)
Nicotiana tabacum	*uidA*	Sytnik et al. (2005)
Nicotiana tabacum	*cry9Aa2*	Chakrabarti et al. (2007)
Nicotiana tabacum	*rbcL*	Kanevski et al. (1999)
Nicotiana tabacum	*aadA* and *gfp*	Davarpanah et al. (2009)
Arabidopsis thaliana	*aadA*	Davarpanah et al. (2009)
Solanum tuberosum	*aadA* and *gfp*	Verma and Daniell (2007)
Brassica napus	*aadA* and *cry1Aa10*	Hou et al. (2003)
Daucus carotadehydrogenase	*badh*	Kumar et al. (2004)
Zea mays	*ManA*	Ahmadabadi et al. (2007)

2.5.2 DISEASE RESISTANCE

Plastid engineering found to be an effective option in development of plants resistant to various bacterial and fungal diseases. In tobacco, MSI-99 gene which codes for antimicrobial peptide introduced into plastid resulted in transplastomic plants [88% (T1) and 96% (T2)] showed resistance to pathogens colletotrichum destructive and pseudomonas syringe (Wang et al., 2015) (Table 2.3).

2.5.3 HERBICIDE RESISTANCE

Many herbicides are plant specific to their inhibitory effect and many of which takes in the plastid. Hence, various herbicides have been used for design of chloroplast selectable markers. For example, the enzyme HPPD involved in biosynthesis of quinones and vitamin E. It is inhibited by Diketonitrile and sulcatrione, which are metabolic derivative of herbicide isoxaflutiole. Chloroplast over expression of HPPD gene from barley showed tolerance to sulcatrione in tobacco (Kang et al., 2003) (Table 2.3).

2.5.4 IMPROVEMENT OF PHOTOSYNTHESIS

Many efforts have been initiated to engineer improvements in photosynthesis to meet the challenges of increasing demands for food and fuel in rapidly changing environmental conditions. Various transgenes have been introduced into either the nuclear or plastid genomes in attempts to increase photosynthetic efficiency. Researchers have exploited the technology to understand how plastid genes are regulated, to determine the function of plastid gene products, to produce large amounts of particular endogenous or foreign proteins, or to alter photosynthesis or metabolism of the alga or plant. Recently, RUBISCO has been manipulated through plastid transformation to achieve higher rate of photosynthesis (Table 2.2). Scientists has also illustrated how plastid operons could be created for expression of the multiple genes needed to introduce new pathways or enzymes to enhance photosynthetic rates or reduce photorespiration (Hanson et al., 2012).

Apart from above-mentioned advantages, plastid also used as pharmaceutical, vaccine, biomaterials bioreactors. Several chloroplast-derived biopharmaceutical proteins have been reported till date. Stable expression of a pharmaceutical protein in chloroplasts was first reported for GVGVP, a protein-based polymer with medical uses such as wound coverings, artificial pericardia, and programed drug delivery. Human ST (hST), a secretory protein, was expressed inside chloroplasts in a soluble, biologically active and disulfide-bonded form. The key use of hST is in the cure of hypopituitary dwarfism in children; additional indications are treatment of Turner syndrome, chronic renal failure, and human immunodeficiency virus wasting syndrome.

2.6 LIMITATIONS OF CHLOROPLAST TRANSFORMATION

For some unknown reason, chloroplast transformation frequencies are much lower than those for nuclear transformation. Prolonged selection procedures, typically, 2–4 regeneration and selection cycles, under high selection pressure are required for the recovery of transformants. The methods of transgene transfer into chloroplasts are limited, and they are either expensive or require regeneration from protoplasts. These transformation systems are far more successful with tobacco than with other plant species. Products of transgenes ordinarily would accumulate in green plant parts only.

2.7 CONCLUSION

The interesting features of plastid compartment and genome are exceptional advantages of plastid genome engineering. Plastid genomes are highly conserved when compared to nuclear genome. A varied repertoire of selectable markers has been developed over two decades since the first report of plastid transformation. Plastid genome sequencing is a starting point to elucidate many process-related plastid gene functions, metabolic processes, and generate high-efficient transformation vectors. Limited number of plastid genomic sequences is the major constraint to add new traits and increasing marker value in commercial crops. The interesting features of plastid compartment and genome are exceptional advantages of plastid genome engineering. Improvement of chloroplast isolation and evolution of genome sequencing technology is useful in implementing plastid transformation technology in various plant species.

KEYWORDS

- **transformation**
- **endosymbiosis**
- **DNA delivery**
- **PEG**
- **biolistic method**
- **cpDNA**
- **homologous recombination**

REFERENCES

Ahmadabadi, M.; Ruf, S.; Bock, R. A Leaf-Based Regeneration and Transformation System for Maize (*Zea mays* L.). *Transgen. Res.* **2007,** *16* (4), 437–448.

Bhattacharya, D.; Archibald, J. M.; Weber, A. P.; Reyes-Prieto, A. How Do Endosymbionts Become Organelles? Understanding Early Events in Plastid Evolution. *Bioessays* **2007,** *29*, 1239–1246.

Boynton, J. E.; Gillham, N. W.; Harris, E. H.; Hosler, J. P.; Johnson, A. M.; Jones, A. R.; Randolph-Anderson, B. L.; Robertson, D.; Klein, T. M.; Shark, K. B. Chloroplast Transformation in *Chlamydomonas* with high velocity microprojectiles. *Science* **1988,** *240*, 1534–1538.

Chakrabarti, S. K.; Lutz, K. A.; Lertwiriyawong, B.; Svab, Z.; Malig, P. Expression of the *cry9Aa2 B.t.* Gene in Tobacco Chloroplasts Confers Resistance to Potato Tuber Moth. *Transgen. Res.* **2007,** *15*, 481.

Daniell, H. Transgene Containment by Maternal Inheritance: Effective or Elusive? *Proc. Nat. Acad. Sci. U.S.A.* **2007,** *104*, 6879–6880.

Daniell, H.; Carmona-Sanchez, O.; Burns, B. Chloroplast Derived Antibodies, Biopharmaceuticals and Edible Vaccines. In: *Molecular Farming*; Fischer, R., Schillberg, S., Eds.; Verlag Publishers: Weinheim, Germany, 2004; pp 113–133.

Davarpanah, S. J.; Jung, S. H.; Kim, Y. J.; Park, Y.; Min, S. R.; Liu, J. R.; Jeong, W. J. Stable Plastid Transformation in *Nicotiana benthamiana*. *J. Plant Biol.* **2009,** *52* (3), 244–250.

Gatehouse, J. A. Biotechnological Prospects for Engineering Insect Resistant Plants. *Plant Physiol.* **2008,** ***146***, 881–887.

Hanson, M. R.; Gray, B. N.; Ahner, B. A. Chloroplast Transformation for Engineering of Photosynthesis. *J. Exp. Bot.* **2012**. doi:10.1093/jxb/ers325.

Hou, B.; Zhou, Y.; Wan, Li.; Zhang, Z.; Shen, G.; Chen, Z.; Hu, Z. Chloroplast Transformation in Oilseed Rape. *Transgen. Res.* **2003,** *12* (1), 111–114.

Igloi, G.; Kossel, H. Transcriptional Apparatus of Chloroplast. *Crit. Rev. Plant Sci.* **1992,** *10*, 525–558.

Kanevski, I.; Maliga, P.; Rhoades, D. F.; Gutteridge, D. Plastome Engineering of Ribulose-1,5-Bisphosphate Carboxylase/Oxygenase in Tobacco to Form a Sunflower Large Subunit and Tobacco Small Subunit Hybrid. *Plant Physiol.* **1999,** *119*, 133–141.

Kang, T. J.; Seo, J. E.; Loc, N. H.; Yang, M. S. Herbicide Resistance of Tobacco Chloroplasts Expressing the Bar Gene. *Mol. Cells* **2003,** *16* (1), 60–66.

Kumar, S.; Dhingra, A.; Daniell, H. Plastid-Expressed *Betaine Aldehyde Dehydrogenase* Gene in Carrot Cultured Cells, Roots, and Leaves Confers Enhanced Salt Tolerance. *Plant Physiol.* **2004,** *136* (1), 2843–2854.

Liu, C. W.; Lin, C. C.; Chen, J.; Tseng, M. J. Stable Plastid Transformation in Cabbage (*Brassica oleracea* L. var. *capitata* L.) by Particle Bombardment. *Plant Cell Rep.* **2007,** *26*, 1733–1744.

Maliga, P. Towards Plastid Transformation in Flowering Plants. *TIBTECH* vol. 11, Elsevier Science Publishers Ltd., 1993; pp 101–107.

Maliga, P. Progress Towards Commercialization of Plastid Transformation Technology. *Trends Biotechnol.* **2003,** *21*, 20–28.

Maliga, P.; Tungsuchat-Huang, T. Plastid Transformation in *Nicotiana tabacum* and *Nicotiana sylvestris* by Biolistic DNA Delivery to Leaves. *Methods Mol Biol.* **2014,** *1132*, 147–163.

Nguyen, T. T.; Nugent, G.; Cardi, T.; Dix, P. J. Generation of Homoplasmic Plastid Transformants of a Commercial Cultivar of Potato (*Solanum tuberosum* L.). *Plant. Sci.* **2005,** *168*, 1495–1500.

Obembe, O,O.; Popoola, J. O.; Leelavathi, S.; Reddy, V.S. Recent advances in plastid transformation. Indian J.Sci.Technol, 2010, 3(2): 1229-1235.

Palmer, J. D. The Molecular Biology of Plastids. In: *Cell Culture and Somatic Cell Genetics of Plants*; Bogorad, L., Vasil, I. K., Eds.; Academic Press: Cambridge, MA, 1991; Vol 7.4, pp 5–53.

Singh, A. K.; Verma, S. S.; Bansal, K. C. Plastid Transformation in Eggplant (*Solanum melongena* L.). *Transgen. Res.* **2010,** *19*, 113–119.

Sytnik, E.; Komarnytsky, I.; Gleba, Y.; Kuchuk, N. Transfer of Transformed Chloroplasts from *Nicotiana tabacum* to the *Lycium barbarum* plants. *Cell Biol. Int.* **2005,** *29* (1), 71–75.

Suzuki, H., Ingersoll, J.; Stern, D. B.; Kindle, K. L. Generation and Maintenance of Tandemly Repeated Extrachromosomal Plastid DNA in *Chlamydomonas* Chloroplast. *Plant J.* 1997, *11* (4), 635–658.

Svab, Z.; Handukieviczç, P.; Maliga. P. Stable Transformation of Plastids in Higher Plants. *Proc. Natl. Acad. Sci. U.S.A.* **1990,** *87*, 8526–8530.

Svab, Z.; Maliga, P. High-Frequency Plastid Transformation in Tobacco by Selection for a Chimeric *aadA* Gene. *Proc. Nat. Acad. Sci. U.S.A.* **1993,** *90*, 913–917.

Turkec, A. Transformation and Expression of *Bacillus sphaericus* Binary Toksin Genes in *Chlamydomonas reinhardtii. Turkish J. Field Crops.* **1999,** *4*, 85–90.

Verma, D.; Daniell, H. Chloroplast Vector System for Biotechnology Applications. *Plant Physiol.* **2007,** *145* (4), 1129–1143.

Wambugu, P. W.; Brozynska, M.; Furtado, A.; Waters, D. L.; Henry, R. J. Relationships of Wild and Fomesticated Rices (*Oryza* AA Genome Species) Based Upon Whole Chloroplast Genome Sequences. *Nat. Sci. Rep.* **2015**. doi:10.1038/srep13957.

Wani, S. H.; Haider, N.; Kumar, H.; Singh, N. B. Plant Plastid Engineering. *Curr. Genomics* **2010,** *11*, 500–512.

Wang, Y.; Wei, Z.; Zhang, Y.; Lin, C.; Zhong, X.; Wang, Y.; Ma, J.; Ma. J.; Xing, S. Chloroplast-Expressed MSI-99 in Tobacco Improves Disease Resistance and Displays Inhibitory Effect against Rice Blast Fungus. *Int. J. Mol. Sci.* **2015,** *16* (3), 4628–4641.

Weber, A. P. M.; Osteryoung, K. W. From Endosymbiosis to Synthetic Photosynthetic Life. *Plant Physiol.***2010,** *154*, 593–59.

CHAPTER 3

FUNGAL TRANSFORMATIONS AND ADVANCEMENTS

VARSHA C. MOHANAN[1*] and SANGITA SAHNI[2]

[1]*Bharat Chattoo Genome Research Centre, Faculty of Science, The M.S. University of Baroda, Vadodara 390002, Gujarat, India*

[2]*Department of Plant Pathology, Tirhut Agriculture College, Dholi, Muzaffarpur, Bihar, India*

[*]*Corresponding author. E-mail: vcm484@gmail.com*

CONTENTS

ABSTRACT

Recently, fungal systems have gained immense importance as model systems for genetic research due to its amenability to genetic manipulation. For genetic manipulation, the establishment of transformation system is the prerequisite. Fungal transformations have come a long way since last few decades and has enabled researchers to develop several model fungal systems. Researchers have developed different transformation techniques and gene functional analytical techniques for fungal systems. Both physical and biological methods of transformations techniques have been developed in fungus. This chapter discusses about the recent advancements in different transformation techniques including some major physical methods like protoplast transformation, lithium acetate method, bolistic method, electroporation and also the most popular biological method, the *Agrobacterium*-mediated transformation.

3.1 INTRODUCTION

As biology advanced with years, the need for understanding organisms in depth increased. With the discovery of DNA (Watson and Crick, 1953) as the basic genetic element of living organisms, the efforts to understand DNA as well as to manipulate it to the benefit of mankind began with interest. Simple microorganisms were used as model organisms for research. With the discovery of bacterial transformation principles (Hershey and Chase, 1952; McCarty, 1944), a new path was opened. A further transformation of bacterial organisms was taken up as a methodology for basic genetic studies. Genetics of the fungi is key since fungi are important in agrobiotech industry, plant pathology, and clinical pathology. Filamentous fungi has grown to be researchers' much preferred model system recently, owing to their cellular assembly and complex mode of existence which can be extrapolated to both prokaryotes and higher eukaryotes. The small and tractable genome makes it easier to conduct genetic studies. Since many of the filamentous fungi have highly efficient secretory system and high production of some industrially important enzymes has interested several researchers all over the world. Clinically and agronomically important filamentous fungi have gained immense importance in last two decades. Molecular technology combined with genetic approaches has helped researchers to improve the fungal system for fundamental studies as well as for applied biology. Developing a genetic transformation system for a new model organism is a major challenge, since immense effort is needed in the application of existing variety of existing transformation methods and choosing the best suitable and efficient for the new system. The basic method of transformation was undertaken both physically and biologically depending on the recipient host biology and efficacy. The physical methods included electroporation, biolistics, vacuum infiltration, glass bead method, and shock wave method. The biological method of transformation was either using protoplasts (naked cells) or *Agrobacterium*-mediated methods. All these transformation methods are key to functional genomics of the fungal system. In fungi, a major problem for genetic analysis is the multinucleate nature of filamentous fungi. The presence of multiple nuclei can fail methods, such as gene replacement and insertional mutagenesis, which rely on the isolation of homokaryotic transformants derived from a single transformation event to study loss of function mutants (Vijn and Govers, 2003). Therefore, different methods of transformation techniques were followed by different gene functional analytical techniques. Researchers adopted both forward and reverse genetics to functional genomics.

3.2 PROTOPLAST TRANSFORMATION OF FUNGAL SPECIES

Protoplast generation is the most critical step in most of the protocols of fungal transformation. Protoplasting has always been a task of unstable success rate since it fairly depends on the nature of the fungal cell wall, its components, and the digestive enzyme used. Understanding filamentous fungal cell wall and defining a suitable digestive enzyme is therefore essential to ensure a good protoplast preparation. Filamentous fungal cell walls are complex and composite whose components largely remain the same but proportion changes through different stages, designed for a variety of functions. Fungal cell wall is composed of fibrillar material bound by sugars, proteins, lipids, and several polysaccharides. The fibrillar material remains the same, while the other components change in need. The ratio of different cell wall components may vary in different fungal species, which is also stage specific and environmentally adapted. These differences in the cell wall composition make standardization of fungal protoplasting a difficult task.

The first attempt for protoplasting was done by using digestive enzymes from various organisms, namely *Helix pomatia* intestine preparations (helicase, glusulase) (Beggs, 1978; Hinnen et al., 1978), and slowly advanced from enzymes of microbiological origin like that from actinobacteria

(*Arthrobacter luteus*, *Streptomyces graminofaciens*, *Micromonospora chalcea*), for example, zymolyase (Hsiao and Carbon, 1979) to commercial enzymes derived from fungi like *Trichoderma viridae* (novozyme) (Beach and Nurse, 1981) and *Trichoderma harzianum* to maintain reproducibility. Obtaining the quality protoplasts for transformations also meant maintaining the protoplast during the methodology with osmotic stabilizers, for a protoplast is defined as a cell without a cell wall. As discussed, the composition of the cell wall is different in different fungal species and different in different life stages. The ignorance of the exact fungal cell wall composition may lead to excessive use of lytic enzymes which is a hindrance in fine protoplasting even in the presence of a suitable osmotic stabilizer. Attaining and maintaining this quality of protoplast was rather difficult; therefore, many researchers preferred to use spheroplasts instead of protoplasts. Spheroplasts are those which are not complete protoplasts but have lost a considerable part of their cell wall making them vulnerable to transformations (Fig. 3.1).

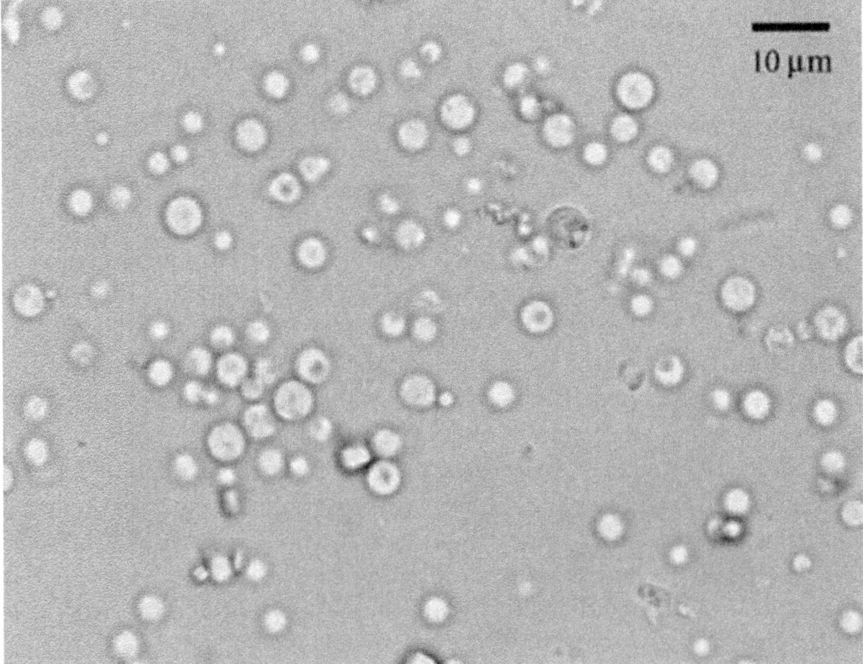

FIGURE 3.1 Microscopic observation of *T. palustris* protoplasts. (Reprinted from http://slideplayer.com/user/11976231/)

The methodology of preparation of spheroplasts with glusulase in *Saccharomyces cerevisiae* proved a breakthrough where stable protoplasts were prepared by using 1-M sorbitol (Hutchison and Hartwell, 1967). Later, Hinnen et al. (1978) reported the transformation of yeast leu− with a chimeric ColE1 carrying leu2+ gene. Later in the same year, first *Escherichia coli*–yeast shuttle vector was developed which could propagate in *S. cerevisiae*. The transformants were obtained in high frequency. With the success of protoplasting and transformation in yeast, the method was extended to filamentous fungi. Transformation of filamentous fungal species was first reported in *Neurospora* (Case et al., 1979). The mutant fungus lacking catabolic and synthetic quinase activity was transformed with a gene encoding catabolic dehydroquinase (qa-2+) (Fincham, 1989). *Neurospora* inositol mutant strain was transformed with wild-type isolated gene and selected on inositol. Supposedly, inositol mutants had weak cell membranes with greater porosity which enabled them to take up the external DNA. The report was looked at with skepticism because a low percentage of the inositol mutants reverted back to the wild type spontaneously. The coming decades saw a revolution in the genetic research with new methods, technology, and strategies to manipulate biological system for development of science and applications to mankind. Protoplasting and transformation of fungal species followed with highest success in *Aspergillus nidulans*, *Magnaporthe grisea*, and many other fungal species. The protoplasts of deletion of acetamidase *A. nidulans* strain were transformed with derivative of pBR322 plasmid containing structural gene for acetamidase (Fincham, 1989). Even though there was a little success in filamentous fungal transformations like *Neurospora* sp. and *Aspergillus* sp., the protocol of transformation was needed to optimize for every other fungi, and highly mutated naturally evolved strains owing to the differences in their acquired cell wall features. Even though through years new protocols advanced and fungal transformations and genetic studies became a routine work, the transformation of filamentous fungus is still not as easy as the other yeast and still faces the problem of low transformation frequencies. Researchers later adopted different modifications to the protoplast transformations to overcome the efficiency crisis. Alternatives included using recipient strains which have much more permeable membranes, using metal ions like magnesium and calcium to facilitate DNA intake and the usage of polyethylene glycol (PEG). Potential disadvantages of PEG-mediated transformation include difficulty in obtaining high concentrations of viable protoplasts, low transformation efficiency, high percentages of transient transformants, and frequent multiple loci integration. However, due to its simplicity in technical operation and equipment required, the PEG-mediated method remains the most commonly used method to conduct transformation in filamentous fungi (Fincham, 1989).

Fungal protoplasts has been successfully made and used for transformations in *Neurospora* (Case et al., 1979), *Aspergillus*, *Penicilluim*, *Podospora*, *Magnaporthe*, and many other fungal species till date. But the transformation frequencies of these fungal transformations vary from species to species.

3.2.1 LITHIUM ACETATE

An alternative method of protoplasting for transformation used by many research labs was lithium acetate method. Both lithium acetate and heat shock enhanced the transformation efficiency of intact cells but not that of spheroplasts. First published in 1983, lithium method is one of the most widely used methods for yeast transformations. Monovalent cations such as Na^+, K^+, Rb^+, Cs^+, and particularly, Li^+ but not divalent cations such as Ca^{2+}(effective for *E. coli* transformation) enhanced the transformation efficiency of intact *S. cerevisiae* cells. The reason for the effectiveness of these monovalent cations might be due to their mild chaotropic effect during the transformation. The lithium method was most effective when: (1) PEG is used; (2) LiAc, and (3) heat shock is given to enhance the transformation efficiency; and (4) when the cells are at the mid-log phase ($OD_{610} = 1.6$) (Ito et al., 1983).

High concentrations of lithium acetate induced cell permeability along with PEG to increase cell agglomeration of the competent cells. The protocol was considerably successful for yeast transformations (Ito et al., 1983) and later adopted for fungal species like *Neurospora crassa* (Dhawale et al., 1984) and *Ustilago violacea* (Binninger et al., 1987).

3.2.2 ELECTROPORATION

Electroporation is a widely used technique for transformation which works on the basic principle that strong electric fields induce reversible membrane permeability enabling the intake of molecular DNA into the cells. The theory of electroporation exploit the nature of the phospholipid membrane which is bound together by the hydrophobic and the hydrophilic interactions can be easily disturbed and deformed temporarily by mild and quick electric pulses and get repaired immediately after the disturbance. DNA enters the cells through the transient pores which are created due to altered polarization on the cell membrane caused by the pulsed electric field and gets trapped in the cell when the voltage is removed. The duration and voltage of the electric field (generally between 0.5 and 2 V approximately) influences the membrane potential created, the extent of membrane poration, the duration of this deformed state, and the mode and duration of the external molecular flow into the cell (Fig. 3.2). Other factors affecting the technique efficiency are concentration of DNA, the tolerance of cell to the electric pulse, and the heterogenicity of the transforming cellular material. Electroporation of conidia, germinating conidia, and protoplasts are common in filamentous fungi. Conidia are pretreated with cell wall weakening agents before electroporating (Ozeki et al., 1994), whereas germinating conidia (Dobrowolska and Staczek, 2009; Sánchez and Aguirre, 1996) and protoplasts can be directly electroporated.

FIGURE 3.2 Gene Pulser Xcell™ electroporation system. (Reprinted from http://www.bio-rad.com/en-ch/product/gene-pulser-xcell-electroporation-systems)

3.2.3 BIOLISTICS OR PARTICLE BOMBARDMENT

The technique was invented by J. C. Sanford of Cornell University in 1987 (Sanford, 1988). Biolistics, otherwise known as biological ballistics, is a technique where heavy metal microparticles are coated with DNA and accelerated with high velocity so as to enter the cell and transfer macromolecules, also known as the gene gun method. It was first used for plant transformation in cereals (Rivera et al., 2014). Later, the technique became widely used in other systems also like algae, yeast, filamentous fungi, and animal tissues (Klein and Fitzpatrick-McElligott, 1993). The salient features of biolistics are that it avoids the then traditional enzymatic treatment which was a limitation to maintain 100% viability; it can be used for the transformation of most types of host material like bacteria, spores, conidia, hyphae, tissue, or even subcellular organelles like mitochondria. The technique has been successful for various species. An additional feature of the technique is the possibility of multiple and chimeric gene transfer, with a less processing time. The factors affecting efficacy of the technique was mainly the acceleration of the microparticle with optimized kinetic energy, the concentration of DNA coated on the particle as well as the number of DNA-coated particles (Rivera et al., 2012). Pneumatic devices (Rinberg et al., 2005), instruments utilizing a mechanical impulse or macroprojectile; centripetal, magnetic, or electrostatic forces; spray or vaccination guns; and apparatus based on electric discharge particle acceleration (Mccabe and Christou, 1993) are different methods used to accelerate the microparticles. The choice of particle for biolististcs is often tungsten or gold. Gold is more preferred due to its inert nature. Tungsten degrades DNA with time (Fig. 3.3). The preferred concentration of DNA per bombardment is 0.3 μg of DNA with about 120 μg of particle. The gene gun consists of a high-pressure and a low-pressure chamber with a diaphragm in the middle that accelerates microparticles of gold, tungsten, or platinum that is covered with DNA. To transform fungi, an inert gas such as Helium with a pressure in the range of 500–2000 psi is used to accelerate microgold particles to speeds of 400 m/s or higher in a partial vacuum of about approximately 30 mmHg for bombardment.

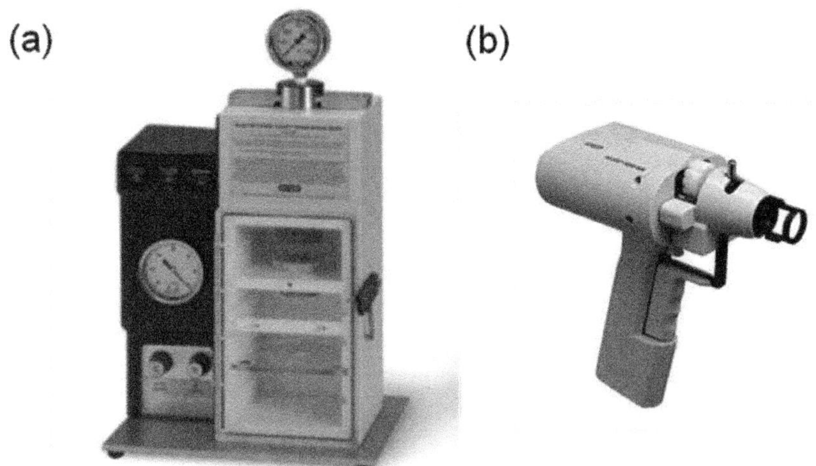

FIGURE 3.3 Instrumentations used for particle bombardment: (a) bolistics unit and (b) gene gun.

The main advantage of biolistics is its technical simplicity and the fact that it can be used to transform any species even if the system is genetically less tractable or less amenable. The disadvantages of the technique are the multiple copy gene insertions leading to undesired effects, low efficacy, high cost of biolistics instrumentation and equipment, and also the difficulty in acquiring special license to use such equipment in some countries (e.g., the United Kingdom) (Rivera et al., 2014).

3.2.4 GLASS BEADS

Vigorous agitation with glass beads in presence of plasmid DNA enables the transformation of yeast cells. The method was easily accepted because of its simplicity and little demand of sophisticated equipment and extremely less cost. The method is efficient in disrupting the yeast cell walls without enzymatic treatment and so makes yeast transformation possible. The mixture of the plasmid DNA, transforming fungal material and 0.3 g sterile glass beads is vigorously vortexed at a speed of 100 m/s for about 15–45 s. Glass bead agitation is the least efficient methods of transformation tried till date. This method was not suitable for filamentous fungi since they had thick cell wall which was not easily breakable as that of yeast cells. The method has been tried out in *Saccharomyces* species, *Picha pastoris* and *Candida* species (Costanzo and Fox, 1988; Lim et al., 2008; Payne et al., 1995).

3.2.5 SHOCK WAVES

Shock waves are pressure pulses with a peak positive pressure in the range of 30–150 MPa. They last about 0.5 and 3 μs, followed by a tensile pulse of up to 20 MPa for about 2–20 μs. They are generally produced by electrohydraulic, electromagnetic, or piezoelectric shock wave generators. The technique was initially used for clinical purposes but later found use in DNA transformations. The exact mechanism responsible for shock-wave-assisted cell permeabilization is still not clear. It is probably because of shock-wave-induced cavitation. Shock-wave-mediated insertion of DNA has been reported in *E. coli* (Divya et al., 2011; Loske et al., 2011), *Pseudomonas aeruginosa*, *Salmonella* (Divya et al., 2011), *Aspergillus niger*, *Trichoderma reesei*, *Phanerochaete chrysosporium*, and *Fusarium oxysporum* (Magaña-Ortíz et al., 2013). Several authors have published articles on shock-wave-mediated DNA delivery in eukaryotic cells and prokaryotes (Covert et al., 2001; de Groot et al., 1998; Maruthachalam et al., 2008; Yang et al., 2011). The application of shock waves to transform cells has several advantages. Expensive enzymatic treatments are not required, the transformation frequency is higher in comparison with other available methods, and the method is fast, easy to perform, and reproducible. Additionally, the same frequency, energy, voltage, and number of shock waves can be used to transform diverse species of fungi (Magaña-Ortíz et al., 2013). At present, the main drawback for the use of shock waves is the need for relatively expensive equipment.

3.2.6 RESTRICTION ENZYME-MEDIATED INTEGRATION

Restriction enzyme-mediated integration (REMI) has grown to be a very effective technique for genetic studies. This technique requires the transformation of the organism with a restriction enzyme linearized DNA along with the restriction enzyme. The DNA then integrates non-homologously into the genomes at cognate restriction sites created with the help of the restriction enzyme. REMI has been widely used for insertional mutagenesis and identification of genes with interesting phenotypes. REMI was first demonstrated in *S. cerevisiae* (Manivasakam and Schiestl, 1998; Orr-Weaver and Szostak, 1983; Schiestl and Petes, 1991) after which it was widely used in many plant pathogenic fungi (Bölker et al., 1995; Linnemannstöns et al., 1999; Lu et al., 1994; Shi et al., 1995; Thon et al., 2000). Most REMI transformants showed single integration with little chromosome rearrangements making it one of the favorite tools of molecular biologists of the time. Biologist first used this technique in *S. cerevisiae* in 1991, when yeast was transformed with *Bam*HI linearized DNA in presence of *Bam*HI enzyme. It is also seen that the DNA of interest gets integrated at *Bam*HI sites in the genome (Schiestl and Petes, 1991). Very soon, REMI became a popular technique for random insertion mutagenesis. It was used for genetic studies in organisms like *Dictyostelium discoideum*, *Magnaporthe oryzae*, *Ustilago maydis*, *Cochliobolus heterostrophus*, and *Colletotrichum granimicola* (Adachi et al., 1994; Balhadère et al., 1999; Bölker et al., 1995; Lu et al., 1994; Thon et al., 2000).

3.2.7 AGROBACTERIUM-MEDIATED TRANSFORMATION

Agrobacterium tumefaciens-mediated transformation is a technique which exploits the infection biology of *Agrobacterium* to transform the fungal recipient host. *Agrobacterium* is a Gram-negative soil bacterium which transfers part of its tumor-inducing plasmid Ti-plasmid T-DNA into the host plant to induce tumor during infection. The "*vir* genes" present on the Ti plasmid get induced in the presence of phenolic compounds like acetosyringone which is produced due to wounding of plant tissues (Rho et al., 2001). The induction of *vir* genes assists the transfer of T-DNA from *Agrobacterium* to the host cells. Together VirA histidine kinase and the cytoplasmic response regulator VirG protein play a central role in regulating *vir* gene expression in response to phenolics. Acetosyringone activates VirA, a membrane bond receptor, which activates the VirG (transcription factor). The activated VirG can then interact with activator elements found in the promoters of the *virA*, *virB*, *virC*, *virD*, *virE*, and *virG* operons, resulting in elevation of their expression levels. VirC and VirD (both nicking endonucleases) bind to the RB/overdrive sequence and cut the ssT-DNA region out of the Ti plasmid. VirE2 binds to the ssT-DNA, protecting it from degeneration by nucleases and self-annealing. VirB2–11 forms a T-pilus through which the VirE2-coated ssT-DNA is transferred from the bacteria into the targeted cell (Zupan et al., 2000). Inside the host cell, a C-terminal located NLS in VirE2 directs the DNA into the nucleus, where host factors are believed to facilitate its integration into the genome, possibly mediated by the DNA repair system (Michielse et al., 2004). If no great sequence similarity exists between the target genome and the introduced T-DNA, the T-DNA integrates randomly into the nuclear genome (Mullins and Kang, 2001). Initially, *Agrobacterium* was used to transform plants and also for insertional mutagenesis as tool for genetic studies (Mullins et al., 2001). Afterword, *Agrobacterium tumefaciens* mediated transformation (ATMT) was used to transform filamentous fungi also. Fungi that have been recalcitrant to transformation by other systems have been successfully transformed by co-cultivation with *Agrobacterium* (Meyer et al., 2003). The transformation of *Aspergillus awamori* using ATMT was 600-fold efficient than the conventional techniques of the time (de Groot et al., 1998). Later, several filamentous fungal transformations were also done using *Agrobacterium* making fungal much easier than before (Covert et al., 2001; Maruthachalam, 2003; Maruthachalam et al., 2008; Nakamura et al., 2011; Vijn and Govers, 2003; Yang et al., 2011). Many binary vectors with different features and utilities have been developed to support molecular genetic studies of fungi via ATMT (Chang et al., 2005; Frandsen, 2011; Mullins et al., 2001; Paz et al., 2011; Sørensen et al., 2014). Both forward and reverse genetics is easily carried out by ATMT. The forward genetic strategy of random insertional mutagenesis of the fungal genome via ATMT has been successfully applied to several fungi to identify many genes essential for their life cycle and pathogenicity (Islam et al., 2012; Jeon et al., 2007; Münch et al., 2011). ATMT also favored reverse genetic strategies by facilitating efficient targeted gene manipulation via homologous recombination (Gouka et al., 1992; Xue et al., 2013; Zwiers and De Waard, 2001), targeted point mutation (Yang et al., 2015), gene silencing, and functional interference using transgenes (Ding et al., 2015).

The advantage of *Agrobacterium*-mediated transformation over other techniques was high efficiency, integrative nature of the T-DNA and the varying range of starting materials which can be used for the method. ATMT could be performed on hyphae, spores, protoplasts, and mycelia. Since the transferred DNA gets integrated into the fungal genome by illegitimate recombination, ATMT can be used for both random mutagenesis and targeted mutagenesis. Generation of single copy recombinants as well as multiple copy recombinants using ATMT has made it an efficient tool for both research and industrial purpose. Methods like inverse PCR, tail PCR, vectorette PCR, plasmid rescue, etc. was used for locus identification of the integration of the T-DNA.

A random mutagenesis for generating mutants with loss of function is not a suitable method to identify genes which are essential or redundant in the genome. Such genes can be studied by promoter and enhancer-trapping techniques. ATMT is also used for successful promoter and enhancer trapping. ATMT vectors are engineered with featured sequences which detect the promoter activity or enhance the transcription of the gene. For instance, reporter genes like luciferase, green fluorescent protein (GFP) have been successfully used for the purpose. In plant pathogenic fungi, the studies related to the fungal establishment require a depth of its invasion into the host tissues. In such cases, reporter genes like GFP, red fluorescent protein (RFP), and β glucuronidase have improved the efficacy of research because of the simplified tagging of these visual markers. For enhancer traps, the reporter gene is cloned under a minimal promoter toward the end of the T-DNA. The insertion of T-DNA near an enhancer element in the genome would give a sufficient expression of the minimal promoter which was much subtle otherwise. To monitor promoter activity, the reporter gene can be tagged under the desired promoter (Weld et al., 2006).

Target directed mutagenesis has recently been widely used for functional genetic studies since the full genome sequence is available for quite a number of model fungal systems. ATMT is also used for target directed mutagenesis by incorporating the mutation cassette (for strategies like knock outs, disruptions, additions, or point mutations) along with the flanking sequences in the cloning cassette within the T-DNA.

The split-marker technique, also known as bipartite gene targeting, has emerged as a widely accepted technique. Two different DNA fragments, each containing two-thirds of the selection marker gene combined with one of the required homologous recombination sequences (HRS) are constructed. A true integration is obtained only if a triple crossing over occurs in the genome one between the homologues overlaps of the selection marker and the other two between the homologous DNA sequences in the genome to that in the two constructs. The technique has been found to increase the homologous recombination in the genome than those with single DNA fragment (Jeong et al., 2007). This technique has been extended to be used with ATMT recently. Wang et al. (2010) used the split-marker strategy and showed it to be compatible with ATMT by co-cultivating the target fungus (*Grosmannia clavigera*) with two different *A. tumefaciens* strains, each carrying a binary plasmid with either the up- or down-stream HRS combined with two-thirds of the selection marker. The split-marker approach increased the gene targeting frequency from 46% to 74%. Because of the triple cross over events required for its efficacy, the number of transformants obtained from split marker is considerably low. But those which were selected had a high probability of the triple cross homologous recombination (Frandsen, 2011). A more recent development in this area is the development of OSCAR (one step construction of *Agrobacterium*-recombination-ready plasmids) gateway vectors for fungal transformations. OSCAR allows single step four fragment fusion by multisite gateway cloning, which depends on four attB recombination sites. The vector is a gateway destination plasmid with a binary back bone and attP2r-ccdB-attP3 cassette (pOSCAR), which is combined with two HRSs PCR amplified with primers having attB2r-attB1r and attB4-attB3 terminal recombination sequences and with marker vector pA-Hyg-OSCAR with attP1r-hygR-aatP4 cassette via BP clonase catalyzed reaction. The resulting binary plasmid will have the selection marker surrounded by two HRS. The *ccdB* killer gene is lost during recombination. The *ccdB* killer gene in the recipient plasmid ensures that only correctly recombined plasmids give rise to viable transformants (Paz et al., 2011).

The availability of genome sequence has enabled direct mutagenesis by knock out, disruption, or point mutations, however, the analysis of essential genes is difficult in fungal systems due to reasons like large genome, lack of haplo-insufficiency libraries, lack of established tight regulatory promoters and also the lack of established mating techniques in many species. Post-transcriptional silencing through RNA interference proves an efficient alternative in such cases. Randomly integrated trans-acting silencing constructs can effectively silence target genes. RNAi silencing constructs have been developed where in a fragment of the target gene is cloned into a binary vector on either sides of an intron spacer in opposite directions (Wang et al., 2009) or any one of the fragments of the target sequence to be cloned in opposite direction are taken to be longer than the other, so that the extended part acts as a spacer and then cloned into the binary vector by restriction enzyme and ligation (Gong et al., 2007).

3.2.8 VACUUM INFILTRATION

Vacuum infiltration is a technique generally accompanied to *Agrobacterium*-mediated transformation to increase the efficiency of transformation. The technique was widely used well before by plant physiologists to study plant pathogen interaction. The pathogenic bacteria were vacuum infiltrated into the cell space so as to trigger a defense response by the host plant. This technique was further applied by researchers to improve transformation. It was first used for plant transformations in Arabidopsis. The principle of infiltration in vacuum was effective for fungal transformation as well because it helped the *Agrobacterium* to get into the intercellular spaces and somehow helped the process of genetic modification. The actual mechanism is still not clearly understood. But this technique is efficient only for some particular *Agrobacterium*-mediated transformations, mostly because all *Agrobacterium* species cannot infect certain species of fungi. Most successfully used *Agrobacterium* species for fungal transformation and gene manipulation studies is the *A. tumefaciens*. Vacuum infiltration was first used in *Agaricus biporus* (Chen et al., 2000), and later, the method was adopted for *F. oxysporum* (Mullins et al., 2001), *Phytophthora infestans* (Vijn and Govers, 2003), *Coprinus cinereus* (Burns et al., 2005), and *Phanerochaete crysosporium* (Sharma et al., 2006).

3.3 MARKERS FOR TRANSFORMATION OF FUNGI

In the last three decades, a number of selectable markers were developed to enable effective transformation in fungi. Both auxotrophic markers and dominant selection markers were effectively used. Auxotrophic markers were those which usually complement the nutritional deficiency in an auxotrophic strain, and dominant markers were mostly those which conferred some special phenotype which is not indigenous to the wild type.

Auxotrophic mutants exist in many fungal species that allow the isolation of wild-type genes using a simple selection. Generally, growth on medium lacking a particular amino acid or nutrient gives the selection. Mutants in orotidylic acid pyrophosphorylase (OMPppase) have been isolated and have been used successfully to transform another species. In both *S. cerevisiae* and *N. crassa*, mutations resulting in loss of orotidine-5′-phosphate decarboxylase (ura3 and pyr-4, respectively) confer resistance to the normally inhibitory analog 5-fluoro-orotic acid. *S. cerevisiae* lys2 (2-aminoadipate reductase deficient) mutants can grow (in the presence of lysine) on 2-aminoadipate as the sole nitrogen source, whereas wild-type *S. cerevisiae* cannot (Fincham, 1989). Successful transformation was also done in *A. nidulans* taking advantage of this gene for selection. Auxotrophic markers have three advantages over the dominant selection in fungi because a single copy of the gene is enough for complementation. Second, the marker itself may be used to direct a chromosomal integration event at its homologous site, so that no other locus is disturbed. And third, auxotrophic selections usually give low backgrounds so that false positives are kept to a minimum. However, the disadvantage of using an auxotrophy as a selectable marker is that one requires a starting mutant strain. While some mutations may be directly selected for, such as ura3 using 5-fluoroorotic acid, or screened visually using color pigmentation, such as ade2 (Mccabe and Christou, 1993), most auxotrophies require screening thousands of mutants.

Other auxotrophic markers used in filamentous fungi are trp-1 which encodes a trifunctional enzyme for tryptophan biosynthesis and arg12 which encodes for ornithine carbamoyl transferase of *N. crassa*; and amdS-encoding acetamidase (Kelly and Hynes, 1985) and niaD nitrate reductase of *A. nidulans* (Malardier et al., 1989). These genes are mostly conserved; for example, *trpC* gene and *pyrG* gene of *A. nidulans* correspond to the trp-1 and pyr-4 of *N. crassa*, and therefore they are widely used in heterologous hosts as well.

Dominant selection markers used today in fungi are mostly genes conferring resistance to any specific antibiotics, herbicides, etc. The *hph* gene (hygromycin) is the most commonly used selection system because of its effectiveness in most, but not all, systems. Other selective agents such as phleomycin, sulfonylurea, nourseothricin, bialophos, carboxin, blasticidin S, and benomyl have also been used (Ruiz-Díez, 2002). Most of these genes are isolated from their native genomes and have been cloned into transformation vectors for selection purpose of heterologous strains. For example, resistance to the antitubulin drug benomyl is provided by the Bml gene of *Neurospora* (Orbach et al., 1986). Many antibiotic resistance genes from bacteria also function suitably as selective markers in fungal transformations: phleomycin, bleomycin, hygrogmycinB, and methotrexate (Austin et al., 1990).

Drug-resistance markers as dominant selection markers are more advantageous that one does not require a recessive mutant strain as a host. In addition, drug resistance markers also permit multiplying and transforming many strains of the same species without genetic crosses or other manipulations. Therefore, these dominant markers are extremely useful in strains whose genetic information is not available. One disadvantage is that, unlike markers where there is homology between the gene present on the plasmid and the gene in the chromosome, these heterologous markers cannot be used to direct integration events within the host genome. Another disadvantage being the possibility of high backgrounds, because single-gene mutations (common during transformation procedures) often lead to elevated levels of resistance. Apart from these two drawbacks, there is a necessity of the parent strain to be checked for minimum inhibitory concentration of the drug, because of the possibility of most wild-type field strains might have natural resistance to the drug or might have evolved to acquire resistance in course of time.

Although fungal transformation has advanced quite far to enable researchers to accomplish in-depth genetic studies, functional genomics of fungi is still far more challenging due the complexities of fungal system and also less fidelity in the targeted genetic modifications achieved due to low transformation frequencies. ATMT has helped to improve fungal genetic studies for sure, but the optimization of the technique in targeted genetic modification still is to be significantly improved. For the same reason, researchers in fungal system at times prefer protoplasting over ATMT for experiments for targeted genetic modifications. Past two decades have seen immense advancement in fungal genetics and the development of new vector constructs, and proper maintenance and sharing of cultures among the scientific community would hasten the advancement of fungal research in the coming years.

KEYWORDS

- **transformation**
- **electroporation**
- **biolistic**
- **gene gun**
- ***Agrobacterium*-mediated transformation**

REFERENCES

Adachi, H. H.; Hasebe, T. T.; Yoshinaga, K. K.; Ohta, T. T.; Sutoh, K. K. Isolation of *Dictyostelium discoideum* Cytokinesis Mutants by Restriction Enzyme-Mediated Integration of the Blasticidin S Resistance Marker. *Biochem. Biophys. Res. Commun.* **1994,** *205*, 1808–1814.

Austin, B.; Hall, R. M.; Tyler, B. M. Optimized Vectors and Selection for Transformation of *Neurospora crassa* and *Aspergillus nidulans* to Bleomycin and Phleomycin Resistance. *Gene* **1990,** *93*, 157–162.

Balhadère, P. V.; Foster, A. J.; Talbot, N. J. Identification of Pathogenicity Mutants of the Rice Blast Fungus *Magnaporthe grisea* by Insertional Mutagenesis. *Mol. Plant. Microbe Interact.* **1999,** *12*, 129–142.

Beach, D.; Nurse, P. High-Frequency Transformation of the Fission Yeast *Schizosaccharomyces pombe*. *Nature* **1981,** *12*, *290* (5802), 140–142.

Beggs, J. D. Transformation of Yeast by a Replicating Hybrid Plasmid. *Nature* **1978,** *275*, 104–109.

Binninger, D. M.; Skrzynia, C.; Pukkila, P. J.; Casselton, L. A. DNA-Mediated Transformation of the Basidiomycete *Coprinus cinereus*. *EMBO J.* **1987,** *6*, 835–840.

Bölker, M.; Böhnert, H. U.; Braun, K. H.; Görl, J.; Kahmann, R. Tagging Pathogenicity Genes in *Ustilago maydis* by Restriction Enzyme-Mediated Integration (REMI). *Mol. Gen. Genet.* **1995,** *248*, 547–552.

Burns, C.; Gregory, K. E.; Kirby, M.; Cheung, M. K.; Riquelme, M.; Elliott, T. J, Challen, M. P.; Bailey, A.; Foster, G. D. Efficient GFP Expression in the Mushrooms *Agaricus bisporus* and *Coprinus cinereus* Requires Introns. *Fungal Genet. Biol.* **2005,** *42*, 191–199.

Case, M. E.; Schweizer, M.; Kushner, S. R.; Giles, N. H.; Kushnert, S. R.; Giles. N. H. Efficient Transformation of *Neurospora crassa* by Utilizing Hybrid Plasmid DNA. *Proc. Nat. Acad. Sci. USA* **1979,** *76*, 5259–5263.

Chang, H. K.; Park, S. Y.; Lee, Y. H.; Kang, S. A Dual Selection Based, Targeted Gene Replacement Tool for *Magnaporthe grisea* and *Fusarium oxysporum*. *Fungal Genet. Biol.* **2005,** *42*, 483–492.

Chen, X.; Stone, M.; Schlagnhaufer, C.; Romaine, C. P. A Fruiting Body Tissue Method for Efficient *Agrobacterium*-mediated Transformation of *Agaricus bisporus*. *Appl. Environ. Microbiol.* **2000,** *66*, 4510–4513.

Costanzo, M. C.; Fox, T. D. Transformation of Yeast by Agitation with Glass Beads. *Genetics* **1988,** *120*, 667–670.

Covert, S. F.; Kapoor, P.; Lee, M.; Briley, A.; Nairn, C. J. *Agrobacterium tumefaciens*-Mediated Transformation of *Fusarium circinatum*. *AMB Express* **2001,** *105*, 259–264.

de Groot, M. J.; Bundock, P.; Hooykaas, P. J.; Beijersbergen, A. G. *Agrobacterium tumefaciens*-Mediated Transformation of Filamentous Fungi. *Nat. Biotechnol.* **1998,** *16*, 839–842.

Dhawale, S. S.; Paietta, J. V.; Marzluf, G. A. A New, Rapid and Efficient Transformation Procedure for *Neurospora*. *Curr. Genet.* **1984,** *8*, 77–79.

Ding, Z. T.; Zhang, Z.; Luo, D.; Zhou, J. Y.; Zhong, J.; Yang, J.; Xiao, L.; Shu, D.; Tan, H. Gene Overexpression and RNA Silencing Tools for the Genetic Manipulation of the S-(+)-Abscisic Acid Producing Ascomycete *Botrytis cinerea*. *Int. J. Mol. Sci.* **2015,** *16*, 10301–10323.

Divya, P. G.; Anish, R. V.; Jagadeesh, G.; Chakravortty, D. Bacterial Transformation Using Micro-shock Waves. *Anal. Biochem.* **2011,** *419*, 292–301.

Dobrowolska, A; Staczek, P. Development of Transformation System for *Trichophyton rubrum* by Electroporation of Germinated Conidia. *Curr. Genet.* **2009,** *55*, 537–542.

Fincham, J. R. Transformation in Fungi. *Microbiol. Rev.* **1989,** *53*, 148–170.

Frandsen, R. J. N. A Guide to Binary Vectors and Strategies for Targeted Genome Modification in Fungi Using *Agrobacterium tumefaciens*-Mediated Transformation. *J. Microbiol. Methods* **2011,** *87*, 247–262.

Gong, X.; Fu, Y.; Jiang, D.; Li, G.; Yi, X.; Peng, Y. L-Arginine is Essential for Conidiation in the Filamentous Fungus *Coniothyrium minitans*. *Fungal Genet. Biol.* **2007,** *44*, 1368–1379.

Gouka, R. J.; Gerk, C.; Hooykaas, P. J.; Bundock, P.; Musters, W.; Verrips, C. T.; de Groot, M. J. Transformation of *Aspergillus awamori* by *Agrobacterium tumefaciens*-Mediated Homologous Recombination. *Nat. Biotechnol.* **1999,** *17*, 598–601.

Hershey, A. D.; Chase, M. Independent Functions of Viral Protein and Nucleic Acid in Growth of Bacteriophage. *J. Gen. Physiol.* **1952,** *36*, 39–56.

Hinnen, A.; Hicks, J. B.; Fink, G. R. Transformation of Yeast. *Proc. Nat. Acad. Sci. USA* **1978,** *75*, 1929–1933.

Hsiao, C. L.; Carbon, J. High-Frequency Transformation of Yeast by Plasmids Containing the Cloned Yeast ARG4 Gene. *Proc. Nat. Acad. Sci. USA* **1979,** *76*, 3829–3833.

Hutchison, H. T.; Hartwell, L. H. Macromolecule Synthesis in Yeast Spheroplasts. *J. Bacteriol.* **1967,** *94*, 1697–705.

Islam, M. N.; Nizam, S.; Verma, P. K. A Highly Efficient *Agrobacterium* Mediated Transformation System for Chickpea Wilt Pathogen *Fusarium oxysporum* f. sp. *ciceri* Using DsRed-Express to Follow Root Colonisation. *Microbiol. Res.* **2012,** *167*, 332–338.

Ito, H.; Fukuda, Y.; Murata, K.; Kimura, A. Transformation of Intact Yeast Cells Treated with Alkali Cations. *J. Bacteriol.* **1983,** *153*, 163–168.

Jeon, J.; Park, S. Y.; Chi, M. H.; Choi, J.; Park, J.; Rho, H. S.; Kim, S.; Goh, J.; Yoo, S.; Choi, J.; Park, J. Y.; Yi, M.; Yang, S.; Kwon, M. J.; Han, S. S.; Kim, B. R.; Khang, C. H.; Park, B.; Lim, S. E.; Jung, K.; Kong, S.; Karunakaran, M.; Oh, H. S.; Kim, H.; Kim, S.; Park, J.; Kang, S.; Choi, W. B.; Kang, S.; Lee, Y. H. Genome-Wide Functional Analysis of Pathogenicity Genes in the Rice Blast Fungus. *Nat. Genet.* **2007,** *39*, 561–565.

Jeong, J. S.; Mitchell, T. K.; Dean, R. A. The *Magnaporthe grisea* Snodprot1 Homolog, MSP1, Is Required For Virulence. *FEMS Microbiol. Lett.* **2007,** *273*, 157–165.

Kelly, J. M.; Hynes, M. J. Transformation of *Aspergillus niger* by the amdS Gene of *Aspergillus nidulans*. *EMBO J.* **1985,** *4*, 475–479.

Klein, T. M.; Fitzpatrick-McElligott, S. Particle Bombardment: A Universal Approach for Gene Transfer to Cells and Tissues. *Curr. Opin. Biotechnol.* **1993,** *4*, 583–590.

Lim, C. S. Y.; Tung, C. H.; Rosli, R.; Chong, P. P. An Alternative *Candida* spp. Cell Wall Disruption Method Using a Basic Sorbitol Lysis Buffer and Glass Beads. *J. Microbiol. Methods* **2008,** *75*, 576–578.

Linnemannstöns, P.; Voss, T.; Hedden, P.; Gaskin, P.; Tudzynski. B. Deletions in the Gibberellin Biosynthesis Gene Cluster of *Gibberella fujikuroi* by Restriction Enzyme-Mediated Integration and Conventional Transformation-Mediated Mutagenesis. *Appl. Environ. Microbiol.* **1999,** *65*, 2558–2564.

Loske, A. M.; Campos-Guillen, J.; Fernández, F.; Castaño-Tostado, E. Enhanced Shock Wave-Assisted Transformation of *Escherichia coli*. *Ultrasound Med. Biol.* **2011,** *37*, 502–510.

Lu, S.; Lyngholm, L.; Yang, G.; Bronson, C.; Yoder, O. C.; Turgeon, B. G. Tagged Mutations at the Tox1 Locus of *Cochliobolus heterostrophus* by Restriction Enzyme-Mediated Integration. *Proc. Nat. Acad. Sci.* **1994,** *91*, 12649–12653.

Magaña-Ortíz, D.; Coconi-Linares, N.; Ortiz-Vazquez, E.; Fernández, F.; Loske, A. M.; Gómez-Lim, M. A. A Novel and Highly Efficient Method for Genetic Transformation of Fungi Employing Shock Waves. *Fungal Genet. Biol.* **2013,** *56,* 9–16.

Malardier, L.; Daboussi, M. J.; Julien, J.; Roussel, F.; Scazzocchio, C.; Brygoo, Y. Cloning of the Nitrate Reductase Gene (niaD) of *Aspergillus nidulans* and Its Use for Transformation of *Fusarium oxysporum. Gene* **1989,** *78,* 147–156.

Manivasakam, P.; Schiestl, R. H. Nonhomologous End Joining during Restriction Enzyme-Mediated DNA Integration in *Saccharomyces cerevisiae. Mol. Cell. Biol.* **1998,** *18,* 1736–1745.

Maruthachalam, K.; Nair, V.; Rho, H. S.; Choi, J.; Kim, S.; Lee, Y. H. *Agrobacterium tumefaciens*-Mediated Transformation in *Colletotrichum falcatum* and *C. acutatum. J. Microbiol. Biotechnol.* **2008,** *18,* 234–241.

Maruthachalam, M. *Agrobacterium tumefaciens*-Mediated Transformation of *Helminthosporium turcicum,* the Maize Leaf-Blight Fungus. *Arch. Microbiol.* **2003,** *180,* 279–284.

Mccabe, D.; Christou, P. Direct DNA Transfer Using Electric Discharge Particle Acceleration. *Plant Cell Tissue Organ Cult.* **1993,** *33,* 227–236.

McCarty, M. Studies on the Chemical Nature of the Substance Inducing Transformation of Pneumococcal Types. *Strain* **1944,** *79,* 137–158.

Meyer, V.; Mueller, D.; Strowig, T.; Stahl, U. Comparison of Different Transformation Methods for *Aspergillus giganteus. Curr. Genet.* **2003,** *43,* 371–377.

Michielse, C. B.; Ram, A. F. J.; Hooykaas, P. J. J.; Van Den Hondel, C. A. M. J. J. Role of Bacterial Virulence Proteins in *Agrobacterium*-Mediated Transformation of *Aspergillus awamori. Fungal Genet. Biol.* **2004,** *41,* 571–578.

Mullins, E. D.; Kang. S. Transformation: A Tool for Studying Fungal Pathogens of Plants. *Cell Mol. Life Sci.* **2001,** *58,* 2043–2052.

Mullins, E. D.; Chen, X.; Romaine, P.; Raina, R.; Geiser, D. M.; Kang, S. *Agrobacterium*-Mediated Transformation of *Fusarium oxysporum*: An Efficient Tool for Insertional Mutagenesis and Gene Transfer. *Phytopathology* **2001,** *91,* 173–180.

Münch, S.; Ludwig, N.; Floss, D. S.; Sugui, J. A.; Koszucka, A. M.; Voll, L. M.; Sonnewald, U.; Deising, H. B. Identification of Virulence Genes in the Corn Pathogen *Colletotrichum graminicola* by *Agrobacterium tumefaciens*-Mediated Transformation. *Mol. Plant. Pathol.* **2011,** *12,* 43–55.

Nakamura, M.; Kuwahara, H.; Onoyama, K.; Iwai, H. *Agrobacterium tumefaciens*-Mediated Transformation for Investigating Pathogenicity Genes of the Phytopathogenic Fungus *Colletotrichum sansevieriae. Curr. Microbiol.* **2012,** *65,* 176–182.

Orbach, M. J.; Porro, E. B.; Yanofsky, C. Cloning and Characterization of the Gene for Beta-Tubulin from a Benomyl-Resistant Mutant of *Neurospora crassa* and Its Use as a Dominant Selectable Marker. *Mol. Cell. Biol.* **1986,** *6,* 2452–2461.

Orr-Weaver, T. L.; Szostak, J. W. Multiple, Tandem Plasmid Integration in *Saccharomyces cerevisiae. Mol. Cell. Biol.* **1983,** *3,* 747–749.

Ozeki, K.; Kyoya, F.; Hizume, K.; Kanda, A.; Hamachi, M.; Nunokawa, Y. Transformation of Intact *Aspergillus niger* by Electroporation. *Biosci. Biotechnol. Biochem.* **1994,** *58,* 2224–2227.

Payne, W. E.; Gannon, P. M.; Kaiser, C. A. An Inducible Acid Phosphatase from the Yeast Pichia pastoris: Characterization of the Gene and Its Product. *Gene* **1995,** *163,* 19–26.

Paz, Z.; García-Pedrajas, M. D.; Andrews, D. L.; Klosterman, S. J.; Baeza-Montañez, L.; Gold, S. E. One Step Construction of *Agrobacterium*-Recombination-Ready-Plasmids

(OSCAR), an Efficient and Robust Tool for ATMT Based Gene Deletion Construction in Fungi. *Fungal Genet. Biol.* **2011,** *48*, 677–684.

Rho, H. S.; Kang, S.; Lee, Y. H. *Agrobacterium tumefaciens*-Mediated Transformation of the Plant Pathogenic Fungus, *Magnaporthe grisea. Mol. Cells* **2001,** *12*, 407–411.

Rinberg, D.; Simonnet, C.; Groisman, A. Pneumatic Capillary Gun for Ballistic Delivery of Microparticles. *Appl. Phys. Lett.* **2005,** *87*, 14103.

Rivera, A. L.; Gómez-Lim, M.; Fernández, F.; Loske, A. M. Physical Methods for Genetic Transformation of Fungi and Yeast. *Phys. Life Rev.* **2012,** *9*, 308–345.

Rivera, A. L.; Magaña-Ortíz, D.; Gómez-Lim, M.; Fernández, F.; Loske, A. M. Physical Methods for Genetic Transformation of Fungi and Yeast. *Phys. Life Rev.* **2014,** *11*, 184–203.

Ruiz-Díez, B. A Review: Strategies for the Transformation of Filamentous Fungi. *J. Appl. Microbiol.* **2002,** *92* (2), 189–195.

Sánchez, O.; Aguirre, J. Efficient Transformation of *Aspergillus nidulans* by Electroporation of Germinated Conidia. *Fungal Genet. Newsl.* **1996,** *43*, 48–51.

Sanford, J. The Biolistic Process. *Trends Biotechnol.* **1988,** *6*, 299–302.

Schiestl, R. H.; Petes, T. D. Integration of DNA Fragments by Illegitimate Recombination in *Saccharomyces cerevisiae. Proc. Nat. Acad. Sci. USA* **1991,** *88*, 7585–7589.

Sharma, K. K.; Gupta, S.; Kuhad, R. C. *Agrobacterium*-Mediated Delivery of Marker Genes to *Phanerochaete chrysosporium* Mycelial Pellets: A Model Transformation System for White-Rot Fungi. *Biotechnol. Appl. Biochem.* **2006,** *43*, 181–186.

Shi, Z.; Christian, D.; Leung, H. Enhanced Transformation in *Magnaporthe grisea* by Restriction Enzyme Mediated Integration of Plasmid DNA. *Phytopathology* **1995,** *85*, 329–333.

Sørensen, L. Q.; Lysøe, E.; Larsen, J. E.; Khorsand-jamal, P.; Nielsen, K. F.; John, R.; Frandsen, N. Genetic Transformation of *Fusarium avenaceum* by *Agrobacterium tumefaciens*-Mediated Transformation and the Development of a USER-Brick Vector Construction System. *BMC Mol. Biol.* **2014,** *15*, 15.

Thon, M. R.; Nuckles, E. M.; Vaillancourt, L. J. Restriction Enzyme-Mediated Integration Used to Produce Pathogenicity Mutants of *Colletotrichum graminicola. Mol. Plant. Microbe Interact.* **2000,** *13*, 1356–1365.

Vijn, I.; Govers, F. *Agrobacterium tumefaciens* Mediated Transformation of the Oomycete Plant Pathogen *Phytophthora infestans. Mol. Plant. Pathol.* **2003,** *4*, 459–467.

Wang, F.; Tao, J.; Qian, Z.; You, S.; Dong, H.; Shen, H.; Chen, X.; Tang, S.; Ren, S. A Histidine Kinase PmHHK1 Regulates Polar Growth, Sporulation and Cell Wall Composition in the Dimorphic Fungus *Penicillium marneffei. Mycol. Res.* **2009,** *113*, 915–923.

Wang, Y.; Diguistini, S.; Bohlmann, J.; Breuil, C. *Agrobacterium*-Meditated Gene Disruption Using Split-Marker in *Grosmannia clavigera*, a Mountain Pine Beetle Associated Pathogen. *Curr. Genet.* **2010,** *56* (3), 297–307.

Watson, J. D.; Crick, F. H. C. Molecular Structure of Nucleic Acids. *Nature* **1953,** *171*, 737.

Weld, R. J.; Plummer, K. M.; Carpenter, M. A.; Ridgway, H. J. Approaches to Functional Genomics in Filamentous Fungi. *Cell. Res.* **2006,** *16*, 31–44.

Xue, C.; Wu, D.; Condon, B. J.; Bi, Q.; Wang, W.; Turgeon, B. G. Efficient Gene Knockout in the Maize Pathogen *Setosphaeria turcica* using *Agrobacterium tumefaciens*-Mediated Transformation. *Phytopathology* **2013,** *103*, 641–647.

Yang, F.; Abdelnabby, H.; Xiao, Y. Novel Point Mutations in β-Tubulin Gene for Carbendazim Resistance Maintaining Nematode Pathogenicity of *Paecilomyces lilacinus. Eur. J. Plant Pathol.* **2015,** *143*, 57–68.

Yang, L.; Yang, Q.; Sun, K.; Tian, Y.; Li, H. *Agrobacterium tumefaciens* Mediated Transformation of ChiV Gene to *Trichoderma harzianum*. *Appl. Biochem. Biotechnol.* **2011,** *163*, 937–945.

Zupan, J.; Muth, T. R.; Draper, O.; Zambryski, P. The transfer of DNA from *Agrobacterium tumefaciens* into Plants: A Feast of Fundamental Insights. *Plant J.* **2000,** *23*, 11–28.

Zwiers, L. H.; De Waard, M. A. Efficient *Agrobacterium tumefaciens*-Mediated Gene Disruption in the Phytopathogen *Mycosphaerella graminicola*. *Curr. Genet.* **2001,** *39*, 388–393.

PART II

Gene Silencing in Plants

CHAPTER 4

RNA INTERFERENCE AND VIRUS-INDUCED GENE SILENCING IN PLANTS

MAHESH KUMAR[1*], TUSHAR RANJAN[2], NAGATEJA NATRA[3], MD. SHAMIM[1], KUMARI RAJANI[4], VINOD KUMAR[1], and TAMOGHNA SAHA[5]

[1]*Department of Molecular Biology and Genetic Engineering, Bihar Agricultural University, Sabour, Bhagalpur 813210, Bihar, India*

[2]*Department of Basic Science & Humanities Genetics, Bihar Agricultural University, Sabour, Bhagalpur 813210, Bihar, India*

[3]*Department of Plant Pathology, Irrigated Agriculture Research & Extension Center, Washington State University, 24106 N. Bunn Road, Prosser, WA 99350, United States*

[4]*Department of Seed Technology, Bihar Agricultural University, Sabour, Bhagalpur 813210, Bihar, India*

[5]*Department of Entomology, Bihar Agricultural University, Sabour, Bhagalpur 813210, Bihar, India*

[*]*Corresponding author. E-mail: maheshkumara2zbau@gmail.com*

CONTENTS

ABSTRACT

RNA interference (RNAi) has modernized the studies to determine the role of a particular/novel gene. RNAi is a biological process where RNA molecule inhibits the expression of a particular gene by targeting and destructing specific mRNA molecules. RNAi is also known as posttranscriptional gene silencing (PTGS), co-suppression, and quelling. Silencing of target genes by RNAi technology came into the limelight just after discovery of plant defense mechanism against virus, where it was believed that plant encode short, noncoding region of viral RNA sequences, which after infection recognizes and degrades viral mRNA. These short and noncoding RNA sequences might be against viral DNA/RNA polymerase and other important genes necessary for viral infection and multiplication. On the theme of the above concept of plant virologist introduced short nucleotides sequence into the viruses and expression of target genes in the infected plants was found to be suppressed. This most popular phenomenon is known as "virus-induced gene silencing" (VIGS) and brings the boom in the era of biotechnologists. This chapter is detailed with the mechanism and recent advancement of RNAi technology.

4.1 INTRODUCTION

RNA interference (RNAi) has revolutionized the studies to determine the role of a particular gene. RNAi is a biological process where RNA molecule inhibits the expression of a particular gene by targeting and destructing specific mRNA molecules. RNAi is also known as posttranscriptional gene silencing (PTGS), cosuppression, and quelling. The discovery of RNAi was totally serendipity. The concept of RNAi for the first time came into the existence while the study of transcriptional inhibition by antisense RNA expressed in transgenic *Petunia* plant conducted by Napoli et al. (1990). These plant scientists were trying to introduce additional copies of chalcone synthase gene responsible for darker pigmentation of flowers. The transgenic copy intended to make more corresponding gene products. But instead of darker flowers, white or less pigmented flowers were observed indicating the suppressed/decreased expression of endogenous chalcone synthase gene (Napoli et al., 1990; Ecker and Davis, 1986). This suggests down regulation of endogenous gene by the event posttranscriptional inhibition due to their mRNA degradation (Romano and Macino, 1992; Van Blokland et al., 1994). Silencing of target genes by RNAi technology came into the limelight just after discovery of plant defense mechanism against virus, where it was believed that plant encode short, noncoding region of viral RNA sequences, which after infection recognizes and degrades viral mRNA. These short and noncoding RNA sequences might be against viral DNA/RNA polymerase and other important genes necessary for viral infection and multiplication. On the theme of the above concept of plant virologist introduced short nucleotides sequence into the viruses and expression of target genes in the infected plants was found to be suppressed (Covey et al., 1997; Ratcliff et al., 1997). This most popular phenomenon is known as "virus-induced gene silencing" (VIGS) and brings the boom in the era of biotechnologists. Just after, a year later in 1998, Craig Mello and Andrew Fire's performed works in the laboratory to study effect of RNAi in *Caenorhabditis elegans*, and interestingly they found that dsRNA effectively silenced the target gene in comparison to antisense ssRNA (100-folds more potent). The term RNAi was coined by these two scientists for the first time and they were awarded Nobel Prize in the field of medicine in 2006 for this breakthrough (Fire et al., 1998). After this great discovery of dsRNA as an extremely potent trigger for gene silencing, it became very realistic to unravel the mechanism of RNAi action in various biological systems (Guo and Kemphues, 1995; Pal-Bhadra et al., 1997). Proteins machinery necessary for gene silencing was discovered in *C. elegans* for the first time in 1999 and comprehensive analysis indicates that common fundamental mechanism must be operated throughout the eukaryotes such as fungi, Drosophila, and plants (Tabara et al., 1999). Scientific community had started realizing that RNAi pathway has ancient origin and coming from primitive eukaryotes to recent human beings. Parallelly in the meanwhile, different groups of scientists working on PTGS system in plant, Drosophila, and worm came up with interesting facts and their results were par with each other. They observed that small RNA ranging in length from 21 to 23 nucleotides is generated from dsRNA in cell extracts and could serve as a de-novo silencing trigger for RNAi in cell extracts free of dsRNA treatments. They concluded that short 21–23 nucleotides siRNA are the outcome of Dicer and RNA-induced silencing complex (RISC) (Hamilton and Baulcombe, 1999; Hammond et al., 2000; Zamore et al., 2000). Now these days, engineered synthetic RNA has been extensively used to induce sequence-specific gene silencing and became a very popular tool for knockdown of eukaryotic genes. Figure 4.1 represents overall timeline from their RNAi discovery to recent advance application for mankind. As with many great discoveries, the history of RNAi is a tale of scientists able to interpret unexpected results in a novel and imaginative way.

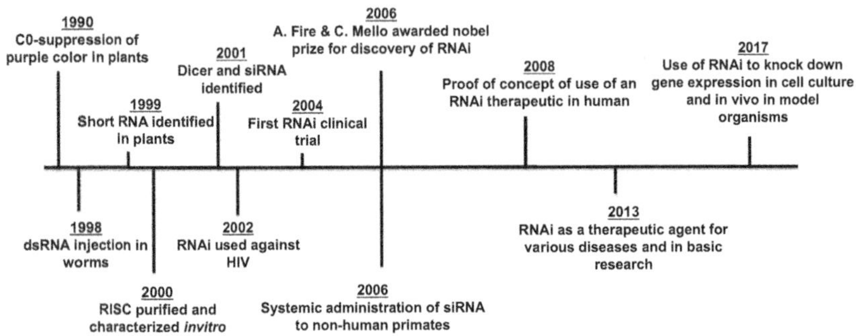

FIGURE 4.1 Timeline for RNA interference discovery and their advanced application.

The short RNA molecules, a key to RNAi technology, are of two types: (1) microRNA (miRNA) and (2) small interfering RNA (siRNA). miRNAs are endogenous or purposefully expressed product (organism's own genome product), whereas siRNAs are derived product of exogenous origin such as virus, transposon. Both have different precursor, for example, miRNA seems to be processed from stem-loop with partial complementary dsRNA, whereas siRNA appears from fully complementary dsRNA (Tomari and Zamore, 2005). Inspite of these differences, both short nucleotides are very much related in terms of their biogenesis and mode of action (Meister and Tuschl, 2004). Likewise, both Dicer and RISC assembly are needed during their synthesis from precursor molecules and targeting as well. Small RNAs are the key mediators of RNA silencing and related pathways in plants and other eukaryotic organisms. Silencing pathways couple the destruction of double-stranded RNA (dsRNA) with the use of the resulting small RNAs to target other nucleic acid molecules that contain the complementary sequence. This discovery has revolutionized our ideas about host defense and genetic regulatory mechanisms in eukaryotes. Small RNAs can direct the degradation of mRNAs and single-stranded viral RNAs, the modification of DNA and histones, and the inhibition of translation. Viruses might even use small RNAs to do some targeting of their own to manipulate host-gene expression.

4.2 PRINCIPAL COMPONENTS LIES AT THE HEART OF RNAI PATHWAY

4.2.1 DICER: A GATEWAY INTO THE RNA INTERFERENCE

Dicer is a member of the RNase III family proteins with dsRNA-specific nuclease activity and it acts as a primary candidate for biogenesis of siRNA during gene silencing (Tomari and Zamore, 2005). These enzymes have several critical motifs spread throughout the polypeptide chain from N-terminus to C-terminus, which is responsible for their efficient performance (Meister and Tuschl, 2004). RNase III enzymes are characterized by the domains in order from N-to-C terminus: a DEXD domain, a DUF283 domain, a PAZ (Piwi/Argonaute/Zwille) domain, two tandem RNaseIII domain, and a dsRNA binding domain (Fig. 4.2A). Apart from ribonuclease-specific PAZ domain, Dicer do possess helicase domain and their function has been implicated in processing long dsRNA substrate (Cenik et al., 2011). Out of these five crucial domains, PAZ and RNase III are very critical for precise excision of siRNA from dsRNA precursor (Zhang et al., 2004). PAZ domain recognizes the duplex RNA end with three nucleotides overhang, resulting in stretching of two helical turn along the surface of the protein. This leads to the cleavage of one out of the two strands at a time by two different RNase III domains separately. The final product after Dicer action is 21–23-nt long fragments with two nucleotides overhang at 3′ end, which now act as a substrate for RISC (Tomari and Zamore, 2005). Current finding suggests that PAZ domain is capable of binding the exactly 2 nucleotide 3′ overhang of dsRNA, while the RNaseIII catalytic domains form a pseudo dimer around the dsRNA to initiate cleavage of the strands. This results in a functional shortening of the dsRNA strand. The distance between the PAZ and RNaseIII domains is determined by the angle of the connector helix and influences the length of the micro RNA product (Macrae et al., 2006). In some of the organism, only one copy of Dicer is responsible for the processing of both miRNA and siRNA but interestingly, in *Drosophila*, Dicer 1 is solely devoted for miRNA biogenesis, whereas Dicer 2 is used for siRNA track (Tomari and Zamore, 2005). The molecular weight of Dicer ranges from 80 to 219 kDa (human Dicer). The difference in size is due to the presence of all five domains in human Dicer and absence of few domains in Dicer characterized from *Giardia intestinalis*. Other variants of Dicer are characterized by absence of ATPase domain or PAZ domain or RNA binding domains. Although functional ATPase domain is not very necessary for the action of Dicer to the substrate molecules, but study also gives clue that ATPase domain is very critical for switching/movement of both RNase III domains and biochemical studies indicates mutation in ATPase domain leads to the abolishment of siRNA procession (Tomari and Zamore, 2005). Because most vertebrates, especially *C. elegans*, express only one Dicer protein, interactions with additional proteins must modulate the specificity of these enzymes. Study indicates R2D2-like protein, RDE-1 and 4 form a complex with Dicer and are essential for RNAi pathway but not miRNA functioning (Tabara et al., 2002).

4.2.2 RISC: AT THE CORE OF RNA INTERFERENCE

RISC is a generic term for a family of heterogeneous molecular complexes that can be programed to target almost any gene for silencing. In general, RISC programing is triggered by the appearance of dsRNA in the cytoplasm of a eukaryotic cell. RISC is a multiprotein complex composed of ribonucleoproteins (Argonaute protein) incorporates one strand of dsRNA fragments (siRNA, miRNA) to the target transcripts. To purify RISC, Tuschl and colleagues used cell extracts derived from human HeLa cells. They partially purify RISC by conjugating the 3′ termini of siRNAs to biotin, which enabled co-immunoprecipation of the siRNA with associated protein complexes. Precipitated complexes were further purified based on size and molecular weight. Two proteins of ~100 kDa were also identified that corresponded to Argonaute 1 and Argonaute 2 (Ago1 and Ago2). Biochemical isolations of RISC have revealed a variety of different RNPs, ranging from modest size (150 kDa) up to 3-MDa particle termed "holo-RISC" and many other intermediate sizes has also been observed (Höck et al., 2007; Martinez et al., 2002; Pham et al., 2004). The complete structure of RISC is still unsolved. Recent research has reported a large number of RISC-associated proteins, which includes mainly, Argonaute proteins and RISC-loading complex. Both these components assembled together to perform its functions efficiently. RISC-loading complex is basically made up of Dicer, Argonaute, and TRBP (protein with three dsRNA-binding domains) (Fig. 4.2B). In 2005, Gregory et al. identified a 500-kDa minimal RISC by characterizing proteins that copurified with human Dicer. Two proteins were found to be associated with Dicer, Ago2, and TRBP (the HIV trans-activating response RNA-binding protein) (Gregory et al., 2005). Parallelly, the minimal RISC, sufficient for target RNA recognition and cleavage efficiently, was demonstrated to be simply an Argonaute protein bound to a small RNA (Rivas et al., 2005). Argonaute proteins are ubiquitously found in plant, animal, many fungi, Protista, and even in few archaea as well. Although all AGO proteins harbor PAZ, MID (middle), and PIWI domains, they are divided into three groups on the basis of both their phylogenetic relationships and their capacity to bind to small RNAs. Group 1 members bind to miRNAs and siRNAs and are referred to as AGO proteins. Group 2 members bind to PIWI-interacting RNAs (piRNAs) and are referred to as PIWI proteins. Group 3 members have been described only in worms, where they bind to secondary siRNAs. AGOs are large proteins (ca. 90–100 kDa) consisting of one variable N-terminal domain and conserved C-terminal PAZ, MID, and PIWI domains. Experiments with bacterial and animal AGO proteins have elucidated the roles of these three domains in small RNA pathways. The MID domain binds to the 5′ phosphate of small RNAs, whereas the PAZ domain recognizes the 3′ end of small RNAs. The PIWI domain adopts a folded structure similar to that of RNaseH enzymes and exhibits endonuclease activity, which is carried out by an active site usually carrying an Asp–Asp–His (DDH) motif (Vaucheret, 2008).

Presence of these proteins has also been reported in prokaryotes but their function in lower organisms is still a mystery. Among eukaryotes, number of Argonaute gene ranging from a single copy to dozens of copies (even more than two dozens) is found to be observed. Multiple copies (paralogous proteins) of Argonaute proteins in *C. elegans* reflects their functionally redundancy and their evolutionary significance remain unknown. Studies suggest genes for Argonaute proteins ample to recompense for one another (Grishok et al., 2001). The Argonaute associated with siRNA binds to the 3′-untranslated region of mRNA and prevents the production of proteins in several ways. The recruitment of Argonaute proteins to targeted mRNA can induce mRNA degradation. The Argonaute–miRNA complex can also effect the formation of functional ribosomes at the 5′-end of the mRNA. The complex competes with the translation initiation factors and/or abrogates ribosome assembly. Also, the Argonaute–miRNA complex can adjust protein production by recruiting cellular factors such as peptides or post-translational modifying enzymes, which degrade the growing of polypeptides (Hutvagner and Simard, 2008).

The Argonaute superfamily can be divided into three separate subgroups: the Piwi clade that binds piRNAs, the Ago clade that associates with miRNAs and siRNAs, and a third clade that has only been found and characterized in nematodes so far (Yigit et al., 2006). All gene-regulatory phenomena involving ~20–30-nt RNAs are thought to require one or more Argonaute proteins, and these proteins are the central, defining components of the various forms of RISC. The double-stranded products of Dicer enter into a RISC assembly pathway that involves duplex unwinding, culminating in the stable association of only one of the two strands with the Ago effector protein (Meister and Tuschl, 2004; Tomari and Zamore, 2005). Thus, one guide strand directs target recognition by Watson–Crick base pairing, whereas the other strand of the original small RNA duplex, known as the passenger strand, is discarded. In human, there are eight AGO family members, some of which are investigated intensively. However, even though AGO1–4 is capable of loading miRNA, endonuclease activity, but RNAi-dependent gene silencing is exclusively found with AGO2. Considering the sequence conservation of PAZ and PIWI domains across the family, the uniqueness of AGO2 is presumed to arise from either the N-terminus or the spacing region linking PAZ and PIWI motifs. Several AGO family in plants also attracts tremendous effort of studying. AGO1 is clearly involved in miRNA-related RNA degradation and plays a central role in morphogenesis. In some organisms, it is strictly required for epigenetic silencing. Interestingly, it is regulated by miRNA itself. AGO4 does not involve in RNAi-directed RNA degradation, but in DNA methylation and other epigenetic regulation, through small RNA (siRNA) pathway. AGO10 is involved in plant development. AGO7 has a function distinct from AGO 1 and 10 and is not found in gene silencing induced by transgenes. Instead, it is related to developmental timing in plants (Meister et al., 2004; Meins et al., 2005). At the cellular level, Ago proteins localize diffusely in the cytoplasm and nucleus and, in some cases, also at distinct foci, which include P-bodies and stress granules. The second clade, Piwi (named after the *Drosophila* protein PIWI, for **P**-element-**i**nduced **wi**mpy testis), is most abundantly expressed in germ-line cells and functions in the silencing of germ-line transposons. A major biochemical difference between Argonaute clades is the means by which members acquire guide RNAs. Ago-guide RNAs which are generated from dsRNA in the cytoplasm by a specialized nuclease named Dicer. Members of the Piwi clade are thought to form guide RNAs in a "ping-pong" mechanism in which the target RNA of one Piwi protein is cleaved and becomes the guide RNA of another Piwi protein. Maternally inherited guide piRNAs are believed to initiate this gene-silencing cascade. Class 3 Argonautes obtain guide RNAs by Dicer-mediated cleavage of exogenous and endogenous long dsRNAs (Aravin et al., 2007; Brennecke et al., 2008; Yigit et al., 2006).

The hallmark domains of Argonaute proteins are: N-terminal PAZ (similar to Dicer enzymes and share common evolutionary origin), mid-domain and C-terminal PIWI domain, a unique to the Argonaute superfamily proteins (Fig. 4.2B and C). The PAZ domain is named after discovery of proteins PIWI, AGO, and Zwille, whereby it is found to conserve. The PAZ domain interacts with 3′ end of both siRNA/miRNA in sequence independent manner and finally it hybridize with the target mRNA via base-pairing interaction, leads to the cleavage or translation inhibition (Tang, 2005). PIWI domain, which is very essential for RNA backbone cleavage has structurally resemblance with RNaseH. The active site is composed of triad amino acids, aspartate–aspartate–glutamate, which coordinates with divalent metal ion and provides binding energy for catalysis. In few Argonaute proteins, PIWI domain participates in interaction with the Dicer via one of the RNaseIII domains (Meister et al., 2004). Between the Mid and PIWI domain, a MC motif is present which is thought to be involved in interaction sites for the 5′ cap of siRNA/miRNA and control their translation (Hutvagner and Simard, 2008). The overall structure of Argonaute is bilobed, with one lobe consisting of the PAZ domain and the other lobe consisting of the PIWI domain flanked by N-terminal (N) and middle (Mid) domains (Fig. 4.2B and C). The Argonaute PAZ domain has RNA 3′ terminus binding activity,

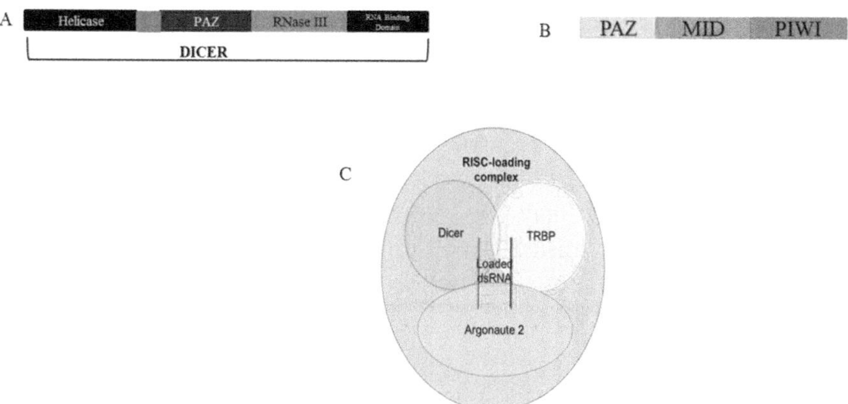

FIGURE 4.2 Principal components of RNA interference. (A) Schematic representation of all predicted domain organization on the polypeptide chain of Dicer protein. Helicase: N-terminal and C-terminal helicase domains. PAZ: Pinwheel–Argonaute–Zwille domain. RNase III: bidentate ribonuclease III domains. (B) Schematic representation of all predicted domain organization on the polypeptide chain of Argonaute protein. (C) Hypothetical complete RISC-loading complex, allows loading of dsRNA fragment generated by Dicer to Argonaute protein by the assistance of TRBP. (A & B, adapted from Naqvi et al., 2009; C, reprinted from wikipedia (https://en.wikipedia.org/wiki/RNA-induced_silencing_complex)

and the co-crystal structures reveal that this function is used in guide-strand binding. The other end of the guide strand engages a 5′-phosphate binding pocket in the mid domain, and the remainder of the guide tracks along a positively charged surface to which each of the domains contributes. The protein–DNA contacts are dominated by sugar-phosphate backbone interactions, as expected for a protein that can accommodate a wide range of guide sequences. Guide strand nucleotides 2–6, which are especially important for target recognition, are stacked with their Watson–Crick faces exposed and available for base pairing (Richard and Sontheimer, 2009).

4.3 GENERAL MECHANISM OF RNAI

The RNAi pathway, ubiquitous to most of the eukaryotes, consisting short RNA molecule binds to specific target mRNA, form a dsRNA hybrid, and inactivate the mRNA by preventing from producing a protein. Apart from their role in defense against viruses, protozoans, it also influences the development of organisms. During RNAi, the dsRNA formed in cells by DNA- or RNA-dependent synthesis of complementary strands or introduced into cells by viral infection or artificial expression is processed to 20-bp double-stranded siRNAs containing 2-nt 3′ overhangs (Filipowicz et al., 2005). The siRNAs are then incorporated into an RISC, which mediates the degradation of mRNAs with sequences fully complementary to the siRNA (Fig. 4.3). In another recent pathway, occurring in the nucleus, siRNAs formed from repeat element transcripts and incorporated into the RNAi-induced transcriptional silencing complex may guide chromatin modification and silencing. The genetics and biochemistry of the latter process are best characterized for the plants and yeast, but related pathways also operate in other organisms (Lippman and Martienssen, 2004).

4.3.1 INITIATION: PROCESSING OF PRECURSOR DSRNA

RNAi pathway, a RNA dependent pathway, can be activated by either exogenous or endogenous short dsRNA molecules in the cytoplasm. The precursor of siRNA, termed as primary siRNA or pri-siRNA, folds back to form a long stem-loop structure (endogenous source dsRNA), leaving two 3′ overhang nucleotide and 5′ phosphate group at the cleavage site (Hannon and Rossi, 2004). In case of miRNA, Drosha and Pasha are responsible for trimming the end of stem-loop like pri-miRNA inside the nucleus, leading to the generation of pre-miRNA. Now, this pre-miRNA is transported to the cytoplasm by the help of Ran-GTP-mediated exportin-5 nuclear transporter, where Dicer chops the dsRNA into mature miRNA (Lund et al., 2004).

Processing of exogenous RNAs is cytoplasmic, which leads to the biogenesis of siRNA, only require Dicer but not Drosha. Dicer contains two RNase III domains, one helicase domain, one dsRNA binding domain, and one Piwi/Argonaute/Zwille domain (PWZ). The PWZ domain is also found in Argonaute family proteins, known to be very essential for RNAi. The current finding suggests the binding of Dicer to the end of dsRNA is far more effective than internal binding. Dicer will associate with an existing terminus of dsRNA and cut ~21 nucleotides away from the end, forming a new end with two 3′ overhangs. As a result of this stepwise cutting, a pool of 21-nt long small RNA with two 3′ overhangs nucleotides will be generated from long dsRNAs (Hammond, 2005). Several organisms contain more than one Dicer genes, with each Dicer preferentially processing dsRNAs from different sources. *Arabidopsis thaliana* has four Dicer-like proteins, out of which, DLC-1 is participated in miRNA maturation; DLC-2 preferentially process dsRNA from plant virus; and DLC-3 is required for generating small RNAs from endogenous repeated sequences. Interestingly, most of the mammals encode only one Dicer gene (Xie et al., 2004).

4.3.2 SELECTION OF SIRNA STRAND AND ASSEMBLY OF RISC

The products of dsRNA and pre-siRNA processing by Dicer are 20-bp duplexes with 3′ overhangs. However, miRNAs and siRNAs present in functional RISCs have to be single stranded for pairing with the target RNA. How are the duplexes converted to single-chain forms and how is a correct (i.e., antisense or "guide") strand selected for loading onto the RISC? The latter question is of practical importance because artificial siRNAs can be directly used to trigger RNAi in order to knock-down genes. Measurements of the potency of different double- and single-stranded siRNAs and sequence analysis of the duplexes formed by pre-siRNA processing by Dicer have indicated that the strand incorporated into the RISC is generally the one whose 5′ terminus is the thermodynamically less stable end of the duplex (Khvorova et al., 2003). Recent studies suggest that, in Drosophila, the Dcr-2–R2D2 heterodimer senses the differential stability of the duplex ends and decides which siRNA strand should get selected. Photocross-linking to siRNAs containing 5-iodouracils at different positions demonstrated that Dicer binds to a less stable and R2D2 to a more stable siRNA end. The most conserved members of RISC are Argonaute proteins, which are essential most for RISC functions. Argonaute proteins are highly rich in basic amino acids and these residues are basically responsible for cross-linking with the guide RNA in plants (Tomari et al., 2004). Argonaute proteins are characterized by the presence of two homology regions, the PAZ domain and the PIWI domain (RNase H like functional motif). PAZ domain also appears in Dicer proteins which specifically recognize the unique structure of two 3′ nucleotides overhangs of siRNAs. 5′ Phosphate group is recognized by the PIWI domain in Argonaute proteins and therefore required for siRNA to assemble into RISC. SiRNA lacking this phosphate group in 5′ end will be rapidly phosphorylated by an endogenous kinas (Nykanen et al., 2001). Transfer of Dicer processed dsRNA to RISC is mediated by several unknown proteins. An ATP-dependent process is needed to activate RISC, which helps in unwinding of siRNA duplex, leaving only single-strand RNA joining the active form of RISC. Studies on comparing stability between functional and nonfunctional siRNA indicates that the 5′ antisense region of the functional siRNAs were less thermodynamically stable than the 5′ sense regions, providing a basis for their selective entry into the RISC. The strand remained within the RISC function as a guild to locate targets mRNA sequence through Watson–Crick base paring, whereas the other stand of duplex siRNA is either cleaved or discarded during the loading process. The endonuclease Argonaute 2, the only member of the Argonaute subfamily of proteins with observed catalytic activity in mammalian cells, is responsible for this slicing activity. Cleaved transcripts will undergo subsequent degradation by cellular exonucleases. The guiding strand of siRNA duplex inside RISC will be intact during this process and therefore permit RISC function catalytically. This robust cleavage pathway makes it a very attractive method of choice for potential therapeutic applications of RNAi (Elbashir et al., 2001). Whether siRNA-mediated regulation has an impact on initiation, elongation, or termination, or whether it acts co-translationally, is still a matter of debate. For example, human Ago2 binds to m7GTP and thus can compete with eukaryotic translation initiation factor 4E (eIF4E) for binding to them 7GTP-cap structure of mRNA; association of human Ago2 with eIF6 and large ribosomal subunits also suggests that miRNAs inhibit an early step of translation. However, miRNAs and AGOs are found associated with polysomes, suggesting that inhibition occurs after initiation, at least in some cases (Vaucheret, 2008).

In plants, the majority of miRNAs hybridize to target mRNA with a near-perfect complementarity and mediate an endonucleolytic cleavage through a similar, if not identical, mechanism used by the siRNA pathway. While in animal, miRNA interacts only with 3′ UTR of mRNA (e.g., *lin-4*) and regulated expression of proteins negatively. The central mismatch between miRNA–mRNA hybridization is believed to be responsible for the lack of RNAi-mediated mRNA cleavage events (i.e., lack of RISC-mediated mRNA degradation). miRNA–mRNA complex associated with Ago proteins finally transfer to processing body (P-body), where mRNA finally degraded by RISC-independent pathway (Liu et al., 2005; Sen and Blau, 2005). RNAi-mediated silencing of genes is not limited to the posttranscriptional level only. In plants, it has been shown that siRNA can also trigger de-novo DNA methylation and transcriptional silencing. Recent evidence suggests that siRNAs can inactivate transcription through direct DNA methylation and other types of covalent modification in the genomes of certain species. Several studies also demonstrated that RNAi machinery in the fission yeast *S. pombe* plays a critical role in formation and maintenance of higher order chromatin structure and function. It is hypothesized that expression of centromeric repeats results in the formation of a dsRNA that is cleaved by Dicer into siRNAs that direct DNA methylation of heterochromatic sites and regulates the expression of genes (Mette et al., 2000; Wassenegger et al., 1994). Many plant and some animal viruses encode suppressors of post-transcriptional RNA silencing that interfere with the accumulation or function of siRNAs. Recent crystallographic studies have revealed how the p19 suppressor protein of *Tombusviridae* elegantly and effectively sequesters siRNAs aimed at destroying viral RNA (Baulcombe, 2004; Vargason et al., 2003).

RNA silencing functions as a natural immunity mechanism in plant defense against pathogen invasion (Ding, 2010), and many viruses have evolved to express virus-silencing repressor (VSR) proteins to counteract host antiviral RNA silencing as mentioned in Figure 4.3. Some of the VSRs were studied at molecular level such as 2b of Cucumber mosaic, P69 of the turnip yellow mosaic virus (TYMV), and HC-Pro of the turnip mosaic virus, in Arabidopsis. The P19 protein of tombusviruses, undoubtedly the best known VSR so far, prevents RNA silencing by siRNA sequestration through binding ds-siRNA with a high affinity (Silhavy et al., 2010). Crystallographic studies have revealed that P19 forms is a tail-to-tail homodimer, which acts like a molecular caliper, measuring the length of siRNA duplexes and binding them in a sequence-independent way, selecting for the 19-bp long dsRNA region of the typical siRNA (Vargason et al., 2003). Latest findings have also confirmed that P19 inhibits the spread of the ds-siRNA duplex identified as the signal of RNA silencing (Dunoyer et al., 2010).

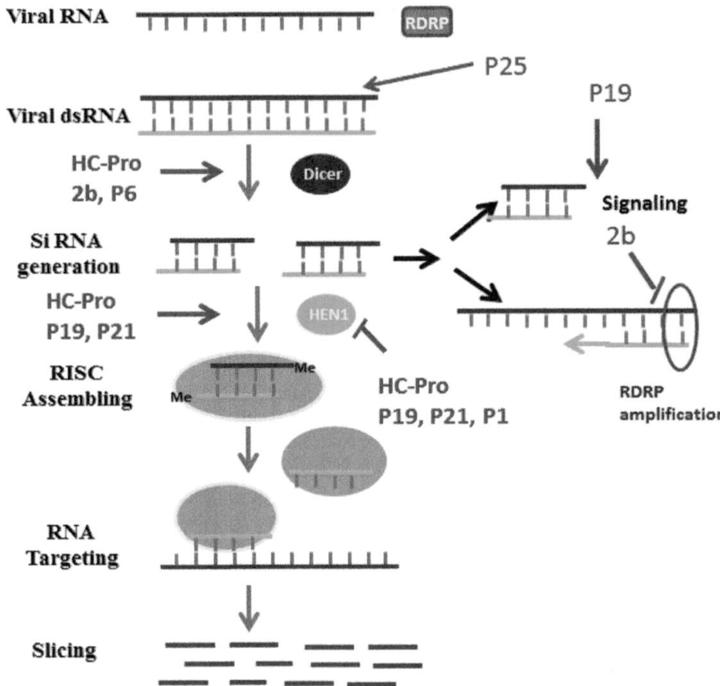

FIGURE 4.3 Viral RNA silencing in plant and its counter defense. (Adapted from Costa et al., 2013)

Other VSRs, such as the Tomato aspermy cucumovirus 2b protein or B2 of the insect-infecting Flock house virus, also bind ds-siRNA in a size-specific manner; nevertheless, structural studies have shown that their modes of binding siRNAs do not share any similarity with P19 (Chen et al., 2008).

Identified two viral proteins were shown to inhibit the processing of dsRNA to siRNAs in agroinfiltration assays: P14 of Pothos latent aureusvirus and P38 of Turnip crinkle virus. Recently, it was discovered that the action of the P38 protein occurs through AGO1 binding and that it interferes with the AGO1-dependent homeostatic network, which leads to the inhibition of Arabidopsis DCLs (Azevedo et al., 2010). In addition to P14 and P38, the P6 VSR of the Cauliflower mosaic virus (CaMV) has been shown to interfere with vsiRNA processing. P6 was previously described as a viral translational trans-activator protein essential for virus biology. Importantly, P6 has two importin-alpha dependent nuclear localization signals, which are mandatory for CaMV infectivity. A recent discovery showed that one of the nuclear functions of P6 is to suppress RNA silencing by interacting with dsRNA-binding protein 4, which is required for the functioning of DCL4.

4.4 VIRUS-INDUCED GENE SILENCING: MECHANISMS AND APPLICATIONS

Van Kammen was first to use the term "virus-induced gene silencing" to describe the phenomenon of recovery from virus infection (van Kammen, 1997), though the term has since been applied almost exclusively to the technique involving recombinant viruses to knockdown expression of endogenous genes (Baulcombe, 1999; Ruiz et al., 1998). RNA silencing has become a major focus of molecular biology and biomedical research around the world. To reduce the losses caused by plant pathogens, plant biologists have adopted numerous methods to engineer resistant plants. Among them, RNA-silencing-based resistance has been a powerful tool that has been used to engineer resistant crops during the last two decades. Based on this mechanism, diverse approaches were developed. VIGS is a virus vector technology that exploits an RNA-mediated antiviral defense mechanism. In plants infected with unmodified viruses, the mechanism is specifically targeted against the viral genome. However, with virus vectors carrying inserts derived from host genes the process can be additionally targeted against the corresponding mRNAs. VIGS has been used widely in plants for analysis of gene function and has been adapted for high-throughput functional genomics. Until now, most applications of VIGS have been studied in *Nicotiana benthamiana*. However, new vector systems and methods are being developed that could be used in other plants, including Arabidopsis. VIGS also helps in the identification of genes required for disease resistance in plants. These methods and the underlying general principles also apply when VIGS is used in the analysis of other aspects of plant biology.

When a plant virus infects a host cell, it activates an RNA-based defense that is targeted against the viral genome. The dsRNA in virus-infected cells is thought to be the replication intermediate that causes the siRNA/RNase complex to target the viral single-stranded RNA. In the initially infected cell, the viral ssRNA would not be a target of the siRNA/RNase complex because this replication intermediate would not have accumulated to a high level. However, in the later stages of the infection, as the rate of viral RNA replication increases, the viral dsRNA and siRNA would become more abundant. Eventually, the viral ssRNA would be targeted intensively and virus accumulation would slow down (Voinnet, 2001). Many plant viruses encode proteins that are suppressors of this RNA silencing process. These suppressor proteins would not be produced until after the virus had started to replicate in the infected cell so they would not cause complete suppression of the RNA based defense mechanism. However, these proteins would influence the final steady-state level of virus accumulation. Strong suppressors would allow virus accumulation to be prolonged and at a high level. Conversely, if a virus accumulates at a low level, it could be due to weak suppressor activity (Brigneti et al., 1998). The dsRNA replication intermediate would be processed so that the siRNA in the infected cell would correspond to parts of the viral vector genome, including any nonviral insert. Thus, if the insert is from a host gene, the siRNAs would target the RNase complex to the corresponding host mRNA and the symptoms in the infected plant would reflect the loss of the function in the encoded protein.

There are several examples that strongly support this approach to suppression of gene expression. Thus, when tobacco mosaic virus (TMV) or potato virus X (PVX) vectors were modified to carry inserts from the plant phytoene desaturase gene the photobleaching symptoms on the infected plant reflected the absence of photoprotective carotenoid pigments that require phytoene desaturase. Similarly, when the virus carried inserts of a chlorophyll biosynthetic enzyme, there were chlorotic symptoms and, with a cellulose synthase insert, the infected plant had modified cell walls (Kjemtrup et al., 1998). Genes other than those encoding metabolic enzymes can also be targeted by VIGS. For example, if the viral insert corresponded to genes required for disease resistance, the plant exhibited enhanced pathogen susceptibility. In one such example, the insert in a tobacco rattle virus (TRV) vector was from a gene (EDS1) that is required for N-mediated resistance to TMV. The virus vector-infected N-genotype plant exhibited compromised TMV resistance. The symptoms of a TRV vector carrying a *leafy* insert demonstrate how VIGS can be used to target genes that regulate development. *Leafy* is a gene required for flower development. Loss-of-function *leafy* mutants produce modified flowers that are phenocopied in the TRV-*leafy*-infected plants. Similarly the effects of tomato golden mosaic virus (TGMV) vectors carrying parts of the gene for a cofactor of DNA polymerase illustrate how VIGS can be used to target essential genes. The plants infected with this geminivirus vector were suppressed for division growth in and around meristematic zones of the shoot (Peele et al., 2001).

To exploit the ability to knockdown, in essence, any gene of interest, RNAi via siRNAs has generated a great deal of interest in both basic and applied biology. There are an increasing number of large-scale RNAi screens that are designed to identify the important genes in various biological pathways. Because disease processes also depend on the combined activity of multiple genes, it is expected that turning off the activity of a gene with specific siRNA could produce a therapeutic benefit to mankind. Based on the siRNAs-mediated RNA silencing (RNAi) mechanism, several transgenic plants has been designed to trigger RNA silencing by targeting pathogen genomes. Diverse targeting approaches have been developed based on the difference in precursor RNA for siRNA production, including sense/antisense RNA, small/long hairpin RNA, and artificial miRNA precursors. Virologists has been designed many transgenic plants expressing viral coat protein (CP), movement protein, and replication-associated proteins, showing resistant against infection by the homologous virus. This type of pathogen-derived resistance has been reported in diverse viruses including tobamo-, potex-, cucumo-, tobra-, Carla-, poty-, and alfalfa mosaic virus groups as well as the luteovirus group (Abel et al., 1986; Ding, 2010). Transgene RNA-silencing-mediated resistance is a process that is highly associated with the accumulation of viral transgene-derived siRNAs. One of the drawbacks of the sense/antisense transgene approach is that the resistance is unstable, and the mechanism often results in delayed resistance or low efficacy/resistance. This may be due to the low accumulations of transgene-derived siRNA in PTGS due to defense mechanism encoded by plants. Moreover, numerous viruses, including potyviruses, cucumoviruses, and tobamoviruses, are able to counteract these mechanisms by inhibiting this type of PTGS. Therefore, the abundant expression of the dsRNA to trigger efficient RNA silencing becomes crucial for effective resistance. To achieve resistance, inverse repeat sequences from viral genomes were widely used to form hairpin dsRNA in vivo, including small hairpin RNA, self-complementary hpRNA, and intron-spliced hpRNA. Among these methods, self-complementary hairpin RNAs separated by an intron likely elicit PTGS with the highest efficiency. The presence of inverted repeats of dsRNA-induced PTGS (IR-PTGS) in plants also showed high resistance against viruses. IR-PTGS is not required for the formation of dsRNA for the processing of primary siRNAs, but the plant RDRs are responsible for the generation of secondary siRNAs derived from nontransgene viral genome, which further intensify the efficacy of RNA silencing induced by hpRNA, a process named RNA-silencing transitivity. Among them, the sequence similarity between the transgene sequence and the challenging virus sequence is the most important. Scientists have engineered several transgenic plants with multiple hpRNA constructs from different viral sources, or with a single hpRNA construct combining different viral sequence, was created. Thus, multiple viruses can be simultaneously targeted, and the resulting transgenic plants show a broader resistance with high efficacy. In addition to the sequence similarity, the length of the transgene sequence also contributes to high resistance. In general, an average length of 100–800 nt of transgene sequence confers effective resistance (Bucher et al., 2006; Himber et al., 2003).

By mimicking the intact secondary structure or hairpin loop of endogenous miRNA precursors, artificial miRNAs (amiRNAs) are designed and processed in vivo to target the genes of interest. The strategy of expressing amiRNAs was first adopted to knockdown endogenous genes for functional analysis. The technology is widely used in engineering antiviral plants and animals. Compared to conventional RNAi strategies, amiRNAs have many advantages: (1) Owing to the short sequence of amiRNAs, a long viral cDNA fragment is not required; thus, the full extent of off-target effects are avoided, and the biosafety of transgenic crops is increased compared to siRNAs from long hairpin RNA; (2) tissue- or cell-specific knockout/downs of genes of interest can be realized because of different tissue- or cell-specific promoters being used; and (3) the relaxed demand on sequence length makes amiRNAs especially useful in targeting a class of conserved genes with high sequence similarities, like tandem arrayed genes, because a short conserved sequence is more easily found in these genes (Schwab, 2006).

Virus which has been modified and used for silencing the gene of interest is summarized in Table 4.1. TMV is one of the modified viruses which were used for effective *pds* gene silencing in *N. benthamiana* plants. TMV is the first modified virus for application of VIGS methods to plants. The viral delivery leads down regulation of transcript of target gene through its homology-dependent degradation so potential of VIGS for analysis of gene function was easily recognized. TRV was also modified to be a tool for gene silencing in plants. VIGS has been effectively applied in *N. benthamiana* and in tomato by using TRV vectors. The significant advantage of TRV-based VIGS in *Solanaceous* species is the ease of introduction of the VIGS vector into plants. The VIGS vector is placed between right border and left border (LB) sites of T-DNA and inserted into *Agrobacterium tumefaciens* (Liu et al., 2002; Ratcliff et al., 2001).

Another property of TRV is the more vigorous spreading all over the entire plant including meristem, and infection symptoms of TRV are mild. Modified TRV vectors such as pYL156 and pYL279 have strong duplicate 35S promoter and a ribozyme at C-terminus for more efficient and faster spreading. These vectors are also able to infect other plant species. TRV-based vector has been used by Liu et al. (2005) for gene silencing in tomato. Very recently, Pflieger et al. have shown that a viral vector derived from TYMV has the ability to induce VIGS in *A. thaliana*. VIGS of *N. benthamiana* using PVX was also achieved. PVX-based vectors have more limited host range (only three families of plants are susceptible to PVX) than TMV-based vectors (nine plant families show susceptibility for TMV) but PVX-based vectors are more stable compared to TMV. Geminivirus-derived

TABLE 4.1 A Range of Plant Viruses Used for Silencing of Target Genes and their Hosts with Targeted Genes.

Virus	Silencing Host	Gene Silenced	References
Tobacco mosaic virus	*Nicotiana benthamiana*	*pds*	Kumagai et al. (1995)
	Nicotiana tabacum		
Potato virus X	*Nicotiana benthamiana*	*pds*	Faivre-Rampant et al. (2004)
	Arabidopsis		
Tobacco rattle virus	*Nicotiana benthamiana*	*Rar1, EDS1, NPR1/NIM1, pds, rbcS*	Ratcliff et al. (2001),
	Tomato, *Solanum* species, chilli pepper		
Pea early browining virus	*Pisum sativum*	*Pspds, uni, kor, pds*	Constantin et al. (2004)
Cabbage leaf curl virus	*Arabidopsis*	*CH42, pds*	Turnage et al. (2002)
Barley stripe mosaic virus	Barley	*Pds, Lr21, Rar1, Sgt 1, Hsp90*	Holzberg et al. (2002), Scofield et al. (2005)
Bean pod mottle virus	*Glycine max*	*pds*	Zhang et al. (2006)
African cassava mosaic virus	*Nicotiana benthamiana*	*Pds, su, cry78d2*	Fofana et al. (2004)
Tomato yellow leaf curl china virus	*Nicotiana benthamiana*	*Pcna, pds, su, gfp*	Kjemtrup et al. (1998)
	Nicotiana tabacum		

vectors can be used for VIGS studies especially to study function of genes involved in meristem function. TGMV was used to silence a meristematic gene, proliferating cell nuclear antigen (PCNA) in *N. benthamiana*. The TGMV-based silencing vector had been used for also silencing of nonmeristematic gene silencing. Satellite-virus-based vectors are also used for efficient gene silencing in plants only with the help of other helper viruses. This two-component system is called Satellite-virus-induced silencing system, SVISS (Fofana et al., 2004; Peele et al., 2001). Previously, barley stripe mosaic virus (BSMV) was developed for efficient silencing of *pds* gene in barley. This system was then used for silencing of wheat genes. BSMV is a positive sense RNA virus containing a tripartite (α, β, γ) genome. The modified γ of BSMV genome replaced by DNA vector was used for plant gene cloning. β-Genome has been deleted for viral CP production defect. Each of the modified DNAs is used to synthesize RNAs by in vitro transcription. Recently, Brome mosaic virus strain has been modified for VIGS of *pds*, *actin*, and *rubisco activase*. These genes were also silenced in important model plants such as rice (Tao and Zhou, 2004). Steps for VIGS have been shown in Figure 4.4. Protocols for VIGS are as follow:

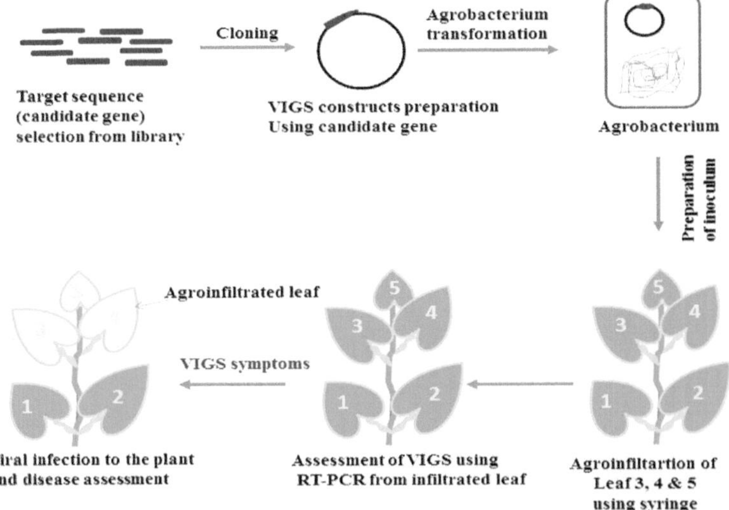

FIGURE 4.4 Steps of virus-induced gene silencing (VIGS). VIGS starts by the cloning of the target gene fragment (200–1300 bp) into a virus infectious cDNA, which is in a binary vector under the control of the CaMV 35S promoter. The recombinant virus construct is then transformed into agrobacterium (*Agrobacterium tumefaciens*) for agrobacterium-mediated virus infection. VIGS will target to the virus carried host gene fragment as to the viral genome and also the endogenous host gene target.

4.4.1 TARGET SEQUENCE SELECTION

si-Fi (siRNA Finder; http://labtools.ipk-gatersleben.de/) software could be used to select 250–400-nt sequence regions that are predicted to produce high numbers of silencing-effective siRNAs. When possible, select at least two preferably nonoverlapping regions of the gene of interest for VIGS analyses. Observation of the same phenotype induced by silencing using each of the two or more independent VIGS constructs is a good indication that the phenotype is due to specific silencing of the intended target gene, therefore, allowing greater confidence in the obtained results. When attempting to silence an individual member of a gene family consider selecting the sequences from the 30- or 50-UTR regions, which are generally more variable than the CDS. This should minimize the risk of off-target silencing. On the other hand, in cases when a great deal of functional redundancy is expected among different gene family members, it should be possible to design VIGS construct(s) from the conserved gene regions in order to target several or even all gene family members simultaneously. Regarding VIGS experimental design, at least one negative control VIGS construct containing a 250–400-nt fragment of a nonplant origin gene, such as the Aequorea victoria Green Fluorescent Protein gene or the *Escherichia coli* β-glucuronidase gene should be included.

4.4.2 VIGS CONSTRUCTS PREPARATION

Clone the VIGS target sequences into for example the BSMV RNAc vector pCa–cbLIC via ligation independent cloning (LIC), in either sense or antisense orientation. Antisense constructs may be slightly more efficient in inducing gene silencing. Transform the sequence verified pCa-cb-LIC VIGS construct into *A. tumefaciens* GV3101 by electroporation. For this MicroPulser (Bio-Rad) electroporator, 0.1-cm gap electroporation cuvettes, and homemade electrocompetent cells could be used: *Agrobacterium* cultures grown to a final OD600 of 1.2 and the cells will be pelleted by centrifugation and washed in ice-cold sterile 10% glycerol seven times in total. Electroporation can be done using the manufacturer's pre-set conditions for *Agrobacterium*, that is, one 2.2-kV pulse. Plate an aliquot of the transformation mixture on LB agar supplemented with 25 µg/ml gentamycin and 50 µg/ml kanamycin. As BSMV requires all three genomic segments, RNAa, RNAb, and RNAc, for successful infection, it is necessary to also produce *A. tumefaciens* GV3101 strains containing pCaBS-α (BSMV RNAα) and pCaBS-β (BSMV RNAβ).

4.4.2.1 PREPARATION OF VIRUS INOCULUM AND INFECTING TARGET PLANTS WITH ENGINEERED VIRUS

Prepared engineered virus introduced into the leaf of dicot plants (e.g., well-studied *N. benthamiana*) via agroinfiltration. For *N. benthamiana* agroinfiltration, grow 5 ml cultures (LB supplemented with 25 μg/ml gentamycin and 50 μg/ml kanamycin) of *A. tumefaciens* strains carrying pCa-cbLIC VIGS constructs overnight at 28°C with constant shaking at 220 rpm. For each BSMV RNAc construct, BSMV RNAα and RNAβ constructs in 5 ml cultures will also be required. Pellet the *A. tumefaciens* cells at 2500 rcf for 20 min, resuspend in infiltration buffer [10 mM $MgCl_2$, 10 mM 2-(*N*-morpholino)ethanesulfonic acid pH 5.6, and 150 μM acetosyringone] to a final optical density at 600 nm (OD600) and incubate at room temperature without shaking for 3 h or longer. Mix *A. tumefaciens* strains carrying BSMV RNAα, RNAβ, and RNAγ strains together in 1:1:1 ratio and pressure infiltrate the bacteria into the abaxial side of fully expanded leaves of approximately 25–30 days old *N. benthamiana* plants using a needleless 1-ml syringe. Use 0.5–1 ml of *Agrobacterium* suspension per leaf and aim to infiltrate the whole area of each leaf.

4.4.2.2 ASSESSMENT OF VIRUS-INDUCED GENE SILENCING

Successful silencing of the targets gene in the VIGS construct-infected plants is assessed using quantitative reverse-transcription PCR (qRT-PCR). The primers used for this purpose should bind outside the region targeted for silencing.

4.4.2.3 VIRAL INFECTION TO THE PLANT AND DISEASE ASSESSMENT

After confirming the turning off of target gene one has to infect the host (plant) from the susceptible virus for the disease assessment.

4.5 CONCLUSION

The discovery of RNAi, the process of sequence-specific gene silencing initiated by dsRNA, has broadened our understanding of gene regulation and has revolutionized methods for genetic analysis. Gene expression is regulated by transcriptional and posttranscriptional pathways, which are crucial for optimizing gene output and for coordinating cellular programs. In plant, 20–24-nltd RNAi regulate gene expression networks necessary for proper development, cell viability, and stress responses. Gene-silencing techniques represent great opportunities for plant breeding. Several practical applications in economically important crops are possible as well as research on gene function and expression. RNAi stability in plants is a very important feature to be accessed in the near future as well as the development of tissue specific and inducible promoters. These are two crucial points for the establishment of this technology as a marketable option. Control of metabolic pathways will also represent a major challenge when trying to obtain plants with altered levels of specific metabolites. The use of artificial miRNA to engineer viral resistant plants also shows great potential. Continuing research on GS in woody plants will probably include plant protection to multiple pathogens (viruses, bacteria), silencing of specific metabolic pathways (lignin synthesis, ethylene, allergens, caffeine, and others), improvement of fruit and wood quality, production of secondary metabolites, and developmental and reproductive trait alteration in plants (induced male sterility and self-compatibility). The ability to switch off genes and interfere with expression patterns in plants, provided by gene silencing techniques, will probably represent a great impact in woody plant breeding.

KEYWORDS

- **RNA interference**
- **gene silencing**
- **precursor molecules**
- **virus**
- **double-stranded RNA**

REFERENCES

Abel, P. P.; Nelson, R. S.; De, B.; Hoffmann, N.; Rogers, S. G.; Fraley, R. T.; Beachy, R. N. Delay of Disease Development in Transgenic Plants that Express the Tobacco Mosaic Virus Coat Protein Gene. *Science* **1986,** *232*, 738–743.

Aravin, A. A.; Hannon, G. J.; Brennecke, J. The Piwi-piRNA Pathway Provides an Adaptive Defense in the Transposon Arms Race. *Science* **2007,** *318*, 761–764.

Azevedo, J.; Garcia, D.; Pontier, D.; Ohnesorge, S.; Yu, A.; Garcia, S.; Braun, L.; Bergdoll, M.; Hakimi, M. A.; Lagrange, T.; Voinnet, O. Argonaute Quenching and Global Changes in Dicer homeostasis Caused by a Pathogen-Encoded GW Repeat Protein. *Genes Dev.* **2010,** *24*, 904–915.

Baulcombe, D. Fast Forward Genetics Based on Virus-Induced Gene Silencing. *Curr. Opin. Plant Biol.* **1999,** *2*, 109–113.

Baulcombe, D. RNA Silencing in Plants. *Nature* **2004,** *431*, 356–363.

Brigneti, G.; Voinnet, O.; Li, W. X.; Ji, L. H.; Ding, S. W.; Baulcombe, D. C. Viral Pathogenicity Determinants Are Suppressors of Transgene Silencing in *Nicotiana benthamiana. EMBO J.* **1998,** *17* (22), 6739–6746.

Brennecke, J.; Malone, C. D.; Aravin, A. A.; Sachidanandam, R.; Stark, A.; Hannon, G. J. An Epigenetic Role for Maternally Inherited piRNAs in Transposon Silencing. *Science* 2008, *322*, 1387–1392.

Bucher, E.; Lohuis, D.; van Poppel, P. M.; Geerts-Dimitriadou, C.; Goldbach, R.; Prins, M. Multiple Virus Resistance at a High Frequency Using a Single Transgene Construct. *J. Gen. Virol.* **2006,** *87*, 3697–3701.

Cenik, E. S.; Fukunaga, R.; Lu, G.; Dutcher, R.; Wang, Y.; Tanaka Hall, T. M.; Zamore, P. D. Phosphate and R2D2 Restrict the Substrate Specificity of Dicer-2, an ATP-Driven Ribonuclease. *Mol. Cell* **2011,** *42* (2), 172–184.

Chen, H. Y.; Yang, J.; Lin, C.; Yuan, Y. A. Structural Basis for RNA-Silencing Suppression by Tomato Aspermy Virus Protein 2b. *EMBO Rep.* **2008,** *9*, 754–760.

Constantin, G. D.; Krath, B. N.; MacFarlane, S. A.; Nicolaisen, M.; Johansen, I. E.; Lund, O. S.; Virus-induced Gene Silencing as a Tool for Functional Genomics in a Legume Species. *Plant J.* **2004,** *40* (4), 622–631.

Costa, AT.; Bravo, JP.; Makiyama, RK.; Nunes, AV.; Maia, IG. Viral Counter Defense X Antiviral Immunity in Plants: Mechanisms for Survival. Current Issues in Molecular Virology - Viral Genetics and Biotechnological Applications, 2013. ed. Victor Romanowski, ISBN 978-953-51-1207-5.

Covey, S.; Al-Kaff, N.; Lángara, A.; Turner, D. Plants Combat Infection by Gene Silencing. *Nature* **1997,** *385* (6619), 781–782.

Ding, S. W. RNA-Based Antiviral Immunity. *Nat. Rev. Immunol.* **2010,** *10*, 632–644.

Dunoyer, P.; Brosnan, C. A.; Schott, G.; Wang, Y.; Jay, F.; Alioua, A.; Himber, C.; Voinnet, O. An Endogenous, Systemic RNAi Pathway in Plants. EMBO J. **2010,** *29*, 1699–1712.

Ecker, J. R.; Davis, R. W. Inhibition of Gene Expression in Plant Cells by Expression of Antisense RNA. *Proc. Natl. Acad. Sci. USA* **1986,** *83* (15), 5372–5376.

Elbashir, S. M.; Lendeckel, W.; Tuschl, T. RNA Interference is Mediated by 21- and 22-Nucleotide RNAs. *Genes Dev.* **2001,** *15* (2), 188–200.

Faivre-Rampant, O.; Gilroy, E. M.; Hrubikova, K.; Hein I.; Millam S.; Loake, G. J.; Birch, P.; Taylor, M.; Lacomme C. Potato Virus X-Induced Gene Silencing in Leaves and Tubers of Potato. Plant Physiol. 2004, *134*(4): 1308–1316.

Filipowicz, W.; Jaskiewicz, L.; Kolb, F. A.; Pillai, S. R. Post-Transcriptional Gene Silencing by siRNAs and miRNAs. *Curr. Opin. Struct. Biol.* **2005,** *15*, 331–341.

Fire, A.; Xu, S.; Montgomery, M. K.; Kostas, S. A.; Driver, S. E.; Mello, C. C. Potent and Specific Genetic Interference by Double-Stranded RNA in *Caenorhabditis elegans. Nature* **1998,** *391* (6669), 806–811.

Fofana, B. F.; Sangaré, A.; Collier, R.; Taylor, C.; Fauquet, C. M. A Geminivirus-Induced Gene Silencing System for Gene Function Validation in Cassava. *Plant Mol. Biol.* **2004,** *56* (4), 613–624.

Gregory, R. I.; Chendrimada, T. P.; Cooch, N.; Shiekhattar, R. Human RISC Couples MicroRNA Biogenesis and Posttranscriptional Gene Silencing. *Cell* **2005,** *123*, 631–640.

Grishok, A.; Pasquinelli, A. E.; Conte, D.; Li, N.; Parrish, S.; Ha, I.; Baillie, D. L.; Fire, A.; Ruvkun, G.; Mello, C. C. Genes and Mechanisms Related to RNA Interference Regulate Expression of the Small Temporal RNAs that Control *C. elegans* Developmental Timing. *Cell* **2001,** *106* (1), 23–34.

Guo, S.; Kemphues, K. Par-1, a Gene Required for Establishing Polarity in *C. elegans* Embryos, Encodes a Putative Ser/Thr Kinase that is Asymmetrically Distributed. *Cell* **1995,** *81* (4), 611–620.

Hamilton, A. J.; Baulcombe, D. C. A Species of Small Antisense RNA in Posttranscriptional Gene Silencing in Plants. *Science* **1999,** *286* (5441), 950–952.

Hammond, S. M.; Bernstein, E.; Beach, D.; Hannon, G. J. An RNA-Directed Nuclease Mediates Post-transcriptional Gene Silencing in *Drosophila* cells. *Nature* **2000,** *404* (6775), 293–296.

Hammond, S. M. Dicing and Slicing: The Core Machinery of the RNA Interference Pathway. *FEBS Lett.* **2005,** *579* (26), 5822–5829.

Hannon, G. J.; Rossi, J. J. Unlocking the potential of the human genome with RNA interference. *Nature* **2004,** 431 (7006), 371–378.

Himber, C.; Dunoyer, P.; Moissiard, G.; Ritzenthaler, C.; Voinnet, O. Transitivity-Dependent and -Independent Cell-to-cell Movement of RNA Silencing. *EMBO J.* **2003,** *22*, 4523–4533.

Höck, J.; Weinmann, L.; Ender, C.; Rüdel, S.; Kremmer, E.; Raabe, M.; Urlaub, H.; Meister, G. Proteomic and Functional Analysis of Argonaute-Containing mRNA–Protein Complexes in Human Cells. *EMBO Rep.* **2007,** *8* (11), 1052–1060.

Holzberg, S.; Brosio, P.; Gross, C.; Pogue, G. P. Barley Stripe Mosaic Virus-induced Gene Silencing in a Monocot Plant. *Plant J.* 2002, *30* (3), 315–327.

Hutvagner, G.; Simard, M. J. Argonaute Proteins: Key Players in RNA Silencing. *Nat. Rev. Mol. Cell Biol.* 2000, *9* (1), 22–32.

Khvorova, A.; Reynolds, A.; Jayasena, S. D. Functional siRNAs and miRNAs Exhibit Strand Bias. *Cell* **2003,** *115*, 209–216.

Kjemtrup, S.; Sampson, K. S.; Peele, C. G.; Nguyen, L. V.; Conkling, M. V.; Thompson, W. F.; Robertson, D. Gene Silencing from Plant DNA Carried by a Geminivirus. *Plant J.* 1998, *14*, 91–100.

Lippman, Z.; Martienssen, R. The Role of RNA Interference in Heterochromatic Silencing. *Nature* **2004,** *431*, 364–370.

Liu, Y.; Schiff, M.; Dinesh-Kumar, S. P. Virus-Induced Gene Silencing in Tomato. *Plant J.* **2002,** *31*(6), 777–786.

Liu, J.; Valencia-Sanchez, M. A.; Hannon, G. J.; Parker, R. MicroRNA-Dependent Localization of Targeted mRNAs to Mammalian P-bodies. *Nat. Cell Biol.* **2005,** *7* (7), 719–723.

Lund, E.; Güttinger, S.; Calado, A.; Dahlberg, J. E.; Kutay, U. Nuclear Export of MicroRNA Precursors. *Science* **2004,** *303* (5654), 95–98.

Macrae, I. J.; Zhou, K.; Li, F.; Repic, A.; Brooks, A. N.; Cande, W. Z.; Adams, P. D.; Doudna, J. A. Structural Basis for Double-Stranded RNA Processing by Dicer. *Science* **2006,** *311* (5758), 195–198.

Martinez, J.; Patkaniowska, A.; Urlaub, H.; Lührmann, R.; Tuschl, T. Single-Stranded Antisense siRNAs Guide Target RNA Cleavage in RNAi. Cell **2002,** *110* (5), 563–574.

Meins, F.; Si-Ammour, A.; Blevins, T. RNA Silencing Systems and their Relevance to Plant Development. *Annu. Rev. Cell Dev. Biol.* **2005,** *21* (1), 297–318.

Meister, G.; Landthaler, M.; Patkaniowska, A.; Dorsett, Yair.; Teng, G.; Tuschl, T. Human Argonaute2 Mediates RNA Cleavage Targeted by miRNAs and siRNAs. *Mol. Cell* **2004,** *15* (2), 185–197.

Meister, G.; Tuschl, T. Mechanisms of Gene Silencing by Double-Stranded RNA. *Nature* **2004,** *431*, 343–349.

Mette, M. F.; Aufsatz, W.; vander-Winden, J.; Matzke, M. A.; Matzke, A. J. Transcriptional Silencing and Promoter Methylation Triggered by Double-Stranded RNA. *EMBO J.* **2000,** *19* (19), 5194–201.

Napoli, C.; Lemieux, C.; Jorgensen, R. Introduction of a Chimeric Chalcone Synthase Gene into Petunia Results in Reversible Co-suppression of Homologous Genes in Trans. *Plant Cell* **1990,** *2* (4), 279–289.

Naqvi, AR.; Islam, MN.; Choudhury, NR.; Haq, Q. The Fascinating World of RNA Interference. Int. J. Biol. Sci. **2009,** *5*(2):97-117.

Nykanen, A.; Haley, B.; Zamore, P. D. ATP Requirements and Small Interfering RNA Structure in the RNA Interference Pathway. *Cell* **2001,** *107* (3), 309–321.

Pal-Bhadra, M.; Bhadra, U.; Birchler, J. Cosuppression in *Drosophila*: Gene Silencing of Alcohol Dehydrogenase by White-Adh Transgenes is Polycomb Dependent. *Cell* **1997,** *90* (3), 479–490.

Peele, C.; Jordan, C. V.; Muangsan, N.; Turnage, M.; Egelkrout, E.; Eagle, P.; Hanley-Bowdoin, L.; Robertson, D. Silencing of a Meristematic Gene Using Geminivirus-Derived Vectors. *Plant J.* **2001,** *24*, 357–366.

Pham, J. W.; Pellino, J. L.; Lee, Y. S.; Carthew, R. W.; Sontheimer, E. J.. A Dicer-2-Dependent 80s Complex Cleaves Targeted mRNAs During RNAi in *Drosophila*. Cell **2004,** *117* (1), 83–94.

Ratcliff, F.; Harrison, B.; Baulcombe, D. A Similarity between Viral Defense and Gene Silencing in Plants. *Science* **1997,** *276* (5318), 1558–1560.

Ratcliff, F.; Martín-Hernáandez, A. M.; Baulcombe, D. C. Tobacco Rattle Virus as a Vector for Analysis of Gene Function by Silencing. *Plant J.* **2001,** *25* (2), 237–245.

Richard, W. C.; Sontheimer, E. J. Origins and Mechanisms of miRNAs and siRNAs. *Cell* 2009, *136* (4), 642–655.

Rivas, F. V.; Tolia, N. H.; Song, J. J.; Aragon, J. P.; Liu, J.; Hannon, G. J.; Joshua-Tor, L. Purified Argonaute 2 and an siRNA Form Recombinant Human RISC. *Nat. Struct. Mol. Biol.* 2005, *12* (4), 340–349.

Romano, N.; Macino, G. Quelling: Transient Inactivation of Gene Expression in *Neurospora crassa* by Transformation with Homologous Sequences. *Mol. Microbiol.* **1992,** *6* (22), 3343–3353.

Ruiz, M. T.; Voinnet, O.; Baulcombe, D. C. Initiation and Maintenance of Virus-Induced Gene Silencing. *Plant Cell* **1998,** *10*, 937–946.

Schwab, R. Highly Specific Gene Silencing by Artificial MicroRNAs in *Arabidopsis*. *Plant Cell* **2006,** *18*, 1121–1133.

Scofield, S. R.; Huang, L.; Brandt, A. S.; Gill, B. S. Development of a Virus-induced Gene-silencing System for Hexaploid Wheat and its Use in Functional Analysis of the Lr21-mediated Leaf Rust Resistance Pathway. *Plant Physiol.* **2005,** *138*, 2165–2173.

Silhavy, D.; Molnár, A.; Lucioli, A.; Szittya, G.; Hornyik, C.; Tavazza, M.; Burgyán, J. A Viral Protein Suppresses RNA Silencing and Binds Silencing-Generated, 21- to 25-Nucleotide Double-Stranded RNAs. *EMBO J.* **2002,** *21*, 3070–3080.

Sen, G. L.; Blau, H. M. Argonaute 2/RISC Resides in Sites of Mammalian mRNA Decay Known as Cytoplasmic Bodies. *Nat. Cell Biol.* **2005,** *7* (6), 633–636.

Tabara, H.; Sarkissian, M.; Kelly, W. G.; Fleenor, J.; Grishok, A.; Timmons, L.; Fire, A.; Mello, C. C. The rde-1 Gene, RNA Interference, and Transposon Silencing in *C. elegans*. *Cell* **1999,** *99* (2), 123–132.

Tabara, H.; Yigit, E.; Siomi, H.; Mello, C. C. The dsRNA Binding Protein RDE-4 Interacts with RDE-1, DCR-1, and a DExH-Box Helicase to Direct RNAi in *C. elegans*. *Cell* **2002,** *109*, 861–871.

Tang, G. siRNA and miRNA: An Insight into RISCs. *Trends Biochem. Sci.* **2005,** *30* (2), 106–114.

Tao, X.; Zhou, X. A Modified Viral Satellite DNA that Suppresses Gene Expression in Plants. *Plant J.* **2004,** *38* (5), 850–860.

Tomari, Y.; Matranga, C.; Haley, B.; Martinez, N.; Zamore, P. D. A Protein Sensor for siRNA Asymmetry. *Science* **2004,** *306*, 1377–1380.

Tomari, Y.; Zamore, P. D. Perspective: Machines for RNAi. *Genes Dev.* **2005,** *19*, 517–529.

Turnage, M. A.; Muangsan, N.; Peele, C. G.; Robertson, D. Geminivirus-based Vectors for Gene Silencing in Arabidopsis. *Plant J.* **2002,** *30* (1), 107–114.

Van Blokland, R.; Vander Geest, N.; Mol, J. N, M.; Kooter, J. M. Transgene-Mediated Suppression of Chalcone Synthase Expression in *Petunia hybrida* Results from an Increase in RNA Turnover. *Plant J.* **1994,** *6* (6), 861–877.

van Kammen, A. Virus-Induced Gene Silencing in Infected and Transgenic Plants. *Trends Plant Sci.* **1997,** *2*, 409–411.

Vargason, J. M.; Szittya, G.; Burgyan, J.; Tanaka, H. T. M. Size Selective Recognition of siRNA by an RNA Silencing Suppressor. *Cell* **2003,** *115*, 799–811.

Vaucheret, H. Plant Argonautes. *Trends Plant Sci.* **2008,** *13* (7), 350–358.

Voinnet, O. RNA Silencing as a Plant Immune System against Viruses. *Trends Genet.* **2001,** *17* (8), 449–459.

Wassenegger, M.; Heimes, S.; Riedel, L.; Sänger, H. L. RNA-Directed De Novo Methylation of Genomic Sequences in Plants. *Cell* **1994,** *76* (3), 567–576.

Yigit, E.; Batista, P. J.; Bei, Y.; Pang, K. M.; Chen, C. C.; Tolia, N. H.; Joshua-Tor, L.; Mitani, S.; Simard, M. J.; Mello, C. C. Analysis of the *C. elegans* Argonaute Family Reveals that Distinct Argonautes Act Sequentially during RNAi. *Cell* **2006,** *127*, 747–757.

Xie, Z.; Johansen, L. K.; Gustafson, A. M.; Kasschau, K. D.; Lellis, A. D.; Zilberman, D.; Jacobsen, S. E.; Carrington, J. C. Genetic and Functional Diversification of Small RNA Pathways in Plants. *PLoS Biol.* **2004,** *2* (5), E104.

Zamore, P. D.; Tuschl, T.; Sharp, P. A.; Bartel, D. P. RNAi: Double-Stranded RNA Directs the ATP Dependent Cleavage of mRNA at 21 to 23 Nucleotide Intervals. *Cell* **2000,** *101* (1), 25–33.

Zhang, H.; Kolb, F.; Jaskiewicz, L.; Westhof, E.; Filipowicz, W. Single Processing Center Models for Human Dicer and Bacterial RNase III. *Cell* **2004,** *118*, 57–68.

Zhang, C.; Ghabrial S. A. Development of Bean Pod Mottle Virus-Based Vectors for Stable Protein Expression and Sequence-Specific Virus-Induced Gene Silencing in Soybean. *Virology* **2006,** *344* (2), 401–411.

CHAPTER 5

VIRUS-INDUCED GENE SILENCING

KISHORE DEY[1*], MICHAEL MELZER[2], and JOHN HU[2]

[1]*Florida Department of Agriculture and Consumer Sciences (FDACS), Gainesville, Florid, FL, United States*

[2]*Department of Plant and Environmental Protection Sciences, University of Hawaii, Honolulu, HI, United States*

[*]*Corresponding author. E-mail: kishore.dey@freshfromflorida.com*

CONTENTS

ABSTRACT

Virus-induced gene silencing (VIGS) is a powerful reverse genetic technology used to unravel the functions of genes. It uses viral vectors carrying a fragment of a gene of interest to generate double-stranded RNA, which initiates the silencing of the target gene. The virus vector is used to induce RNA-mediated silencing of a gene or genes in the host plant. A wide range of viruses have been modified for use as VIGS vectors. As the name suggests, VIGS uses the host plant's natural defense mechanisms against viral infection to silence plant genes. VIGS is methodologically simple and is widely used to determine gene functions, including disease resistance, abiotic stress, biosynthesis of secondary metabolites, and signal transduction pathways. The growing amount of plant genomic and transcriptome data provide a quick way to select important genes for functional analysis. Recently, the VIGS system, besides its ability to silence genes has found an important application in the CRISPR/Cas editing system, a most recent and promising genetic tools for targeted genome editing and precise knocking out of entire genes.

5.1 INTRODUCTION

The field of plant biology is expanding at a tremendous pace. The complete genomes of a large number of plants have been sequenced and more are in progress. With advancements in DNA sequencing, such as high-throughput technology, it is important to unravel the functions of the billions of sequence reads they produce. The next major challenges are to identify the functions of the sequenced genes and engineer the applicable crop traits to meet human needs.

Virus-induced gene silencing (VIGS) is a powerful reverse genetic technology used to unravel the functions of genes. It uses viruses as vectors to carry out targeted gene silencing. The virus vector is used to induce RNA-mediated silencing of a gene or genes in the host plant. The process of silencing is triggered by dsRNA molecules, the mechanism of which is explained in this chapter. Over the years, a large number of viruses have been modified for use as VIGS vectors and a list of these vectors is also included. As the name suggests, VIGS uses the host plant's natural defense mechanisms against viral infection to silence plant genes. VIGS is methodologically simple and is widely used to determine gene functions, including disease resistance, abiotic stress, biosynthesis of secondary metabolites, and signal transduction pathways.

Van Kammen (1997) first used the term VIGS to describe a naturally occurring plant-resistance mechanism against viral infection. To understand how VIGS functions, it is helpful to know the various RNA-mediated pathways in plants. The first indication that plants possessed this innate defense mechanism came from studies on the genetic engineering of plants for increased yield and protection from pests. However, the expression of an engineered gene often was unpredictable, with instances of the silencing of endogenous genes when homologous transgenes were expressed. This phenomenon was initially termed co-suppression, since the introduced transgene triggered silencing of endogenous gene-sharing sequences with the homologous transgene (De Carvalho Niebel et al., 1995). This phenomenon of co-suppression was later identified as posttranscriptional gene silencing (PTGS).

PTGS was first reported in *Petunia* during experiments designed to increase the floral pigmentation by the overexpression of the *chalcone synthase* (*chs*) gene. Intriguingly, the experiment did not result in dark-purple flowers as expected but instead produced colorless flowers (Jorgensen et al., 1996; Napoli et al., 1990; Van der Krol et al., 1990). Earlier examples of PTGS do exist in the literature, although the mechanism was not apparent at the time. For example, in 1928, S. A. Wingard observed that the upper leaves of tobacco plants infected with *Tobacco ring spot virus* became resistant to successive inoculations with the same virus. Moreover, the virus was present even after 10 generations of vegetative propagation and the recovered tissues showed a substantial reduction in virus concentration. This resistance was due to RNA silencing, or PTGS, which was unknown at that time. Researchers later discovered that they could produce this immunity with different viruses when certain viral genes under the control of the CaMV constitutive promoter (35S) were introduced into the host plants. When viral cDNA was used as a transgene then plants acquired immunity against the same virus or related viruses (Baulcombe and English, 1996). This phenomenon was attributed to the mechanism of PTGS or gene silencing, more generally known as RNA silencing or RNA interference (RNAi) (Baulcombe and English, 1996; Fire et al., 1998). Unraveling the molecular mechanism of PTGS required considerable effort from various disciplines, and many review articles have been published (Cogoni and Macino, 2000; Depicker and Van Montagu, 1997; Hammond et al., 2001; Hannon, 2002). Early examples of PTGS came from plants displaying recovery phenotypes from viral infection and that later became resistant to future infection from the same virus. This recovery phenomenon led to the discovery of the systemic component of PTGS, the spreading of a diffusible signal following virus infection that signaled the host of the invading virus. Readers are encouraged to review the elegant research done to discover this mobile silencing signal (Mlotshwa et al., 2002). Subsequent studies demonstrated that PTGS is present in all higher eukaryotes and acts as a natural defense pathway in plants against viral infection. When a plant targets specific introduced genes and develops resistance to other plant viruses, it is called pathogen-derived resistance (Ratcliff et al., 1997; Vance and Vaucheret, 2001).

Although the basic knowledge of RNA silencing was still accumulating, the sense and antisense technology used to downregulate gene expression in many experimental hosts, such as *Caenorhabditis elegans* and *Nicotiana benthamiana*, was already in common use on a large scale. Its efficiency, however, was variable. It was only after an accidental discovery that a hairpin construct, rather than a sense or antisense RNA construct, was more effective in downregulating gene expression, that the role of RNAi as a natural defense mechanism in eukaryotes was gradually accepted. Another key breakthrough was the discovery of small RNA (smRNA) 21–28-nt long in the silenced tissue. This was homologous to the endogenous genes in both mammalian and plant system.

In this chapter, we will discuss discoveries that were instrumental in the development of VIGS. We describe the basic underlying molecular mechanism of VIGS, the methodology and various experimental requirements, and its advantages and disadvantages. Finally, we consider the future prospects of VIGS in relation to CRISPR (clustered regulatory interspaced short palindromic repeats)/Cas9 technology. Besides using it to overexpress or silence genes, VIGS has emerged as the preferred delivery system for the cutting edge CRISPR/Cas9 genome-editing technology.

5.2 OVERVIEW OF AN RNA-SILENCING MECHANISM IN PLANTS

Since VIGS utilizes a virus vector that exploits the host's gene-silencing mechanism, understanding the underlying virus–host interactions is essential for understanding VIGS technology. At least three different pathways of RNA silencing are recognized in plants. The first pathway involves the silencing of the plant's own messenger RNA (mRNA). This pathway regulates various gene expression functions using endogenous microRNAs (miRNAs) encoded by distinct nucleotide sequences. These miRNAs bind in a gene-specific manner to mRNAs, cleaving them or inhibiting protein translation (Baulcombe, 2004; Vaucheret, 2006). The second pathway involves DNA methylation and gene transcription suppression. It operates at the chromatin level and mainly in a response against transposons (Lippman and Martienssen, 2004; Waterhouse et al., 2001). RNA-directed DNA methylation (RdDM) cleaves dsRNA into smRNAs that guide the RNA-induced transcriptional silencing complex with its guide sequence complementary to its targets, modifies them, and renders them transcriptionally inactive (Lippman and Martienssen, 2004; Waterhouse et al., 2001). DNA modification by methylation is an important method of targeting transposable elements (Liu et al., 2010). Promoters and other regions that are usually not transcribed are also targets for methylation in transcriptional silencing (Matzke et al., 2001). This methylation usually corresponds to transcriptional silencing (Hirochika et al., 2000). The third pathway identified in plants is cytoplasmic small-interfering RNA (siRNA) silencing (Hamilton and Baulcombe, 1999), also known as PTGS. This pathway operates at a posttranscriptional level after the gene is transcribed and transported into the cytoplasm. This is the primary pathway a plant uses against viral infection and is the underlying mechanism of VIGS technology.

In plant systems, the silencing mechanism is triggered by the cleavage of dsRNA by an RNAaseIII-like Dicer enzyme into siRNAs or miRNAs 21–25 nt in length (Hamilton and Baulcombe, 1999; Lakatos et al., 2006). These smRNAs are first incorporated into a multisubunit ribonuclease complex, known as RISC (RNA-induced silencing complex). The siRNAs guide the RISC to homologous RNA sequences, promoting a sequence-specific RNA degradation (Hamilton et al., 2002). One of the important components of RISC is the Argonaute (Ago) protein, which cleaves single-stranded (ss)RNA complementary to siRNAs into smRNA fragments (Tomari and Zamore, 2005). This gene inactivation process protects plants and animals against transposons and viral RNA or DNA (Mello and Conte, 2004).

The general RNA silencing pathway (Fig. 5.1) shows how replicating viruses are targeted after infection by the host's silencing machinery. The ssRNA of a positive-strand RNA virus is converted to dsRNA, in the cytoplasm of the plant cell by its own RNA-dependent RNA polymerase (RdRp) or by plant-encoded RdRp (Dalmay et al., 2000; Waterhouse et al., 2001). The dsRNA derived from the viral ssRNA is believed to be the main target of the host-silencing machinery and involves the siRNA/RNase complex. The dsRNAs are the replication intermediates and accumulate in the host tissues after the virus establishes itself. In the first infected cells, it is assumed that these dsRNA molecules do not accumulate in large numbers. In the later stages, however, the dsRNA increases and the resulting siRNA accumulates in large amounts. The siRNA then targets the viral ssRNA, reducing the number of virus particles in the cytoplasm.

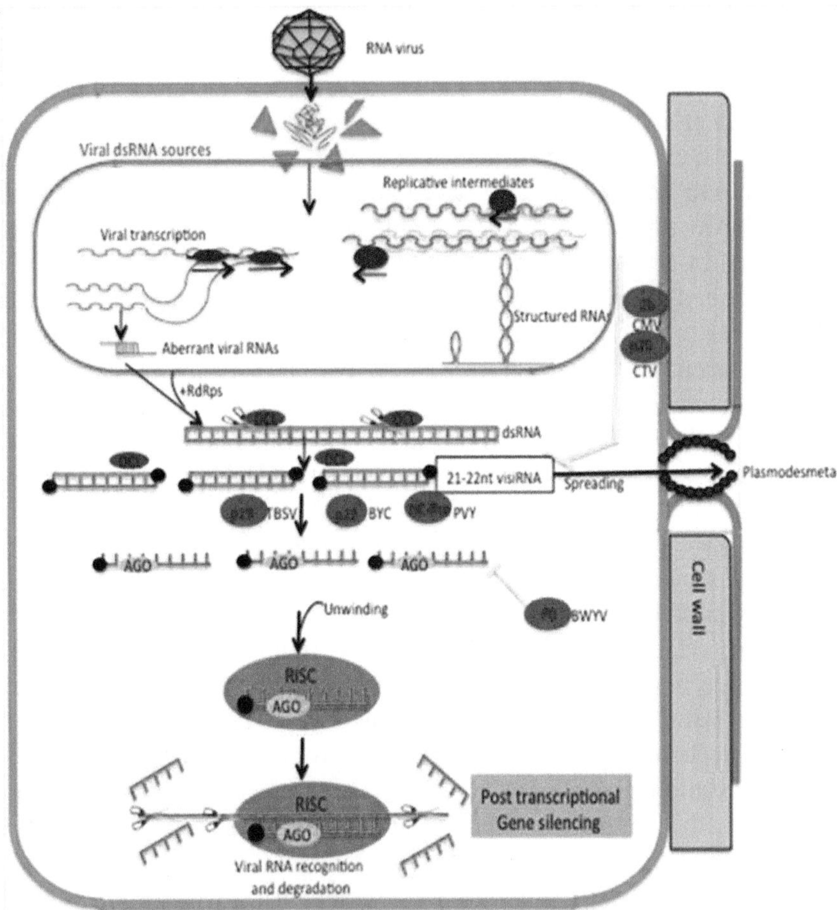

FIGURE 5.1 *Schematic showing the generalized RNA-silencing pathway.* RNA silencing is initiated by the recognition of viral dsRNAs, which are processed to viral siRNA (vsiRNAs) of 21–24 nt by dsRNA-specific RNases called DCLs. In the next step, vsiRNAs are recruited into an AGO-containing complex, also called an RNA induced-silencing complex (RISC). Here, the duplex is unwound and only one of the strands, called a guide strand providing specificity, is recruited into AGO-containing complexes to target viral RNA. AGO mediates the slicing. Also shown is the short-distance spread of the RNA silencing signal, the 21–22-nt vsiRNA spreads to adjacent cells of the plant through the plasmodesmata, and multiple points of disruption by the plant virus suppressor are sequestering the vsiRNA by p19 protein of *Tomato bushy stunt virus* (TBSV), preventing the assembly of different effectors such as P0 protein of *Beet western yellow virus* (BWYV), and blocking the spread of systemic silencing such as p20 protein of *Citrus tristeza virus* (CTV) or 2b protein of *Cauliflower mosaic virus* (CMV). The long-distance silencing signal may be due to secondary vsiRNA produced in an amplification loop by the action of plant RDRps and their cofactors. (Adapted from ViralZone/ExPASy.com and used with permission from ViralZone, Swiss Institute of Bioinformatics.)

At the biochemical level, the Dicer processes dsRNA into 21–23-nt dsRNA. ATP, RNA helicase, and the Ago protein then incorporate these short dsRNAs into the RISC complex (Meister and Tuschl, 2004). Inside the RISC complex, double-stranded siRNA is unwound and only one strand used for locating the complementary viral RNAs for degradation. This is occurs in a sequence-specific manner by the Ago slicer component of the RISC complex. All RNA viruses form dsRNA-like structures during replication. The recognition of this molecule by the host PTGS pathway is a key step in the process (Baulcombe, 2004). Since the siRNAs that result from degradation of the virus, target more virus particles, this process is also known as *cis*-acting siRNAs (Vaucheret, 2006), or VIGS.

Although most plant viruses are RNA viruses, some DNA viruses are economically important. These include the geminiviruses, nanoviruses, and caulimoviruses, which have unique replication strategies. Geminiviruses and nanoviruses are ssDNA viruses and replicate by rolling circle replication with a dsDNA intermediate (Laufs et al., 1995). Caulimoviruses have a dsDNA without genome and replicate by RNA intermediates using reverse transcription (Hull et al., 1987). Despite having no dsRNA intermediate in their replication, geminiviruses can trigger the host RNA silencing mechanism. Geminiviruses can reportedly trigger transcriptional gene silencing as well (Vanitharani et al., 2005). To take advantage of this, it is important that effective geminivirus-based VIGS be constructed. Since this chapter deals with RNA silencing by RNA viruses, readers interested in the RNA silencing response in DNA viruses should refer to the excellent review by Vanitharani et al. (2005).

It was soon discovered that most plant viruses encode proteins to suppress the host's PTGS (Anandalakshmi et al., 1998; Brigneti et al., 1998; Kasschau and Carrington, 1998). The indication that a counter-defense mechanism was present in plant viruses came from the observation that mild disease symptoms caused by a single virus infection were often more severe in plants infected with multiple viruses. This suggested a synergistic effect among the viruses, resulting in an increase in viral replication and higher virus titers (Bance, 1991; Pruss et al., 1997; Rochow and Ross, 1955). This synergism was widely observed, especially within the genus *Potyvirus* (Bance, 1991; Calvert and Ghabrial, 1983; Goldberg and Brakke, 1987; Rochow and Ross, 1955). All these observations led to the discovery of viral encoded proteins called as suppressor proteins, which were found to strongly influence the counteracting RNA-silencing mechanism of the host. Viruses with strong suppressors accumulate at higher levels in their hosts and are more persistent.

An intriguing aspect of RNA silencing is its noncell autonomous nature. Viral RNA molecules introduced at the inoculation site are able to cause silencing at a distance from this site (Mlotshwa et al., 2002; Voinnet, 2005). This phenomenon was first observed in *N. benthamiana* plants, when a scion expressing a *nitrite reductase* transgene (35S-*Nia*) was grafted on a tobacco rootstock expressing a *Nia*-silenced transgene (Palauqui et al., 1997), the systemic leaves were silenced for the introduced transgenes. These experiments indicated that a mobile signal is involved in systemic silencing traveling from the stock to the scion (Mlotshwa et al., 2002).

Gene silencing includes both short-range, cell-to-cell silencing, and long-distance, systemic cell silencing (Kalantidis et al., 2008). Short-range movement is usually restricted to a few cells, whereas long-distance movement of the silencing signal affects the whole plant (Kalantidis et al., 2008). The systemic silencing signal moves through the phloem, similar to the movement of dye (Turgeon and Wolf, 2009), from source to sink (Dunoyer et al., 2010; Molnar et al., 2010; Schwach et al., 2005).

Because of the potential usefulness of systemic RNA silencing, research is still attempting to discover the nature of the mobile signal and clarify its molecular mechanism. Although various lines of investigation suggest that smRNAs are the most likely candidates for the mobile silencing signal (Mallory et al., 2001; Pyott and Molnar, 2015; Zhang et al., 2014), there are no clear experimental evidences demonstrating these findings yet. Another intriguing question is why the systemic silencing phenomenon is only initiated by introduced transgenes and not by endogenous genes. Despite these unanswered questions, it is generally established that in a virus–plant interaction, the systemic silencing component of RNA silencing is the basic mechanism a plant employs in its defense against viruses. More recent studies produced new evidence about the identity of the mobile signal (Dunoyer et al., 2005, 2007; Himber et al., 2003), RNA-dependent RNA polymerase 6 (RDRP6), conjugated with the suppressor of gene silencing (SGS3) appears to be an integral component of the systemic spread of RNA silencing (Dalmay et al., 2000; Himber et al., 2003; Schwach et al., 2005). Our current understanding of the systemic movement of RNA silencing suggests that: (1) it involves two components, short-distance spread among neighboring cells and long-distance spread through the vasculature; (2) 21-nt siRNAs are implicated in the short-distance movement of silencing; and (3) secondary siRNAs and RdRP are involved in its long-distance spread (Brosnan et al., 2007).

In summary, this sequence-specific mechanism of RNA silencing by knocking down or silencing endogenous genes has been exploited in VIGS technology. Hairpin-like constructs or short gene sequences similar to the endogenous genes are cloned into viral genomes to trigger RNA silencing of the targeted homologous plant genes. Because of this method's effectiveness in expressing foreign proteins and muting endogenous genes through RNA silencing (Kurth et al., 2012), using these modified plant viruses became known as VIGS (Baulcombe, 1999).

5.3 OVERVIEW OF VIGS TECHNOLOGY

To develop a robust and effective VIGS system, it is important to choose the right plant virus to modify. In the field of plant virology, viruses are cloned and used as a tool in reverse genetics to gain knowledge of functions of the viral genome and its mechanisms of replication (Boyer and Haenni, 1994; Nagyová and Subr, 2006). These modified viruses are commonly known as infectious clones. These clones enable plant virologist to study the role of viral gene products in host–pathogen interactions. They also improve understanding of virus disease cycles, providing information useful in developing effective disease management strategies.

Early attempts to validate VIGS technology used *Tobacco mosaic virus* (TMV) and *Potato virus X* (PVX). Genes were targeted that produced distinctive phenotypes, such as silencing of green fluorescent protein (GFP) in transgenic tobacco expressing GFP (Fig. 5.2), the photo-bleaching of leaves caused by a loss of carotenoid pigments when *phytoene desaturase* (*pds*) was disrupted (Kumagai et al., 1995; Ruiz et al., 1998). Other examples targeted the chlorophyll biosynthetic enzyme, resulting in plant chlorosis (Kjemtrup et al., 1998), and the *cellulose synthase* gene, resulting in a modification of plant cell walls (Burton et al., 2000). With the initial success of VIGS, researchers began targeting essential genes (Peele et al., 2001) such as those involved in plant resistance (Peele et al., 2001) encoding metabolic enzymes, increasing crop yield, or plant growth and development. For example, when a VIGS vector constructed with *Tobacco rattle virus* (TRV) was modified with the *EDS1* gene required for N-mediated resistance to TMV (Peart et al., 2002), the inoculated plants had an enhanced susceptibility to TMV.

Most of the early studies were on the model plant, *N. benthamiana*, and then shifted to other species like *Arabidopsis*, tomato (Liu et al., 2002b), petunia, and barley (Burch-Smith et al., 2004; Lu et al., 2003b; Romeis et al., 2001). However, *N. benthamiana* has been the mainstay of VIGS research because of the clear phenotypic effects it produces. The reason for the increased susceptibility of the host to a variety of viruses is still not understood but there are two possible explanations. First, the plasmodesmatal exclusion limit is much wider in *N. benthamiana* compared to other plant species (Waigmann et al., 2000) allowing easier viral movement between cells. Second, *N. benthamiana* may be defective in one of the components of RNA silencing, compromising its antiviral response to viral attacks (Yang et al., 2004).

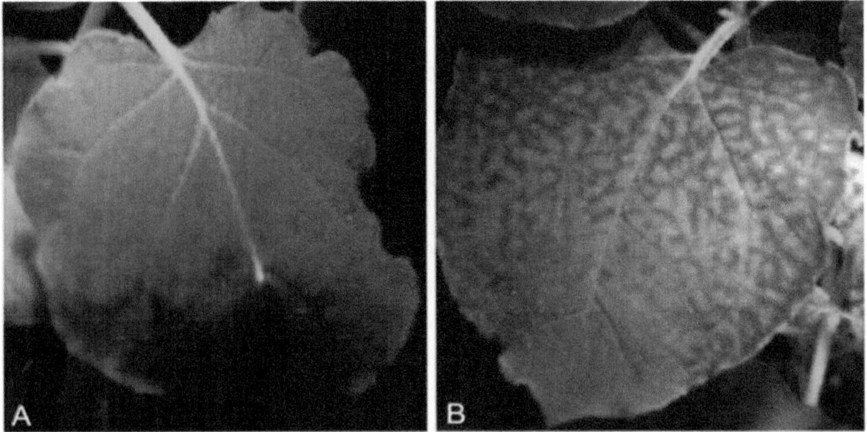

FIGURE 5.2 *Virus-induced silencing in 16C trasgenic N. benthamiana for GFP.* Leaves examined under a long-wavelength UV light at 7 weeks postinoculation. (A) Un-inoculated leaves showing GFP fluorescence. (B) Leaves co-infiltrated with 35S-sGFP and a pBIC-35S-empty vector induced silencing. The noninoculated upper leaves showing development of red trails due to systemic silencing of GFP.

Following the pioneering work with VIGS on TMV, TRV, and PVX, other viruses were modified as VIGS vectors for different economically important crops (Table 5.1). For example, in cassava, an important staple crop throughout sub-Saharan Africa, a VIGS vector was developed using a DNA virus, African cassava mosaic virus (ACMV). The validity of this vector was demonstrated by silencing sulfur gene (su) encoding one component of magnesium chelatase, a chlorophyll-synthesis enzyme, which resulted in yellowish white spots on the leaves (Fofana et al., 2004). This ACMV-VIGS vector was then used to silence the enzyme involved in the biosynthesis of the toxic substance known as the cyanogenic glycoside linamarin. The results were promising, as VIGS significantly reduced the level of this chemical, offering the prospect of using cassava leaves as a future fodder crop for livestock.

TABLE 5.1 Plant Viruses Used as VIGS Vectors, the Nature of their Genomes and their Important Hosts.

Virus/Type	Group	Natural Hosts	Silenced Host Species	Gene Silenced	References
African cassava mosaic virus DNA virus, bipartite	Begomovirus	*Manihot esculenta*	*N. benthamiana*, *M. esculenta*	*pds*, *su*, *cyp79d2*	Fofana et al. (2004)
Apple latent spherical virus RNA virus, bipartite	Cheravirus	Apple	*N. tabacum*, *N. occidentalis*, *N. benthamiana*, *N. glutinosa*, *Solanum lycopersicon*, *A. thaliana*, cucurbit species, several legume species	*pds*, *su*, *pcna*	Igarashi et al. (2009)
Barley stripe mosaic virus RNA virus, tripartite	Hordeivirus	Barley, wheat, oat, maize, spinach	*Hordeum vulgare*, *Triticum aestivum*	*Pds*, *TaEra1*	Holzberg et al. (2002), Manmathan et al. (2013)
Bean pod mottle virus RNA virus, bipartite	Cucumovirus	*Phaseolus vulgaris*, *Glycine max*	*G. max*	*Pds*, *GmRPA3*	Atwood et al. (2014), Zhang and Ghabrial (2006)
Brome mosaic virus RNA virus, tripartite	Bromovirus	Barley	*Hordeum vulgare*, *Oryza sativa*, and *Zea mays*	*pds*, *actin 1*, *rubisco activase*	Ding et al. (2006)
Cabbage leaf curl virus DNA virus, bipartite	Begomovirus	Cabbage, broccoli, cauliflower	*A. thaliana*	*gfp*, *CH42*, *pds*	Turnage et al. (2002)
Cucumber mosaic virus RNA virus, tripartite	Cucumovirus	Cucurbits, *S. lycopersicon*, *Spinacia oleracea*	*G. max*	*chs*, *sf30h1*	Nagamatsu et al. (2007)

TABLE 5.1 *(Continued)*

Virus/Type	Group	Natural Hosts	Silenced Host Species	Gene Silenced	References
Pea early browning virus	Tobravirus	*Pisum sativum*, *Phaseolus vulgaris*	*P. sativum*	*pds*, *uni*, *kor*	Constantin et al. (2004)
RNA virus, Bipartite					
Poplar mosaic virus	Carlavirus	Poplar	*N. benthamiana*	*gfp*	Naylor et al. (2005)
RNA virus, monopartite					
Potato virus X	Potexvirus	*Solanum tuberosum*, *Brassica campestris* ssp. *rapa*	*N. benthamiana*, *A. thaliana*	*gus*, *pds*, *DWARF*, *SSU*, *NFL*, *LFY*	Ruiz et al. (1998)
RNA virus, monopartite					
Satellite tobacco mosaic virus	RNA satellite virus	*Nicotiana glauca*	*N. tabacum*	*Several genes*	Gosselé et al. (2002)
RNA virus, satellite					
Tomato bushy stunt virus	Tombusvirus	*S. lycopersicon*, *N. benthamiana*	*N. benthamiana*	*gfp*	Hou and Qiu (2003)
RNA virus					
Tobacco curly shoot virus	DNA satellite-like virus	*N. tabacum*	*N. tabacum*, *Solanum lycopersicon*, *Petunia hybrida*, *N benthamiana*	*gfp*, *su*, *chs*, *pcna*	Huang et al. (2009)
DNA satellite-like virus					
Tobacco mosaic virus	Tobamovirus	*N. tabacum*	*N. benthamiana*, *N. tabacum*	*pds*, *psy*	Kumagai et al. (1995)
RNA virus, monopartite					

TABLE 5.1 *(Continued)*

Virus/Type	Group	Natural Hosts	Silenced Host Species	Gene Silenced	References
Tobacco rattle virus RNA virus, bipartite	Tobravirus	Wide host range	*N. benthamiana*, *A. thaliana*, *S. lycopersicon*	*pds*, *rbcS*, *FLO/LFY (NFL)* *Sllea4*	Liu et al. (2002b), Ratcliff et al. (2001), Senthil-Kumar and Udayakumar (2006)
Tomato golden mosaic virus DNA virus, bipartite	Begomovirus	*S. lycopersicon*	*N. benthamiana*	*su*, *luc*	Peele et al. (2001)
Tomato yellow leaf curl China, virus-associated b DNA satellite	Begomovirus	*S. lycopersicon*	*N. benthamiana*, *S. lycopersicon*, *N. glutinosa*, *N. tabacum*	*pcna*, *pds*, *su*, *gfp*	Tao and Zhou (2004)
Turnip yellow mosaic virus RNA virus, monopartite	Tymovirus	Brassicaceae	*A. thaliana*	*pds*, *lfy*	Pflieger et al. (2008)

Similarly, a breakthrough was achieved with *Brome mosaic virus* (BMV), which infects rice, maize, and barley, three important monocotyledonous plants. The VIGS–BMV complex was modified to silence phytoene desaturase (*pds*) in barley (Adams et al., 2013; Ding et al., 2006), and later in rice for silencing the *actin* and *rubisco activase* genes. It is expected that extension of VIGS technology to cereal crops will have a tremendous impact on increasing yield and on disease management strategies.

There are two examples of using viruses that infect trees as VIGS vectors: *Poplar mosaic virus* (PopMV) (Naylor et al., 2005) and *Apple latent spherical virus* (ALSV) (Igarashi et al., 2009; Sasaki et al., 2011). ALSV-VIGS soon became a versatile vector able to silence a broad range of plants including tobacco, tomato, and legumes (Yaegashi et al., 2007). Two other viruses revolutionizing VIGS technology in woody plants are *Citrus tristeza virus* (CTV) and *Grapevine leaf roll associated virus-2* (GLRaV-2). Both viruses have relatively large genomes (Dolja et al., 2006), and unique advantages over viruses in the VIGS system with smaller genomes. Because of their importance, they are discussed in more detail in the following section. Other important viruses used as vectors of VIGS include, *Pea early browning virus* (PEBV), *Bean pod mottle virus*, and *Cucumber mosaic virus*, targeting *pds* and *chs* (Nagamatsu et al., 2007; Zhang et al., 2009). These vectors effectively targeted essential genes of the host. For a list of currently developed VIGS vectors, see Table 5.1.

5.4 ADVANTAGES AND DISADVANTAGES OF VIGS

5.4.1 ADVANTAGES

VIGS technology is superior to other gene disruption methods such as sense, antisense, or transgenic technology for several reasons. (1) Development of the VIGS vector is relatively fast because of higher technical capability. Nowadays, creating a VIGS vector for small viral genomes has become routine, and viruses with larger RNA genomes have been greatly simplified. For example, the availability of high-fidelity polymerases have made it possible to cloning large viral fragments, whereas alternatives to completely avoid cell-free based cloning using various technique like circular polymerase extension cloning technique can ameliorate the toxicity problems often encountered in cell-based cloning (Quan and Tian, 2011). (2) There are viral vectors available for most of the important crops. (3) Only a partial sequence of the target gene is needed to silence a functional gene, so complete knowledge of the gene is not required. Even if sequences as small as 20–30 nucleotides can be effective for VIGS, as long as they have a perfect complementarity with the target sequence. (4) VIGS can silence multiple genes using the same construct. This may include genes with multiple copies with high homology, especially in plants with polyploidy genomes such as wheat (*Triticum aestivum*). (5) VIGS is easy to use and flexible enough to be introduced by agro-infiltration, biolistic methods, or mechanical inoculation (Fig. 5.3). Further improvements in VIGS delivery systems, such as Agrodrench or a quick inoculation by toothpick or viral sap, have made

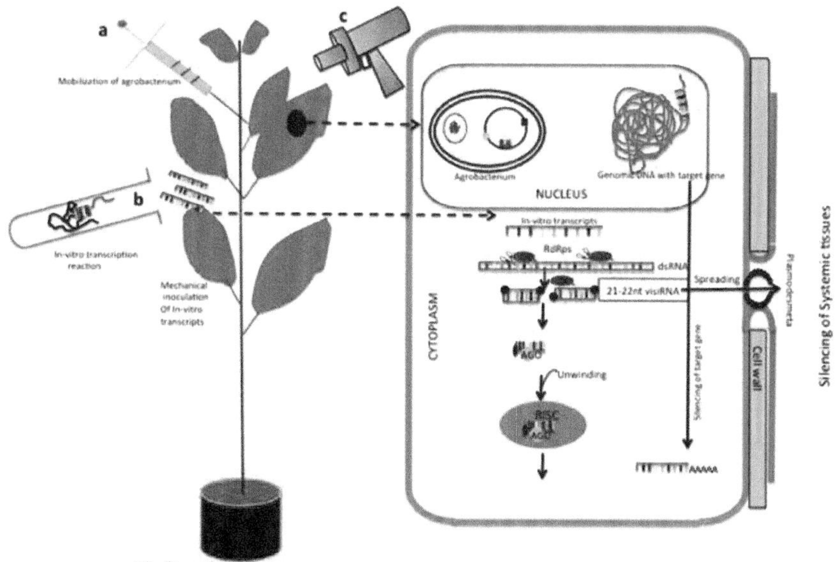

FIGURE 5.3 Schematic representation of various methods used to inoculate plants with VIGS into plants and subsequent RNA silencing inside a plant cell: (A) target genes from a DNA or RNA virus are cloned in a binary vector, transformed into *Agrobacterium*, and introduced into plants with a syringe; (B) mechanical inoculation by rubbing synthesized transcripts onto a plant leaf; or (C) direct introduction of the VIGS vector using a gene gun. Right inset shows the mechanism of virus-induced gene silencing. Silencing is initiated by the recognition of viral dsRNAs and the replication of intermediate RNA viruses. The transcripts from DNA virus also gets amplified by host RdRP to dsRNA, which are all processed to vsiRNA of 21–24 nt by dsRNA-specific RNAses called DCLs. In the next step, vsiRNA are recruited into RNA induced silencing complex (RISC) where the duplex is unwound and only one of the strand, called as guide strand (providing the specificity) is recruited to target viral RNA. The component, AGO mediates the slicing. Also shown is the short distance spread of RNA silencing signal, the 21–22 nt vsiRNA spreads to adjacent cells of the plant tissues through the plasmodesmata. The long-distance silencing signal is believed to be due to secondary vsiRNA produced in an amplification loop by the action of plant RDRs and their cofactors.

the technology more robust (Lu et al., 2003a; Vaghchhipawala et al., 2012). (6) VIGS rapidly initiates gene silencing. VIGS-induced phenotypes are visible in a relatively short time, usually 2–3 weeks. Any downstream application involving metabolic or transcript profiling from the silencing effect can be carried out in a relatively short time. (7) VIGS can operate at the transcriptional level, silencing the target gene by extensive methylation. As described previously, this RNA-mediated, sequence-specific mechanism is known as RdDM (Wassenegger, 2000). RdDM has been demonstrated with various RNA-VIGS. When plants simultaneously engineered for the same viral transgene are challenged with RNA-VIGS, it always results in extensive methylation of the transgene (Pélissier et al., 1999; Wassenegger et al., 1994). The VIGS vector also can cause extensive transcriptional methylation of endogenous promoters, providing an indirect method of regulating gene expression. Interestingly, the silencing of target genes by this mechanism is hereditable, causing an epigenetic change in the genome of plants infected by the virus (Jones et al., 2001). (8) VIGS can be used with plant species that are difficult to transform. Transformation is a labor- and time-intensive procedure. Any crop plant susceptible to viruses potentially can be transformed using a VIGS vector.

5.4.2 CHALLENGES

There are several challenges to conventional VIGS technology. One of the challenges is the method of viral delivery. The majority of VIGS vectors were developed for dicots with only a few available for monocots. Although various modifications of the *Agrobacterium*-mediated delivery system have been made, it is still not efficient for monocotyledonous plants. The viral sap inoculation method, however, provides an alternative means of inoculation but requires prior propagation of the virus in an appropriate host such as *N. benthamiana* to increase its titer before inoculation with the extracted sap into the target plants.

There are several inherent challenges associated with the virus used for constructing the VIGS vector. (1) The usefulness of a VIGS vector often depends on the host range of its virus. A virus cannot be used outside of its experimental host range. (2) The virus may replicate too slowly or be unstable in the VIGS construct. This instability can result in deletion of sequences the viral gene during virus replication. (3) The size of the viral genome can also be a problem. Viral constructs made from smaller viral genomes usually replicate faster than viruses with large genomes and considerably affect the outcome of the silencing response. (4) Another difficulty with the virus chosen for the silencing constructs would be the presence of antisilencing suppressor proteins in their genomes. Strong viral suppressors can overcome the host's silencing pathways and significantly alter the desired silencing response. (5) The inability of some viruses to move systemically throughout the plant and infect every cell. TMV, PVX, and TGMV are unable to infect the plant meristems, including its emerging leaves (Matthews, 1991).

Another challenge is the difficulty in observing the effect of silencing a particular gene. The VIGS phenotype could be obscured by natural symptoms induced by viral infection, complicating assessment of the results. For this reason, viruses that produce severe symptoms in their host plants or accumulate to high titers should be avoided. Another challenge is the need for sound technical knowledge when selecting the gene for silencing. A poor selection of gene may result in undesired off-target silencing. Also, position of the gene, its length, or its orientation in the vector can cause inefficient silencing. Lastly, given the current public opinion against genetically modified organisms, the use of genetically modified plant viruses in VIGS technology underscores the requirement for strict containment of their operations.

5.5 DEVELOPMENT OF VIGS METHODOLOGY

5.5.1 TYPES OF VIGS VECTORS

Many viruses have been modified to serve as VIGS vectors. The selection of a VIGS vector depends on many factors: DNA or RNA, infectivity, severity of symptoms, method of transmission, presence of suppressor proteins in the genome that could counteract the host PTGS mechanism, and others.

There are important differences in the biology of RNA and DNA viruses. RNA viruses use their own polymerase to replicate, forming an intermediate dsRNA-like structure in the cytoplasm of the host plant. This replication also depends on the host's cytoplasmic membrane and ribosomes. DNA viruses replicate in the nucleus of the host plant and mainly depend on the replication machinery of the host. The dsRNA structure of RNA viruses is a strong inducer of gene silencing, so the host's PTGS mechanism may directly target the RNA VIGS vector. Some early VIGS vectors constructed with RNA viruses were TMV, PVX, and TRV (Liu et al., 2002b) (Fig. 5.4). The list of VIGS vectors has increased over the years and now includes: ALSV (Igarashi et al., 2009; Sasaki et al., 2011) and *Barley stripe virus* (Bruun-Rasmussen et al., 2007).

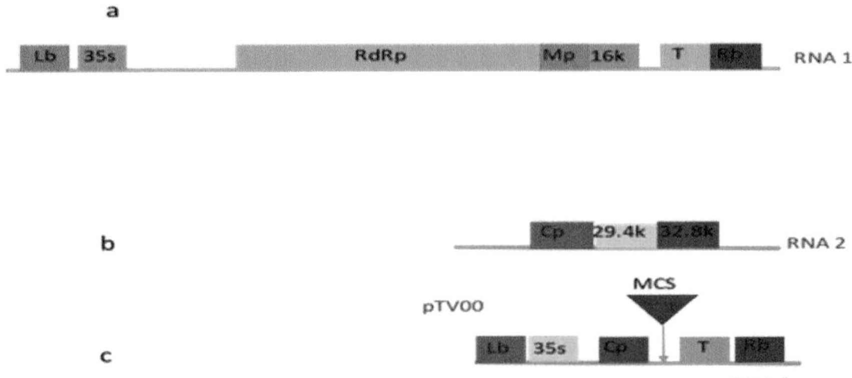

FIGURE 5.4 *Genomic organization of TRV.* (A) The complete TRV RNA 1 gene cloned into the T-DNA region of the binary vector pBINTRA6. (B) Organization of the TRV RNA 2 genome. (C) A binary vector pTV00 used for cloning the TRV–RNA2 genome component in a multiple cloning site (MCS) between the 35S promoter and the terminator (T) sequence. The whole cassette is inserted between the left border (LB) and right border (RB) of the T-DNA of *Agrobacterium*. The different open reading frames (ORF) are RNA-dependent RNA polymerase (RdRP), coat protein (Cp), and 29.4 and 32.8 kb proteins. (Adapted from Ratcliff et al., 2001)

5.5.2 CONSTRUCTION OF VIGS VECTORS

Many studies suggest that the ideal insert length for a VIGS vector should be 200–300 nt (Burch-Smith et al., 2004, 2006; Liu and Page, 2008; Rodrigo et al., 2011; Zhang et al., 2010). Software is available that can predict the silencing efficiencies of the 21-nt siRNA generated from the gene in vivo (Xu et al., 2006).

There are of two types of VIGS vectors constructed with RNA viruses: infectious RNAs (in vitro) and infectious cDNAs (in vivo). The in vitro technique uses bacterial phage promoters such as T7, SP6, or T3 inserted at the 5′ end of the genome. The promoter regulates transcription of the cloned viral genome (Nagyová and Subr, 2006). The complete viral cDNA is then transcribed to produce the desired quantities of viral RNA. The synthesized RNA is mechanically introduced into plants or transfected into protoplasts using the polyethylene glycol. With the in vitro approach, there is an inherent risk of RNA degradation during in vitro manipulations. Once inside the plant, however, the transcribed RNA system functions directly as mRNA and does not need to enter the nucleus as translation occurs in the cytoplasm (Nagyová and Subr, 2006).

The in vivo approach on the other hand, uses *Agrobacterium* to deliver the synthesized virus into the plant. This is a robust method of introducing infectious clones into plants (Grimsley et al., 1986) and is especially suited for insect-transmitted viruses, that otherwise are recalcitrant to in vitro procedures. In the in vivo technique, full-length viral cDNA is cloned under control of a promoter such as the CaMV-35S promoter. It is then inserted into the T-DNA of the binary vector, mobilized in *Agrobacterium* cells, and then transferred to plant cells by agro-infiltration. Transcription of full-length viral cDNA occurs in vivo in the nucleus of the plant cells. Compared to the in vitro method, however, the cDNA is dependent on the host machinery for transcription, so its needs to enter the cell nucleus (Boyer and Haenni, 1994).

DNA-based vectors have the advantage of bypassing RNA transcription and so are amenable to high throughput delivery system such as biolistics or agro-inoculations (Liu et al., 2002a; Zhang et al., 2009). VIGS vectors constructed with DNA viruses include: *Rice tungro bacilliform virus* (Purkayastha et al., 2010), *Tomato golden mosaic virus* (Kjemtrup et al., 1998; Peele et al., 2001), *Cabbage leaf curl virus* (CaLCuV) (Turnage et al., 2002), ACMV (Fofana et al., 2004), and viral satellites such as *Tobacco mosaic satellite virus* (Gosselé et al., 2002), *Tomato yellow leaf curl China virus*, DNAβ (Cui et al., 2004), *Tobacco curly shoot virus*, DNAβ (Qian et al., 2006), and others. Satellite DNAs are DNA molecules which are only associated with viruses from the genus *Begomovirus*. Satellites can also be used as VIGS vectors but need to be co-inoculated with their helper virus (Huang et al., 2009; Tao and Zhou, 2004).

5.5.3 INOCULATION OF PLANTS WITH VIGS VECTORS

There are several commonly used methods for introducing VIGS vectors into plants, including manual rub inoculation using a wounding agent such as carborundum, agro-inoculation, and microprojectile biolistic bombardment (Fig. 5.3). To use manual rub inoculation, in vitro transcripts of the viral genome are first generated by the transcription reaction and the RNA transcripts mechanically rubbed on the upper leaf surface. This method is time consuming and challenging due to the labile nature of RNA, but it works well with certain virus–host systems such as TMV and *N. benthamiana* (Shivprasad et al., 1999), or TRV and *Arabidopsis thaliana* (Ratcliff et al., 2001). Agro-inoculation on the other hand, is easy to use and lacks the problem of RNA degradation. It is becoming the method of choice for use with the VIGS system. Agro-inoculation also can be used with both DNA and RNA viruses. There are two different inoculation methods, toothpick inoculation and inoculation by agro-infiltration. With the first method, a toothpick is loaded with *Agrobacterium* from a culture containing the viral construct; the plant is then stab-inoculated with the bacterium. For agro-infiltration, a syringe is used to infiltrate the leaf mesophyll with the *Agrobacterium* culture. A detailed demonstration of these methods is available at http://www.sainsbury-laboratory.ac.uk/david-baulcombe.

Microprojectile bombardment is gaining popularity because it can be used on a wide range of plant tissues. In this technique, a high-velocity gene gun is used to deliver biologically inert, high-density microparticles (usually gold). The particles are coated with nucleic acids and penetrate the plant cell walls, entering the nucleus. The DNA is integrated directly into the host genome. Although this method is used for both DNA and RNA viruses, it is most successful with DNA viruses (Muangsan et al., 2004). A major advantage of microparticle bombardment is that it does not depend on host compatibility. With the particle gun, the backbone sequences of the vector may be excluded, eliminating integration into the host genome (Sudowe and Reske-Kunz, 2013). It also eliminates a possible host response to the pathogen, *Agrobacterium*. One disadvantage of this method is that the transgenes might get damaged during particle bombardment, resulting in the integration of broken transgenes into the host genome.

There are also examples of these techniques had to be modified individually or in combinations depending on the individual virus–host combinations. For instance, a major accomplishment of agro-inoculation was the successful introduction of the bipartite genome of TRV into plants (Fig. 5.4). The two RNA components of the virus were cloned separately into Ti plasmids in two different strains of *Agrobacterium.* The strains were then simultaneously introduced into the plant by agro-infection, using either toothpicks or a syringe (Lu et al., 2003b).

In another instance of virus–host combinations, the virus titer needs to be increased by propagation in another host before it can be inoculated into the target host. It was noticed, for example, that TRV could not be inoculated directly into *Arabidopsis* either by mechanical inoculation or by agro-inoculation. If the TRV was agroinoculated into *N. benthamiana* first to increase the virus titer, however, and then mechanically inoculated using viral sap from *N. benthamiana*, infection of *Arabidopsis* was successful (Lu et al., 2003b). Similarly, an ingenious method was developed for inoculating CTV into citrus plants (Fig. 5.5). *In-vitro* RNA from infectious cDNA clones of CTV could infect protoplasts, but since the virus titer was found to be very low in the protoplasts, neither the *in vitro* transcripts nor the viral sap from protoplasts could infect citrus trees. The researchers overcame the problem by first amplifying the virus by successive passage of the virions through protoplasts until the crude sap of the protoplasts was able to infect the citrus plants (Satyanarayana et al., 2001).

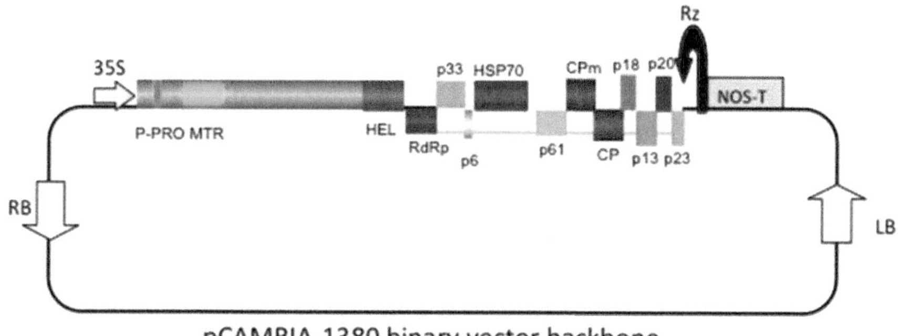

FIGURE 5.5 *Schematic of VIGS–Citrus tristeza virus (CTV) constructed in a binary vector.* Schematic representation of a full-length infectious cDNA clone of CTV with an open reading frames (ORF) between the enhanced 35S promoter of *Cauliflower mosaic virus* at the 5′ end, and ribozyme (Rz) of *Subterranean* clover mottle virus satellite RNA and *nopaline synthase terminator* (NOS-T) at the 3′ end in the binary vector pCAMBIA-1380. The vector plasmid is referred to as wild-type CTV (CTV-wt) is based on CTV isolate T36.

5.5.4 ASSESSMENT OF VIGS

Assessment of the VIGS phenotype is usually visible several days postinoculation. Ideally, the response would be distinct, but often it is poorly developed or absent. Environmental factors and the physiological condition of the plant may affect the outcome of a VIGS experiment, so pilot experiments with a well-characterized vector under optimum conditions may be necessary. To improve the assessment, marker genes, such as *pds* that produce photobleaching, may be helpful. Selecting genes that more closely resemble the target gene may be appropriate. For example, to identify disease resistance genes, pilot experiments with control vectors that could target well-characterized defense-related genes of the host plant should be completed first. The outcome of which would predict the progression of VIGS expression in an actual experiment. If the targeted gene is expressed at a low level, the plant may not show a response. Therefore, the absence of visible change should be interpreted cautiously. There is also a possibility of misinterpreting the result of VIGS due to the presence of a random gene in the plant that is identical to the target gene. This may be overcome by choosing a second, nonoverlapping gene from the same target gene (Lu et al., 2003b). If expression of the second gene is the same as the target gene, this supports selection of the target gene. Another consideration in interpreting the results of VIGS is due to pleiotropy. It is well known that VIGS can produce unintended secondary effects in the host that might complicate the intended outcome. Lu et al. (2003b) reported that out of 5000 genes screened for resistance, fewer than 10 genes directly showed a loss of disease resistance due to VIGS, while 100 genes resulted in a loss of cell death phenotype unrelated to VIGS. This result indicates that the results from VIGS experimentation may sometimes show pleiotropic effects and needs to be interpreted carefully.

5.6 TOBACCO RATTLE VIRUS AS A VIGS VECTOR

TRV is a useful VIGS vector with a host range of over 60 species in more than 12 families, including the Solanaceae. It induces only mild symptoms and easily infects adjacent cells and plant meristems. TRV is an RNA virus with a bipartite genome. The RNA1 encodes genes for replication and movement, whereas genes in RNA2 are involved in virion formation and transmission by nematodes. Since TRV has a small genome it is amenable to genetic manipulation such as cloning and multiplexing. Most importantly, TRV does not integrate into plant genomes and is capable of infecting germline cells (Martín-Hernández and Baulcombe, 2008).

Ratcliff et al. (2001) successfully constructed a TRV-VIGS vector by making separate cDNA clones of RNA1 and RNA2 in binary vectors under the control of the *Cauliflower mosaic virus* (CaMV) 35S promoter (Fig. 5.4). The infectious cDNA of TRV RNA1 incorporated intron 3 of the *A. thaliana* Col-0 *nitrate reductase NIA1* gene inserted to interrupt the RdRp. The RNA1 was then cloned into a PBINTRA6 vector. This step was needed to stabilize the cDNA in *E. coli*, otherwise *E. coli* expressed certain ORFs toxic to the cells harboring the clones. The RNA2 was modified by replacing the 29.4 kb and 32.8 kb genes and cloned into the pTV00 binary vector. To be infectious, the *Agrobacterium* cultures carrying the RNA1 and RNA2 clones were mixed and infiltrated into *N. benthamiana* as described above. The VIGS was also tagged with the GFP reporter gene so its movement could be followed in the plant under UV light. This TRV-VIGS was found to be very robust and easily spread throughout the inoculated plants. In comparison to PVX, this GFP tagged TRV-VIGS vector persisted for more than 16 weeks postinoculation, whereas the PVX vector persisted but only 28 days postinoculation (dpi). To show its effectiveness in silencing the endogenous gene, the authors compared TRV with PVX for silencing the *pds* gene, *pds* in *N. benthamiana.* It was observed that the photo-bleaching effect was more pronounced in case of TRV than PVX. PDS prevents plant from photobleaching and is involved in the production of carotenoids (Demmig-Adams and Adams, 1992).

To assess TRV infection, leaves infected systemically were detected 10 days postinoculation, for its subgenomic RNAs. Although the virus caused a mild mosaic of the infected leaves, the mosaic usually disappeared after a few weeks due to a recovery mechanism exhibited in later stage of gene silencing. During this recovery phase, the plant was immune to secondary infection by TRV and virus RNA levels drop as replication declined. Interestingly, the persistence of TRV was apparently low because the virus was able to evade the host's PTGS (Ratcliff et al., 2001). In comparison, researchers found that VIGS constructed from PVX were eliminated eventually by the host's PTGS, indicated by a halt in its replication. The authors have also extended this study to *Arabidopsis*, with similar results. The TRV RNA1 and RNA2 which were modified with PDS caused photobleaching in systemic leaves 10 dpi and later in the stems, axillary shoots, and sepals.

5.7 VIGS WITH LARGER VIRAL GENOMES FOR USE IN WOODY PLANTS

The use of VIGS in woody plants is especially useful, but the technology is still being developed. There are several bottlenecks, such as a resistance to traditional genetic modification, a lack of knowledge about viruses infecting perennial plants, and the technical challenges of genetically modifying the large viruses known to infect woody plants.

Some of the economically important perennial fruit crops, such as grapevines and citrus (Dolja and Koonin, 2013) are known to be infected by viruses in the family *Closteroviridae*. Two of the most important viruses infecting the citrus and grapevines are CTV and GLRaVs, respectively. These viruses have the largest genomes among plant viruses. Much of the knowledge about the possibility of using CTV or GLRaVs as VIGS vectors came from work on Closterovirus vectors as expression vectors. Closterovirus-derived vectors provide an attractive means of long-term recombinant protein production in their natural hosts by the extended gene expression in their perennial hosts. One study demonstrated that a CTV vector in citrus plants retained a foreign gene for a decade (Dawson and Folimanova, 2013).

Therefore, this virus could be engineered to express multiple foreign genes either by inserting the genes in a suitable position or by substituting them in place of nonessential genes (Dawson and Folimonova, 2013; El-Mohtar and Dawson, 2014). The disadvantages of using viruses with large genomes, however, are their slow replication, tissue specificity (Bar-Joseph and Murant, 1982), and the inability to produce them on a large scale (Dolja and Koonin, 2013).

Vectors based on CTV are constructed from infectious clones of CTV from the isolate T36 (GenBank accession no. AY170468). This isolate was engineered into the binary vector pCAMBIA-1380 (Satyanarayana et al., 1999, 2001). The CTV genomic RNA is driven by the duplicated 35S promoter of CaMV at the 5′ end and a ribozyme sequence obtained from *Subterranean clover mottle virus* at the 3′ end. For ease of cloning, unique restriction sites *PacI* and *StuI* of a foreign gene were inserted between ORF-p23 and the 3′-untranslated region that is strategically placed so the coat protein subgenomic RNA controller element (CE) will drive expression of any amplicon (Fig. 5.5).

CTV produces substantial amounts of genomic and subgenomic replicative intermediates as double-stranded RNAs (Dodds and Bar-Joseph, 1983; Hilf et al., 1995). Since these molecules are known to be strong inducers of host antiviral gene silencing activity, CTV may be an effective vector for gene silencing. It is known that the ORFs nearest the 3′ end are under the control of their respective subgenomic RNAs CEs, so they are expressed in large amounts during the earliest stage of infection (Navas-Castillo et al., 1997). Therefore, a silencing construct to be used should be inserted into its extreme 3′ end. Although some of these ORFS at the 3′ end were also shown to be suppressors of gene silencing, but interestingly, they do not prevent gene silencing from being induced in the host plant. It is hypothesized that for the virus to establish, they are expressed early on to thwart the host antiviral gene silencing response.

A very useful application has emerged by utilizing the CTV as a VIGS vector to protect the citrus plants against huanglongbing (HLB, citrus greening), associated with the bacterium *Candidatus* Liberibacter asiaticus (*C*Las). The novelty of the approach relies on the indirect use of the CTV-based VIGS to control the Asian citrus psyllid (*Diaphorina citri*) that vectors the bacteria. Since *C*Las and CTV both infect the plant's phloem-associated cells, *D. citri* would acquire the silencing triggers against itself, causing abnormalities that would prevent it from spreading the bacterium to new hosts. To test this concept, Hajeri et al. (2014) used CTV to express the *Awd* gene derived from *D. citri* as silencing components in citrus phloem-associated cells. As anticipated, *D. citri* acquired the silencing triggers during feeding. This caused abnormalities, such as a malformed wing phenotype due to expression of the *Awd* gene (Fig. 5.6). This concept of utilizing the gene silencing mechanism of the host to target insect and pests is called host-induced gene silencing (Nunes and Dean, 2012).

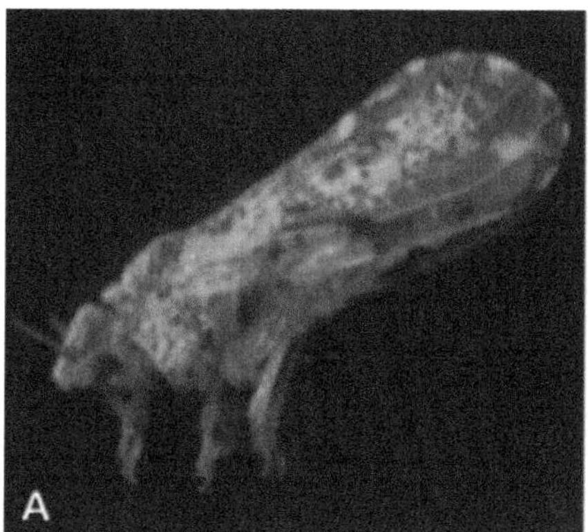

FIGURE 5.6 *Effect of CTV-based plant-mediated RNAi in Diaphorina citri.* Images showing the effects on wing morphology of *Diaphorina citri* after ingesting phloem sap of *Citrus macrophylla* with (A) wild-type CTV (controls) and (B) with CTV-VIGS expressing truncated abnormal wings (CTV-tAWd) targeting *Diaphorina citri*. (Image courtesy Subhas Hajeria, William O. Dawson, and Siddarame Gowda and the Journal of Biotech.)

Using the CTV vector to mitigate HLB was faster than obtaining resistance through transgenic technologies. It is anticipated that in future this RNAi-based VIGS technology could help control other economically important phloem-feeders, such as aphids, whiteflies, and scale insects, reducing insecticide use in citrus orchards (Gatehouse and Price, 2011; Walker and Allen, 2010).

Another significant milestone in the use of large viral genomes was the development of a VIGS vector based on *Grapevine leafroll-associated virus 2* (GLRaV-2) (Fig. 5.7). Grape growers are continually trying to improve the traditional grape varieties because of their economic importance in making wine. Improvements through genetic engineering have had limited success, however, and resistance by the public to genetically modified organism has slowed progress further. With the advent of powerful RNAi technology, VIGs vectors have the capability of introducing the desired traits without making any heritable changes in the plant genome. Like CTV, GLRaV-2 is a large closterovirus that accumulates in the phloem tissue of its host. It is capable of expressing recombinant proteins in the phloem, thereby ensuring easy movement through the sugar transport mechanism. There are several advantages to using GLRaV-2 as a VIGS vector. GLRaV-2 is present in all grape-growing regions of the world, so a GLRaV-2-derived VIGS vector could be used across geographical boundaries. Further, the molecular biology of this virus has been studied extensively; creating a foundation for genetic manipulation using this virus. GLRaV-2 is also very stable, similar to CTV, and as a VIGS vector can tolerate relatively large insertions. This virus also has the ability to introduce multiple RNAi triggers in a single cassette, enabling the silencing of various pathogens infecting grapevines. Another unique advantage of using GLRaV-2 is that, unlike other viruses from woody plants, GLRaV-2 can infect the herbaceous host, *N. benthamiana*, an experimental host adaptable to genetic modifications.

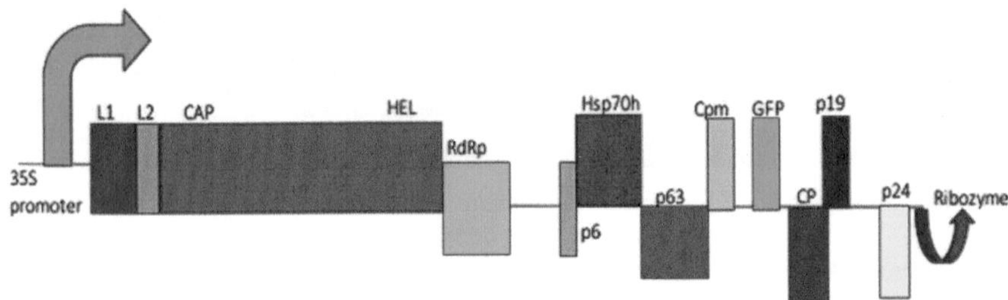

FIGURE 5.7 *GLRaV-2-derived vector tagged with a GFP reporter gene.* The various open reading frames are, L1 and L2, papain-like leader proteases; CAP, capping enzyme; HEL, RNA helicase; RdRp, RNA-dependent RNA polymerase; p6, 6-kDa movement protein; Hsp70h, heat shock protein (70 kDa), homolog; p63, 63-kDa virion protein; CPm, minor capsid protein; CP, major capsid protein; p19, 19-kDa protein; p24.

It is anticipated that in coming years, GLRaV-2 could be an important tool in functional genomics of the grapevine and revolutionize the wine-making industry.

5.8 NEXT GENERATION VIGS WITH CRISPR/CAS SYSTEM

VIGS has made a tremendous impact in plant biology by silencing and then identifying endogenous genes. However, with one of the most recent and promising genetic tools, the CRISPR/Cas DNA system, it is now possible for targeted genome editing and precise knocking out of entire genes. In recent studies, CRISPR/Cas9 was used to edit plant genomes such as rice, *N. benthamiana* and *Arabidopsis* for heritable changes (Nekrasov et al., 2013; Shan et al., 2013). The procedure is simple, requiring only transgenic plants expressing cas9 and guide RNA (gRNA). (The technical terms are explained below.) Additionally, the genetic modifications are present in subsequent generations. The VIGS system, besides its ability to silence genes has found an important application in the CRISPR/Cas editing system. It can be used as a vehicle to transport the CRISPR/Cas editing system into plant system.

It is expected that CRISPR/Cas will transform the way plant traits are modified in the future. Although this technology is new, a number of proof of concept studies in model plants have shown its potential as a powerful gene-editing technology. The efficiency, accuracy, and flexibility of the CRISPR/Cas9 genome engineering system has been demonstrated in various eukaryotes such as yeast, zebrafish, and worms (DiCarlo et al., 2013; Friedland et al., 2013; Hwang et al., 2013; Mali et al., 2013). The potential applications have been growing rapidly and include the cutting-edge application of gene editing in the germlines of humans and other organisms (Mali et al., 2013). This method was recently adopted in plant systems in various transient experiments or in transgenic plants and is becoming the method of choice for plant scientists.

Like RNAi, the CRISPR/Cas gene-editing technology was derived from a naturally occurring plant-defense mechanism. It provides a form of acquired immunity to the cleavage of DNA present in certain prokaryotes and confers resistance against foreign genetic elements such as phages and plasmids. It is based on the type II CRISPR (Fig. 5.8). CRISPR is a sequence of short, repetitious segments followed by a short segment of spacer DNA. The spacer DNA could be from previous exposures to a virus, plasmid, or bacterium. Evidence that the source of the spacers was a bacterial genome was the first hint of the CRISPR's role in an adaptive immunity analogous to RNAi. It was soon proposed that the spacers identified in bacterial genomes served as templates for RNA molecules that the bacteria transcribed immediately after an exposure to an invading phage. Further studies revealed that an important protein called Cas9 was involved, together with the transcribed RNA, to recognize the invading phage and cut the RNA into small pieces (crRNA) in the CRISPR system (Fig. 5.8) (Horvath and Barrangou, 2010; Jiang et al., 2013; Ran et al., 2013). CRISPRs are found in almost 90% of the sequenced Archaea and up to 40% of bacterial genomes (Horvath and Barrangou, 2010).

Native bacterial CRISPR RNAs also can be altered into a single gene known as a single-guide RNA (sgRNA) (Jinek et al., 2012; Schaeffer and Nakata, 2015). Using sgRNA has made the system more flexible, allowing it to simplify genome editing by combining sgRNA and Cas 9 in a heterologous system.

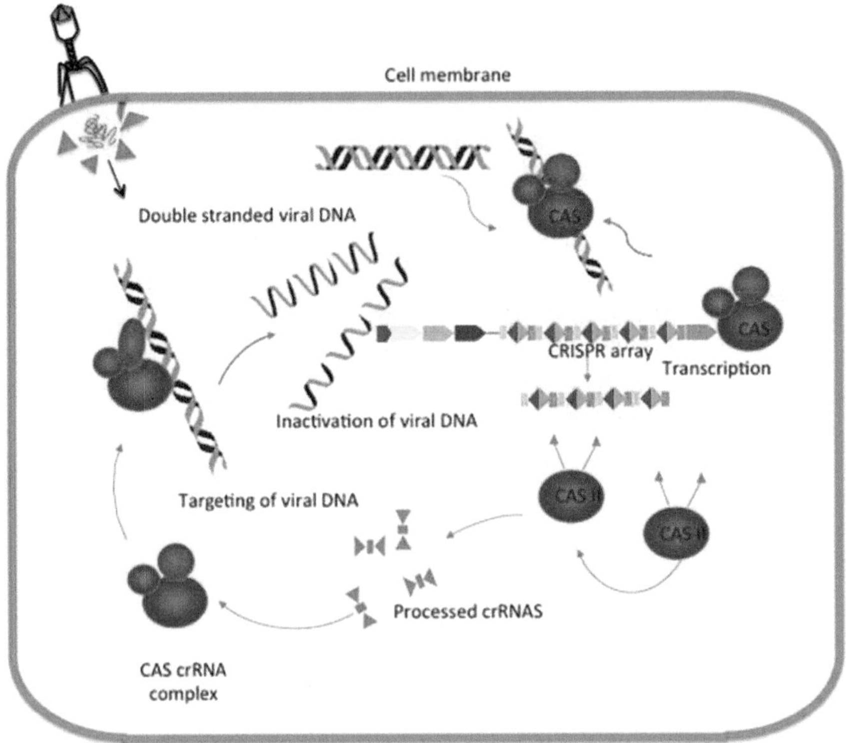

FIGURE 5.8 *Simplified overview of the CRISPR/Cas mechanism.* In the first step, once the viral DNA enters into the bacterial cell, the Cas complex immediately recognizes it as foreign DNA and introduces a novel repeatable spacer unit at one end of the CRISPR unit. In the second step, the repeat array of the CRISPR is transcribed into a pre-CRISPR targeting RNA (crRNA) that is subsequently processed into mature crRNAs. These crRNA acts as guide by a Cas complex to target the exogenous viral nucleic acid. The repeats of CRISPR array is shown in diamond whereas, the spacers are shown in rectangles. (Adapted from Horvath and Barrangou, 2010).

Applying the CRISPR/Cas9 system in plants uses both components; the Cas9 enzyme catalyzes DNA cleavage and the sgRNA recruits Cas9 to the target site. This site is usually located about 20 nucleotides before the protospacer motif and cleaves the DNA. The natural mechanism plants use to reattach the cleaved ends of DNA is called nonhomologous end joining (Xie et al., 2014) and usually results in a mutation either by frameshift, insertion/deletion, or insertion of a stop codon. Therefore, by simply designing a sgRNA with a complementary sequence, virtually any gene can be edited with this heterologous system.

5.8.1 INTEGRATION OF VIGS AND CRISPR/CAS9

As mentioned in the previous section, recognition of the usefulness of the TRV-based VIGS vector in functional genomics was followed by its use to deliver the components for genome editing into plants. TRV is ideally suited since it can systemically infect a wide range of important crop plants. Moreover, TRV is widely used to transiently infect any plant using the TRV-VIGS system, so the protocols are well established. The ability of TRV to infect the plant meristems makes it an ideal candidate for delivery of CRISPR/Cas9 since any seeds derived from these plants will have the induced modifications that are heritable. This bypasses the need for time-consuming transformations or tissue culture to obtain mutant seeds.

In a recent study, TRV delivered sgRNA molecules to edit the *pds* gene in *N. benthamiana* (Ali et al., 2015) (Fig. 5.9). To develop the system, researchers used *Agrobacterium*-mediated transformation protocol to generate transgenic lines of *N. benthamiana* that overexpressed Cas9. Next, they modified the RNA2 genome of TRV for sgRNA delivery. The sgRNA directed to target the PDS was expressed by a promoter derived from Pea early browning virus (PEBV). Subsequently, they reconstituted

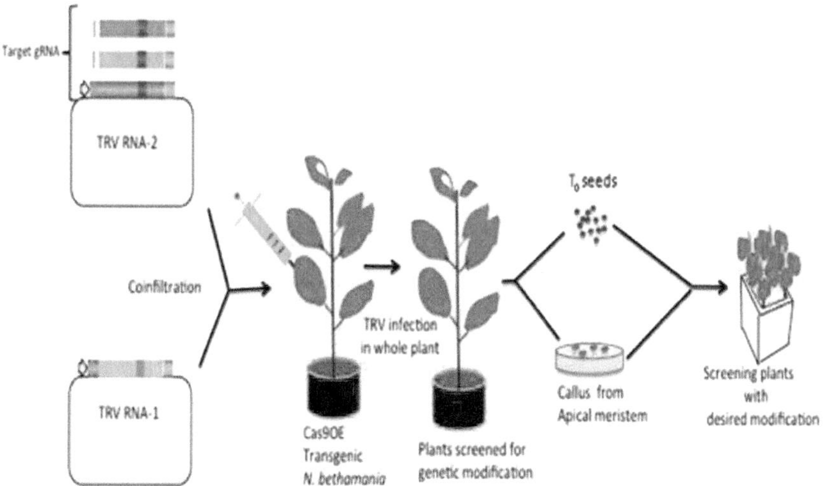

FIGURE 5.9 *Experimental scheme used by Ali et al. (2015) molecular plant.* A 20-nucleotide target gene under control of the PEBV promoter was integrated into the RNA 2 genome and used as gRNA and transformed into *Agrobacterium*. The RNA 1 genome of TRV mobilized into *Agrobacterium* and co-infiltrated into transgenic *N. benthamiana* expressing Cas9. The RNA 2 genome can be used on multiple targets. Plants are then analyzed for the desired modifications using various assays. The leaves of plants with the genetic modification can be regenerated or the seeds harvested and screened for the mutants.

the functional TRV virus by introducing RNA1 of its bipartite genome into tobacco leaves by agro-infiltration. After 2 weeks, they assayed the plants and found the genomic modifications in systemically infected leaves. Importantly, the genetic modification for the PDS gene was present in the progeny due to infection of the meristematic cells and subsequent seed transmission. The demonstration of TRV for virus-mediated genome editing suggests the possibility of modifying a wide variety of plant species by using other RNA viruses as vectors.

Recently, the use of CRISPR/Cas9 was extended to include a DNA virus, Cabbage leaf curl virus (CaLCuV) in the genus *Geminivirus*. Since DNA viruses replicate in the nuclei of plant cells, expression of sgRNA should be more efficient since genome editing occurs in the nucleus (Yin et al., 2015). Moreover, CaLCuV has a number of hosts in the Brassicaceae including cabbage, cauliflower, and *Arabidopsis*. It also infects *N. benthamiana* and other solanaceous crops.

5.9 CONCLUSION AND FUTURE PROSPECTS

VIGS is a powerful reverse genetics technology. The number of genes discovered since the technique was developed has increased greatly (Pang et al., 2013; Qu et al., 2012; Yin et al., 2015). The future of VIGS may depend on the high-throughput gene function capability of the different virus vectors. This capability is established for species susceptible to VIGS and is now used in more than 30 plant species. Most recently, ALSV (Igarashi et al., 2009), *Barely stripe mosaic virus* (Yuan et al., 2011), and CaLCuV (Turnage et al., 2002) have been developed, respectively, for apple, barley, and cabbage. Identification of viruses with broad host ranges will enable functional characterization of useful genes from those crop plants. Because of its large host range, TRV has already demonstrated its potential for functional gene analysis. DNA viruses like the geminiviruses are also being considered as many infect soybeans and other economically important vegetable crops.

The growing amount of plant genomic and transcriptome data provide a quick way to select important genes for functional analysis. Most of these studies include various plants but are limited to analyzing marker genes. The real test will be extending research to experiments on the novel genes constantly being discovered. The unique advantage of VIGS technology over the well-established transgenic technology is that the altered phenotypes are not due to a stable transformation of the plant genome. Moreover, silencing with VIGS is possible in plants with diverse genetic backgrounds.

For example, which are lethal to the plants, is only possible by downregulating the gene using a VIGS vector and not by any other traditional gene knockout technique. Similarly, plant phenotypes dependent on specific environmental conditions are easier to screen using the VIGS system. Finally, public resistance to GMOs should not apply to VIGS. If this system is publicly accepted, many crops would not face the long regulatory approvals or boycotts in the marketplace that hinder GMOs.

The parallel advancement of Crisper-Cas and VIGS technology and the proven versatility of TRV and CaLCuV could be used with CRISPR–Cas system for either gene editing or gene knocking out. It seems obvious that in near future there will be a lot of flexibility in scientific research in terms of the desired outcome, the merging of VIGS with CRISPR–Cas technology will therefore revolutionize all aspects of functional genomics in the field of plant biology.

The use of TRV and CaLCuV in the Crisper-Cas/VIGS system has demonstrated the versatility of this approach to gene editing and gene knockout. The demands of today's research and the flexibility of this technology could revolutionize all aspects of functional genomics in plant biology.

KEYWORDS

- **gene silencing/RNA silencing**
- **PTGS**
- **miRNA**
- **siRNA**
- **DICER**
- **RISC assembly**

REFERENCES

Adams, M. J.; King, Q.; Carstens, E. B. Ratification Vote on Taxonomic Proposals to the International Committee on Taxonomy of Viruses. *Arch. Virol.* **2013,** *158*, 2023.

Ali, Z.; Abul-Faraj, A.; Li, L.; Ghosh, N.; Piatek, M.; Mahjoub, A.; Aouida, M.; Piatek, A.; Baltes, N. J.; Voytas, D. F. Efficient Virus-Mediated Genome Editing in Plants Using the CRISPR/Cas9 System. *Mol. Plant* **2015,** *8*, 1288–1291.

Anandalakshmi, R.; Pruss, G. J.; Ge, X.; Marathe, R.; Mallory, A. C.; Smith, T. H.; Vance, V. B. A viral suppressor of gene silencing in plants. *Proc. Nat. Acad. Sci.* **1998,** *95*, 13079–13084.

Atwood, S. E.; O'Rourke, J. A.; Peiffer, G. A.; Yin, T.; Majumder, M.; Zhang, C.; Cianzio, S. R.; Hill, J. H.; Cook, D.; Whitham, S. A. Replication Protein A Subunit 3 and the Iron Efficiency Response in Soybean. *Plant, Cell Environ.* **2014,** *37*, 213–234.

Bance, V. B. Replication of Potato Virus X RNA is Altered in Coinfections with Potato Virus Y. *Virology* **1991,** *182*, 486–494.

Bar-Joseph, M.; Murant, A. Closterovirus Group. *Descriptions of Plant Viruses*. Commonwealth Mycological Institute, 1982.

Baulcombe, D. RNA Silencing in Plants. *Nature* **2004,** *431*, 356–363.

Baulcombe, D. C. Fast Forward Genetics Based on Virus-Induced Gene Silencing. *Curr. Opin. Plant Biol.* **1999,** *2*, 109–113.

Baulcombe, D. C.; English, J. J. Ectopic Pairing of Homologous DNA and Post-transcriptional Gene Silencing in Transgenic Plants. *Curr. Opin. Biotechnol.* **1996,** *7*, 173–180.

Boyer, J.-C.; Haenni, A.-L. Infectious Transcripts and cDNA Clones of RNA Viruses. *Virology* **1994,** *198*, 415–426.

Brigneti, G.; Voinnet, O.; Li, W. X.; Ji, L. H.; Ding, S. W.; Baulcombe, D. C. Viral Pathogenicity Determinants Are Suppressors of Transgene Silencing in *Nicotiana benthamiana*. *EMBO J.* **1998,** *17*, 6739–6746.

Brosnan, C.; Mitter, N.; Christie, M.; Smith, N.; Waterhouse, P.; Carroll, B. Nuclear Gene Silencing Directs Reception of Long-distance mRNA Silencing in Arabidopsis. *Proc. Nat. Acad. Sci.* **2007,** *104*, 14741–14746.

Bruun-Rasmussen, M.; Madsen, C. T.; Jessing, S.; Albrechtsen, M. Stability of Barley Stripe Mosaic Virus-Induced Gene Silencing in Barley. *Mol. Plant-Microbe Interact.* **2007,** *20*, 1323–1331.

Burch-Smith, T. M.; Schiff, M.; Liu, Y.; Dinesh-Kumar, S. P. Efficient Virus-Induced Gene Silencing in Arabidopsis. *Plant Physiol.* **2006,** *142*, 21–27.

Burch-Smith, T. M.; Anderson, J. C.; Martin, G. B.; Dinesh-Kumar, S. P. Applications and Advantages of Virus-Induced Gene Silencing for Gene Function Studies in Plants. *Plant J.* **2004,** *39*, 734–746.

Burton, R. A.; Gibeaut, D. M.; Bacic, A.; Findlay, K.; Roberts, K.; Hamilton, A.; Baulcombe, D. C.; Fincher, G. B. Virus-Induced Silencing of a Plant Cellulose Synthase Gene. *Plant Cell* **2000,** *12*, 691–705.

Calvert, L.; Ghabrial, S. Enhancement by Soybean Mosaic Virus of Bean Pod Mottle Virus Titre in Doubly Infected Soybeans. *Phytopathology* **1983,** *73*, 992–997.

Cogoni, C.; Macino, G. Post-transcriptional Gene Silencing Across Kingdoms. *Curr. Opin. Genet. Dev.* **2000,** *10*, 638–643.

Constantin, G. D.; Krath, B. N.; MacFarlane, S. A.; Nicolaisen, M.; Elisabeth Johansen, I.; Lund, O. S. Virus-Induced Gene Silencing as a Tool for Functional Genomics in a Legume Species. *Plant J.* **2004,** *40*, 622–631.

Cui, X.; Tao, X.; Xie, Y.; Fauquet, C. M.; Zhou, X. A DNAβ associated with Tomato Yellow Leaf Curl China Virus Is Required for Symptom Induction. *J. Virol.* **2004,** *78*, 13966–13974.

Dalmay, T.; Hamilton, A.; Rudd, S.; Angell, S.; Baulcombe, D. C. An RNA-Dependent RNA Polymerase Gene in *Arabidopsis* is Required for Posttranscriptional Gene Silencing Mediated by a Transgene but Not by a Virus. *Cell* **2000,** *101*, 543–553.

Dawson, W. O.; Folimonova, S. Y. Virus-Based Transient Expression Vectors for Woody Crops: A New Frontier for Vector Design and Use. *Annu. Rev. Phytopathol.* **2013,** *51*, 321–337.

De Carvalho Niebel, F.; Frendo, P.; Van Montagu, M.; Cornelissen, M. Post-transcriptional Cosuppression of Beta-1,3-glucanase Genes Does Not Affect Accumulation of Transgene Nuclear mRNA. *Plant Cell* **1995,** *7*, 347–358.

Demmig-Adams, B.; Adams, W. W. Carotenoid Composition in Sun and Shade Leaves of Plants with Different Life Forms. *Plant, Cell Environ.* **1992,** *15*, 411–419.

Depicker, A.; Van Montagu, M. Post-transcriptional Gene Silencing in Plants. *Curr. Opin. Cell Biol.* **1997,** *9*, 373–382.

DiCarlo, J. E.; Norville, J. E.; Mali, P.; Rios, X.; Aach, J.; Church, G. M. Genome Engineering in *Saccharomyces cerevisiae* Using CRISPR-Cas Systems. *Nucleic Acids Res.* **2013,** gkt135.

Ding, X. S.; Schneider, W. L.; Chaluvadi, S. R.; Mian, M. A. R.; Nelson, R. S. Characterization of a Brome Mosaic Virus Strain and Its Use as a Vector for Gene Silencing in Monocotyledonous Hosts. *Mol. Plant-Microbe Interact.* **2006,** *19*, 1229–1239.

Dodds, J. A.; Bar-Joseph, M. Double-Stranded RNA from Plants Infected with Closteroviruses. *Phytopathology* **1983,** *73*, 419–423.

Dolja, V. V.; Koonin, E. V. The Closterovirus-Derived Gene Expression and RNA Interference Vectors as Tools for Research and Plant Biotechnology. *Front. Microbiol.* **2013,** *4*, 83.

Dolja, V. V.; Kreuze, J. F.; Valkonen, J. P. T. Comparative and Functional Genomics of Closteroviruses. *Virus Res.* **2006,** *117*, 38–51.

Dunoyer, P.; Himber, C.; Ruiz-Ferrer, V.; Alioua, A.; Voinnet, O. Intra- and Intercellular RNA Interference in *Arabidopsis thaliana* Requires Components of the MicroRNA and Heterochromatic Silencing Pathways. *Nat. Genet.* **2007,** *39*, 848–856.

Dunoyer, P.; Himber, C.; Voinnet, O. DICER-LIKE 4 is Required for RNA Interference and Produces the 21-Nucleotide Small Interfering RNA Component of the Plant Cell-to-Cell Silencing Signal. *Nat. Genet.* **2005,** *37*, 1356–1360.

Dunoyer, P.; Schott, G.; Himber, C.; Meyer, D.; Takeda, A.; Carrington, J. C.; Voinnet, O. Small RNA Duplexes Function as Mobile Silencing Signals between Plant Cells. *Science* **2010,** *328*, 912–916.

El-Mohtar, C.; Dawson, W. O. Exploring the Limits of Vector Construction Based on *Citrus tristeza* Virus. *Virology* **2014,** *448*, 274–283.

Fire, A.; Xu, S.; Montgomery, M. K.; Kostas, S. A.; Driver, S. E.; Mello, C. C. Potent and Specific Genetic Interference by Double-Stranded RNA in *Caenorhabditis elegans. Nature* **1998,** *391*, 806–811.

Fofana, I. B. F.; Sangare, A.; Collier, R.; Taylor, C.; Fauquet, C. M. A Geminivirus-Induced Gene Silencing System for Gene Function Validation in Cassava. *Plant Mol. Biol.* **2004,** *56*, 613–624.

Friedland, A. E.; Tzur, Y. B.; Esvelt, K. M.; Colaiácovo, M. P.; Church, G. M.; Calarco, J. A. Heritable Genome Editing in *C. elegans* Via a CRISPR-Cas9 System. *Nature Methods* **2013,** *10*, 741–743.

Gatehouse, J. A.; Price, D. R. G. Protection of Crops against Insect Pests Using RNA Interference. *Insect Biotechnology*; Springer: Berlin, 2011; pp. 145–168.

Goldberg, K.-B.; Brakke, M. K. Concentration of Maize Chlorotic Mottle Virus Increased in Mixed Infections with Maize Dwarf Mosaic Virus, Strain B. *Phytopathology* **1987,** *77*, 162–167.

Gosselé, V.; Faché, I.; Meulewaeter, F.; Cornelissen, M.; Metzlaff, M. SVISS—A Novel Transient Gene Silencing System for Gene Function Discovery and Validation in Tobacco Plants. *Plant J.* **2002,** *32*, 859–866.

Grimsley, N.; Hohn, T.; Hohn, B. Recombination in a Plant Virus: Template-Switching in Cauliflower Mosaic Virus. *EMBO J.* **1986,** *5*, 641.

Hajeri, S.; Killiny, N.; El-Mohtar, C.; Dawson, W. O.; Gowda, S. *Citrus tristeza* Virus-Based RNAi in Citrus Plants Induces Gene Silencing in *Diaphorina citri*, a Phloem-Sap Sucking Insect Vector of Citrus Greening Disease (Huanglongbing). *J. Biotechnol.* **2014,** *176*, 42–49.

Hamilton, A.; Voinnet, O.; Chappell, L.; Baulcombe, D. Two Classes of Short Interfering RNA in RNA Silencing. *EMBO J.* **2002,** *21*, 4671–4679.

Hamilton, A. J.; Baulcombe, D. C. A Species of Small Antisense RNA in Posttranscriptional Gene Silencing in Plants. *Science* **1999,** *286*, 950–952.

Hammond, S. M.; Caudy, A. A.; Hannon, G. J. Post-transcriptional Gene Silencing by Double-stranded RNA. *Nat. Rev. Genet.* **2001,** *2*, 110–119.

Hannon, G. J. RNA Interference. *Nature* **2002,** *418*, 244–251.

Hilf, M. E.; Karasev, A. V.; Pappu, H. R.; Gumpf, D. J.; Niblett, C. L.; Garnsey, S. M. Characterization of Citrus Tristeza Virus Subgenomic RNAs in Infected Tissue. *Virology* **1995,** *208*, 576–582.

Himber, C.; Dunoyer, P.; Moissiard, G.; Ritzenthaler, C.; Voinnet, O. Transitivity-Dependent and Independent Cell-to-Cell Movement of RNA Silencing. *EMBO J.* **2003,** *22*, 4523–4533.

Hirochika, H.; Okamoto, H.; Kakutani, T. Silencing of Retrotransposons in Arabidopsis and Reactivation by the ddm1 Mutation. *Plant Cell Online* **2000,** *12*, 357–368.

Holzberg, S.; Brosio, P.; Gross, C.; Pogue, G. P. Barley Stripe Mosaic Virus-Induced Gene Silencing in a Monocot Plant. *Plant J.* **2002,** *30*, 315–327.

Horvath, P.; Barrangou, R. CRISPR/Cas, the Immune System of Bacteria and Archaea. *Science* **2010,** *327*, 167–170.

Hou, H.; Qiu, W. A Novel Co-delivery System Consisting of a Tomato Bushy Stunt Virus and a Defective Interfering RNA for Studying Gene Silencing. *J. Virol. Methods* **2003,** *111*, 37–42.

Huang, C.; Xie, Y.; Zhou, X. Efficient Virus-Induced Gene Silencing in Plants Using a Modified Geminivirus DNA1 Component. *Plant Biotechnol. J.* **2009,** *7*, 254–265.

Hull, R.; Covey, S. N.; Maule, A. J. Structure and Replication of Caulimovirus Genomes. *J. Cell Sci.* **1987,** *1987*, 213–229.

Hwang, W. Y.; Fu, Y.; Reyon, D.; Maeder, M. L.; Tsai, S. Q.; Sander, J. D.; Peterson, R. T.; Yeh, J. R. J.; Joung, J. K. Efficient Genome Editing in Zebrafish Using a CRISPR-Cas System. *Nat. Biotechnol.* **2013,** *31*, 227–229.

Igarashi, A.; Yamagata, K.; Sugai, T.; Takahashi, Y.; Sugawara, E.; Tamura, A.; Yaegashi, H.; Yamagishi, N.; Takahashi, T.; Isogai, M. Apple Latent Spherical Virus Vectors for Reliable and Effective Virus-Induced Gene Silencing among a Broad Range of Plants Including Tobacco, Tomato, *Arabidopsis thaliana*, Cucurbits, and Legumes. *Virology* **2009,** *386*, 407–416.

Jiang, W.; Bikard, D.; Cox, D.; Zhang, F.; Marraffini, L. A. RNA-Guided Editing of Bacterial Genomes Using CRISPR-Cas Systems. *Nat. Biotechnol.* **2013,** *31*, 233–239.

Jinek, M.; Chylinski, K.; Fonfara, I.; Hauer, M.; Doudna, J. A.; Charpentier, E. A Programmable Dual-RNA-Guided DNA Endonuclease in Adaptive Bacterial Immunity. *Science* **2012,** *337*, 816–821.

Jones, L.; Ratcliff, F.; Baulcombe, D. C. RNA-Directed Transcriptional Gene Silencing in Plants can be Inherited Independently of the RNA Trigger and Requires Met1 for Maintenance. *Curr. Biol.* **2001,** *11*, 747–757.

Jorgensen, R. A.; Cluster, P. D.; English, J.; Que, Q.; Napoli, C. A. Chalcone Synthase Cosuppression Phenotypes in Petunia Flowers: Comparison of Sense vs. Antisense Constructs and Single-Copy vs. Complex T-DNA Sequences. *Plant Mol. Biol.* **1996,** *31*, 957–973.

Kalantidis, K.; Schumacher, H. T.; Alexiadis, T.; Helm, J. M. RNA Silencing Movement in Plants. *Biol. Cell* **2008,** *100*, 13–26.

Kasschau, K. D.; Carrington, J. C. A Counterdefensive Strategy of Plant Viruses: Suppression of Posttranscriptional Gene Silencing. *Cell* **1998,** *95*, 461–470.

Kjemtrup, S.; Sampson, K. S.; Peele, C. G.; Nguyen, L. V.; Conkling, M. A.; Thompson, W. F.; Robertson, D. Gene Silencing from Plant DNA Carried by a Geminivirus. *Plant J.* **1998,** *14*, 91–100.

Kumagai, M. H.; Donson, J.; Della-Cioppa, G.; Harvey, D.; Hanley, K.; Grill, L. K. Cytoplasmic Inhibition of Carotenoid Biosynthesis with Virus-Derived RNA. *Proc. Nat. Acad. Sci.* **1995,** *92*, 1679–1683.

Kurth, E. G.; Peremyslov, V. V.; Prokhnevsky, A. I.; Kasschau, K. D.; Miller, M.; Carrington, J. C.; Dolja, V. V. Virus-Derived Gene Expression and RNA Interference Vector for Grapevine. *J. Virol.* **2012,** *86*, 6002–6009.

Lakatos, L.; Csorba, T.; Pantaleo, V.; Chapman, E. J.; Carrington, J. C.; Liu, Y. P.; Dolja, V. V.; Calvino, L. F.; López-Moya, J. J.; Burgyán, J. Small RNA Binding is a Common Strategy to Suppress RNA Silencing by Several Viral Suppressors. *EMBO J.* **2006,** *25*, 2768–2780.

Laufs, J.; Traut, W.; Heyraud, F.; Matzeit, V.; Rogers, S. G.; Schell, J.; Gronenborn, B. In Vitro Cleavage and Joining at the Viral Origin of Replication by the Replication Initiator Protein of Tomato Yellow Leaf Curl Virus. *Proc. Nat. Acad. Sci.* **1995,** *92*, 3879–3883.

Lippman, Z.; Martienssen, R. The Role of RNA Interference in Heterochromatic Silencing. *Nature* **2004,** *431*, 364–370.

Liu, C.; Lu, F.; Cui, X.; Cao, X. Histone Methylation in Higher Plants. *Annu. Rev. Plant Biol.* **2010,** *61*, 395–420.

Liu, E.; Page, J. E. Optimized cDNA Libraries for Virus-Induced Gene Silencing (VIGS) Using Tobacco Rattle Virus. *Plant Methods* **2008,** *4*, 1.

Liu, Y.; Schiff, M.; Dinesh-Kumar, S. P. Virus-Induced Gene Silencing in Tomato. *Plant J.* **2002a,** *31*, 777–786.

Liu, Y.; Schiff, M.; Marathe, R.; Dinesh-Kumar, S. P. Tobacco Rar1, EDS1 and NPR1/NIM1 Like Genes Are Required for N-Mediated Resistance to Tobacco Mosaic Virus. *Plant J.* **2002b,** *30*, 415–429.

Lu, R.; Malcuit, I.; Moffett, P.; Ruiz, M. T.; Peart, J.; Wu, A. J.; Rathjen, J. P.; Bendahmane, A.; Day, L.; Baulcombe, D. C. High Throughput Virus-Induced Gene Silencing Implicates Heat Shock Protein 90 in Plant Disease Resistance. *EMBO J.* **2003a,** *22*, 5690–5699.

Lu, R.; Martin-Hernandez, A. M.; Peart, J. R.; Malcuit, I.; Baulcombe, D. C. Virus-Induced Gene Silencing in Plants. *Methods* **2003b,** *30*, 296–303.

Mali, P.; Yang, L.; Esvelt, K. M.; Aach, J.; Guell, M.; DiCarlo, J. E.; Norville, J. E.; Church, G. M. RNA-Guided Human Genome Engineering Via Cas9. *Science* **2013,** *339*, 823–826.

Mallory, A. C.; Ely, L.; Smith, T. H.; Marathe, R.; Anandalakshmi, R.; Fagard, M.; Vaucheret, H.; Pruss, G.; Bowman, L.; Vance, V. B. HC-Pro Suppression of Transgene Silencing Eliminates the Small RNAs But Not Transgene Methylation or the Mobile Signal. *Plant Cell Online* **2001,** *13*, 571–583.

Manmathan, H.; Shaner, D.; Snelling, J.; Tisserat, N.; Lapitan, N. Virus-Induced Gene Silencing of *Arabidopsis thaliana* Gene Homologues in Wheat Identifies Genes Conferring Improved Drought Tolerance. *J. Exp. Bot.* **2013,** *64*, 1381–1392.

Martín-Hernández, A. M.; Baulcombe, D. C. Tobacco Rattle Virus 16-Kilodalton Protein Encodes a Suppressor of RNA Silencing that Allows Transient Viral Entry in Meristems. *J. Virol.* **2008,** *82*, 4064–4071.

Matthews, R. E. F. The Nature of Dark Green Tissue. *Plant Virology*, 3rd ed. Academic Press: San Diego, CA, 1991; pp 448–449.

Matzke, M. A.; Matzke, A. J.; Pruss, G. J.; Vance, V. B. RNA-Based Silencing Strategies in Plants. *Curr. Opin. Genet. Dev.* **2001,** *11*, 221–227.

Meister, G.; Tuschl, T. Mechanisms of Gene Silencing by Double-Stranded RNA. *Nature* **2004,** *431*, 343–349.

Mello, C. C.; Conte, D. Revealing the World of RNA Interference. *Nature* **2004,** *431*, 338–342.

Mlotshwa, S.; Voinnet, O.; Mette, M. F.; Matzke, M.; Vaucheret, H.; Ding, S. W.; Pruss, G.; Vance, V. B. RNA Silencing and the Mobile Silencing Signal. *Plant Cell Online* **2002,** *14*, S289–S301.

Molnar, A.; Melnyk, C. W.; Bassett, A.; Hardcastle, T. J.; Dunn, R.; Baulcombe, D. C. Small Silencing RNAs in Plants are Mobile and Direct Epigenetic Modification in Recipient Cells. *Science* **2010,** *328*, 872–875.

Muangsan, N.; Beclin, C.; Vaucheret, H.; Robertson, D. Geminivirus VIGS of Endogenous Genes Requires SGS2/SDE1 and SGS3 and Defines a New Branch in the Genetic Pathway for Silencing in Plants. *Plant J.* **2004,** *38*, 1004–1014.

Nagamatsu, A.; Masuta, C.; Senda, M.; Matsuura, H.; Kasai, A.; Hong, J. S.; Kitamura, K.; Abe, J.; Kanazawa, A. Functional Analysis of Soybean Genes Involved in Flavonoid Biosynthesis by Virus-Induced Gene Silencing. *Plant Biotechnol. J.* **2007,** *5*, 778–790.

Nagyová, A.; Subr, Z. Infectious Full-Length Clones of Plant Viruses and their Use for Construction of Viral Vectors. *Acta Virol.* **2006,** *51*, 223–237.

Napoli, C.; Lemieux, C.; Jorgensen, R. Introduction of a Chimeric Chalcone Synthase Gene into Petunia Results in Reversible Co-suppression of Homologous Genes in Trans. *Plant Cell Online* **1990,** *2*, 279–289.

Navas-Castillo, J.; Albiach-Martĺ, M. R.; Gowda, S.; Hilf, M. E.; Garnsey, S. M.; Dawson, W. O. Kinetics of Accumulation of Citrus Tristeza Virus RNAs. *Virology* **1997,** *228*, 92–97.

Naylor, M.; Reeves, J.; Cooper, J. I.; Edwards, M. L.; Wang, H. Construction and Properties of a Gene-Silencing Vector Based on Poplar Mosaic Virus (genus *Carlavirus*). *J. Virol. Methods* **2005,** *124*, 27–36.

Nekrasov, V.; Staskawicz, B.; Weigel, D.; Jones, J. D. G.; Kamoun, S. Targeted Mutagenesis in the Model Plant *Nicotiana benthamiana* Using Cas9 RNA-Guided Endonuclease. *Nat. Biotechnol.* **2013,** *31*, 691–693.

Nunes, C. C.; Dean, R. A. Host-Induced Gene Silencing: A Tool for Understanding Fungal Host Interaction and for Developing Novel Disease Control Strategies. *Mol. Plant Pathol.* **2012,** *13*, 519–529.

Palauqui, J. C.; Elmayan, T.; Pollien, J. M.; Vaucheret, H. Systemic Acquired Silencing: Transgene-Specific Post-transcriptional Silencing is Transmitted by Grafting from Silenced Stocks to Non-silenced Scions. *EMBO J.* **1997,** *16*, 4738–4745.

Pang, J.; Zhu, Y.; Li, Q.; Liu, J.; Tian, Y.; Liu, Y.; Wu, J. Development of *Agrobacterium*-Mediated Virus-Induced Gene Silencing and Performance Evaluation of Four Marker Genes in *Gossypium barbadense*. *PLoS ONE* **2013,** *8*, e73211.

Peart, J. R.; Cook, G.; Feys, B. J.; Parker, J. E.; Baulcombe, D. C. An EDS1 Orthologue Is Required for N-Mediated Resistance against Tobacco Mosaic Virus. *Plant J.* **2002,** *29*, 569–579.

Peele, C.; Jordan, C. V.; Muangsan, N.; Turnage, M.; Egelkrout, E.; Eagle, P.; Hanley-Bowdoin, L.; Robertson, D. Silencing of a Meristematic Gene Using Geminivirus-Derived Vectors. *Plant J.* **2001,** *27,* 357–366.

Pélissier, T.; Thalmeir, S.; Kempe, D.; Sänger, H.-L.; Wassenegger, M. Heavy De Novo Methylation at Symmetrical and Non-Symmetrical Sites is a Hallmark of RNA-Directed DNA Methylation. *Nucleic Acids Res.* **1999,** *27,* 1625–1634.

Pflieger, S.; Blanchet, S.; Camborde, L.; Drugeon, G.; Rousseau, A.; Noizet, M.; Planchais, S.; Jupin, I. Efficient Virus-Induced Gene Silencing in Arabidopsis Using a 'One-Step' TYMV-Derived Vector. *Plant J.* **2008,** *56,* 678–690.

Pruss, G.; Ge, X.; Shi, X. M.; Carrington, J. C.; Vance, V. B. Plant Viral Synergism: The Potyviral Genome Encodes a Broad-Range Pathogenicity Enhancer that Transactivates Replication of Heterologous Viruses. *Plant Cell Online* **1997,** *9,* 859–868.

Purkayastha, A.; Mathur, S.; Verma, V.; Sharma, S.; Dasgupta, I. Virus-Induced Gene Silencing in Rice Using a Vector Derived from a DNA Virus. *Planta* **2010,** *232,* 1531–1540.

Pyott, D. E.; Molnar, A. Going Mobile: Non-Cell-Autonomous Small RNAs Shape the Genetic Landscape of Plants. *Plant Biotechnol. J.* **2015,** *13,* 306–318.

Qian, Y.; Mugiira, R. B.; Zhou, X. A Modified Viral Satellite DNA-Based Gene Silencing Vector Is Effective in Association with Heterologous Begomoviruses. *Virus Res.* **2006,** *118,* 136–142.

Qu, J.; Ye, J.; Geng, Y.-F.; Sun, Y.-W.; Gao, S.-Q.; Zhang, B.-P.; Chen, W.; Chua, N.-H. Dissecting Functions of KATANIN and WRINKLED1 in Cotton Fiber Development by Virus-Induced Gene Silencing. *Plant Physiol.* **2012,** *160,* 738–748.

Quan, J.; Tian, J. Circular Polymerase Extension Cloning for High-Throughput Cloning of Complex and Combinatorial DNA Libraries. *Nat. Protocols* **2011,** *6,* 242–251.

Ran, F. A.; Hsu, P. D.; Wright, J.; Agarwala, V.; Scott, D. A.; Zhang, F. Genome Engineering Using the CRISPR–Cas9 System. *Nat. Protocols* **2013,** *8,* 2281–2308.

Ratcliff, F.; Harrison, B. D.; Baulcombe, D. C. A Similarity between Viral Defense and Gene Silencing in Plants. *Science* **1997,** *276,* 1558–1560.

Ratcliff, F.; Martin-Hernandez, A. M.; Baulcombe, D. C. Technical Advance: Tobacco Rattle Virus as a Vector for Analysis of Gene Function by Silencing. *Plant J.* **2001,** *25,* 237–245.

Rochow, W.; Ross, A. F. Virus Multiplication in Plants Doubly Infected by Potato Viruses X and Y. *Virology* **1955,** *1,* 10–27.

Rodrigo, G.; Carrera, J.; Jaramillo, A.; Elena, S. F. Optimal Viral Strategies for Bypassing RNA Silencing. *J. R. Soc. Interface* **2011,** *8,* 257–268.

Romeis, T.; Ludwig, A. A.; Martin, R.; Jones, J. D. G. Calcium-Dependent Protein Kinases Play an Essential Role in a Plant Defence Response. *EMBO J.* **2001,** *20,* 5556–5567.

Ruiz, M. T.; Voinnet, O.; Baulcombe, D. C. Initiation and Maintenance of Virus-Induced Gene Silencing. *Plant Cell* **1998,** *10,* 937–946.

Sasaki, S.; Yamagishi, N.; Yoshikawa, N. Efficient Virus-Induced Gene Silencing in Apple, Pear and Japanese Pear Using Apple Latent Spherical Virus Vectors. *Plant Methods* **2011,** *7,* 1.

Satyanarayana, T.; Bar-Joseph, M.; Mawassi, M.; Albiach-Martı, M. R.; Ayllón, M. A.; Gowda, S.; Hilf, M. E.; Moreno, P.; Garnsey, S. M.; Dawson, W. O. Amplification of Citrus Tristeza Virus from a cDNA Clone and Infection of Citrus Trees. *Virology* **2001,** *280,* 87–96.

Satyanarayana, T.; Gowda, S.; Boyko, V.; Albiach-Marti, M.; Mawassi, M.; Navas-Castillo, J.; Karasev, A.; Dolja, V.; Hilf, M.; Lewandowski, D. An Engineered Closterovirus RNA

Replicon and Analysis of Heterologous Terminal Sequences for Replication. *Proc. Nat. Acad. Sci.* **1999,** *96*, 7433–7438.

Schaeffer, S. M.; Nakata, P. A. CRISPR/Cas9-Mediated Genome Editing and Gene Replacement in Plants: Transitioning from Lab to Field. *Plant Sci.* **2015,** *240*, 130–142.

Schwach, F.; Vaistij, F. E.; Jones, L.; Baulcombe, D. C. An RNA-Dependent RNA Polymerase Prevents Meristem Invasion by Potato Virus X and Is Required for the Activity But not the Production of a Systemic Silencing Signal. *Plant Physiol.* **2005,** *138*, 1842–1852.

Senthil-Kumar, M.; Udayakumar, M. High-Throughput Virus-Induced Gene-Silencing Approach to Assess the Functional Relevance of a Moisture Stress-Induced cDNA Homologous to Lea4. *J. Exp. Bot.* **2006,** *57*, 2291–2302.

Shan, Q.; Wang, Y.; Li, J.; Zhang, Y.; Chen, K.; Liang, Z.; Zhang, K.; Liu, J.; Xi, J. J.; Qiu, J.-L. Targeted Genome Modification of Crop Plants Using a CRISPR–Cas System. *Nat. Biotechnol.* **2013,** *31*, 686–688.

Shivprasad, S.; Pogue, G. P.; Lewandowski, D. J.; Hidalgo, J.; Donson, J.; Grill, L. K.; Dawson, W. O. Heterologous Sequences Greatly Affect Foreign Gene Expression in Tobacco Mosaic Virus-Based Vectors. *Virology* **1999,** *255*, 312–323.

Sudowe, S.; Reske-Kunz, A. B. *Biolistic DNA Delivery: Methods and Protocols*. Humana Press: New York, NY, 2013.

Tao, X.; Zhou, X. A Modified Viral Satellite DNA that Suppresses Gene Expression in plants. *Plant J.* **2004,** *38*, 850–860.

Tomari, Y.; Zamore, P. D. Perspective: Machines for RNAi. *Genes Dev.* **2005,** *19*, 517–529.

Turgeon, R.; Wolf, S. Phloem Transport: Cellular Pathways and Molecular Trafficking. *Annu. Rev. Plant Biol.* **2009,** *60*, 207–221.

Turnage, M. A.; Muangsan, N.; Peele, C. G.; Robertson, D. Geminivirus-Based Vectors for Gene Silencing in Arabidopsis. *Plant J.* **2002,** *30*, 107–114.

Vaghchhipawala, Z. E.; Vasudevan, B.; Lee, S.; Morsy, M. R.; Mysore, K. S. *Agrobacterium* May Delay Plant Nonhomologous End-Joining DNA Repair via XRCC4 to Favor T-DNA Integration. *Plant Cell* **2012,** *24*, 4110–4123.

Van der Krol, A. R.; Mur, L. A.; Beld, M.; Mol, J.; Stuitje, A. R. Flavonoid Genes in Petunia: Addition of a Limited Number of Gene Copies May Lead to a Suppression of Gene Expression. *Plant Cell Online* **1990,** *2*, 291–299.

Van Kammen, A. Virus-Induced Gene Silencing in Infected and Transgenic Plants. *Trends Plant Sci.* **1997,** *2*, 409–411.

Vance, V.; Vaucheret, H. RNA Silencing in Plants—Defense and Counterdefense. *Science* **2001,** *292*, 2277–2280.

Vanitharani, R.; Chellappan, P.; Fauquet, C. M. Geminiviruses and RNA Silencing. *Trends Plant Sci.* **2005,** *10*, 144–151.

Vaucheret, H. Post-transcriptional Small RNA Pathways in Plants: Mechanisms and Regulations. *Genes Dev.* **2006,** *20*, 759–771.

Voinnet, O. Non-Cell Autonomous RNA Silencing. *FEBS Lett.* **2005,** *579*, 5858–5871.

Waigmann, E.; Chen, M. H.; Bachmaier, R.; Ghoshroy, S.; Citovsky, V. Regulation of Plasmodesmal Transport by Phosphorylation of Tobacco Mosaic Virus Cell-to-Cell Movement Protein. *EMBO J.* **2000,** *19*, 4875–4884.

Walker, W. B.; Allen, M. L. Expression and RNA Interference of Salivary Polygalacturonase Genes in the Tarnished Plant Bug, *Lygus lineolaris*. *J. Insect Sci.* **2010,** *10*, 173.

Wassenegger, M. RNA-Directed DNA Methylation. *Plant Gene Silencing*. Springer: Berlin, 2000; pp. 83–100.

Wassenegger, M.; Heimes, S.; Riedel, L.; Sänger, H. L. RNA-Directed De Novo Methylation of Genomic Sequences in Plants. *Cell* **1994,** *76*, 567–576.

Waterhouse, P. M.; Wang, M.-B.; Lough, T. Gene Silencing as an Adaptive Defence against Viruses. *Nature* **2001,** *411*, 834–842.

Xie, K.; Zhang, J.; Yang, Y. Genome-Wide Prediction of Highly Specific Guide RNA Spacers for CRISPR–Cas9-Mediated Genome Editing in Model Plants and Major Crops. *Mol. Plant* **2014,** *7*, 923–926.

Xu, P.; Zhang, Y.; Kang, L.; Roossinck, M. J.; Mysore, K. S. Computational Estimation and Experimental Verification of Off-Target Silencing during Posttranscriptional Gene Silencing in Plants. *Plant Physiol.* **2006,** *142*, 429–440.

Yaegashi, H.; Yamatsuta, T.; Takahashi, T.; Li, C.; Isogai, M.; Kobori, T.; Ohki, S.; Yoshikawa, N. Characterization of Virus-induced Gene Silencing in Tobacco Plants Infected with Apple Latent Spherical Virus. *Arch. Virol.* **2007,** *152*, 1839–1849.

Yang, S.-J.; Carter, S. A.; Cole, A. B.; Cheng, N.-H.; Nelson, R. S. A Natural Variant of a Host RNA-Dependent RNA Polymerase is Associated with Increased Susceptibility to Viruses by *Nicotiana benthamiana. Proc. Nat. Acad. Sci. USA* **2004,** *101*, 6297–6302.

Yin, K.; Han, T.; Liu, G.; Chen, T.; Wang, Y.; Yu, A. Y. L.; Liu, Y. A Geminivirus-Based Guide RNA Delivery System for CRISPR/Cas9 Mediated Plant Genome Editing. *Sci. Rep.* **2015,** *5,* 14926, 1-10.

Yuan, C.; Li, C.; Yan, L.; Jackson, A. O.; Liu, Z.; Han, C.; Yu, J.; Li, D. A High Throughput Barley Stripe Mosaic Virus Vector for Virus Induced Gene Silencing in Monocots and Dicots. *PLoS ONE* **2011,** *6*, e26468.

Zhang, C.; Bradshaw, J. D.; Whitham, S. A.; Hill, J. H. The Development of an Efficient Multipurpose Bean Pod Mottle Virus Viral Vector Set for Foreign Gene Expression and RNA Silencing. *Plant Physiol.* **2010,** *153*, 52–65.

Zhang, C.; Ghabrial, S. A. Development of Bean Pod Mottle Virus-Based Vectors for Stable Protein Expression and Sequence-Specific Virus-Induced Gene Silencing in Soybean. *Virology* **2006,** *344*, 401–411.

Zhang, C.; Yang, C.; Whitham, S. A.; Hill, J. H. Development and Use of An Efficient DNA-Based Viral Gene Silencing Vector for Soybean. *Mol. Plant-Microbe Interact.* **2009,** *22*, 123–131.

Zhang, W.; Kollwig, G.; Stecyk, E.; Apelt, F.; Dirks, R.; Kragler, F. Graft-Transmissible Movement of Inverted-Repeat-Induced siRNA Signals into Flowers. *Plant J.* **2014,** *80*, 106–121.

PART III

Transgenic Crops and Biosafety

CHAPTER 6

TRANSGENIC PLANTS AND THEIR APPLICATION IN CROP IMPROVEMENT

VINOD KUMAR[1*], NIMMY M. S.[2], TUSHAR RANJAN[1], RAVI RANJAN[1], BISHUN DEO PRASAD[1], and CHANDAN KISHORE[3]

[1]*Department of Molecular Biology and Biotechnology, BAU, Sabour, Bhagalpur 813210, Bihar, India*

[2]*ICAR-NRC on Plant Biotechnology, IARI, Pusa Campus, New Delhi 110012, India*

[3]*Department of Plant Breeding and Genetics, BAU, Sabour, Bhagalpur 813210, Bihar, India*

[*]*Corresponding author. E-mail: biotech.vinod@gmail.com*

CONTENTS

ABSTRACT

The cultivable land available for crop production is shrinking day by day because of fast urban growth as well as land degradation, and the trend is expected to be much more drastic in developing countries than in the developed ones. Yields of several crops have already reached a plateau in developed countries, and therefore, most of the productivity gains in the future will have to be achieved in developing countries through better natural resources management and crop improvement. Productivity gains are essential for long-term economic growth, but in the short term, these are even more important for maintaining adequate food supplies for the increasing global population. Therefore, in this context, biotechnology will play an important role in food production in the near future. In this regard, we attempt to take a critical but practical look at the prospects and constraints of various types of biotechnologies and their application for increasing crop production and improving nutritional quality. Within this, we also address the critical issues of biosafety and impact of the genetically engineered crops on the environment. Genetic engineering offers plant breeders access to an infinitely wide array of novel genes and traits, which can be inserted through a single event into high-yielding and locally adapted cultivars.

6.1 INTRODUCTION

The United Nations have projected that world population will increase by 25% to 7.5 billion by 2020. On an average, an additional 73 million people are added annually, of which 97% will live in the developing countries. At the moment, nearly 1.2 billion people live in a state of "absolute poverty," of which 800 million people live under uncertain food security, and 160 million preschool children suffer from malnutrition (FAO, 1986). A large number of people are deficient from micronutrients such as iron, zinc, and vitamin A. Food insecurity and malnutrition result in serious health problems in common people, with least human potential. The cultivable land available for crop production is shrinking day by day because of fast urban growth as well as land degradation, and the trend is expected to be much more drastic in developing countries than in the developed ones. In 1990, per capita crop land availability was less than 0.25 ha in Egypt, Kenya, Bangladesh, Vietnam, and China. However, by 2025, countries such as Peru, Tanzania, Pakistan, Indonesia, and Philippines are likely to join this group (Engelman et al., 1995). These decreases in the amount of land available for crop production and increase in human population will have major implications for food security over the next 2–3 decades. There had been a remarkable increase in total grain production between 1950 and 1980 but only a marginal increase was realized during 1980–1990. Much of the early increase rise in grain production resulted from an increase in area under cultivation, irrigation, better agronomic practices, and improved cultivars. Yields of several crops have already reached a plateau in developed countries, and therefore, most of the productivity gains in the future will have to be achieved in developing countries through better natural resources management and crop improvement. Productivity gains are essential for long-term economic growth, but in the short-term, these are even more important for maintaining adequate food supplies for the increasing global population. Therefore in this context, biotechnology will play an important role in food production in the near future. In this regard, we attempt to take a critical but practical look at the prospects and constraints of various types of biotechnologies and their application for increasing crop production and improving nutritional quality. Within this, we also address the critical issues of biosafety and impact of the genetically engineered crops on the environment.

Genetic engineering offers plant breeders access to an infinitely wide array of novel genes and traits, which can be inserted through a single event into high-yielding and locally adapted cultivars. Genetic engineering has an edge over conventional crop breeding program in various aspects (Table 6.1). Decision should be made in regards to transfer of gene of interest for crop improvement by conventional breeding approaches or genetic engineering (Fig. 6.1). In comparison to conventional breeding approach in respect to time and revalidation of gene of interest, genetic engineering approach offers rapid introgression of novel genes and traits into elite agronomic backgrounds (Fig. 6.2). Future impacts of biotechnology in crop production will

TABLE 6.1 Comparison between Conventional Breeding and Genetic Engineering.

Conventional Breeding	Genetic Engineering
Limited to exchanges between the same or very closely species	Allows the direct transfer of one or just a few genes between either closely or distantly related organisms
Little or no guarantee of any particular gene combination from the millions of crosses generated	Crop improvement can be achieved in shorter time as compared to conventional
Undesirable gene can be transferred along with desirable	Allows plants to be modified by removing or switching off particular genes
Takes long time to achieve desired result	

(Reprinted from http://slideplayer.com/slide/4650387/)

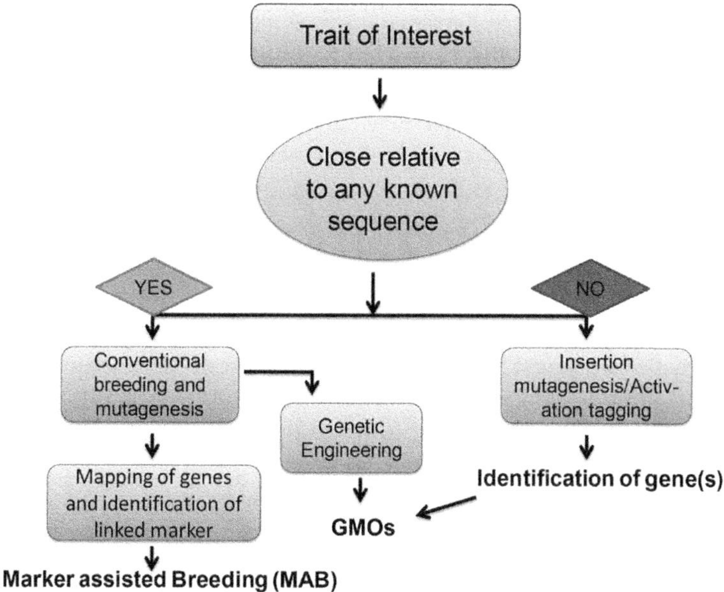

FIGURE 6.1 Decision-making box by using both conventional and modern biotechnology approaches for crop breeding. (Adapted from DANIDA 2002).

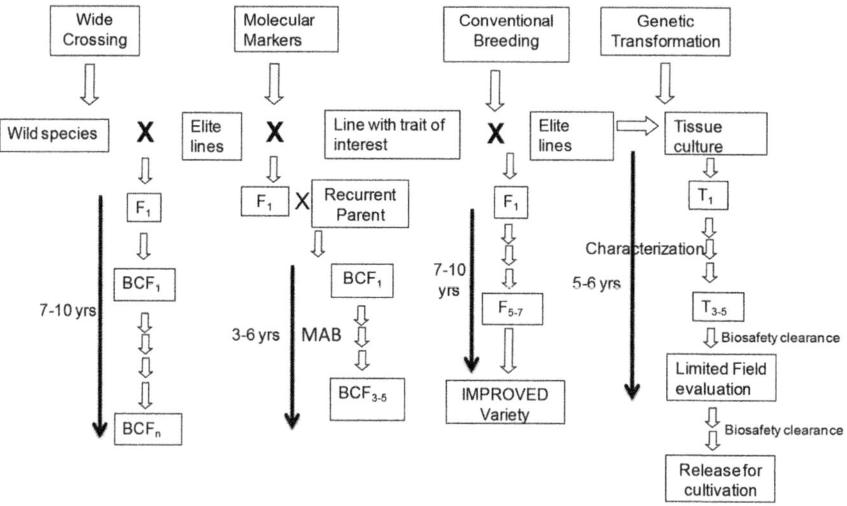

FIGURE 6.2 A schematic outline of biotechnological approaches in crop improvement. Lines derived through genetic transformation can be released as varieties or used as a donor parent in the conventional breeding. The lines derived from wide crossing can take many generations (BCFn) to obtain homozygous and stable lines, and such material can either be used as improved lines or as a donor parent in conventional breeding or marker-assisted selection. (Adapted from Sharma et al., 2002).

be in the areas of: (1) developing new hybrid crops based on genetic male sterility; (2) exploit transgenic apomixes to fix hybrid vigor in inbred crops; (3) increase resistance to insect pests, diseases, and abiotic stress factors; (4) improve effectiveness of biocontrol agents; (5) enhance nutritional value (vitamin A and iron) of crops and postharvest quality; (6) increase efficiency of soil phosphorus uptake and nitrogen fixation; (7) improve adaptation to soil salinity and aluminum toxicity; (8) understanding nature of gene action and metabolic pathways; (9) increase photosynthetic activity, sugar, and starch production; and (10) production of pharmaceuticals and vaccines. New crop cultivars with resistance to insect pests and diseases combined with biocontrol agents should lead to a reduced reliance on pesticides, and thereby reduce farmers' crop protection costs, while benefiting both the environment and public health. Similarly, genetic modification for herbicide resistance to achieve efficient and cost-effective weed control can increase farm incomes, while reducing the labor demand for weeding and herbicide application which leads to further toxicity and degradation to soil health. Labor released from agriculture can then be used for other profitable endeavors. In addition, there is an urgent need for less labor-intensive agricultural practices in countries significantly affected by human immune deficiency virus (HIV). By increasing crop productivity, agricultural biotechnology can substitute the need to cultivate new land and thereby conserve biodiversity in areas that are marginal for crop production. The potential of these technologies as compared to conventional method (Table 6.1) has been extensively tested in the model crop species of temperate and subtropical agriculture. However, there is an urgent need for an increased focus on crops relevant to the small farm holders and poor consumers in the developing countries of the humid and semiarid tropics. The promise of biotechnology can be realized by utilizing the information and products generated through research on genomics and transgenics to increase the productivity of crops through enhanced resistance to biotic and abiotic stress factors and improved nutritional quality.

In the scenario of global climate change and increasing population growth rate, we need sustainable crop production and an urgent need for plant breeding along with modern tools and techniques rather than solely depending on conventional methods. As we already discussed that globally, population is expected to rise to more than 9 billion by 2050 (Raven, 2014). Crop improvement can be achieved by different methods of gene transfer and genetic engineering has emerged as a powerful tool with transferring and expressing foreign genes into plants and transgenic plants resistant to insect pests, pathogens, and other abiotic stresses as well as producing novel compounds of pharmaceutical and industrial value have been developed. Transgenic plant development is the result of integrated application of rDNA technology and tissue-culture techniques. The progress in this area is tremendous with advances in plant molecular biology with majority of crop genomes has been sequenced and gene functions have been annotated. Even though there are legal and ethical issues to the full implementation of plant biotechnology and transgenic technology, advances in this field have led to crop improvement.

The era of genetically engineered plant begins with the tobacco plants transformed by *Agrobacterium tumefaciens* in the early 1980s. *A. tumefaciens* evolved from being a mere plant pathogen to a powerful genetic transformation agent for biotechnology research. The list of plant species that can be transformed by *Agrobacterium* seems to grow daily. However, *Agrobacterium*-mediated gene transfer into monocots was not possible at the beginning, but later on reproducible protocols were standardized for rice (Hiei et al., 1994), maize (Ishida et al., 1996) wheat (Cheng et al., 1997), and sugarcane (Arencibia et al., 1998). Genetic modification of crops has enabled plant breeders to modify plants in novel ways and has the potential to overcome important problems of modern agriculture. Introduction of genes into plants has been made possible using *Agrobacterium* as a biological vector, and direct gene-transfer techniques. *Agrobacterium*-based methods are more efficient and simple but have the disadvantages that are not applicable in every plant species. Recent developments indicate that these host-range limitations can be overcome by developing specific plant-cell culture procedures and defining inoculation and co-cultivation conditions (Park et al., 1996). Some important nonhost species such as maize and rice have now been stably transformed by *Agrobacterium*. Although plant transformation was initially experimental, the potential of commercialization of new improved varieties was early realized and a fast-growing international Agricultural Biotech market has already been formed. New transgenic varieties have been produced that are resistant to pathogens, insects, herbicides, or express novel characters that improve product quality and agronomic traits. The new opportunities to modify plants in novel ways with genetic modification present new responsibilities for safe use to avoid adverse effects on human health and the environment (Dale and Irwin, 1998). Risk assessment studies are integral part in the production and placing to the market a transgenic variety. Different countries have adopted different approaches in biosafety assessment. International harmonization of biosafety standards is an important challenge as we face the international trade of transgenic plant products.

The role of international organizations such as OECD and United Nations may be critical toward this goal. In the following sections, we will present the methods and techniques that are utilized in the production of transgenic plants emphasizing recent developments for multiple genes transfers at once; we will consider the major achievements of the new technology and the future prospects and challenges, as well as remaining technological gaps; we will enumerate and discuss some of the possible risks involved in the unrestrained use of transgenic plants and their products; and finally, we will present the current status of the regulatory framework pertaining the field release of transgenic plants.

6.2 GLOBAL STATUS OF COMMERCIALIZED BIOTECH/GM CROPS

The global area of biotech crops continued to increase for the 19th year at a sustained growth rate of 3–4% or 6.3 million hectares (~16 million acres), reaching 181.5 million hectares or 448 million acres in 2014. Biotech crops have set a precedent in that the biotech area has grown impressively every single year for the past 19 years, with a remarkable 100-fold increase since the commercialization began in 1996. Thus, biotech crops are considered as the fastest adopted crop technology in the history of modern agriculture. In 2014, a total of 18 million farmers planted biotech crops in 28 countries, wherein over 94.1% or greater than 16.9 million were small and resource-poor farmers from developing countries. The highest increase in any country, in absolute hectarage growth was in the United States with 3 million hectares. In summary, during the period of 1996–2014, biotech crops have been successfully grown in accumulated hectarage of 1.78 billion hectares (4.4 billion acres).

6.3 TRANSGENIC PLANTS AND AGRICULTURE

A number of documents as evidence increased for crop yields, higher farm income, and health and environment benefits associated with GM crops. In 1996, when GM crops were first officially commercialized, six countries around the world planted a total of 1.7 million hectares of GM crops. In 2010, GM crop area reached up to 148 million hectares in 29 countries (of which 19 countries were in the developing world). This 87-fold growth makes GM the fastest growing crop technology adopted in modern agriculture. Total

15.4 million farmers who planted GM crops in 2010, >90% (14.4 million) were small farmers in developing countries, including in three African countries: Burkina Faso, South Africa, and Egypt. Almost 100,000 farmers in Burkina Faso cultivated GM cotton on 260,000 ha in 2010 (representing a 126% increase from 2009), and GM crops are estimated to have benefited Burkina Faso's economy by over US$100 million per year.

Similarly, in South Africa, the first and biggest producer of GM crops in Africa, GM technology is reported to have enhanced farm income by US$156 million in the period 1998–2006. South Africa is the only African country among the five principal GM-producing countries (along with India, Argentina, Brazil, and China), and farmers there planted 63 million hectares of GM crops in 2010 alone. GM crops are always beneficial in terms of health and income compared to traditional farming in which huge pesticide, herbicide, and labor are required.

6.4 ISOLATION AND CLONING THE GENE OF INTEREST

Identification and cloning of the gene of interest is a first limiting step in the transgenic development process. Selection, identification, characterization, and cloning of agriculturally important genes require a huge effort both in terms of human as well as financial capital. One of the earliest developments is the introduction of insect resistance by transgenic technology. The discovery of *Bt* genes has revolutionized plant transgenic. Spores of the soil bacterium *Bacillus thuringiensis* (Bt) contain a crystal (cry) protein (δ-endotoxin). Inside insect gut, the crystals break apart and release a toxin that binds to and creates pores in the intestinal lining. Instead of whole gene truncated *cry* gene is used in Bt crops. Figure 6.3 shows the truncated cry gene structure.

6.4.1 MECHANISM OF TOXICITY

Bt gene (also known as *cry* gene) was identified and isolated from Gram-positive bacteria *Bacillus thuringiensis*. The *cry* gene is used in the production of insect resistant crops (genetically modified) and biological insecticides as well. *Bacillus thuringiensis*, during sporulation produces a toxic protein which possesses insecticidal activity against Lepidopteron, Coleopteron, Hymenopterans, Dipterans, and Nematodes. This crystal protein is called as Cry protein, encoded by cry gene present on the plasmid (nonchromosomal gene). As soon as the Cry protein crystals reach the digestive tract of the insect, the prevailing alkaline condition, there causes denaturation of the insoluble crystals. This denaturation makes the crystal soluble and prone to proteases activity in the gut of insect. Proteolysis of Cry crystal leads to release of cry toxin, which forms pore in the cell membrane of the gut by inserting themselves into it. The pore causes cell lysis and ultimately death of insects.

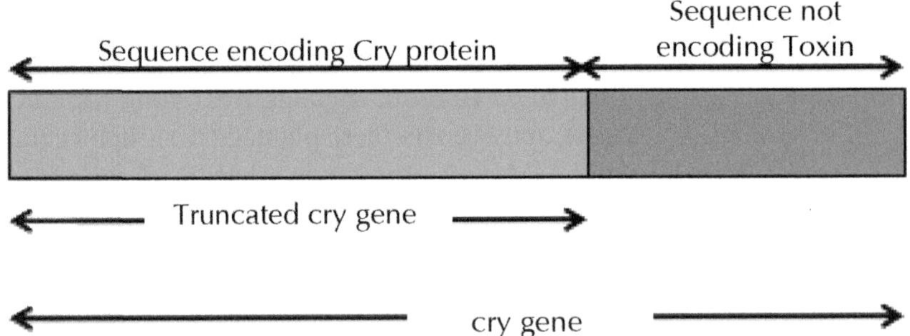

FIGURE 6.3 Gene sequence showing the truncated cry gene. (Reprinted from http://nptel.ac.in/courses/102103013/module6/lec1/3.html)

6.4.2 *CONTROL OF GENE EXPRESSION*

The level of gene expression is determined by regulatory sequences such as promoters as well as 5′ UTR elements. Transgene promoters: Most commonly used is the CaMV 35S promoter of cauliflower mosaic virus. It is a constitutive promoter (turned on all the time in all tissues) that gives high levels of expression in plants. Most commonly used terminator sequence is the nopaline synthase (*nos*) gene from *A. tumefaciens.* Figure 6.4 shows expression cassette of *Bt* gene along with promoter and terminator.

6.5 METHODS OF GENE TRANSFER IN PLANTS

Development of transgenic plants depends on the availability of procedures of plant transformation. Mainly two types of effective gene transfer methods to plants are there, the first one is based on the use of *Agrobacterium* as a biological vector also called as natural genetic engineer and the second is based on the use of physical, electrical, or chemical treatments to introduce isolated DNA into cells alleviating the need for vector use. The latter techniques are commonly termed direct gene transfer methods. Flow chart of gene transfer is given in Figure 6.5.

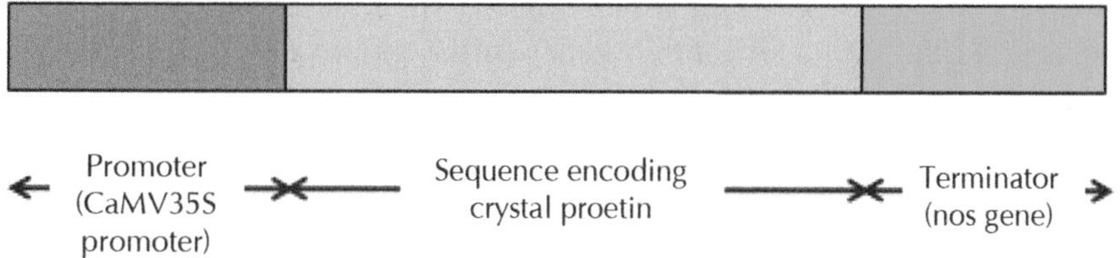

FIGURE 6.4 Expression cassette of *Bt* gene along with promoter and terminator. (Reprinted from http://nptel.ac.in/courses/102103013/module6/lec1/4.html)

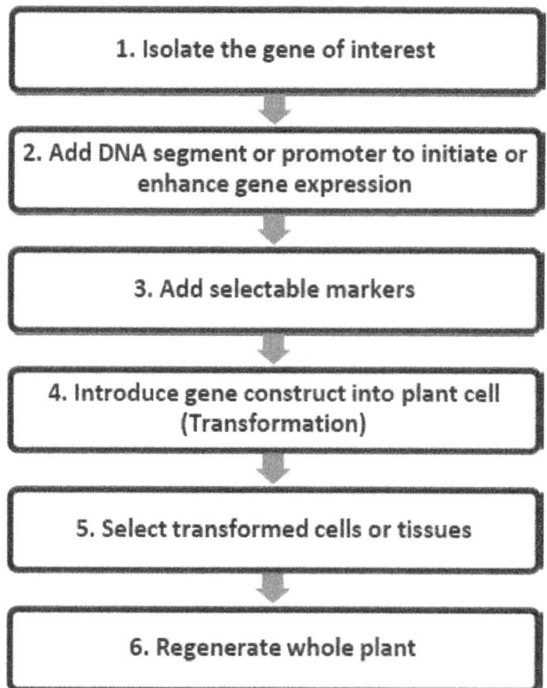

FIGURE 6.5 Steps involved in production of transgenic plants. (Reprinted from http://nptel.ac.in/courses/102103013/module6/lec1/3.html)

6.5.1 INDIRECT GENE TRANSFER USING AGROBACTERIUM AS VECTOR

The most widely used method for the introduction of new genes into plants is based on the natural DNA transfer capacity of *A. tumefaciens*. In nature, this soil bacterium causes tumor formation (called crown gall) on a large number of dicotyledonous plant species. During this infection, a part of the Ti-plasmid of *Agrobacterium*, called T-DNA, is transferred and integrated into the plant genome. The natural capacity of these bacteria made us to use as a natural vector of foreign genes (inserted into the Ti-plasmid) into plant chromosomes. *Agrobacterium*-based and direct gene transfer techniques were developed in parallel, but the former is today the most widely used method because of its simplicity and efficiency in many plants, although it still having limitations in terms of the range of species which are amenable to transformation. These limitations are due to the natural host range of *Agrobacterium*, which generally infects herbaceous dicotyledonous species most efficiently and is less effective on monocotyledonous and woody species. In these plants, direct gene transfer techniques offer the means to establish transformation systems but many of these techniques suffer from a relatively low efficiency of transformation. Attempts are therefore being made to exploit and adapt the relatively simple and convenient *Agrobacterium* system to transform recalcitrant plant species. Recent work has shown that these host-range limitations are not absolute and by developing specific plant cell culture procedures and defining inoculation and co-cultivation conditions, some important nonhost species have now been stably transformed by *Agrobacterium*, although there are still many plant species for which *Agrobacterium* transformation is not usable. The development of reliable transformation protocols for recalcitrant species depends on the establishment of an efficient regeneration procedure, a high transformation rate of the regenerable cells, and an effective selection for regenerating transformed cells (Gheysen et al., 1998). The plant genotype is an important factor, which determines both the regeneration capacity and the efficiency of *Agrobacterium* transformation. Equally important is the choice of the bacterial strain and the external conditions during the preculture and co-cultivation of agrobacteria and plant material. The *Agrobacterium* transformation methods are using two different procedures. The first transformation is dependent upon regeneration of the callus into plants while the second method is free of regeneration. The purpose of the regeneration procedure is twofold: it allows the recovery of uniformly transformed shoots and the selection of such shoots. For many plant species, the lack of suitable regeneration method is one of the main bottlenecks in developing a transforming procedure. A particular regeneration method is usually only efficient with a limited number of genotypes even within a species. Somaclonal variation may also be problematic with some regeneration procedures. Therefore, many efforts have been devoted to the development of regeneration-independent transformation procedures, such as meristem transformation and in planta transformation techniques.

The shoot apex has been used in meristem transformation as an attractive target for transformation since it contains the meristematic cells from which all the aerial parts of the plant are derived. Because meristems are multicellular organs, primary transformants are expected to be chimeric, consisting of transformed and untransformed sectors. This has two important consequences. First, it does not always result in germ-line transformation and transmission of the transgenes to the offspring and second, a stringent selection procedure cannot be applied (Gheysen et al., 1998). These have, as a result, this transformation method to be labor intensive and very inefficient. Several reports clearly give evidence for stable transformation that has been achieved through meristem transformation with *Agrobacterium* infection of important crops, such as *Musa acuminata* (May et al., 1995), *Oryza sativa* (Park et al., 1996). *O. sativa* is the first cereal species that has been stably transformed via *Agrobacterium*. Targeting cells of meristem for transformation has, therefore, the advantage that transformed cell lineages, can be obtained without the involvement of a regeneration pathway which involves dedifferentiation and reorganization of cells, so somaclonal variation is not a problem and the transformation is rather genotype-independent. These are the main advantages of an approach like that. Nevertheless, additional manipulations (e.g., hormonal treatments) are necessary to obtain transformants with acceptable frequencies, reintroducing a factor of genotype dependence in the procedure. More than a decade, several in planta methods for *Agrobacterium*-mediated transformation of *Arabidopsis thaliana* have been developed that do not involve any tissue culture steps. In the first-described procedure (Feldmann and Marks, 1987), imbibed seeds are infected with *Agrobacterium*, allowed to grow into mature plants, and finally transformants were identified among the seeds harvested from these plants. Bechtold et al. (1993) inoculated flowering *A. thaliana* plants by vacuum infiltration with an *Agrobacterium* suspension and managed to get transformants at even higher frequencies. Another technique which has been developed recently (Clough and Bent, 1998) is floral dip. It is a simple dipping of developing floral tissues into an *Agrobacterium* suspension. The absence of any tissue culture step (so somaclonal variation does not occur), the simplicity, and the relatively high efficiency of the transformation procedure would make such techniques attractive to adapt the technique to other plant species, recalcitrant to regeneration procedure.

6.5.2 DIRECT GENE TRANSFER

The development of novel direct gene transfer methodology, bypassing limitations imposed by *Agrobacterium*-host specificity and cell culture constraints, has allowed the engineering of almost all major crops, including formerly recalcitrant cereals, legumes, and woody species. Direct gene transfer transformation methods are independent of species and genotype in terms of DNA delivery, but their efficiency is influenced by the type of target cell, and their utility for the production of transgenic plants in most cases depends on the ease of regeneration from the targeted cells, as most methods operate on cells cultured in vitro. As direct gene transfer referred methods such as particle bombardment, DNA uptake into protoplasts, treatment of protoplasts with DNA in the presence of polyvalent cations, fusion of protoplasts with bacterial spheroplasts, fusion of protoplasts with liposomes containing foreign DNA, electroporation-induced DNA uptake into intact cells and tissues, silicon carbide fiber-induced DNA uptake, ultrasound-induced DNA uptake, microinjection of tissues and cells, electrophoretic DNA transfer, exogenous DNA application and imbibition, macroinjection of DNA (Barcelo and Lazzeri, 1998; Walden and Schell, 1990). The most significant direct gene transfer methods are presented in Table 6.2. Most workers in transgenic plant research are interested primarily in applying a transformation technique rather than in its mechanism of operation, so there is a general wish for technically simple methods which are easily transferred between laboratories and which ideally do not require expensive, specialized equipment. Of the above direct gene transfer techniques, particle bombardment and protoplast transformation are today the most widely used. The former most closely satisfies the criteria of technical simplicity and reproducibility, although it requires a specialized particle gun, the commercial version of which uses relatively expensive consumables. Protoplast transformation can be highly efficient, but demands more complicated cell culture techniques and is limited by the difficulty of regenerating plants. Tissue electroporation is relatively simple, applicable to regenerable tissues and has produced stably transformed plants in several systems after only a relatively short period of development. These results suggest the method should receive further attention to evaluate its potential for wider application. Ultrasound and silicon carbide fiber-mediated techniques are newer methods, which are again technically quite simple.

Microinjection and laser-mediated transformation are specialized techniques, which are at present inefficient. Electrophoretic transfer to date does not give us evidence that the gene transfer actually occurs. Whole-plant direct gene transfer methods would be methods of choice for most users, but despite several claims of high transformation efficiencies most critical studies have not produced evidence for integrative transformation (Barcelo and Lazzeri, 1998).

TABLE 6.2 The Most Significant Methods of Direct Gene Transfer.

Methods	Description
Particle bombardment	Delivery of DNA into cells using microscopic gold or tungsten particles coated with DNA as carriers accelerated into target cells by gunpowder, gas or air pressure or by electrical discharge (Christou, 1995)
Protoplast transformation	DNA introduction into protoplasts using PEG-mediated DNA uptake and electroporation, liposome (containing plasmid DNA) fusion (Caboche, 1990; Krens et al., 1982)
Tissue electroporation	Transformation of plant organs or regenerable cell cultures (D' Halluin et al., 1992; Li et al., 1991)
Ultrasound-induced transformation	DNA uptake into protoplasts, suspension cells, and tissues induced by ultrasound waves (Joersbo, 1990; Joersbo and Brunstedt, 1990; Zhang et al., 1991)
Silicon carbide fiber or whisker transformation	The fibers perforate cell walls and allow DNA to penetrate the cells (Frame et al., 1994)
Laser-mediated transformation	Laser beams are used to create openings in cell components and organelles allowing DNA insertion (Weber et al., 1989)
Microinjection	Direct delivery of DNA into plant cells using a microsyringe (Schnorf et al., 1991)
Macroinjection	The injection of plasmid DNA (uncloned native DNA) into the lumen of developing inflorescence using a hypodermic syringe is called macroinjection
Lipofection	Introduction of DNA into cells via liposomes is known as lipofection. Lipofection is the method of choice for DNA delivery into animal cells cultured in vitro

Particle bombardment delivery of DNA into cells using microscopic gold or tungsten particles coated with DNA as carriers accelerated into target cells by gunpowder, gas or air pressure, or by electrical discharge (Christou, 1995). Protoplast transformation DNA introduction into protoplasts using PEG-mediated DNA uptake and electroporation, liposome (containing plasmid DNA) fusion (Caboche, 1990; Krens et al., 1982). Tissue electroporation:

transformation of plant organs or regenerable cell cultures (D' Halluin et al., 1992; Li et al., 1991), ultrasound-induced transformation DNA uptake into protoplasts, suspension cells, and tissues induced by ultrasound waves (Joersbo, 1990; Joersbo and Brunstedt, 1990; Zhang et al., 1991); silicon-carbide fiber or whisker transformation: the fibers perforate cell walls and allow DNA to penetrate the cells (Frame et al., 1994); laser-mediated transformation: laser beams are used to create openings in cell components and organelles allowing DNA insertion (Weber et al., 1989).

Microinjection direct delivery of DNA into plant cells using a microsyringe (Schnorf et al., 1991). Plant genetic engineering is now at a crucial crossroad. The gene transfer constraints appear to have been removed from a number of important crops. Technical problems still remain, but they are not insurmountable. The attention of the scientific community is gradually shifting to other areas such as identification and cloning of genes responsible for multigenic traits. The study of genomes (known as "genomics") involves the mapping, sequencing, and analysis of genomes in order to determine the structure and function of every gene in an organism. This has already been accomplished in several microorganisms and much effort has been devoted to the complete sequencing of the genome of higher eukaryotes including plants. Information derived from analysis of such data will be used to map entire biochemical pathways, which will be then easier to transfer and incorporate in transgenic organisms. Thus, genomic information can be used to improve important plant traits through genetic engineering, such as high and stable yield and product quality. Utilization of genomics in transgenic technology will also require establishment of routine techniques for simultaneous multiple gene transfer in a single transformation event. In most cases, one or a few genes are transferred to the plant genome along with a selectable marker that facilitates selection of transgenic tissues. Genetic transformation with a single target gene has been used for the production of transgenic crop plants that expressing herbicide tolerance, resistance to fungal, viral, and bacterial diseases and insect pests. In addition, improved agronomic characteristics have been achieved by manipulating metabolic pathways through overexpression of a specific gene or the use of antisense sequences. As most agronomic characteristics are polygenic in nature plant genetic engineering will require manipulation of complex metabolic or regulatory pathways involving multiple genes or gene complexes. Redirecting complex biosynthetic pathways and modifying polygenic agronomic traits requires the integration of multiple transgenes into the plant genome, while ensuring their stable transgenic crops: recent developments and prospects, inheritance, and expression in succeeding generations. Transfer of multiple genes via *Agrobacterium*-mediated transformation although possible is technically demanding and becomes increasingly problematic as the number of genes and the size of the transferred DNA increases. But this transfer could be achieved through cobombardment in a simple process in which genes carried on separate plasmids are mixed prior to transfer by particle bombardment. In this manner, numerous genes can be transferred simultaneously using a single selectable marker (Chen et al., 1998). This will certainly be one of the major future goals of transgenic technology in plants.

6.6 CHARACTERIZATION OF TRANSFORMANTS

The presence and activity of introduced gene are confirmed by observation of phenotype and by advanced methods as listed below.

Southern blot: In Southern blotting, separated DNA fragments obtained after electrophoresis are transferred to a filter membrane and subsequent fragment detection is accompanied by probe hybridization.

Northern blot: It is used to study gene expression by detection of mRNA (or isolated mRNA) in a plant sample.

Western blot: It is used to study the gene expression by detection of the protein produced by the transformed regenerated plants.

Real-time PCR: It is used to detect the copy number of transgene as well as expression level of transformed gene.

6.6.1 EVALUATION OF TRANSFORMED PLANTS

Plant is evaluated for the presence and activity of introduced gene. The effect of various environmental factors on the transgenic plant is also evaluated. Evaluation for food or feed safety and evaluation of basic containment level is also very important.

6.7 ACHIEVEMENTS AND FUTURE PROSPECTS

Through conventional plant breeding, genes can be transmitted only by crossing in the same or closely related species. Transgenic techniques have allowed genetic material to be transferred between completely unrelated organisms, so that breeders can incorporate characteristics that are not normally available within a species. The modified organisms exhibit properties that would be impossible to obtain by conventional breeding techniques. Modern biotechnology makes plant breeding programs more effective in two important ways. First, it allows transfer of specific genes, incorporating into the new variety only those traits that are wanted. This makes the process of trait transfer faster, more exact, cheaper, and less likely to fail than traditional crossbreeding methods. Second, it gives breeders the freedom to incorporate genes from unrelated species into the target plant, a possibility that is unprecedented in plant breeding. Transgenic methods have been employed over the last 15 years in a number of important crop plants such as maize, cotton, soybean, oilseed rape, and a variety of vegetable crops like tomato, potato, cabbage, and lettuce. In European Union, 1255 field tests involving transgenic plant varieties have been approved. This number has surpassed 5000 field releases (permits and notifications) in the United States. The commercial production of transgenic crops shows a rapid increase the last few years. The global area (excluding China) of transgenic crops from zero in 1995 has reached 27.8 million hectares in 1998.

Global area cultivated with commercial transgenic crops (excluding China) from 1995 to 1998. The distribution of the area between industrial and developing countries is also shown. During this period, the larger part of global transgenic crops has been grown in industrial countries with significantly less in developing countries. The proportion of transgenic crops grown in industrial countries in 1998 was 84%, slightly less than 1997 (86%), and only 16% grown in the developing countries, with most of that area in Argentina, and the balance in Mexico and South Africa.

6.8 TRANSGENIC CROPS: RECENT DEVELOPMENTS AND PROSPECTS

The major achievements of transgenic plant technology up to now concern tolerance to insect or disease pests, herbicide tolerance, and improved product quality. A description of the major categories of modified traits with characteristic examples will follow.

6.8.1 ABIOTIC STRESSES

Abiotic stresses such as drought, salinity, high or low temperature are major environmental stress factors that adversely affect plant growth and productivity. Genetic engineering is an attractive approach with the potential to improve plant abiotic stress tolerance. The overexpression of enzymes required for glycine betaine (GlyBet) biosynthesis in transgenic plants improve tolerance to various abiotic stresses. Chloroplast transformation has been used to transfer choline monooxygenase (BvCMO), an enzyme that catalyzes the conversion of choline into betaine aldehyde from beet (*Beta vulgaris)* into the plastid genome of tobacco. Transplastomic carrot plants expressing betaine aldehyde dehydrogenase (BADH) gene have also been developed using chloroplast engineering showing highest level of salt tolerance. A large fraction of the world's irrigated crop land is so laden with salt that it cannot be used to grow most important crops. However, researchers at the University of California Davis campus have created transgenic tomatoes that grow well in saline soils. The transgene was highly expressed sodium/proton antiport pump-sequestered excess sodium in the vacuole of leaf cells. There was no sodium buildup in the fruit.

6.8.2 INSECT RESISTANCE

New varieties of maize, cotton, and tobacco, for example, have been developed utilizing a gene from the bacterium *Bacillus thuringiensis* to produce a protein (the Bt protein) that is specifically toxic to certain insect pests including bollworm, but not to animals or humans (Carozzi et al., 1992; Liang et al., 1994). This protein has been used as a pesticide spray for many years. Cultivation of these transgenic plants should help reduce the use of chemical pesticides in cotton production, as well as in the production of many other crops, which could be engineered to contain the *Bacillus thuringiens* gene. *Bt* crops *Bacillus thuringiensis* is a soil-dwelling bacterium which was isolated by Ernst Berliner and it produces proteinaceous crystalline (Cry) inclusions during sporulation which are insecticidal (mainly against lepidopterans), but nontoxic to nontarget organisms including human and animals.

Cry proteins once ingested are solubilized in the alkaline environment of insect mid-gut and then undergoes proteolytic cleavage generating an active toxin of 65–70 kDa which binds specifically to insect mid-gut epithelial cell receptors resulting in cell lysis and finally death (Gahan et al., 2010). The prerequisite for alkaline environment, specific proteases, and receptors explains the nontoxicity of Bt to mammals (which have an acidic gut and lack the corresponding receptors) and why each toxin has a narrow host range (Sanahuja et al., 2011). Since bacterial cry genes are rich in A/T content compared to plant genes, they have had to undergo considerable modification of codon usage and removal of polyadenylation sites before successful expression in plants (de Maagd et al., 1999). Plants engineered to express Cry proteins were first reported in 1987 (Vaeck et al., 1987). *Bt* genes that encode insecticidal Cry proteins have been successfully transferred to important crop plants and a list of few transgenic plants with resistant to insect pests. In 2013, the global area of GM crops planted for commercial purposes was 175.2 million hectares, out of which, 23 million hectares were allocated to *Bt* crops and 47 million hectares to stacked traits (herbicide resistant and *Bt* crops) (James, 2013).

Bt maize has been transformed with either *cry1Ab*, *cry1Ac*, or *cry9C* to protect it against *Ostrinia nubilalis* and *Sesamia nonagriodes*, or with *cry1F* to protect it against *Spodoptera frugiperda*, and with *cry3Bb*, *cry34Ab*, and *cry35Ab* to protect it against the rootworms of the genus *Diabrotica* (James, 2012). By the end of the year 2012, more than 18 million hectares were under the cultivation of *Bt* cotton plants. Most commercially planted Bt cotton contains *cry1Ac* or a fusion gene of *cry1Ac* and *cry1Ab* (James, 2013). *Bt* potatoes protected against *Leptinotarsa decemlineata* have also been planted commercially in North America and Europe and contain the *cry3Aa* gene.

Bt eggplant is another crop which was targeted for control of *Leucinodes orbonalis* and commercialized in India in 2008. *Bt* crucifer vegetables are under development and are targeted against *Plutella xylostella* (James, 2012). *Bt* rice expressing the *Bacillus thuringiensis* toxin is expected to be commercially released in the future (James, 2012). Several GM rice varieties have entered and passed field and environmental release trials, and four varieties entered preproduction trials in farmers' fields in 2001. Also, *Bt* alfalfa has been produced using *cry3a* gene against *Hypera postica* for the first time in Iran (Tohidfar et al., 2013). Finally, the *Bt* trait has been introduced in soybean through either one or two cry genes among *cry1A*b, *cry1Ac*, and *cry1F* (James, 2013).

6.8.3 DISEASE RESISTANCE

Tobacco mosaic virus causes the leaves of some important crop plants to wither and die. Incorporation into the plant of a gene that encodes the coat protein of the virus protects it from disease (Clark et al., 1995). This approach has also been applied to other viral diseases in crops. More progress in development of disease resistant transgenic plants will be seen in the near future. Over the past decade, many efforts were focused on understanding plant–pathogen interactions in molecular terms. This led to the identification of disease resistant plant genes that specify race-specific resistance to pathogens. The tomato disease resistant gene *Pto*, for example, confers resistance to the bacterial pathogen. *Pseudomonas syringae* pv. *tomato* carrying the *avrPto* gene. Recently, Tang et al. (1999) reported that overexpression of the *Pto* gene in transgenic tomato plants activated defense responses and conferred broad resistance to several bacterial pathogens.

6.8.4 HERBICIDE TOLERANCE

Engineering herbicide tolerance in transgenic plants has been accomplished exploiting at least three different mechanisms: overexpression of the target enzyme, modification of the target enzyme, and herbicide detoxification (Tsaftaris, 1996). Examples of transgenic plants developed based on each mechanism are following. Glyphosate is an environmentally more benign, widely used broad-spectrum herbicide. It is easily degraded in the agricultural environment and works by interfering with the EPSPS enzyme system that is present only in plants. Unfortunately, the herbicide kills crop plants as well as weeds. Transgenic plants including maize, soybean, and cotton have been developed, overexpressing an additional copy of the EPSPS gene from *Petunia hybrid* under the strong 35S promoter and exhibiting increased tolerance to glyphosate. Alternatively, expression of a mutant *Aro A* gene from *Salmonella typhimurium* (which encodes EPSPS) in transgenic tobacco resulted in even higher tolerance to the herbicide than overexpression of the wild-type petunia EPSPS gene (for review, see Tsaftaris, 1996). This allows farmers to control weeds in transgenic cultivars spraying with glyphosate alone. A different approach has been applied for development of resistance to the herbicide phosphinothricin (basta). The *bar* gene from *Streptomyces hygroscopicus* or *S. uiridochromogenes* encodes the enzyme phosphinothricin acetyl transferase (PAT), which converts the herbicide to a nontoxic acetylated form. Expression of the *bar* gene in transgenic tobacco, potato, and tomato plants conferred phosphinothricin resistance at up to 10 normal application rate of the hebicide in the field (Wohlleben et al., 1988). Questions have been raised about the safety both to humans and to the environment of some of the broad-leaved weed killers like 2,4-D. Alternatives are available, but they may damage the crop as well as the weeds growing in it. However, genes for resistance to some of the newer herbicides have been introduced into some crop plants and enable them to thrive even when exposed to the weed killer.

6.8.5 PRODUCT QUALITY

Transgenic technologies have been used to modify other important characteristics of plants such as starch composition in potato (Lorberth et al., 1998; Takaha et al., 1998), ripening in tomato (Smith et al., 1990), lignin content in arabidopsis (Ni et al., 1994), flower vase-life in carnation (Bovy et al., 1995), and explore many new possibilities for uses in agriculture as well as in industry. Milled rice is the staple food for a large fraction of the world's human population. Milling rice removes the husk and any beta-carotene it contained. Beta-carotene is a precursor to vitamin A, so it is not surprising that vitamin A deficiency is widespread, especially in the countries of Southeast Asia. The synthesis of beta-carotene requires a number of enzyme-catalyzed steps. In January 2000, a group of European researchers reported that they had succeeded in incorporating three transgenes into rice that enabled the plants to manufacture beta-carotene in their endosperm.

6.8.6 BIOPHARMACEUTICALS

The genes for proteins to be used in human (and animal) medicine can be inserted into plants and expressed by them.

Advantages:

- Glycoproteins can be made (bacteria like *E. coli* cannot do this).
- Virtually unlimited amounts can be grown in the field rather than in expensive fermentation tanks.
- It avoids the danger from using mammalian cells and tissue culture medium that might be contaminated with infectious agents.
- Purification is often easier.

Corn is the most popular plant for these purposes, but tobacco, tomatoes, potatoes, rice, and carrot cells grown in tissue culture are also being used.

Some of the proteins that have been produced by transgenic crop plants:

- human growth hormone with the gene inserted into the chloroplast DNA of tobacco plants;
- humanized antibodies against such infectious agents as

- HIV
- respiratory syncytial virus
- sperm (a possible contraceptive)
- herpes simplex virus, HSV, the cause of "cold sores"
- Ebola virus, the cause of the often-fatal Ebola hemorrhagic fever.

• Protein antigens to be used in vaccines

- an example, patient specific antilymphoma (a cancer) vaccines. B-cell lymphomas are clones of malignant B cells expressing on their surface a unique antibody molecule. Making tobacco plants transgenic for the RNA of the variable (unique) regions of this antibody enables them to produce the corresponding protein. This can then be incorporated into a vaccine in the hopes (early trials look promising) of boosting the patient's immune system, especially the cell-mediated branch to combat the cancer
- other useful proteins like lysozyme and trypsin.
- However, as of April 2012, the only protein to receive approval for human use is glucocerebrosidase, an enzyme lacking in Gaucher's disease. It is synthesized by transgenic carrot cells grown in tissue culture.

6.9 FUTURE PROSPECTS

Major achievements of plant biotechnology are presently limited to traits involving one or a few genes. It will probably require more research before we can manipulate complex traits (such as yield) that are influenced by many genes. However, with newly developed techniques, we can now incorporate multiple genes in plant genomes integrating multiple traits. In addition, advances in structural and functional analysis of higher plant genomes will provide substantial knowledge on important biochemical pathways that are involved in the regulation of more complex characters. This could even enable scientists to identify and transfer entire biochemical pathways from one species to another and incorporate them into new hosts for the benefit of agriculture and/or industry. Eventually, it may also be possible to develop crops for nonfood uses by modifying traits to make them more suitable for industrial purposes, or to use plants rather than animals to make antibodies for medical and agricultural diagnostic purposes, and delivering vaccines with food in developing countries. Current research will see the improvement and development of crops for specific purposes. Plants that require less water could be developed for countries with arid climates. Crop plants engineered to be tolerant to salt could be farmed in salt-damaged farmland or could be irrigated with salty water. Crops with higher yields and higher protein values are also possible. Much current research focuses on understanding and developing useful promoter sequences to control transgenes, and establishing precise methods to insert and place the transgene at specific locations in the recipient chromosomes. Much still needs to be done to improve our knowledge of specific genes and their actions, the potential side effects of adding foreign DNA and of manipulating genes within an organism and the problems associated with transgene silencing.

6.10 PHENOTYPIC STABILITY OF TRANSGENIC VARIETIES

Transgenic crop plants will only be of value if their phenotype is stable in the field and transmitted faithfully in subsequent generations. Although it is possible to study transgenes with a high level of precision, there is often uncertainty related with inactivation and structural instability of the transgenes. This inactivation is well documented and is most frequently correlated with gene silencing and not loss of the transgene. Gene silencing in transgenic plants has been identified as a major obstacle in transgenic technology. Reversal to herbicide sensitivity, for example, of transgenic plants bred for tolerance to herbicides could lead to significant loses. From the applied side, gene silencing has come as an unwelcome surprise and is turning out to be a substantial problem. According to Finnegan and McElroy (1994), of 30 companies polled, nearly all reported some problems with unwanted silencing of transgenes. It has been shown that this unwelcome sensitivity due to inactivation of the transgene mediated by methylation is triggered by stresses like the common agronomic practice of seedling transplantation in the field (Brandle et al., 1995). Thus, it requires closer attention since many crops need transplantation in the field and this imposes a severe stress for the young plantlet. It is clear that some steps that were taken for granted may need to be further investigated for successful commercialization of transgenic crops.

6.11 RISK ASSESSMENT AND THE REGULATORY FRAMEWORK

6.11.1 RISK ASSESSMENT

Advances in transgenic technology bring new responsibilities for safe use of transgenic plants for the benefit of humanity and the environment. The objective in risk assessment is to develop safety procedures, proactively rather than reactively. Safety issues are scale dependent and are probably different in small-scale experimental field trials than in large-scale commercial releases. Long-term effects of transgenic plants and their products may be only detectable after large scale or even commercial production of transgenic crops. Thus, one major challenge concerning safety of transgenic organisms is to develop procedures to assess long-term effects on human health and the environment. Other issues have also been emerged that could be considered in risk assessment. Competitiveness, jobs, and investment are thought to be at risk if the technology is not adopted, and potential benefits lost. Another dimension is the issue of consumer choice and rights to reject the technology at the point of sale. In large part, this has been the focus of the growing consumer movement in all parts of the world, and especially in Europe. In addition, much of the commercialization of the new technology relies on a few international companies with capital and power to dominate in the forming market of transgenic plants. The future may well find just a few key multinational industries active in producing recombinant plants to manage plant diseases and to produce agricultural and industrial products leaving the developing countries (which could benefit the most from the new technology) well behind. This can be alleviated if laboratories from developing countries along with advanced agro-biotech companies share knowledge and technology in a network, aiming for the development of improved transgenic varieties of crops that may be of minor commercial value for the companies, but critical as source of food or agricultural income for the developing countries. A summary of possible risks for human health and the environment, associated with transgenic plants is given below.

6.11.2 RISKS FOR HUMAN HEALTH

- Formation of new allergens from the novel proteins expressed in the transgenic organism, which could trigger allergic reactions at some stage.

- Creation of new toxins through unexpected interactions between the product of the genetic modification and other endogenous constituents of the organism.
- Dispersion of antibiotic resistance genes used as markers from the genetically modified organism derived food to gut microorganisms and intensification of problems with antibiotic resistant pathogens.

6.11.3 RISKS FOR THE ENVIRONMENT

- Gene transfer from the transgenic plant to related species as a result of hybridization that could lead to new pests.
- The transgenic plant escapes its intended use and becomes an invader to the natural environment.
- Harmful effects on nontarget species with the expression, for example, of insecticide toxins that can kill beneficial as well as targeted insects.
- Development of resistance from the continuous use of the same agent on the target organism.
- Harmful effects on ecosystems when transgenic plant products interfere with natural biochemical cycles.
- Harmful effects on biodiversity if a transgene offers an adaptive advantage in transgenic plants escaped in the area of cultivation or in wild relatives where it could be transferred by cross-fertilization. This is practically important if occurring at the centers of genetic variation of cultivated plants. In addition, biodiversity concerns have been raised for current cultivation systems including many locally adopted varieties if they will be substituted by a few new transgenics. The process of examining the above risks from the release of a transgenic plant to the environment can provide a framework for risk assessment. Of course, enumeration and listing all the above major questions or possible risks expressed from scientists, consumers, and ecological groups for field testing and commercialization of transgenic plants does not imply that all the above are a concern for all the different transgenics and in different environments. For example, a risk for possible new allergenicity to the consumer could be meaningful, thus requiring testing prior to release, in cases where new genes coding for possibly new allergenic compounds have been cloned into plants. This question should not concern transgenics without such kind of genes cloned.

6.12 TERMINATOR GENES TECHNOLOGY

This term is used (by opponents of the practice) for transgenes introduced into crop plants to make them produce sterile seeds (and thus force the farmer to buy fresh seeds for the following season rather than saving seeds from the current crop).

The process involves introducing three transgenes into the plant:

- A gene encoding a **toxin** which is lethal to developing seeds but not to mature seeds or the plant. This gene is normally inactive because of a stretch of DNA inserted between it and its promoter.
- A gene encoding a **recombinase**—an enzyme that can remove the spacer in the toxin gene thus allowing to be expressed.
- A **repressor** gene whose protein product binds to the promoter of the recombinase thus keeping it inactive.

How it works

When the seeds are soaked (before their sale) in a solution of tetracycline

- Synthesis of the repressor is blocked.
- The recombinase gene becomes active.
- The spacer is removed from the toxin gene and it can now be turned on.

Because the toxin does not harm the growing plant, only its developing seed, the crop can be grown normally except that its seeds are sterile.

The use of terminator genes has created much controversy:

- Farmers especially those in developing countries want to be able to save some seed from their crop to plant the next season.
- Seed companies want to be able to keep selling seed.

6.13 THE REGULATORY FRAMEWORK

The criteria and factors that determine biosafety assessment of transgenic plants vary in different countries. In European Union, all plants produced by genetic modification must be assessed (technology based assessment), whereas in the United States and Canada only plants modified with particular genes are regulated (product based assessment). There is considerable debate on the safety guaranteed by the two approaches. The use of *Agrobacterium* as a vector for the transformation process implies that the transgenic plants produced will be regulated under both approaches. However, as other methods avoid *Agrobacterium* and sequences derived from plant pathogens, there is considerable difference on the regulatory requirements between North America and Europe. Time and scale differences in commercialization of transgenic crops will also have an effect making one system more "experienced" than the other. In United States, three agencies share the primary responsibility for regulating the genetically modified organisms, whether they be designed for closed systems or for environmental uses. These are the USEPA, the FDA, and the USDA. In addition to federal regulation, several states and municipalities have enacted biotechnology-related legislation, including provisions related to the environmental release of genetically modified organisms. Each of the federal agencies regulating biotechnology is guided in its analysis and decision-making criteria by its specific legislation, that is, the lows passed by Congress charging each agency with specific responsibilities. These laws differ in their mandate as to what populations to consider with regard to adverse effects (e.g., humans, crops, the environment), as well as in their mandate as how to strike a balance between risks and benefits. In addition to its specific legislation, each agency must also adhere to the National Environmental Policy Act, which is binding on all federal agencies.

6.13.1 TRANSGENIC CROPS: RECENT DEVELOPMENTS AND PROSPECTS

In European Union, the regulatory framework on agricultural biotechnology is made of a few European Directives and Regulations (Vega et al., 1999). Directive 90/219 issues the regulations covering the contained use of genetically modified organisms and Directive 90/220 issues the regulations covering the deliberate release into the environment of genetically modified organisms. Directive 90/219 has recently been totally revised. The European Commission also presented a proposal (COM/98/0085) for amending the Directive 90/220 so as to harmonize European approaches to the issue. Several Member States have refused to approve commercialization of transgenic plants approved in other Member States in their territories and others have called for a moratorium. The European Parliament's Committee on the Environment, Public Health, and Consumers proposed a Europe-wide moratorium on all transgenic crops awaiting authorization to be placed in the market. The Council Regulation (EC) No. 258/97 regulates the compulsory indication of the labeling of certain foodstuff produced from genetically modified organisms. The Directive on the legal protection of biotechnological inventions 98/44/EEC regulates issues of intellectual property. For more details on risk assessment and the regulatory framework pertaining transgenic plants, see the chapter by Dr. John Beringer in this volume. Risk assessment needs to have an international dimension and extent beyond the primary country of release or a shared international boundary or a market. International harmonization of regulations and procedures for production, testing, and handling transgenic plants must be a major challenge. Obtaining good scientific data for the long-term effects to the environment will be critical for passing these products from the regulatory framework. This will be even more critical for plants tolerant to different biotic and abiotic stresses. Progress in meeting this challenge will be highly dependent on:

1. how the questions will be formulated and
2. the amount and kind of effort that will be devoted toward this goal.

The role of international organizations toward that goal can be critical and meetings and discussions to facilitate this are of immense importance.

KEYWORDS

- **transgenic**
- **gene expression**
- **genetic engineering**
- **conventional breeding approach**

REFERENCES

Arencibia, A. D.; Carmona, E. R. C.; Tellez, P.; Chan, M. T.; Yu, S. M.; Trujillo, L. E.; Oramas, P. An Efficient Protocol for Sugarcane (*Saccharum* spp. L) Transformation Mediated by *Agrobacterium tumefaciens*. *Transgenic Res.* **1998,** *7*, 213–222.

Barcelo, P.; Lazzeri, P. Direct Gene Transfer: Chemical, Electrical and Physical Methods. In: *Transgenic Plant Research*; Lindsey, K., Ed.; Harwood Academic Publishers: Amsterdam, 1998.

Bechtold, N.; Ellis, J.; Pelletier, G. In Planta Transformation Mediated Gene Transfer by Infiltration of Adult *Arabidopsis* Plants. *C. R. Acad. Sci., Ser. III* **1993,** *316*, 1194–1199.

Bovy, A. G.; van Altvorst, A. C.; Angenent, G. C.; Dons, J. J. M. Genetic Modification of the Vase-Life of Carnation. *Acta Hortic.* **1995,** *405*, 179–189.

Brandle, J. E.; McHugh, S. G.; James, L.; Labbe, H.; Miki, B. L. Instability of Transgene Expression in Field Grown Tobacco Carrying the *csr1-1* Gene for Sulfonylurea Herbicide Resistance. *Biotechnology* **1995,** *13*, 994–998.

Caboche, M. Liposome-Mediated Transfer of Nucleic Acids in Plant Protoplasts. *Physiol. Plant.* **1990,** *79*, 173–176.

Carozzi, N. B.; Warren, G. W.; Desai, N.; Jayne, S. M.; Lotstein, R.; Rice, D. A.; Evola, S.; Koziel, M. G. Expression of a Chimeric CaMV 35S *Bacillus thuringiensis* Insecticidal Protein Gene in Transgenic Tobacco. *Plant Mol. Biol.* **1992,** *20*, 539–548.

Chen, L.; Marmey, P.; Taylor, N. J.; Brizard, J.-P.; Espinoza, C. D; Cruz, P.; Huet, H.; Zhang, S.; de Kochko, A.; Beachy, R. N.; Fauquet, C. M. Expression and Inheritance of Multiple Transgenes in Rice Plants. *Nat. Biotechnol.* **1998,** *16*, 1060–1064.

Cheng, M.; Fry, J. E.; Pang, S.; Zhou, I.; Hironaka, C.; Duncan, D. R. I.; Conner, T. W. L.; Wang, Y. Genetic Transformation of Wheat Mediated by *Agrobacterium tumefaciens. Plant Physiol.* **1997,** *115*, 971–980.

Christou, P. Strategies for Variety-Independent Genetic Transformation of Important Cereals, Legumes and Woody Species Utilizing Particle Bombardment. *Euphytica* **1995,** *85*, 13–27.

Clark, W. G.; Fitchen, J. H.; Beachy, R. N. Studies of Coat Protein-Mediated Resistance to TMV. I. The PM2 Assembly Defective Mutant Conifers Resistance to TMV. *Virology* **1995,** *208*, 485–491.

Clough, S. J.; Bent, A. F. Floral Dip: A Simplified Method for *Agrobacterium*-Mediated Transformation of *Arabidopsis thaliana. Plant J.* **1998,** *16* (6), 735–743.

D' Halluin, K.; Bonne, E.; Bossut, M.; De Beuckleer, M.; Leemans, J. Transgenic Maize Plants by Tissue Electroporation. *Plant Cell* **1992,** *4*, 1495–1505.

Dale, P. J.; Irwin, J. A. Environmental Impact of Transgenic Plants. In: *Transgenic Plant Research*; Lindsey, K., Ed.; Harwood Academic Publishers: Amsterdam, 1998.

DANIDA. Assessment of potentials and constraints for development and use of plant biotechnology in relation to plant breeding and crop production in developing countries. Working paper. Ministry of Foreign Affairs, Denmark. 2002.

de Maagd, R. A.; Bosch, D.; Stiekema, W. *Bacillus thuringiensis* Toxin-Mediated Insect Resistance in Plants. *Trends Plant Sci.* **1999,** *4* (1), 9–13.

Engelman, R.; LeRoy, P. *Conserving Land: Population and Sustainable Food Production.* Population Action International: Washington, DC, 1995.

Feldmann, K. A.; Marks, M. D. *Agrobacterium*-Mediated Transformation of Germinating Seeds of *Arabidopsis thalliana*. A Non-Tissue Culture Approach. *Mol. Gen. Genet.* **1987,** *208*, 1–9.

Finnegan, J.; McElroy, D. Transgene Inactivation-Plants Fight Back. *Biotechnology* **1994,** (12) 883–888.

Food and Agriculture Organization (FAO). *International Code of Conduct on the Distribution and Use of Pesticides*. Food and Agriculture Organization: Rome, Italy, 1986.

Frame, B. R.; Drayton, P. R.; Bagnall, S. V.; Lewnau, C. J.; Bullock, W. P.; Wilson, H. M. Production of Fertile Transgenic Maize Plants by Silicon Carbide Fiber-Mediated Transformation. *Plant J.* **1994,** *6*, 941–948.

Gahan, L. J.; Pauchet, Y.; Vogel, H.; Heckel, D. G. An ABC Transporter Mutation is Correlated with Insect Resistance to *Bacillus thuringiensis* Cry1Ac Toxin. *PLoS Genet.* **2010,** *6* (12), e1001248.

Gheysen, G.; Angenon, G.; Van Montagu, M. *Agrobacterium*-Mediated Plant Transformation: A Scientifically Intriguing Story with Significant Applications. In: *Transgenic Plant Research*; Lindsey, K., Ed.; Harwood Academic Publishers: Amsterdam, 1998.

Hiei, Y.; Ohta, S.; Komari, T.; Kumashiro, T. Efficient Transformation of Rice (*Oryza sativa* L.) Mediated by *Agrobacterium* and Sequence Analysis of the Boundaries of the T-DNA. *Plant J.* **1994,** *6*, 271–282.

Ishida, Y.; Saito, H.; Ohta, S.; Hiei, Y.; Komari, T.; Kumashiro, T. High Efficiency Transformation of Maize (*Zea mays* L.) Mediated by *Agrobacterium tumefaciens*. *Nat. Biotechnol.* **1996,** 14, 745–750.

James, C. Global Status of Commercialized Biotech/GM Crops. *Brief No. 44*. ISAAA: Ithaca, NY, 2012.

James, C. Global Status of Commercialized Biotech/GM Crops: 2014. *ISAAA Briefs No. 49*. James, C. Preview: Global Status of Commercialized Biotech/GM Crops: 2013. *ISAAA Briefs No. 46*. ISAAA: Ithaca, NY, 2013.

Joersbo, M.; Brunstedt, J. Direct Gene Transfer to Plant Protoplasts by Mild Sonication. *Plant Cell Rep.* **1990,** *9*, 207–210.

Krens, F. A.; Molendijk, L.; Wullems, G. J.; Schilperoort, R. A. *In Vitro* Transformation of Plant Protoplasts with Ti-Plasmid DNA. *Nature* **1982,** *296*, 72–74.

Li, B.; Xu, X.; Shi, H.; Ke, X. Introduction of Foreign Genes into the Seed Embryo Cells of Rice by Electro-injection and the Regeneration of Transgenic Rice Plants. *Sci. China* **1991,** *34* (8), 923–931.

Liang, X.; Zhu, Y.; Mi, J.; Chen, Z. Production of Virus Resistant and Insect Tolerant Transgenic Tobacco Plants. *Plant Cell Rep.* **1994,** *14*, 141–144.

Lorberth, R.; Ritte, G.; Willmitzer, L.; Kossmann, J. Inhibition of a Starch-Granule Bound Protein Leads to Modified Starch and Repression of Cold Sweetening. *Nat. Biotechnol.* **1998,** *16*, 473–477.

May, G. D.; Afza, R.; Mason, H. S.; Wiecko, A.; Novak, F. J.; Arntzen, C. J. Generation of Transgenic Banana (*Musa acuminata*) Plants via *Agrobacterium*-Mediated Transformation. *Biotechnology* **1995,** *13*, 486–492.

Ni, W. T.; Paiva, N. L.; Dixon, R. A. Reduced Lignin in Transgenic Plants Containing a Caffeic Acid *O*-Methyltransferase Antisense Gene. *Transgenic Res.* **1994,** *3*, 120–126.

Park, S. H.; Pinson, S. R. M.; Smith, R. H. T-DNA Integration into Genomic DNA of Rice Following *Agrobacterium* Inoculation of Isolated Shoot Apices. *Plant Mol. Biol.* **1996,** *32*, 1135–1148.

Raven, P. H. GM Crops, the Environment and Sustainable Food Production. *Transgenic Res.* **2014,** *23* (6), 915–921.

Sanahuja, G.; Banakar, R.; Twyman, R. M.; Capell, T.; Christou, P. *Bacillus thuringiensis*: A Century of Research, Development and Commercial Applications. *Plant Biotechnol. J.* **2011,** *9* (3), 283–300.

Schnorf, M.; Neuhaus-Url, G.; Galli, A.; Iida, S.; Potrykus, I.; Neuhaus, G. An Improved Approach for Transformation of Plant Cells by Microinjection: Molecular and Genetic Analysis. *Transgenic Res.* **1991,** *1*, 23–30.

Sharma, H. C.; Crouch, J. H.; Sharma, K. K.; Seetharama, N.; Hash, C. T. Applications of biotechnology for crop improvement: Prospects and constraints. *Plant Sci.* **2002,** 163. 381–395.

Smith, C. J. S.; Watson, C. F.; Morris, P. C.; Bird, C. R.; Seymour, G. B.; Gray, J. E.; Arnold, C.; Tucker, G. A.; Schuch, W.; Harding, S. Inheritance and Effect on Ripening of Antisense Polygalacturonase Genes in Transgenic Tomatoes. *Plant Mol. Biol.* **1990,** *14*, 369–379.

Takaha, T.; Critchley, J.; Okada, S.; Smith, S. M. Normal Starch Content and Composition in Tubers of Antisense Potato Plants Lacking D-Enzyme (4-alphaglucanotransferase). *Planta* **1998,** *205*, 445–451.

Tang, X.; Xie, M.; Kim, Y-J.; Zhou, J.; Klessing, D.; Martin, G. B. Overexpression of *Pto* Activates Defense Responses and Confers Broad Resistance. *Plant Cell* **1999,** *11*, 15–29.

Tohidfar, M.; Zare, N.; Jouzani, G. S., Eftekhari, S. M. Agrobacterium-mediated transformation of alfalfa (*Medicago sativa*) using a synthetic cry3a gene to enhance resistance against alfalfa weevil. Plant Cell Tiss. Organ Cult. 2013, 113:227–234.

Tsaftaris, A. The Development of Herbicide-Tolerant Transgenic Crops. *Field Crops Res.* **1996,** *45*, 115–123.

Vaeck, M.; Reynaerts, Á.; Hofte, H.; Jansens, S.; De Beukleer, M. D.; Dean, C. Transgenic Plants Protected from Insect Attack. *Nature* **1987,** *328*, 33–37.

Vega, M.; Bontoux, L.; Llobell, A. Biotechnology for Environmentally Safe Agriculture. *IPTS Rep.* **1999,** *31*, 9–13.

Walden, R.; Schell, J. Techniques in Plant Molecular Biology—Progress and Problems. *Eur. J. Biochem.* **1990,** *192*, 563–576.

Weber, G. Monajembashi, S.; Greulish, K. O.; Wolfrum, J. Uptake of DNA in Chloroplasts of *Brassica napus* (L.) by Means of a UV-Laser Microbeam. *Eur. J. Cell. Biol.* **1989,** *49*, 73–79.

Wohlleben, W.; Arnold, W.; Broer, I.; Hillemann, D.; Strauch, E.; Puhler, A. Nucleotide Sequence of the Phosphinothricin *N*-Acetyltransferase Gene from *Streptomyces uiridochromogenes* Tu 494 and its Expression in *Nicotiana tabacum. Gene* **1988,** *70*, 25–37.

Zhang, L.; Cheng, L. Xu, N.; Zhao, N.; Li, C.; Yuan, J.; Jia, S. Efficient Transformation of Tobacco by Ultrasonication. *Biotechnology* **1991,** *9*, 996–997.

CHAPTER 7

GENETICALLY MODIFIED FOODS AND BIODIESEL

SUHAIL MUZAFFAR*

National Centre for Biological Sciences, GKVK Campus, Bellary Road, Bangalore 560065, India

**E-mail: suhail.bt@gmail.com*

CONTENTS

ABSTRACT

Genetically modified foods are derived from microbes, plants or animals whose genetic material has been modified using recombinant DNA technology to enhance the desired traits. Genetic engineering creates novel combinations of plant, animal, and bacterial genomes that are not normally present in nature. Most of the genetically modified crops have been engineered for resistance to pathogens and herbicides and for enhanced nutrient values. Commercial sale of GM foods started in 1994 when a US based company Calgene (now Monsanto) first marketed its Flavr Savr delayed ripening tomato. Most genetic modifications have basically focused on cash crops in high demand including corn, rice, soybean, and cotton. Although there is a scientific consensus that the currently available GM crops pose no serious threat to human or animal health but the public and various non-scientific organisations are much less likely to perceive GM foods as safe. The major disagreements between scientists and public are related to the safety of GM food for humans and the possible environment and ecological imbalances. In addition to GM foods scientists are preparing for fossil fuel independent technologies for automobile fuels. Biodiesel refers to a vegetable oil based diesel fuel consisting of long-chain esters. It is typically made by chemically reacting lipids with an alcohol.

7.1 GENETICALLY MODIFIED FOODS

7.1.1 OVERVIEW

It has been more than three decades since the first genetically modified (GM) organisms were introduced, but there have been apprehensions over the applications of this technology. The arguments have been focused over the deliberate release of GM crops for agricultural purposes. There have been conflicting opinions by the scientists, experts, and stakeholders over the introduction of GM crops and the debate have run into a deadlock. Because of the concerns over the environmental and biological safety, most of the developing countries have not permitted the farmers to plant any GM crops. Also, due to new regulations by the European Union calling for strict traceability and labeling of GM foods has further discouraged the planting of GM crops in poor countries due to cost-effectiveness (Paarlberg, 2002). These international regulations on the food situation have impacted the people in the developing countries where approximately 800 million people are highly malnourished, which includes around 250 million children (Uzogara, 2000).

7.1.2 HISTORY

Humans have been practicing genetic engineering of crops and livestock from early ages. Farmers have been using different forms of genetic engineering to emphasize certain attributes in them by selective breeding of animals and cross-fertilizing specific species of plants to create new varieties with more desirable characteristics (Schardt, 1994).

Direct involvement of genetic engineering techniques in mainstream agriculture started gaining momentum in the 1960s has continued to evolve over many decades. A new variety of potato called Lenape potato with high solid content for making potato chips was developed in 1967. After few years, this potato developed a toxin called solanine and was therefore withdrawn from the market by the USDA. This development showed that genetic alteration may have many unwanted effects (McMillan and Thompson, 1979).

In 1979, a synthetic growth hormone bovine somatotropin (rBST) was developed at Cornell University, New York. This hormone was successfully injected into dairy cows to increase the capacity of milk production. During 1980, *Agrobacterium tumefaciens*-mediated transformation system was developed for generating transgenic plants by independent research groups in different countries (Fraley et al., 1983; Zambryski et al., 1983).

First genetically engineered foods were made available to the public in the 1990s, as Pfizer Corporation's genetically engineered form of rennet for cheese production was approved. National Institute of Health and American Medical Association reported that meat and milk from cows treated with rBST were safe. Scientists at the Cornell University produced recombinant porcine somatotropin for more meat production in pigs without increasing the feed intake. In 1994, the FDA finally approved for Flavr Savr tomato, the first genetically engineered whole food by the Calgene Corporation (Thayer, 1994).

In the 1990s, there were several landmark developments in the area of genetic engineering, which include cloning of sheep from embryonic cells and adult mammalian cells (Wilmut et al., 1999), introduction of the "biolistic gun" technology for gene transfer (Klein et al., 1987), introduction of the "terminator seeds" (Uzogara, 2000), herbicide and pest resistant plants (Liu, 1999).

7.1.3 BENEFITS OF GM FOODS

The major advantage of GM food crops is their potential of future food security, especially for the small-scale agriculture in the underdeveloped countries. Genetic engineering can provide improved nutritional quality, year-round food availability, and increased shelf-life of different foods which will benefit the farmers, consumers, as well as the environment. Following are some of the potential benefits of GM foods.

7.1.3.1 RESISTANCE AGAINST DISEASES, WEEDS, AND HERBICIDES

Crop plants like tomatoes, tobacco, and corn have also been genetically engineered to enhance resistance against viruses (Wood, 1995). Plants have been GM for insect and herbicide resistance (Wilkinson, 1997). These insecticides and herbicides include *Bt* toxin, glufosinate, glyphosate, imidazolinone, and sulphonyl urea (Uzogara, 2000). Worldwide, herbicides account for 50% of sales, insecticides 30%, and fungicides account for 20% in the agrochemical market and may pose serious health and environmental ill effects (Thayer, 1999). Genetically modified plants resistant to weed-killing herbicides seem to pose negligible risks to human health, but the environmental concerns are hard to dismiss.

7.1.3.2 INCREASED CROP YIELD

Genetic modifications can be employed to reduce the crop loss by making the plants tolerant toward high salinity, pH, temperature, pests, weeds, herbicides, insects, and drought. This will increase the global food production by reducing the crop loss without increasing the area under agriculture (Uzogara, 2000). Genetic manipulations have also been used to generate crops with superior nitrogen fixation ability and will reduce dependence on the utilization of chemical fertilizers (Paoletti and Pimentel, 1996). Some of the success stories of GM foods include the production of herbicide tolerant corn, insect-resistant apple, tomatoes, and soybeans, and virus-resistant cucumbers (Paoletti and Pimentel, 1996; Wood, 1995).

7.1.3.3 IMPROVED NUTRITIONAL QUALITY

Biotechnology can be employed to enhance the levels of vitamins (vitamins A, C, and E), which act as antioxidants and prevent the development of night blindness, heart diseases, and cancers (Ames, 1998; Smaglik, 1999). Genetic engineering can be used to increase certain minerals (such as zinc, iron, iodine) into common staple foods for optimal levels of key nutrients or supplementing some nutritional deficiencies endemic in some developing countries in the world (Wambugu, 1999). GMs have been used to increase the levels of unsaturated fatty acids in soybean, sunflower, and peanuts (Liu and Brown, 1996). Oils low in high unsaturated fatty acids and reduced levels of saturated and *trans*-fatty acids have various important health benefits.

7.1.3.4 IMPROVEMENT IN SHELF-LIFE

The Flavr Savr tomato was the first GM crop and whole food approved by the FDA. It was bioengineered to have a longer shelf-life by delaying ripening and rotting processes. Delayed ripening of fruits and vegetables by suppression of cell wall destroying degrading enzyme polygalacturonase can lead to better flavor, texture, and longer shelf-life (Uzogara, 2000). Like the tomato, the delayed ripening characteristics could also be created in pineapple, strawberry, and raspberry to extend the crop's shelf-life.

7.1.3.5 PROTEIN CONTENT AND QUALITY

Molecular tools can be employed to improve protein quality of different foods and feeds (De Lumen et al., 1997; Harlander and Roller, 2012). It may involve an increasing of the essential amino acid content of the food, like enhancing the content of methionine and lysine residues of certain proteins (Uzogara, 2000).

7.1.3.6 IMPROVEMENTS IN MEAT AND MILK

Genetic engineering like animal cloning can generate large-scale production of farm animals to meet the high demand for meat and other protein foods (Bishop, 1996). This technology can be employed to produce high quantities of meat and milk with low-cost investment. Dairy cows have been treated with BST to enhance milk production in cows, a technology approved by the FDA in 1993. Transgenic animals can be modified for enhanced production of milk and meat with added qualities like lactose-free milk, low-fat milk, and low-cholesterol meat in a cost-effective manner (Koch, 1998).

7.1.3.7 POTENTIAL RISKS OF GENETICALLY MODIFIED FOODS

Nevertheless, the critics of GM foods have concerns regarding genetically engineered foods include altered nutritional quality, toxicity, carcinogenicity, and environmental hazards. The transgene sometimes might get inserted in a wrong spot in the target DNA of the organism, or the new gene may accidentally activate or suppress the expression of other genes, thereby producing undesirable effects (Phillips, 1994). Some of the potential risks of GM crops include the following.

7.1.3.7.1 Environmental Concerns

One of the major concerns regarding the widespread plantation of GM crops is that they can present environmental risks. Transgenic plants having herbicide and insect resistance can cross-pollinate with wild grasses to produce resistant weeds, which would be hard to eradicate from the environment (Kaiser, 1996). Critics of GM crops also fear that commercialization of these crops can pose a threat to crop genetic diversity which is already endangered by present agricultural practices favouring the worldwide adoption of a few crop varieties (Phillips, 1994). Challenges of GM crops want regulations for appropriate studies to assess the risks of GM crops on the balance of the ecosystem, as it has been reported that *Bt* toxin threatens the nonhazardous insects by entering the food chain.

7.1.3.7.2 Toxicity

Genetic engineering could inadvertently induce production of some natural plant toxins by switching on involved in the toxin production. Naturally occurring toxins such as cyanogens in lima beans, pressor amines in bananas, and protease inhibitors in legumes which are normally inactive can get activated to produce enhanced levels of these toxins and pose a hazard to the consumers. This issue has led to a serious debate in different countries especially the EU and the USA to ban the GM foods until enough research is carried out to rule out any possibilities of hazardous toxins in GM foods.

7.1.3.7.3 Antibiotic Resistance

Antibiotic resistance is often used for the selection of transgenic organisms during genetic transformations. There has been a rising concern over the potential risks and unintended consequences these foods. There have been reports suggesting that the antibiotic resistant marker genes GM crops could be transferred into microbial pathogens inside and outside the human body, resulting in antibiotic resistant pathogens.

7.1.3.7.4 Allergic Reactions

Allergenic properties of the donor source (like bacteria) could transfer into the GM recipient plant or animal. Also, GM foods containing known allergens (like peanuts, egg, legumes, and shellfish proteins) could initiate allergic reactions in the vulnerable consumers. Pioneer Hi-bred International expressed Brazil nut proteins into soybeans to enhance its protein content. This led to allergic reactions in consumers who were allergic to Brazil nut, so this product was recalled voluntarily (Nordlee et al., 1996).

7.1.3.7.5 Alteration in Nutritional Quality

Foreign genes might change the nutritional value of GM foods by decreasing levels of certain nutrients while enhancing others causing an alteration of the balance of nutrients in foods. Because of the lack of research in the nutrient quality, nutrient bioavailability, and nutrient metabolism of GM foods, there are growing concerns from the experts for consumption of these foods by children (Young and Lewis, 1995).

7.1.3.7.6 Religious and Ethical Concerns

There have been growing religious, ethical, and cultural concerns regarding genetically engineered foods. For example, some people may avoid eating foods containing the human genes. Jews and Muslims may object to consume foods that contain pig genes. Similarly, vegetarians may be reluctant to vegetables and fruits that contain any animal genes (Crist, 1996).

7.1.4 THE GM FOOD CONTROVERSY

GM food manufacturers subject these foods to a rigorous testing before they can reach the consumers. However, genetic modifications of foods have been surrounded by wide controversies since the early 1990s. The cloning of Dolly the sheep in Scotland sparked several controversies and debates related to and other aspects of genetic engineering (Wilmut et al., 1997). Some people panic that genetic engineering may one day lead to cloning of humans, a concept which is highly opposed in the United States and Great Britain (Uzogara, 2000). Many critics oppose any form of genetic modifications in plants or animals and advocate a complete ban on GM foods. Some of the controversies that sparked opposition to using the genetic engineering in plants and animals include the following:

- The cloning of Dolly the sheep in Great Britain (Dyer, 1996; Wilmut et al., 1997).
- The introduction of "terminator seed" technology (Koch, 1998).
- *Bt* toxin versus the Monarch butterflies (Hileman, 1999).
- Basmati rice patent.
- Mad cow disease in Great Britain (Patterson and Painter, 1999).
- Effect of herbicide resistance on the environment (Longman, 1999).

Some critics in the EU countries think genetic engineering as an adverse technology that threatens health, agriculture, and ecology. Confrontation to GM foods in Britain rose due to incidents like "mad cow disease" as well as *Salmonella* outbreaks. Public further lost confidence in safety regulations of GM foods after a controversial study by a food scientist, Arpad Pusztai, in Scotland claimed that growth of rats was stunted when GM potatoes were fed to them (Enserink, 1999; Ewen and Pusztai, 1999).

The critics of GM foods oppose this technology and believe that scientists should not be allowed to cross nature's boundaries to perform genetic engineering in any of the organisms including humans (Woodard and Underwood, 1997). They also oppose the concept of transfer of genetic material from one organism to another which they think is a challenge to nature. They believe that applying GM techniques to food production may pose several undesirable consequences. For them, safety, religious, ethical, and environmental concerns outweigh the interest in increased food production and improved agriculture generated by genetic engineering.

7.2 EDIBLE VACCINES

As the great physician Hippocrates stated: "Let thy food be thy medicine." Edible vaccines are genetically engineered antigenic proteins expressed in a consumable plant or crop (Fig. 7.1). As the crop is consumed, some of the protein makes its way into the blood stream to initiate an immune response. This immune response has a potential to neutralize the future pathogen encounter. Development of edible vaccines is an emerging technology for cost-effective vaccine delivery system, particularly in the developing nations. It involves the introduction of certain desired genes into the plants to manufacture the encoded proteins. Edible vaccines have the potential to overcome the problems associated with traditional vaccines like cost, administration, and storage of vaccines. Edible vaccines have found applications in the prevention of infectious diseases, autoimmune diseases, cancer, and birth control. Although surmountable, there is a range of technical obstacles including the regulatory and nonscientific issues that need to be addressed before the introducing this technology in the field.

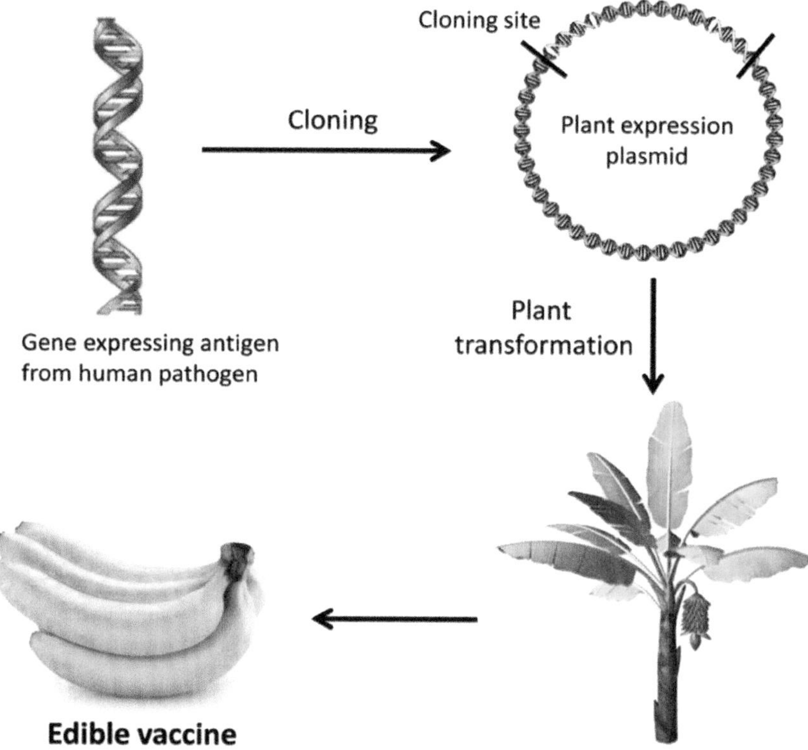

FIGURE 7.1 *Production of edible vaccine.* Edible vaccines are generated by expressing the genes encoding the antigens of bacterial and viral pathogens. The genes encoding the antigens are first cloned in a plant-specific vector using the recombinant DNA technology. The recombinant plasmid is transformed into the plant by various gene delivery methods for the expression of antigen in the edible part of plant.

Vaccination is one of the greatest success stories of modern medicine. Experiments by Edward Jenner and Louis Pasteur suggested that diseases can be prevented just by exposing a patient to a weakened or inactivated pathogen. Jenner successfully prevented children from smallpox by deliberately exposed them to the pus (containing the pathogen) from cowpox. Modern day vaccines contain a specific protein or a group of proteins from a particular pathogen and not the pathogen itself. A protective immune response can be generated from this more specific and less risky exposure. The fundamental principle of this kind of treatment lies in the fact that if the immune system is exposed to the pathogen before actual infection, the host body can readily respond to the pathogen and disease can be prevented.

In spite of worldwide immunization of children against the infectious diseases, almost 20% of infants are still unimmunized, resulting in almost two million avoidable deaths every year (Lal et al., 2007). For complete eradication of these diseases, an absolute coverage of immunisation is desirable but not practical due to various constraints like vaccine production, optimum storage, and distribution. Immunisations for some infectious diseases either are not available everywhere due to the cost or high maintenance storage. Therefore, there is a need for a cost-effective, storage friendly, easy-to-administer, and more acceptable form of vaccines. One of the potential strategies to develop such vaccines is to genetically engineer plants or plant viruses against some of the life-threatening diseases (Moffat, 1995).

Hiatt and co-workers in 1989 were the first scientists to generate antibodies in plants which could be used for the passive immunisation (Hiatt, 1989). The first report of the edible vaccine in plants appeared in 1990 in the form of a patent, when a surface protein from *Streptococcus* was expressed in tobacco (Mason and Arntzen, 1995). However, the notion of edible vaccine gained momentum a few years later when hepatitis B surface antigen was successfully expressed in tobacco via genetic engineering (Mason et al., 1992). Since then, there have been remarkable developments in the area of edible vaccine via recombinant DNA technology. Currently, efforts are being made to produce efficient edible vaccines to combat different genets causing diarrhoea like *Vibrio cholera*, enterotoxigenic *Escherichia coli*, and Rotavirus which are responsible for millions of infant deaths every year. Also, efforts are underway to generate various antigens and antibodies in plants, which can be administered as the edible part of the plant or can be injected into the body directly after isolation and purification (Ma and Hein, 1995). Antigens have been successfully expressed for rabies virus G-protein in tomato, and Norwalk virus capsid protein in potato and tobacco (Mason et al., 2002; McGarvey et al., 1995; Thanavala et al., 1995). Monoclonal antibodies can be generated in plants as an edible vaccine for cancer therapy. Monoclonal antibody (BR-96) has been generated in soybean for targeting lung, breast, and colon cancers (Moffat, 1995; Prakash, 1996). Edible vaccines have also been used to suppress autoimmune disorders like type-1 diabetes, multiple sclerosis, rheumatoid arthritis, etc. (Prakash, 1996).

Some of the advantages of edible vaccines over conventional medicines are as follows:

- low production costs of edible vaccines;
- can be produced in the large quantities to serve more people;
- no refrigeration is required for the storage of edible vaccines;
- no skilled medical personnel is required for the delivery;
- no chances of unknown human pathogens;
- elicits mucosal as well as systemic immunity; and
- no injections required for delivery.

7.3 BIODIESEL

7.3.1 OVERVIEW

Constant utilization of petroleum-based fuels is mostly nonsustainable due to a gradual depletion of the natural sources and emission of pollutants like carbon dioxide in the environment. Biodiesel derived from oil crops is a probable renewable alternative to petroleum fuels. Biodiesel is a diesel fuel consisting of long-chain alkyl esters generated from the lipids of plants and animals. Biodiesel is prepared by chemically reacting lipids with an alcohol to produce fatty acid esters. The calorific value of biodiesel has been estimated to be 37.27 MJ/kg (Elsayed et al., 2003). Though economically viable, biodiesel cannot assure even a fraction of the existing demands. The need of the hour is to find new sustainable and renewable sources of biodiesel. Microalgae emerge as the only possible source of renewable biodiesel can fulfil the global demands for transport fuels.

7.3.2 PROCESS OF BIODIESEL PRODUCTION

Oils consist of triglycerides in which three fatty acids are esterified with a molecule of glycerol. During the biodiesel production, the triglycerides are reacted with methanol by a process called as transesterification or alcoholysis (Fig. 7.2). Transesterification produces methyl esters of fatty acids (biodiesel) and glycerol. The reaction takes place in a stepwise manner where triglycerides are converted to diglycerides, then to monoglycerides, and at last to glycerol.

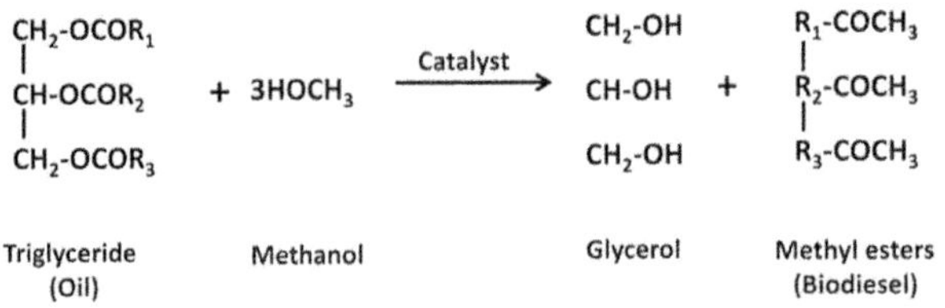

$$\begin{array}{lclcl} CH_2\text{-}OCOR_1 & & & CH_2\text{-}OH & & R_1\text{-}COCH_3 \\ | & & & & & | \\ CH\text{-}OCOR_2 & + \; 3HOCH_3 & \xrightarrow{\text{Catalyst}} & CH\text{-}OH & + & R_2\text{-}COCH_3 \\ | & & & & & | \\ CH_2\text{-}OCOR_3 & & & CH_2\text{-}OH & & R_3\text{-}COCH_3 \\ \text{Triglyceride (Oil)} & \text{Methanol} & & \text{Glycerol} & & \text{Methyl esters (Biodiesel)} \end{array}$$

FIGURE 7.2 *Transesterification.* Biodiesel can be produced from plant oils, animals based oils and fats as well as waste oils with a process known as transesterification. The reaction between the oil and the alcohol is a reversible reaction so the methanol must be added in excess to ensure the complete conversion of oil into biodiesel. R_1–R_3 indicate the hydrocarbon groups.

Transesterification can be catalyzed by alkalis, acids, and lipase enzymes. However, the alkali-catalyzed transesterification is about much faster than the acid catalyzed the reaction. Lipases offer significant advantages but are not feasible at industrial level due to the high cost of the catalyst (Fukuda et al., 2001). Alkali-based catalytic transesterification is carried out at 60°C, as methanol boils off at 65°C at atmospheric pressure and the reaction takes 90 min to complete. Methanol and oil do not mix and form two different layers consequently making the separation easier. Biodiesel is recovered by washing with pure water to remove salts, glycerol, and methanol.

7.3.3 MICROALGAE AS THE SOURCE OF BIODIESEL

Microalgae, also known as microphytes, are unicellular microscopic algae found in freshwater as well as marine water (Thurman and Burton, 1997). Microalgae are important for life on earth as they are capable of performing photosynthesis. They produce roughly half of the total atmospheric oxygen. Microalgae represent an enormous biodiversity, and it has been estimated that there are 200,000–800,000 species out of which almost 50,000 species are described (Starckx, 2012). Since there is an enormous consumption of transport fuel around the globe and for cooking oil and animal fat cannot practically meet this demand, microalgae can be an alternative source of biodiesel which has potential to completely replace the fossil oil due to its rapid growth and oil content. Microalgae have been used by humans for centuries for various purposes. However, it has been only a few decades since humans have started cultivating microalgae (Borowitzka, 1999). The oil content of microalgae can be up to 80% of its dry weight and the organism doubles its biomass within 3.5 h during the exponential growth, thus making it an efficient candidate for the production of fuel oils (Metting Jr., 1996; Spolaore et al., 2006). Different species of microalgae produce different types of lipids and other complex oils and not all kinds of oils are preferred for the production of biodiesel (Chisti, 2007). However, there are enormous resources of preferable lipids present in various species of microalgae. These properties make microalgae a potential candidate for biodiesel production.

KEYWORDS

- **genetically modified organisms**
- **risk assessment**
- **genetic engineering**
- **edible vaccines**
- **biodiesel**

REFERENCES

Ames, B. N. Micronutrients Prevent Cancer and Delay Aging. *Toxicol. Lett.* **1998,** *102*, 5–18.

Bishop, J. Technology and Health: Sheep Cloning Methods Hold Promise of Fast Introduction of Livestock Traits. *Wall Street J.* **1996,** *Thursday March 7*, B6.

Borowitzka, M. A. Commercial Production of Microalgae: Ponds, Tanks, Tubes and Fermenters. *J. Biotechnol.* **1999,** *70*, 313–321.

Chisti, Y. Biodiesel from Microalgae. *Biotechnol. Adv.* **2007,** *25*, 294–306.

Crist, W. Waiter, There's a Flounder in My Fruit. Bio-engineered Fruits and Vegetables with Animal Genetic Materials are Not So Labeled). *Veg. Times* **1996,** *231*, 22.

De Lumen, B.; Krenz, D. C.; Revilleza, M. J. Molecular Strategies to Improve the Protein Quality of Legumes. *Food Technol. (USA)*, **1997,** *464*, 117–126.

Dyer, O. Sheep Cloned by Nuclear Transfer. *BMJ (Clin. Res. Ed.)* **1996,** *312*, 658–658.

Elsayed, M.; Matthews, R.; Mortimer, N. *Carbon and Energy Balances for a Range of Biofuels Options*. Resources Research Unit, Sheffield Hallam Univ.: Sheffield, 2003.

Enserink, M. The Lancet Scolded over Pusztai Paper. *Science* **1999,** *286*, 656.

Ewen, S. W.; Pusztai, A. Effect of Diets Containing Genetically Modified Potatoes Expressing *Galanthus nivalis* Lectin on Rat Small Intestine. *Lancet* **1999,** *354*, 1353–1354.

Fraley, R. T.; Rogers, S. G.; Horsch, R. B.; Sanders, P. R.; Flick, J. S.; Adams, S. P.; Bittner, M. L.; Brand, L. A.; Fink, C. L.; Fry, J. S. Expression of Bacterial Genes in Plant Cells. *Proc. Nat. Acad. Sci.* **1983,** *80*, 4803–4807.

Fukuda, H.; Kondo, A.; Noda, H. Biodiesel Fuel Production by Transesterification of Oils. *J. Biosci. Bioeng.* **2001**, *92* (5):405–416.

Harlander, S.; Roller, S. *Genetic Modification in the Food Industry: A Strategy for Food Quality Improvement*. Springer Science & Business Media: Berlin, 2012.

Hiatt, A.; Cafferkey, R.; Bowdish, K. Production of antibodies in transgenic plants. *Nature* **1989**, 342(6245):76–8.

Hileman, B. Bt Corn Pollen Kills Monarch Caterpillars. *Chem. Eng. News* **1999,** *77*, 7.

Kaiser, J. Pests Overwhelm Bt Cotton Crop. *Science* **1996,** *273*, 423.

Klein, T. M.; Wolf, E.; Wu, R.; Sanford, J. High-Velocity Microprojectiles for Delivering Nucleic Acids into Living Cells. *Nature* **1987,** *327*, 70–73.

Koch, K. *Food Safety Battle: Organic vs. Biotech*. CQ Press: Washington, DC, 1998.

Lal, P.; Ramachandran, V.; Goyal, R.; Sharma, R. Edible Vaccines: Current Status and Future. *Indian J. Med. Microbiol.* **2007,** *25*, 93.

Liu, K. Biotech Crops: Products, Properties, and Prospects. *Food Technol.* **1999,** *53*, 42–49.

Liu, K.; Brown, E. A. Enhancing Vegetable Oil Quality Trough Plant Breeding and Genetic Engineering. *Food Technol.* **1996,** *50*, 67–71.

Longman, P. J. The Curse of Frankenfood. Genetically Modified Crops Stir Up Controversy at Home and Abroad. *US News World Rep.* **1999,** *127*, 38.

Ma, J. K.; Hein, M. B. Immunotherapeutic Potential of Antibodies Produced in Plants. *Trends Biotechnol.* **1995,** *13*, 522–527.

Mason, H. S.; Arntzen, C. J. Transgenic Plants as Vaccine Production Systems. *Trends Biotechnol.* **1995,** *13*, 388–392.

Mason, H. S.; Lam, D.; Arntzen, C. J. Expression of Hepatitis B Surface Antigen in Transgenic Plants. *Proc. Nat. Acad. Sci.* **1992,** *89*, 11745–11749.

Mason, H. S.; Warzecha, H.; Mor, T.; Arntzen, C. J. Edible Plant Vaccines: Applications for Prophylactic and Therapeutic Molecular Medicine. *Trends Mol. Med.* **2002,** *8*, 324–329.

McGarvey, P. B.; Hammond, J.; Dienelt, M. M.; Hooper, D. C.; Fu, Z. F.; Dietzschold, B.; Koprowski, H.; Michaels, F. H. Expression of the Rabies Virus Glycoprotein in Transgenic Tomatoes. *Nat. Biotechnol.* **1995,** *13*, 1484–1487.

McMillan, M.; Thompson, J. An Outbreak of Suspected Solanine Poisoning in Schoolboys. *QJM: Int. J. Med.* **1979,** *48*, 227–243.

Metting, Jr., F. Biodiversity and Application of Microalgae. *J. Ind. Microbiol.* **1996,** *17*, 477–489.

Moffat, A. S. Exploring Transgenic Plants as a New Vaccine Source. *Science* **1995,** *268*, 658–658.

Nordlee, J. A.; Taylor, S. L.; Townsend, J. A.; Thomas, L. A.; Bush, R. K. Identification of a Brazil-Nut Allergen in Transgenic Soybeans. *N. Engl. J. Med.* **1996,** *334*, 688–692.

Paarlberg, R. L. The Real Threat to GM Crops in Poor Countries: Consumer and Policy Resistance to GM Foods in Rich Countries. *Food Policy* **2002,** *27*, 247–250.

Paoletti, M. G.; Pimentel, D. Genetic Engineering in Agriculture and the Environment. *BioScience* **1996,** 665–673.

Patterson, W.; Painter, M. Bovine Spongiform Encephalopathy and New Variant Creutzfeldt-Jakob Disease: An Overview. *Commun. Dis. Publ. Health/PHLS* **1999,** *2*, 5–13.

Phillips, S. C. *Genetically Engineered Foods*. CQ Press: Washington, DC, 1994.

Prakash, C. Edible Vaccines and Antibody Producing Plants. *Biotechnol. Dev. Monit.* **1996,** *27*, 10–13.

Schardt, D. Brave New Foods (Genetically Engineered Foods). *Am. Health* **1994,** *13*, 60.

Smaglik, P. Food as Medicine: Nutritionists, Clinicians Disagree on Role of Chemopreventive Supplements. *Scientist* **1999,** *13 (11), 14.*

Spolaore, P.; Joannis-Cassan, C.; Duran, E.; Isambert, A. Commercial Applications of Microalgae. *J. Biosci. Bioeng.* **2006,** *101*, 87–96.

Starckx, S. A Place in the Sun—Algae is the Crop of the Future, According to Researchers in Geel, *Flanders Today*, 2012.

Thanavala, Y.; Yang, Y.; Lyons, P.; Mason, H.; Arntzen, C. Immunogenicity of Transgenic Plant-Derived Hepatitis B Surface Antigen. *Proc. Nat. Acad. Sci.* **1995,** *92*, 3358–3361.

Thayer, A. M. FDA Gives Go-ahead to Bioengineered Tomato. *Chem. Eng. News* **1994,** *72*, 7–8.

Thayer, A. M. Transforming Agriculture. *Chem. Eng. News* **1999,** *77*, 21–35.

Thurman, H. V.; Burton, E. A. *Introductory Oceanography*. Prentice Hall: Upper Saddle River, NJ, 1997.

Uzogara, S. G. The Impact of Genetic Modification of Human Foods in the 21st Century: A Review. *Biotechnol. Adv.* **2000,** *18*, 179–206.

Wambugu, F. Why Africa Needs Agricultural Biotech. *Nature* **1999,** *400*, 15–16.

Wilkinson, J. Q. Biotech Plants: From Lab Bench to Supermarket Shelf. *Food Technol. USA*, **1997**, 51 (12), 37–42.

Wilmut, I.; Schnieke, A.; McWhir, J.; Kind, A.; Campbell, K. Viable offspring derived from fetal and adult mammalian cells. *Nature* **1997,** *385*, 810–813.

Wilmut, I.; Schnieke, A.; McWhir, J.; Kind, A.; Campbell, K. Viable Offspring Derived from Fetal and Adult Mammalian Cells, **2007**, 9, 3–7

Wood, M. Boosting Plants' Virus Resistance. *Agric. Res.* **1995,** *43*, 18.

Woodard, K.; Underwood, A. Today the Sheep, Tomorrow the Shepherd? Before Scientists Get There, Ethicists Want Some Hard Questions Asked and Answered. *Newsweek* **1997,** *129*, 60.

Young, A. L.; Lewis, C. G. Biotechnology and Potential Nutritional Implications for Children. *Pediatr. Clin. North Am.* **1995,** *42*, 917–930.

Zambryski, P.; Joos, H.; Genetello, C.; Leemans, J.; Van Montagu, M.; Schell, J. Ti Plasmid Vector for the Introduction of DNA into Plant Cells Without Alteration of their Normal Regeneration Capacity. *EMBO J.* **1983,** *2*, 2143.

CHAPTER 8

IMPROVEMENT OF HORTICULTURE CROPS THROUGH BIOTECHNOLOGICAL APPROACHES

ANUPAM ADARSH[1], PANKAJ KUMAR[2], BISHUN DEO PRASAD[2*], RANDHIR KUMAR[1], and TUSHAR RANJAN[3]

[1]*Department of Horticulture (Vegetable and Floriculture), Bihar Agricultural University, Sabour, Bhagalpur, Bihar, India*

[2]*Department of Plant Molecular Biology and Genetic Engineering, Bihar Agricultural University, Sabour, Bhagalpur, Bihar, India*

[3]*Department of Basic Science and Humanities Genetics, Bihar Agricultural University, Sabour, Bhagalpur, Bihar, India*

[*]*Corresponding author. E-mail: dev.bishnu@gmail.com*

CONTENTS

ABSTRACT

Horticultural crops are grown worldwide which provide fibers, nutrients, and vitamins in the human diet. It can be consumed fresh or may be eaten after cooking, processing and constitute important part of meals of billions of people worldwide. Most vegetable crops are annual or biennial and few are perennials. The current level of production is 90 mt and the total area under vegetable cultivation is around 6.2 million hectares which is about 3% of the total area under cultivation in the country. Agricultural biotechnology deals with the practical application of biological organisms or their subcellular components in agriculture. The techniques currently in use include tissue culture, conventional breeding, molecular marker-assisted breeding, and most advanced genetic engineering. Tissue culture is the cultivation of plant cells or tissues on specifically designed nutrient media. Under optimal conditions, a whole plant can be regenerated from a single cell, a rapid and essential tool for mass propagation and production of disease-free plants. Advances in breeding help agriculture achieve higher yields and meet the needs of expanding population with limited land and water resources. In molecular-assisted breeding, molecular markers [identifiable deoxyribonucleic acid (DNA) sequences found at specific location of the genome] are being used. By determining location and likely actions of genes, scientists can quickly and accurately identify plants carrying desirable characteristics; hence, conventional breeding can be conducted with greater precision. Molecular markers can be used in plant breeding to increase the speed and efficiency of the introduction of new genes diversity, taxonomic relationships between plant species and biological processes such as mating systems, pollen, or disease dispersal. Biotechnology enables development of disease diagnostic kits for use in laboratory and field. These kits are able to detect plant diseases early, by testing for the presence of pathogen's DNA or proteins which are produced by pathogens or plants during infection. Conventional agricultural biotechnologies work better when combined with modern biotechnological approaches. Modern agricultural biotechnology refers to biotechnological techniques for the manipulation of genetic material and the fusion of cells beyond normal breeding barriers.

8.1 INTRODUCTION

Horticultural crops grown are worldwide, which provide fiber, nutrients, and vitamins in the human diet. It can be consumed fresh or may be eaten after cooking, processing and constitute important part of meals of billions of people worldwide. Most vegetable crops are annual or biennial and few are perennials. The current level of production is 90 mt and the total area under vegetable cultivation is around 6.2 million hectares which is about 3% of the total area under cultivation in the country. Environmental stress is the primary cause of crop losses worldwide, though it reduces average yields of major crops by more than 50% (Bray et al., 2000). The response of plants to environmental stresses depends on the plant developmental stage and the length or severity of the stress (Bray, 2002). Plants may respond similarly to avoid one or more stresses through morphological or biochemical mechanisms (Capiati et al., 2006). In the 21st century, vegetable crops are affected by global warming and climate change because the global temperature has increased by 0.8°C and is expected to reach 1.1–5.4°C by the end of next century (Fand et al., 2012). The concentrations of CO_2 in the atmosphere have increased from 280 to 370 ppm and are expected to be doubled in 2100 (IPCC, 2007). This change is attributed to the overexploitation and misuse of natural resources for various anthropogenic developmental activities such as increased urbanization, deforestation, and industrialization resulting in aberrant weather events like changes in rainfall patterns, frequent droughts and floods, increased intensity and frequency of heat and cold waves, outbreaks of insect-pests and diseases, etc. affecting profoundly, many biological systems and ultimately the human beings (IPCC, 2007).

High temperatures can cause severe significant losses in tomato productivity in terms of reduced fruit set, abnormal flower development, poor pollen production, dehiscence, viability, ovule abortion and poor viability, reduced carbohydrate availability, and other reproductive abnormalities and smaller and lower quality fruits (Hazra et al., 2007). In pepper, high-temperature exposure at the pre-anthesis stage did not affect pistil or stamen viability, but high post-pollination temperatures inhibited fruit set (Erickson and Markhart, 2002). Plant sensitivity to salt stress results in loss of turgor, growth reduction, wilting, leaf curling and epinasty, leaf abscission, decreased photosynthesis, respiratory changes, loss of cellular integrity, tissue necrosis, and ultimately potentially death of the plant. Most vegetable crops are highly sensitive to flooding. Flooded crops such as tomato accumulate endogenous ethylene that causes damage to the plants (Datta, 2013). Low oxygen levels stimulate an increased production of an ethylene precursor, 1-aminocyclopropane-1-carboxylic acid, in the roots. During the last 40–50 years, air pollution, sulfur dioxide, nitrogen oxide, hydrofluoride, ozone, and acid rain has adverse effect on vegetable production in terms of reducing growth, yield, and quality. A recent study indicated that climate change significantly decreased yield up to more than 50% in case of *Brassica oleracaea*, *Lactuca sativa*, and *Raphanus sativus*. Air pollution these days has become one of the major threats for vegetable crops. Many vegetable crops namely tomato, water melon, potato, squash, soybeans, cantaloupe, peas, carrot, beet, turnip, etc. are more susceptible to air pollution damage. Yield of vegetable can be reduced by 5–15% when daily ozone concentrations reach to greater than 50 ppb (Narayan, 2009).

Adapting to such harsh calamities and adverse conditions, we apply agricultural biotechnologies for improving crop productivities per unit area of land cultivated. Biotechnological application of advanced techniques in breeding can help floriculturist to achieve higher yields and meet needs of expanding population with limited land and water resources (Treasury, 2009). Biotechnology is the science of engineering the genetic makeup of an organism to achieve desired traits in the organism. Agricultural biotechnology deals with the practical application of biological organisms or their subcellular components in agriculture. The techniques currently in use include tissue culture, conventional breeding, molecular marker-assisted breeding, and most advanced genetic engineering. Tissue culture is the cultivation of plant cells or tissues on specifically designed nutrient media. Under optimal conditions, a whole plant can be regenerated from a single cell, a rapid and essential tool for mass propagation and production of disease-free plants (Kumar and Naidu, 2006). Advances in breeding help agriculture achieve higher yields and meet the needs of expanding population with limited land and water resources. As a result of improved plant breeding techniques, the productivity gains in worldwide production of primary crops, including maize, wheat, rice, and oilseed have increased by 21% since 1995, while total land devoted to these crops has increased by only 2% (Treasury, 2009). In molecular-assisted breeding, molecular markers [identifiable deoxyribonucleic acid (DNA) sequences found at specific location of the genome] are being used. By determining location and likely actions of genes, scientists can quickly and accurately identify plants carrying desirable characteristics; hence, conventional breeding can be conducted with greater precision (Mneney et al., 2001; Sharma et al., 2002). Molecular markers can be used in plant breeding to increase the speed and efficiency of the introduction of new genes diversity, taxonomic relationships between plant species and biological processes such as mating systems, pollen, or disease dispersal (Johanson and Ives, 2001). Biotechnology enables development of disease diagnostic kits for use in laboratory and field. These kits are able to detect plant diseases early, by testing for the presence of pathogen's DNA or proteins which are produced by pathogens or plants during infection (Kumar and

Naidu, 2006). Conventional agricultural biotechnologies works better when combined with modern biotechnological approaches. Modern agricultural biotechnology refers to biotechnological techniques for the manipulation of genetic material and the fusion of cells beyond normal breeding barriers. The most obvious example is genetic engineering to create genetically modified organisms (GMOs) through "transgenic" technology involving the insertion or deletion of genes. In genetic engineering or genetic transformation, the genetic material is modified by artificial means. It involves isolation and cutting of a gene at a precise location by using specific restriction enzymes. Selected DNA fragments can then be transferred into the cells of the target organism. The common practice in genetic engineering is the use of a bacterium *Agrobacterium tumefaciens* as a vector to transfer the genetic trait (Johanson and Ives, 2001). A more recent technology is ballistic impregnation method whereby a DNA is attached to a tiny gold or tungsten particle and then "fired" into the plant tissue (Morris, 2011). Crops may be modified for improved flavor, increased resistance to pests and diseases, or enhanced growth in adverse weather conditions. In recent years, biosafety and genetic engineering projects have been initiated in Africa, with the aim of introducing GMOs into Africa's agricultural systems. Already, countries like South Africa, Egypt, and Burkina Faso have commercialized GMOs while many others have developed the capacity to conduct research and development in modern agricultural biotechnology (Mayet, 2007). Green biotechnology is the term referring to the use of environmentally friendly solutions in agriculture, horticulture, and animal breeding processes (Treasury, 2009).

Recombinant DNA (rDNA) technology has significantly augmented the conventional crop improvement and has the potential role to assist plant breeders to meet the increased food demand predicted for the 21st century. Dramatic progress has been made over the past two decades in manipulating genes from diverse and exotic sources and inserting them into microorganisms and crops to confer resistance to biotic or abiotic (pests and diseases, tolerance to herbicides, drought, soil salinity, aluminum and arsenic toxicity, improve postharvest quality, enhance nutrient uptake and nutritional quality; increase photosynthetic rate, sugar and starch production, increase effectiveness of biocontrol agents, improved understanding of gene action and metabolic pathways; and production of drugs and vaccines in crops) (Mtui, 2011). Biotechnology has provided powerful and useful tools ranging from traditional biotechnology such as plant tissue culture to modern biotechnology such as transgenic plants and genetically engineered animals that contribute in improvement of crop production, food quality, and safety, while preserving the environment. It also addresses the complex regulatory framework surrounding modern biotechnology, as well as tools in the pipeline, and intellectual property aspects related to the technology. Biotechnology has had limited commercial success in horticultural crops, including fruits, vegetables, flowers, and landscape plants. Even though the first transgenic crop to reach the market was the *Flavr Savr* tomato, and sweet corn, potato (Freedom, New leaf), squash, and papaya varieties engineered to resist insects and viruses have been approved for commercial use and marketed.

8.2 TISSUE CULTURE

Tissue culture is an important biotechnological tool applied in olericulture for crop improvement. Tissue culture is of in vitro aseptic culture of cells, tissues, organs, or whole plant under controlled nutritional and environmental conditions to produce the clones of plants (Hussain et al., 2012). It can be used in fundamental research to study cell division, plant growth, in the area of plant propagation, disease elimination, plant improvement, and production of secondary metabolites. The key to the successful application of tissue culture is the manipulation of media compositions to achieve desired outcomes (Touchell et al., 2008). The most common application of tissue culture is micropropagation, which usually involves growing of plants in in vitro (agar solidified nutrient media). Micropropagation facilitates production of virus-free planting material and propagation of plant species from undifferentiated callus. Micropropagation technology has a vast potential to produce plants of superior quality, isolation of useful variants in well-adapted high-yielding genotypes with better disease resistance and stress tolerance capacities. Explant is primary requirement for micropropagation technique. Explant is the piece of any part of tissues that is put into culture (Fig. 8.1). Explant may be piece of organ (leaf, stems, roots, cotyledons, and embryo) or specific cell types (leaf tissue, pollen, endosperm, nucellus) (Fig. 8.1). A single explant can be multiplied into several thousand plants in relatively short time period and space under controlled conditions, irrespective of the season and weather on a year round basis (Hussain et al., 2012).

The technique of plant tissue culture depends upon the principle of totipotency (ability of a single cell to express the full genome by cell division) of plant cells (Hussain et al., 2012). Plant tissue culture medium contains all the nutrients required for the normal growth and development of plants. It is mainly composed of macronutrients, micronutrients, vitamins, other organic components, plant growth regulators (PGRs), carbon source, and some gelling agents in case of solid medium. Murashige and Skoog medium

FIGURE 8.1 Different explants used in tissue culture.

(MS medium) is most extensively used for the vegetative propagation of many plant species in vitro. The pH of the media is also important that affects both the growth of plants and activity of PGRs so, it is adjusted to the value between 5.4 and 5.8. Both the solid and liquid medium can be used for culturing. PGRs play key role in the development pathway of plant cells and tissues in culture medium. The type and the concentration of hormones used depend mainly on the species of the plant, the tissue, or organ cultured and the objective of the experiment. Auxins and cytokinins are most widely used PGRs in plant tissue culture and their amount determined the type of culture established or regenerated. The high concentration of auxins generally favors root formation, whereas the high concentration of cytokinins promotes shoot regeneration. A balance of both auxin and cytokinin leads to the development of mass of undifferentiated cells known as callus. The regenerated plants have following advantages.

Production of improved crop varieties:

- Production of disease-free plants (virus)
- Production of genetically identical clones
- Overcoming difficult crosses of postzygotic embryo abortion
- Genetic transformation
- Production of secondary metabolites
- Production of varieties tolerant to salinity, drought, and heat stresses.

8.3 MERISTEM AND BUD CULTURE

Meristem-tip culture is the excision of organized apex of the shoot from a selected donor plant for subsequent in vitro culture. The conditions of culture are regulated to allow organized outgrowth of the apex directly into a shoot, without the intervention of any adventitious organs. (The excised meristem tip is typically small and is removed by sterile dissection under the microscope.) The explants comprise apical dome and a limited number of the youngest leaf primordia and exclude any differentiated provascular or vascular tissues. A major advantage of such small explant is the potential that holds for excluding pathogenic organisms that may have been present in the donor plants from in vitro culture. A second advantage is the genetic stability of inherent technique, since plantlet is produced from an already differentiated apical meristem and propagation from adventitious meristems. Shoot develop directly from the meristem to avoid callus tissue formation and adventitious organogenesis, ensuring that genetic instability and somaclonal variation are minimized. Typically, these explants are between 3 and 20 mm in length, and development of in vitro can still be regulated to allow for direct outgrowth of the organized apex. The axillary buds of in vitro plantlets derived from meristem-tip culture may also be used as a secondary propagule. When the in vitro plantlet has developed expanded internodes, it may be divided into segments, each containing small leaf and an even smaller axillary bud. When these nodal explants are placed on fresh culture medium, the axillary bud will grow directly into a new plantlet, at which time the process can be repeated. This technique has high propagation rate to the original meristem-tip culture technique, and together the techniques form the basis of micropropagation, which is so important to the horticulture industry. Axillary shoots issue from preexisting buds and are normally true to type, indeed the meristematic cells are genetically very stable. In these processes, tissues or cells, either as suspensions or as solids, are maintained under conducive conditions which include proper temperature, proper gaseous and liquid environment, and proper supply of nutrient for their growth and multiplication. Plant tissue culture relies on the fact that many plant cells have the ability to regenerate a whole plant (totipotency). Single cells, plant cells without cell walls (protoplasts), pieces of leaves, or (less commonly) roots can often be used to generate a new plant on culture media given the required nutrients and plant hormones (Vidyasagar, 2006).

Meristem culture is a unique technique to produce virus-free plants. Virus elimination through meristem culture is a popular horticulture practice nowadays (Bhojwani and Razdan, 1983). The application of meristem culture is either to eliminate virus infection in clonal plant or its large-scale production of asexual seedling. A standard method for establishment of meristem culture and subsequent regeneration in eggplant cultivars of Bangladesh to raise meristem-derived virus-free eggplant clones, shoot tips were collected from 30 to 35-day-old field grown plants (Akhtar et al., 2008). Among different hormonal treatments in MS liquid medium, 2.0 mg/L BAP was proved to be best medium for primary establishment of meristem culture in all the cultivars. Best shoot development was found in cv. Islampuri containing MS semisolid medium supplemented with 2.0 mg/L BAP and 1.0 mg/L NAA. Rooting of meristem-derived shoots was found best by using MS medium containing 1.0 mg/L IBA in both cv. Islampuri and cv. Khatkhatia. The successful acclimatization of the in vitro grown plantlets proved the validity of the developed protocol of using biotechnological techniques for improvement of eggplant. Virus eradication is dependent on several parameters. But to take advantage of the nonuniform and imperfect virus distribution in the host plant body, the size of the excised meristem should be as small as possible. The explants smaller than 0.2 mm can't survive and those larger than 0.7 produce plants that still contain mottle virus.

8.4 PROPAGATION BY AXILLARY SHOOTING

This technique has proved to be the most applicable and reliable method of in vitro propagation. Axillary shoot growth is stimulated by overcoming apical meristem dominance. Commercial tissue culture laboratories are now able to propagate a large number of herbaceous ornamental species and several woody plants in this way. However, the propagation of *Pelargonium* and few other horticultural plants are always difficult to propagate by axillary branching.

8.5 PROPAGATION BY DIRECT OR INDIRECT ORGANOGENESIS

Adventitious shoots could arise directly from the tissue of explants without callus formation. Several plants of the family gesneriaceae (Saintpaulia, Streptocarp) regenerate directly buds on leaf explants, likewise *Lilium rnerate* directly buds on leaf explants, likewise *Lilium* regenerates on scales. However more often, like for *Ficus lyrata*, adventitious buds appear on callus. While coffee, cocoa trees, and many conifers are produced by somatic embryogenesis developed on callus or cell suspensions.

8.5.1 IMPROVEMENT OF AXILLARY BRANCHING

To reduce manpower costs, several improvements have been proposed. The more simple method was in vitro layering developed by Wang (1977) to clone PVX-free potato plants. The first plantlets placed on the medium in a horizontal position developed axillary shoots. They are harvested by cutting 1 cm above the medium surface, at 3 weeks intervals. A similar technique called "hedging system" was later used to produce *Pinus radiata*. Ziv (1990) proposed for corn plants, Gladiolus and Nerine, a very rapid propagation system. She reduces the internodes and leaves by introduction of an antigibberellin agent in the medium. Finally, only aggregates of buds are formed, then they finally, only aggregates of buds are formed, and then they are divided and introduced in bioreactors for mass production. Similar systems were developed in Gembloux, to propagate carub trees and asparagus. Since 1988, Duhem was producing very large quantities of Eucalyptus plantlets in Petri dishes without antigibberellin but in complete darkness. Transfers from one Petri to another are made by a simple squashing (Boxus, 1991). The major advantages of meristem culture are that it provides:

1. clonal propagation in vitro with maximal genetic stability;
2. the potential for removal of viral, bacterial, and fungal pathogens from donor plants;
3. the meristem tip as a practical propagule for cryopreservation and other techniques of culture storage;
4. a technique for accurate micropropagation of chimeric material; and
5. cultures those are often acceptable for international transport with respect to quarantine regulations.

8.5.1.1 METHODS/PROCEDURE

- Select a suitable donor plant, in this case, any of the *Solanum tuberosum* ssp. *tuberosum* types following any desired temperature pretreatments. Excise stem segments containing at least one node from the donor plant.
- Remove mature and expanding foliage to expose the terminal and axillary buds. Cut donor segments to 4-cm lengths, and presterilize by immersion in absolute ethanol for 30 s.
- Sterilize by immersing the donor tissues in the sodium hypochlorite solution, with added detergent, for 8 min.
- Following surface sterilization, rinse the tissues three times with sterile distilled water
- Mount the stem segment on the stage of the dissection microscope, and use the tips of hypodermic needles to dissect away progressively smaller, developing leaves to expose the apical meristem of the bud, with the few youngest part of the leaf primordial.
- Excise the explant tissue that should comprise the apical dome and the required number of the youngest leaf primordia.
- After excision, the explant is transferred directly onto the selected growth medium, and the culture vessel is closed.
- Transfer the completed men stem-tip culture to the growth room.
- If the explant is viable, then enlargement, development of chlorophyll, and some elongation will be visible within 7–14 days.
- Maintain the developing plantlet in vitro until the internodes are sufficiently elongated to allow dissection into nodal explants.
- To prepare nodal explants, remove the plantlets from the culture vessels under sterile conditions and separate into nodal segments. Each of these transferred directly onto fresh growth medium to allow axillary bud outgrowth. Extension of this bud should be evident within 7–14 days of culture initiation.

8.5.2 ZYGOTIC EMBRYO CULTURE

Biotechnology could be an alternative approach to conventional breeding methods, effective way to improve plant varieties by means of selection optimization, shortening breeding schemes, ability to tolerate soil-borne disease (Phytophthora root-rot) and abiotic stress (salinity) and therefore diminishing the cost of breeding efforts. A genetic breeding program has been recently initiated and aimed to improve tolerance to biotic and abiotic stresses by combining the use of mutation induction and biotechnological techniques. Considering the eventual use of zygotic embryo culture technique in this program are the in vitro germination, rooting of zygotic embryos, sprout multiplication, and plantlet adaptation. Intervarietal and interspecific crosses, followed by selection, have accounted for the improvement in quality and yield potential of practically all major crops (Raghavan, 1986). Embryo culture involves isolating and growing an immature or mature zygotic embryo under sterile conditions on an aseptic nutrient medium with the goal of obtaining a viable plant. The basic premise for this technique is that the integrity of the hybrid genome is retained in a developmentally arrested or an abortive embryo and that its potential to resume normal growth may be realized if supplied with the proper growth substances. The technique depends on isolating the embryo without injury, formulating a suitable nutrient medium, and inducing continued embryogenic growth and seedling formation. The culture of immature embryos is used to rescue embryos that would normally abort or that would not undergo the progressive sequence of ontogeny. The culture of mature embryos from ripened seeds is used to eliminate seed germination inhibitors or to shorten the breeding cycle if, for example, dormancy is a problem. This culture is easy and only requires a simple nutrient medium with agar, sugar, and minerals.

8.5.2.1 METHODOLOGY

Seeds with different developmental states were used. An embryo was considered mature when it was extracted from ripe fruits, which depended on the genotype. Seeds were dipped into 90% (v/v) ethanol and flamed to surface sterilized as previously indicated. Aseptic seeds were divided by halves into separated cotyledons, excising the plumule–radicle axes together with 1-cm-thick sections of cotyledon, and transferring them into tubes of nutrient medium. For all experiments, zygotic embryos were put on filter paper bridges into glass tubes containing 5 mL of MS salt medium diluted to half strength (1/2 MS) supplemented with 30,000 mg/L of sucrose, 100 mg/L of *i*-inositol, pH 5.7 ± 0.1; except for multiplication experiments, where bencilaminepurine (BA) and giberelic acid (GA3) at 0.5 mg/L were also added. Four week-old entire plantlets were transferred to glass pots containing 10 mL of fresh medium without hormones and grown for eight more weeks. Cultures were grown in a climate-room with a relative humidity of 60%, temperature of 25 ± 2°C and light intensity of 2500 lx provided by *Chiyoda lux* fluorescent lamps and measured using a Yu116 Luxometer. A 16-h light photoperiod was used in this setup. Three-month-old plantlets were transferred to pot containing a mix of soil, organic matters, and charcoal breeze for acclimatization, before these were transferred to normal greenhouse conditions. At this acclimatization state, plants were covered using transparent nylon for 2 weeks and watered three times weekly. First watering was made using MS (1/2) salt medium. This step resulted critical during material adaptation. In tomato from immature embryo culture technique, new plants germinated into from 20-day-old embryos with a low success rate. The germination percentage reached 100% when the embryo age reached 28–32 days. Shoot germination rate was not affected by growth regulators or genotypes. Using the immature embryo culture provided an advantage in rapid generation advancement in comparison with the conventional breeding practice. Immature embryo culture technique offered up to three generations, in contrast to conventional breeding systems which has maximum of 1–2 generations/year.

8.6 APPLICATIONS

8.6.1 RESCUING EMBRYOS FROM INCOMPATIBLE CROSSES

In interspecific and intergeneric hybridization programs, incompatibility barriers often prevent normal seed development and production of hybrids. Although there may be normal fertilization in some incompatible crosses, embryo abortion results in the formation of shriveled seeds. Poor and abnormal development of the endosperm caused embryo starvation and eventual abortion. Isolation of hybrid embryos before abortion and their in vitro culture may prevent these strong postzygotic barriers. The most useful and popular application of embryo cultures is to raise rare hybrids by rescuing embryos of incompatible crosses.

8.6.2 OVERCOMING DORMANCY AND SHORTENING BREEDING CYCLE

Long periods of dormancy in seeds delay breeding works especially in horticultural and crop plants. Using embryo culture techniques, the breeding cycle can be shortened in these plants. For example, the life cycle of Iris was reduced from 2 to 3 years to less than 1 year. Similarly, it was possible to obtain two generations of flowering against one in *Rosa* sp. Germination of excised embryo is regarded as a more reliable test for rapid testing of viability in seeds, especially during dormancy period.

8.6.3 OVERCOMING SEED STERILITY

In early ripening fruit cultivars, seeds do not germinate because their embryos are still immature. Using the embryo culture method, it become promising to raise seedling from sterile seeds of early ripening stone fruits, peach, apricot, plum, etc. "Makapuno" coconuts are very expensive and most relished for their characteristics soft fatty endosperms in place of liquid endosperm. Under normal conditions, the coconut seeds fail to germinate. De Guzman et al. (1971) obtained 85% successes in raising field-grown makapuno trees with the aid of embryo cultures.

8.6.4 PRODUCTION OF MONOPLOID

An embryo culture has been used in production of monoploids of barley. With the cross *Hordeum vulgare*, fertilization occurs normally but thereafter chromosomes of *H. bulbosum* are eliminated, resulting in formation of Monoploid *H. vulgare* embryo which can be rescued by embryo cultures.

8.6.5 CLONAL MICROPROPAGATION

The regenerative potentials are an essential prerequisite in nonconventional methods of plant genetic manipulations. Because of their juvenile nature, embryos have a high potential for regeneration and hence may be for in vitro clonal propagation. This is especially true of conifers and graminaceous members. Both organogenesis and somatic embryogenesis have been induced in major cereals and forage grasses form embryonic tissues. Generally, callus derived from immature embryos of cereals has the desired morphogenetic potential for regeneration and clonal propagation.

8.7 ANTHER AND MICROSPORE CULTURE

Anther culture is a technique by which the developing anthers at a precise and critical stage are excised aseptically from unopened flower bud and are cultured on a nutrient medium where the microspores within the cultured anther develop into callus tissue or embryoids that give rise to haploid plantlets either though organogenesis or embryogenesis. It has been observed that mature pollen grains of *Ginkgo biloba* (a gymnosperm) can be induced to prolifrate in culture to form haploid callus (Tulecke, 1953). After that, direct development of embryos from microspores of *Datura innoxia* by the culture of excised anther was reported by Guha and Maheswari (1964).

8.7.1 POLLEN CULTURE/MICROSPORE CULTURE

Pollen or microspore culture is an in vitro technique by which the pollen grains, preferably at the uninucleated stage, are squeezed out aseptically from the intact anther and then cultured on nutrient medium where the microspores, without producing male gametes, develop into haploid embryoids or callus tissue that give rise to haploid plantlets by embryogenesis or organogenesis. Commercial varieties developed through DH protocols have been reported for many crops, such as wheat (*Triticum aestivum* L.), barley (*H. vulgare* L.), triticale (*Tritico secale* Wittm.), rice (*Oryza sativa* L.), *Brassica* spp., eggplant (*Solanum melongema* L.), pepper (*Capsicum annuum* L.), asparagus (*Asparagus officinalis* L.), and tobacco (*N. tabacum* L.) (Thomas et al., 2003). Tomato (*Solanum lycopersicum* L.) is considered to be recalcitrant to DH technology; however, plants have been derived from anther tissue and globular embryos have been reported from isolated microspores (Segui-Simarro and Nuez, 2007). Cultured tomato anthers produced both gametophytic and sporophytic calli, and regenerated plants were mostly mixoploid, although there were also small numbers of haploid and diploid plants. Supena et al. (2006) developed a successful shed microspore protocol for Indonesian hot pepper (*Capsicum annuum* L.), which utilized a 2-layer culture medium consisting of a liquid upper layer and a solid lower layer containing activated charcoal. Cultured anthers floated to the surface of the upper liquid layer, whereupon microspores dehisced and embryos formed from microspores which were free from the anther wall tissue. In contrast, isolated microspore culture produced higher numbers of sporophytically dividing microspores, but very few embryos. Kim et al. (2008) achieved embryogenesis from hot peppers using NLN medium with 9% sucrose. They also noted that microspore plating density was extremely important with the optimum plating density being 8×10^{-4} – 10×10^{-4} microspores/mL. Recently, improvements in isolated microspore culture of pepper were achieved utilizing co-culture with wheat or pepper ovaries (Lantos et al., 2009). Cultures with pepper ovaries produced multicellular structures but development stopped at this stage, while microspore cultures with wheat ovaries continued embryo development.

8.7.2 FACTORS INFLUENCING ANTHER CULTURE

8.7.2.1 GENOTYPE OF DONOR PLANTS

The genotype of the donor plant plays a significant role in determining the frequency of pollen production. For example, Hordeum of each genotype differs with respect to androgenic response in anther culture.

8.7.2.2 ANTHER WALL FACTOR

The anther wall provides the nourishment in the development of isolated pollen of a number of species. There are reports that glutamine alone or in combination with serine and myoinositol could replace the anther wall factor for isolated cultures.

8.7.2.3 CULTURE MEDIUM

The anther culture medium requirements vary with genotype and probably the age of the anther as well as condition under which donor plants are grown. Incorporation of activated charcoal into the medium has stimulated the induction of androgenesis. The iron in the medium plays a very important role for the induction of haploids. Potato extracts, coconut milk, and growth regulators like auxin and cytokininare used for anther and pollen culture.

8.7.2.4 STAGE OF MICROSPORES

In most of the cases, anthers are most productive when cultured at the uninucleate microspore stage, for example, barely, wheat, rice, etc. Anther of some species gives the best response if pollen is cultured at first mitosis or later stage such as Datura and tobacco.

8.7.2.5 EFFECT OF TEMPERATURE

Temperature enhances the induction frequency of microspore androgenesis (it is the in vitro development of haploid plants originating from totipotent pollen grains through a series of cell division and differentiation). The low-temperature treatment to anther or flower bud enhances the haploid formation. The low temperature affects the number of factors such as dissolution of microtubules lowering of absicisic acid maintenance of higher ratio of viable pollen capable of embryogenesis.

8.7.2.6 PHYSIOLOGICAL STATUS OF DONOR PLANT

Physiological status of donor plant such as water-stress nitrogen requirement and age of donor plant highly affect the pollen embryogenesis. Plants starved of nitrogen may give more responsive anthers compared to those that are well fed with nitrogenous fertilizers.

8.7.3 ADVANTAGE OF POLLEN CULTURE OVER ANTHER CULTURE

1. During anther culture there is always the possibility that somatic cells of the anther that are diploid will also respond to the culture condition and so produce unwanted diploid calli or plantlets.
2. Sometimes the development of microspores inside the anther may be interrupted due to growth inhibiting substances leaking out of the anther wall in contact with nutrient medium.

8.7.4 IMPORTANCE OF POLLEN AND ANTHER CULTURE

1. Utility of anther and pollen culture for basic research:
    a) Cytogenetic studies.
    b) Study of genetic recombination in higher plants.
    c) Study of mode of differentiation from single cell to whole organism.
    d) Study of factor controlling pollen embryogenesis of higher plants.
    e) Formation of double haploid (DH) that is homozygous and fertile.

2. Anther and pollen culture are used for mutation study. For example, nitrate reductae mutants are reported in *N. tabacum*.
3. Anther and pollen-culture use for plant breeding and crop improvement.
4. Anther culture is used to obtain the alkaloid. For example, homozygous recombination *Hyoscyamus niger* having higher alkaloid content is obtain by anther culture.
5. Haploid are used in molecular biology and genetic engineering. For example, Haploid tissue of Arbidopsis and lycopersicon has been used for the transfer and expression of three genes from *Escherchia coli*.

8.8 CELL AND TISSUE CULTURE

Plant tissue culture or cell culture is a technique of growing plant cells, tissues, organs, seeds, or other plant parts in a sterile environment on a nutrient medium. Totipotency: ability of a cell or tissue or organ to grow and develop into a fully differentiated organism. Somatic embryogenesis is the process in which embryo-like structures are formed from somatic tissues and develop into a whole plant. It is of two types, that is, direct somatic embryogenesis: the embryo is formed directly from a cell or small group of cells such as the nucleus, styles, or pollen without the production of an intervening callus. Direct somatic embryogenesis is generally rare, whereas indirect somatic embryogenesis is the process in which callus is first produced from the explant and then embryos are produced from the callus tissue or from a cell suspension cultures. When friable callus is placed into the appropriate liquid medium and agitated, single cells, and/or small clumps of cells are released into the medium and continue to grow and divide, producing a cell-suspension culture. The inoculum used to initiate cell suspension culture should neither be too small to affect cells numbers nor too large to allow the buildup of toxic products or stressed cells to lethal levels. Cell suspension culture techniques are very important for plant biotransformation and plant genetic engineering.

Somatic or asexual embryogenesis is the production of embryo-like structures from somatic cells without gamete fusion. During their development, somatic embryos pass through several stages similar to those observed in zygotic embryogenesis. Somatic embryos are independent of the surrounding tissues and accumulate embryo-specific proteins and mRNAs (Zimmermann, 1993). Somatic embryos arise from in vitro cultured cells in the process called indirect somatic embryogenesis. This process requires the induction of embryogenic competence. Indirect somatic embryogenesis is the most common method to generate somatic embryos for practical uses (Redenbaugh, 1993).

Commercial cultivars of cucumber were explored for embryogenesis and plant regeneration in somatic tissues on PGRs. Maximum callus induction 94.16% and 76% was observed in leaf disc explants on MS medium supplemented with 2,4-dichlorophenoxyacetic acid (2,4-D) (2 mg/L), NAA, and BAP (1.5 mg/L, each), respectively (Table 8.1). Seed cotyledon explants induced maximum calli (77%) on 4.0 + 0.75 mg/L (BAP + NAA). Calli induced in leaf disc on the highest level of 2,4-D (5 mg/L) yielded the highest embryo formation (23%) whereas calli induced on BAP and BAP + NAA (5 + 1 mg/L) regenerated into 14% and 12% shoots, respectively. These shoots were excised and rooted on MSO medium. The plantlets were transplanted in pots and transferred to field after acclimatization. The developed plant material will be morphologically and genetically characterized for homozygosity (Usman et al., 2011).

An efficient protocol of direct somatic embryogenesis (without involving intermediate callus) has been developed from hypocotyl explants of two *Capsicum annuum* L. genotypes with potential for high frequency production of this important horticultural crop. MS medium supplemented with different concentrations of thidiazuron (TDZ) or 2,4-D were used. Two types of media (woody plant medium [WPM] and MS) as well as sucrose concentrations were examined for induction of direct somatic embryogenesis. WPM was significantly more effective on number of somatic embryos formation (9.30) as compared with that of MS results (7.22). The highest number of embryos (14.60) was obtained by using WPM with 80 g/L sucrose. The addition of 1.0 mg/L of $AgNO_3$ enhanced the induction of direct somatic embryogenesis affecting both the percentage of explants forming somatic embryos and the number of somatic embryos per explant, while higher doses (1.5 and 2.0 mg/L) negatively affected the regenerative capacity. MS at half strength contained 30 g/L sucrose was more effective in conversion somatic embryos and producing normal plants. However, increasing sucrose concentration had a negative effect on normal germination of somatic embryos. Finally, plantlets were transferred to a mixture of peatmoss and vermiculite at equal volume with survival rate 54% after 21 days with respect to morphology and growth characteristics (Aboshama, 2011).

The embryogenic capacity of seven cucumber (*Cucumis sativus* L.) cultivars was examined by tissue culture of cotyledon, young first-leaf, and internode explants. Somatic embryogenesis frequencies differed significantly

TABLE 8.1 Media Composition for Callus Induction and Regeneration of Different Vegetable Crops.

Species	Type of Explants Used	Developmental Stage of Ovule/ Ovary	Pretreatment	Induction Medium	Regeneration Medium	Recovery
Allium cepa L. and *A. roylei* Stearn	Female flower bud	Umbel with 20–30% opened flower	4°C	BDS medium + 2 mg/L 2,4-D + 2 mg/L 6-BA	BI medium + 1 mg/L NAA + 2 mg/L 2 ip	Haploid and double haploid plantlets
Allium L.	Flower bud	Just before anthesis	4°C	B5 + 2 mg/L 2,4-D + 2 mg/L BA + 100 g/L sucrose	MS + 1 mg/L NAA + 2 mg/L 2 ip + 100 g/L sucrose	Haploid plantlet
Cucumber (*Cucumis sativus* L.)	Ovary slices	6 h before anthesis	35°C in dark (2–10 days)	CBM + 0.02 mg/L TDZ + 4% sucrose	CBM + 0.02 mg/L NAA + 0.2 mg/L BA	Haploid plantlet
Cucumber (*Cucumis sativus* L.)	Ovary slices	1 day before anthesis	4°C	MS + 0.04 mg/L TDZ + 3% sucrose + 0.8% agar	MS + 0.3mg/L BA	Haploid and double haploid plantlet
Onion (*Allium cepa* L.)	Flower bud	3–5 days before anthesis	4°C	BDS macro + BS micro + MS vitamin + 100g/L sucrose + 0.5 mM putrescine	BDS macro + BS micro + MS vitamin + 100 g/L sucrose + 0.1 mM spermidine	Haploid plantlet
Shallot (*Allium cepa* L. Aggregatum group)	Flower bud	3–5 days before anthesis	4°C	B5 + 1–2 mg/L 2,4-D + 1–2 mg/L BA + 75% sucrose + 0.7% agar	MS + 3% sucrose + 0.75% agar	Haploid, double haploid, and tetraploid
Squash (*Cucurbita pepo* L.)	Ovules	1 day before anthesis	4°C	MS + 0.1, 1.0, 5.0 mg/L 2,4-D	MS medium	Haploid and double haploid

TABLE 8.1 *(Continued)*

Species	Type of Explants Used	Developmental Stage of Ovule/ Ovary	Pretreatment	Induction Medium	Regeneration Medium	Recovery
Sugarbeet (*Beta vulgaris*)	Ovules	Unopened flower at bulbing stages	4°C	MS + 0.04 mg/L TDZ + 3% sucrose + 0.8% agar	MS + 2.0 mg/L BAP 2.5 or 5.0 mg/L $AGNO_3$ + 0.5% activated charcoal	Haploid and double haploid
Summer squash (*Cucurbita pepo* L.)	Ovules	1 day before anthesis	4°C	MS + 0.1, 1.0, 5.0 mg/L 2,4-D	MS + 3% sucrose + 1-mg/L KNO_3 + 1-mg/L 2,4-D	Haploid and double haploid

among the tested cultivars, and "Fushinarimidori" produced the highest number of embryos from either cotyledons or young first leaves. Cotyledon- and first-leaf-derived calluses produced more embryos than calluses from internodes. Somatic embryos were induced from "Aonaga F1" internodes. With relatively high sucrose levels (6% and 9%) in the initiation medium, the frequency of embryogenic callus formation from "Fushinarimidori" cotyledon explants was >90%. The highest yield of somatic embryos occurred in cultures initiated with high sucrose levels (9% or 12%), although 12% sucrose inhibited callus formation and growth. Somatic embryos germinated in a basal liquid medium supplemented with 0.5% activated charcoal, and they developed into well-shaped, healthy plantlets on semisolid medium with 1% sucrose (Lou and Kako, 1994).

In vitro culture conditions represent an unusual combination of stress factors that plant cells encounter (e.g., oxidative stress as a result of wounding at excision of the explant tissue, PGRs, low or high salt concentration in solution, low or high light intensities). The stress associated with in vitro induction of SE may result in an overall stress response as expressed as chromatin reorganization. An extended chromatin reorganization is believed to cause an "accidental" release of the embryogenic program, the latter normally being repressed by a chromatin-mediated gene silencing mechanism (Fehér, 2005). Direct evidence for changes of DNA methylation during SE is well documented (Chakrabarty et al., 2003; Leljak-Levanic et al., 2004; Santos and Fevereiro, 2002). However, the ability of in vitro cultures to generate embryos is limited to a group of cells or a discrete zone of embryogenic callus.

8.9 SOMACLONAL VARIATION

It is the genetic variations in plants that have been produced by plant tissue culture and can be detected as genetic or phenotypic traits.

8.9.1 BASIC FEATURES OF SOMACLONAL VARIATIONS

- Variations for karyotype, isozyme characteristics, and morphology in somaclones may also observe.
- Calliclone (clones of callus), mericlone (clones of meristem), and protoclone (clones of protoplast) were produced.
- Generally heritable mutation and persist in plant population even after plantation into the field.

8.9.2 MECHANISM OF SOMACLONAL VARIATIONS

1. Genetic (heritable variations)
   - Preexisting variations in the somatic cells of explant
   - Caused by mutations and other DNA changes
   - Occur at high frequency

2. Epigenetic (nonheritable variations)
   - Variations generated during tissue culture
   - Caused by temporary phenotypic changes
   - Occur at low frequency.

8.9.3 DETECTION AND ISOLATION OF SOMACLONAL VARIANTS

1. Analysis of morphological characters
   Qualitative characters: Plant height, maturity date, flowering date, and leaf size.
   Quantitative characters: yield of flower, seeds, and wax contents in different plant parts.
2. Variant detection by cytological studies
   Staining of meristematic tissues like root tip, leaf tip with feulgen and acetocarmine provide the number and morphology of chromosomes.
3. Variant detection by DNA contents
   Cytophotometer detection of feulgen-stained nuclei can be used to measure the DNA contents.
4. Variant detection by gel electrophoresis
   Change in concentration of enzymes, proteins, and hemical products like pigments, alkaloids, and amino acids can be detected by their electrophoretic pattern.
5. Detection of disease-resistance variant
   Pathogen or toxin responsible for disease resistance can be used as selection agent during culture.
6. Detection of herbicide resistance variant
   Plantlets generated by the addition of herbicide to the cell culture system can be used as herbicide resistance plant.

7. Detection of environmental stress tolerant variant
    - Selection of high salt tolerant cell lines in tobacco
    - Selection of water-logging and drought resistance cell lines in tomato
    - Selection of temperature stress tolerant in cell lines in pear
    - Selection of mineral toxicities tolerant in sorghum plant (mainly for aluminum toxicity).

8.9.4 ADVANTAGES OF SOMACLONAL VARIATIONS

- Help in crop improvement
- Creation of additional genetic variations
- Increased and improved production of secondary metabolites
- Selection of plants resistant to various toxins, herbicides, high salt concentration, and mineral toxicity
- Suitable for breeding of tree species.

8.9.5 DISADVANTAGES OF SOMACLONAL VARIATIONS

- A serious disadvantage occurs in operations which require clonal uniformity, as in the horticulture and forestry industries where tissue culture is employed for rapid propagation of elite genotypes.
- Sometime leads to undesirable results.
- Selected variants are random and genetically unstable.
- Require extensive and extended field trials.
- Not suitable for complex agronomic traits like yield, quality, etc.
- May develop variants with pleiotropic effects which are not true.

Somatic or asexual embryogenesis is the process by which somatic cells develop into plants. The rapid improvement in somatic embryogenesis methods allows the use of somatic embryos in plant micropropagation as synthetic seeds. However, practical applications of somatic embryogenesis are not limited to synthetic seed technology. Somatic embryogenesis can be used in the regeneration of genetically transformed plants, polyploidy plants, or somatic hybrids. Moreover, promising results indicate the possibility to use somatic embryogenesis in cell selection programs and germplasm cryopreservation. The application of somatic embryogenesis to plant virus elimination, metabolite production, and in vitro mychorrhizal initiation has been investigated (Vincent and Martínez, 1998).

8.10 CHROMOSOME ENGINEERING

It refers to the technologies in which there is manipulation of chromosome to change their mode of genetic inheritance for example haploid *Arabidopsis thaliana* produced by altering the kinetochore protein yielding homozygous line. It will facilitate reverse breeding that downregulate recombination to ensure progeny contain intact parental chromosomes. This technique aims to create artificial chromosomes or to change basic genetic process by manipulating chromosomes proteins (Chan, 2010). About 70% of plant species for example potato, oat, kiwi fruit, etc. have produced from spontaneous interspecific and intergeneric hybridization although extent of natural hybridization differs among different genera and families. The stabilization of these hybrids results in the formation of new biological species (Masterson, 1994).

Three species of *Cyamopsis* were studied to find out barriers to interspecific crosses between *C. tetragonoloba* × *C. serrate* and *C. tetragonoloba* × *C. senegalensis* which serve as a stepping stone for guar improvement. Quantitative production of pollen was identical in all the species. Pollen grains of *C. tetragonoloba* and *C. senegalensis* showed more than 95% of viability and *C. serrata* have 87% viability. Nutritive requirement for in vitro germination of pollen revealed that *C. tetragonoloba* required 25% sucrose + 100 ppm boric acid + 300 pm calcium nitrate and *C. senegalensis* needed 35% sucrose with same basal medium, while *C. serrata* required 35% maltose + 6% PEG 6000 along with above dose of boric acid and calcium nitrate. Moreover, pollen germination in *C. serrata* was initiated after 30 h of incubation and its pollen tubes were slow growing attaining 174.7 μm length in 48 h. The length of style of *C. tetragonoloba* and *C. serrata* was nearly identical (2.6 mm) while *C. senegalensis* possess longest style (3.8 mm). Protein content of stigma + style was nearly identical in all the species and total soluble carbohydrate content in *C. tetragonoloba* and *C. serrata* was nearly identical (5–6 mg/100 mg FW) but lower content was in *C. senegalensis* (2.4 mg/100 mg FW). It was observed that interspecific hybridization between *C. tetragonoloba* × *C. serrata* was successful by use of stub smeared with PGM and 10.43% of pod setting. Color and shape of hybrid seeds was akin to the female parent (*C. tetragonoloba*), hybrid plants showed early flowering just like male parent (*C. serrata*) whereas the plant height was intermediate between the two parents (Ahlawat et al., 2013).

Tomato is highly prone to biotic stresses, especially diseases, insects, and nematodes. Genes are available in different wild species, but it has not been easy to transfer these genes in cultivated species due to problems in crossability. *S. lycopersicum* was crossed with *S. peruvianum* and *Solanum pimpinellofolium*. Twenty-five days after pollination was found to be the optimum time for rescuing the embryos. MS medium supplemented with 1 mg/L GA3, 0.1 mg/L NAA, and 0.5 mg/L BAP was found to be the most effective for germination of the immature putative hybrid embryos. The confirmation of hybridity of the embryo rescued plants from the interspecific crosses of both *S. lycopersicum* var. MT-3 and *S. lycopersicum* var. Kashi Amrit with *S. peruvianum* (WIR-3957) was done using RAPD markers (Kharkongar et al., 2012).

The pioneered plant chromosome engineering research done from 50 years ago by directed transfer of a leaf rust resistance gene from an alien chromosome to a wheat chromosome using X-ray irradiation and an elegant cytogenetic scheme (Qi et al., 2007) but dealing with induced homoeologous pairing and recombination is the most powerful and has been extensively used in wheat. Here, we review the current status of homoeologous recombination-based chromosome engineering research in plants with a focus on wheat and demonstrate that integrated use of cytogenetic stocks and molecular resources can enhance the efficiency and precision of homoeologous-based chromosome engineering for based transfer of virus resistance from an alien chromosome to a wheat chromosome, its characterization, and the prospects for further engineering by a second round of recombination. Wide or distant hybridization has been widely used as an important tool of chromosome manipulation for crop improvement. The chromosome behaviors in F_1 hybrids provide us with the essential genetic basis for chromosome manipulation. The induction of homoeologous pairing in F_1 hybrid plants followed by the incorporation of a single-chromosome fragment from an alien or a wild species into an existing crop species by translocating chromosomes has been used in the production of translocation lines. Chromosome doubling in somatic cells or gametes of F_1 hybrids followed by the incorporation of all alien chromosomes has been used in the production of amphidiploids. Amphidiploidy can be used for a bridge to move a single chromosome from one species to another or for the development of new crops. Chromosome elimination of a uniparental genome during the development of F_1 hybrid embryos has been used in the production of haploids. Haploids are very useful in double-haploid breeding of a true-breeding crop such as wheat and rice since this method can quickly replace genetic recombination while enhancing breeding efficiency or facilitating genetic analysis.

8.11 MOLECULAR MARKER

Marker is a tag which is conspicuous or apparent or which helps in identification of traits. Molecular marker is a DNA sequence that is readily based on the basic strategy, and some major ones are detected and whose inheritance can easily be monitored. A marker must be polymorphic, that is, it must exist in the following:

1. It must be polymorphic as it is the polymorphism is measured for genetic diversity studies.
2. Codominant inheritance: Molecular marker should be detectable in diploid organisms to allow discrimination of homo- and heterozygotes.
3. A marker should be evenly and frequently distributed studies.
4. It should be easy, fast, and cheap to detect.
5. It should be reproducible.
6. High exchange of data between laboratories.

These have been grouped into the following categories:

8.11.1 NON-PCR-BASED APPROACHES

8.11.1.1 RESTRICTION FRAGMENT LENGTH POLYMORPHISM

Restriction fragment length polymorphism (RFLP) was the first technology that enabled the detection of polymorphism at the DNA sequence level. Genetic information, which makes up the genes of higher plants, is stored in the DNA sequences. Variation in this DNA sequence is the basis for the genetic diversity within a species.

8.11.1.1.1 Advantages of RFLP

1. It permits direct identification of a genotype or cultivar in any tissue at any developmental stage in an environment-independent manner.
2. RFLPs are codominant markers, enabling heterozygotes to be distinguished from homozygotes.
3. It has a discriminating power that can be at the species/population (single-locus probes) or individual level (multi-locus probes).
4. The method is simple as no sequence-specific information is required.

8.11.1.1.2 Disadvantages of RFLPs

1. Conventional RFLP analysis requires relatively large amount of highly pure DNA.
2. A constant good supply of probes that can reliably detect variation are needed.
3. It is laborious and expensive to identify suitable marker/restriction enzyme combinations from genomic or cDNA libraries where no suitable single-locus probes are known to exist.
4. RFLPs are time-consuming as they are not amenable to automation.
5. RFLP work is carried out using radioactively labeled probes and therefore requires expertise is autoradiography.

8.11.1.1.3 Procedure of RFLP Analysis

1. DNA isolation,
2. cutting DNA into smaller fragments using restriction enzyme(s),
3. separation of DNA fragments by gel electrophoresis,
4. transferring DNA fragments to a nylon or nitrocellulose membrane filter,
5. visualization of specific DNA fragments using labeled probes, and
6. analysis of results.

8.11.2 PCR-BASED APPROACHES

Random-amplified polymorphic DNA (RAPD), microsatellite or simple sequence repeat polymorphism (SSRP), amplified fragment length polymorphism (AFLP), arbitrarily primed PCR, etc.

8.11.2.1 RANDOM AMPLIFIED POLYMORPHIC DNA MARKERS

RAPD analysis is a PCR-based molecular marker technique. Here, single short oligonucleotide primers are arbitrarily selected to amplify a set of DNA segments distributed randomly throughout the genome. Williams et al. (1990) showed that the differences as polymorphisms in the pattern of bands amplified from genetically distinct individuals behaved as Mendelian genetic markers.

8.11.2.1.1 Advantages

1. Need for a small amount of DNA (15–25 ng) makes it possible to work with populations which are inaccessible for RFLP analysis.
2. It involves nonradioactive assays.
3. It needs a simple experimental set-up requiring only a thermocycler and an agarose assembly.
4. It does not require species-specific probe libraries thus, work can be conducted on a large variety of species where such probe libraries are not available.
5. It provides a quick and efficient screening for DNA-sequence-based polymorphism at many loci.
6. It does not involve blotting or hybridization steps.

8.11.2.1.2 Limitations

1. RAPD polymorphisms are inherited as dominant-recessive characters. This causes a loss of information relative to markers which show codominance.
2. RAPD primers are relatively short, a mismatch of even a single nucleotide can often prevent the primer from annealing; hence, there is loss of band.
3. RAPD is sensitive to changes in PCR conditions, resulting in changes to some of the amplified fragments.

Experiment on 38 genotypes of *M. charantia* including few commercially cultivars collected from different parts of India based on agro-ecological zones were analyzed for diversity study both at morphological and molecular levels (Dey et al., 2006). Diversity based on yield-related traits and molecular analysis was not in consonance with ecological distribution. Among 116 random decamer primers, screened 29 were polymorphic and informative enough to analyze these genotypes. A total of 208 markers generated of which 76 (36.50%) were polymorphic and the number of bands per primer was 7.17 out of them 2.62 were polymorphic. Pair-wise genetic distance (GD) based on molecular analysis ranged from 0.07 to 0.50 suggesting a wide genetic base for the genotypes. The clustering pattern based on yield-related traits and molecular variation was different. So, it may be sufficient and more efficient RAPD primers showing maximum number of polymorphic bands or other available marker systems could be utilized for analysis of germplasm.

8.11.2.2 AMPLIFIED FRAGMENT LENGTH POLYMORPHISM

This is a highly sensitive method for detecting polymorphism throughout the genome, and it is becoming increasingly popular. It is essentially a combination of RFLP and RAPD methods, and it is applicable universally and is highly reproducible. It is based on PCR amplification of genomic restriction fragments generated by specific restriction enzymes and oligonucleotide adapters of few nucleotide bases (Vos et al., 1995).

8.11.2.2.1 AFLP Involves the Following Steps

DNA is cut with restriction enzymes (generally by two enzymes), and double-stranded (ds) oligonucleotide adapters are ligated to the ends of the DNA fragments. Selective amplification of sets of restriction fragments is usually carried with 32 P-labeled primers designed according to the sequence of adapters plus 1–3 additional nucleotides. Only fragments containing the restriction site sequence plus the additional nucleotide will be amplified. Gel analysis of the amplified fragments: The amplification products are separated on highly resolving sequencing gels and visualized using autoradiography. Fluorescent or silver staining techniques can be used to visualize the products in cases where radiolabelled nucleotides are not used in the PCR.

8.11.2.2.2 Advantages

1. This technique is extremely sensitive.
2. It has high reproducibility, rendering it superior to RAPD.
3. It has wide-scale applicability, proving extremely proficient in revealing diversity.
4. It discriminates heterozygotes from homozygotes when a gel scanner is used.
5. It is not only a simple fingerprinting technique, but can also be used for mapping.

8.11.2.2.3 Disadvantages

1. It is highly expensive and requires more DNA than is needed in RAPD (1 mg per reaction).
2. It is technically more demanding than RAPDs, as it requires experience of sequencing gels.

4. AFLPs are expensive to generate as silver staining, fluorescent dye, or radioactivity detect the bands.

Molecular data from mitochondrial, nuclear, and chloroplast DNA RFLPs, nuclear microsatellites, isozymes, and gene sequences of internal transcribed spacers of nuclear ribosomal DNA (ITS; multiple-copy), the single-copy nuclear encoded granule-bound starch synthase gene (GBSSI or waxy), and morphology, have been used to examine hypotheses of species relationships. This study is a companion to the previous GBSSI gene sequence study and to the morphological study of relationships of all 10 wild tomato species (including the recently described *S. galapagense* with a concentration on the most widespread and variable species *S. peruvianum*. These new AFLP data are largely concordant with the GBSSI and morphological data and in general support the species outlined in the latest treatment by C. M. Rick, but demonstrate the distinct nature of northern and southern Peruvian populations of *S. peruvianum*, and suggest that their taxonomy needs revision. *Solanum ochranthum* is supported as sister to wild tomatoes, and *S. habrochaites* and *S. pennellii* reside in a basal polytomy in the tomato clade (Spooner et al., 2005).

8.11.2.3 SIMPLE SEQUENCE REPEATS (MICROSATELLITES)

The term microsatellites were coined by Litt and Lutty (1989). SSRs, also known as microsatellites, are present in the genomes of all eukaryotes. These are ideal DNA markers for genetic mapping and population studies because of their abundance. These SSR length polymorphisms at individual loci are detected by PCR, using locus-specific flanking region primers where the sequence is known. Thus, STMs require precise DNA sequence information for each marker locus from which a pair of identifying flanking markers is designed. This is impractical for many plant and animal species that are not well-characterized genetic systems. Some of these SSR-based methods have been collectively termed microsatellite-primed PCR.

8.11.2.3.1 Steps of SSRs Analysis

Isolate the DNA of representative cultivar/line. Restrict with four base pair cutter. Size fractionation (0.5–0.7 kb) ligates to a suitable vector and transform into *E. coli*. Following hybridization identify the desired transformation.

Go for end sequencing of the selected clones and designing the primers for amplification.

8.11.2.3.2 Advantages

- Codominant markers.
- Highly polymorphic and highly reproducible.

8.11.2.3.3 Disadvantages

- Costly in term of primer designing.

8.11.2.4 SEQUENCE TAGGED SITES

STS is a short unique sequence (60–1000 bp) that can be amplified by PCR, which identifies a known location on a chromosome (Olsen et al., 1989). Specific PCR markers that match the nucleotide sequence of the ends of DNA fragment can be derived from primers, for example, an RFLP probe or an expressed sequence tag. To date, all STSs that have been used in mapping projects have been derived from well-characterized probes or sequences. STSs are the physical DNA landmarks and PCR is the experimental method used to detect them. STS maps simply represent the relative order and spacing of STSs within a region of DNA. Using this technique, tedious hybridization procedures involved in RFLP analysis can be overcome. STSs have been extensively used for physical mapping of genome.

8.11.2.5 SEQUENCE-TAGGED MICROSATELLITES

The term microsatellites were coined by Litt and Lutty (1989). Simple sequence repeats, also known as microsatellites, are present in the genomes of all eukaryotes. These are ideal DNA markers for genetic mapping and population studies because of their abundance. These are tandemly arranged repeats of mono-, di-, tri-, and tetra-nucleotides with different lengths of repeat motifs (e.g., A, T, AT, GA, AGG, AAC, etc.). Motifs are A, AT, AGG, etc. and repeat number is denoted by n. Thus a repeat (AT) nine means AT nucleotides is tandemly arranged one after another nine times. In a genome of a particular species when this repeat is identified in a gene, which constitutes a microsatellite, the gene is sequenced with its flanking sequences to design primers for amplification of microsatellites. The regions flanking the microsatellite are generally conserved among genotypes of the same species. PCR primers to the flanking regions are used to amplify the SSR-containing DNA fragment. Length polymorphism is created when PCR products from different individuals vary in length as a result of variation in the number of repeat units in the SSR. Genebank sequence data have also been used for designing primers for amplification of microsatellites. Thus, SSRP reflects polymorphism based on the number of repeat units in a defined region of the genome.

8.11.2.5.1 Procedure

A specific microsatellite contained within a stretch of DNA can be amplified by PCR using flanking primer sequences, and then analyzed on metaphor agarose or polyacrylamide gels. The gels are stained with ethidium bromide and seen under UV light. The variation in length of the PCR product is a function of the number of SSR units. This is a relatively new technique and is especially useful in inbreeding crops such as wheat and barley, which are characterized by low levels of RFLP variation.

8.11.2.6 SEQUENCE CHARACTERIZED AMPLIFIED REGIONS

A sequence characterized amplified region (SCAR) is a genomic DNA fragment at a single genetically defined locus that is identified by PCR amplification using a pair of specific oligonucleotide primers. Williams et al. (1990) converted RFLP markers into SCARs by sequencing two ends of genomic DNA clones and designing oligonucleotide primers based on the end sequences. These primers were used directly on genomic DNA in a PCR reaction to amplify the polymorphic region. If no AFLP is noticed, then the PCR fragments can be subjected to restriction digestion to detect RFLPs within the amplified fragment. SCARs are inherited in a codominant fashion in contrast to RAPDs, which are inherited in a dominant manner. Paran and Michelmore (1993) converted RAPD markers into SCARs. Amplified RAPD products are cloned and sequenced. The sequence of primers derived from the termini of a band is identified as a RAPD marker. Two 24-base oligonucleotide primers corresponding to the ends of the fragment (the 5′ 10 bases are the same as the original 10-mer used in the RAPD reaction and 14 internal bases from the end) have been synthesized. These primers with their increased specificity generally amplify a single highly repeatable band, not the 5–10 bands for the progenitor 10 base primers. SCARs are similar to STSs, but do not involve DNA hybridization for detection and can therefore contain repeated DNA sequences. SCARs have several advantages over RAPD markers. RAPDs show dominant nature, amplification of multiple loci, and are sensitive to reaction conditions. The mapping efficiency of RAPD markers in F2 populations is decreased by their dominant nature. The conversion of dominant RAPDs to codominant SCARs increases the amount of information per F2 individual. As the annealing conditions for SCARs are more stringent than for RAPDs, SCAR primers detect only one locus. Also, the use of longer oligonucleotide primers for SCARs allows a more reproducible assay than the one obtained with the short primers used for RAPD analysis. SCARs can readily be applied to commercial breeding programs as they do not require the use of radioactive isotopes.

In the past, genetic maps were based mainly on morphological and isozyme markers. But these markers are limited and are influenced by environment and developmental stage. Molecular marker on the other hand is large in number and is not influenced by environment and development stage. Saturated linkage maps are prerequisite for gene tagging, marker-assisted selection, and map-based gene cloning. Yayeh (2005) identified first genetic linkages in male fertile garlic accessions based on single nucleotide polymorphism simple sequence repeats and RAPDs. Thirty seven markers formed nine linkage groups covering 415 centrimorgans (cM) with average distance of 15 cM between loci. A male fertility locus was placed on the map. A 109 point linkage map consisting of three phenotypic loci (P1, Y2, and Rs), 6 restriction fragment length polymorphic DNA (RFLPs), 2 RAPDs, 96 AFLPs, and 2 selective amplification of microsatellite polymorphic loci was constructed in carrot by Vivek and Simon (1999). A genetic map of an interspecific cross in Allium based on amplified length polymorphism markers constructed by Van Heusden et al. (2000). The map based on *A. cepa* markers consisted of eight linkage groups whereas map based an *A. roylei* markers comprised 15 linkage groups. Zhang et al. (2004) constructed linkage map for watermelon using recombinant inbred lines (RILs) from a cross between the high-quality inbred line 97103 and the fusarium wilt resistant plant introduction using RAPD and SCAR markers. This map is useful for further development of quantilative trait loci affecting fruit quality and for identification of genes conferring resistance to fusarium wilt. Resistance to *Verticillium dahliae* race 1 is conferred by a single dominant gene in tomato, *i.e.* locating it on different chromosomes, which subsequently raised the possibility that Verticillium resistance may be controlled by a number of loci. Mapping populations was positioned on the short arm of chromosome 9 tightly linked to the RFLP marker *GP39*. This linkage was confirmed by screening for *GP39* in different breeding lines with known resistance or susceptibility to *Verticillium* indicating the potential use of *GP39* in the rapid detection of *Verticillium* resistance (Diwan *et al*., 1999).

8.11.2.6.1 Assessment of Genetic Diversity

Molecular markers have proved to be excellent tools for assessment of genetic diversity in a wide range of plant species. The information is often of direct utility to plant breeders, since it is indicative of the performance, adaptation, or other agronomic qualities of the germplasm. Molecular markers have provided very useful information about the overall genetic range of crop germplasm. For breeders, this information is important to take decisions regarding the utility of germplasm particularly in search for rare and unique genes. Germplasm of narrow genetic base is obviously unlikely to harbor novel genes, for example, those conferring resistance to biotic and abiotic stresses. RAPD analysis of pepper breeding lines (Heras et al., 1996) revealed very narrow genetic base with more than 50% of the DNA bands being common among all the lines. In an assessment of the world collections of tomato, Villand et al. (1998) found South American accessions to have greater diversity than old world accessions. Shim and Jorgensen (2000) carried out AFLP analysis in wild and cultivated carrots and found that the old varieties released between 1974 and 1976 were more heterogeneous than newly developed F1 hybrids varieties.

8.12 GENE TAGGING

The most interesting application of molecular markers at present time is the ability to facilitate the method of "conventional" gene transfer. Gene tagging refer to mapping of genes of economic importance close to known markers. Thus, a molecular marker very closely linked to gene act as a tag that can be used for indirect selection of gene in breeding programs with the construction of molecular map, especially the RFLP maps, several genes of economic importance like disease resistance, stress tolerance, insect resistance, fertility restoration genes, yield-attributing traits have been tagged. Gene tagging is a prerequisite for marker-assisted selection and map-based gene cloning. In case of tomato TMV resistance Tm-2 locus, nematode resistance, *Mi* gene, *Fusarium oxysporum* resistance gene, powdery mildew resistance gene, has been tagged. Huang *et al.* (2000) tagged powdery mildew resistance gene ol-1 on chromosome 6 of tomato using RAPD and SCAR markers.

8.13 DNA FINGERPRINTING FOR VARIETAL IDENTIFICATION

DNA fingerprinting can be used for varietal identification as well as for ascertaining variability in the germplasm. Although any type of marker can be used but RAPDs, microsatellite and RFLPs are marker of choice for the purpose because all these are PCR based and does not require any prior information on nucleotides. The fingerprinting information is useful for quantification of genetic diversity, characterization of accessions in plant germplasm collections, and for protection of property of germplasm especially the cms lines. Molecular marker has been used widely for DNA fingerprinting of cultivars and breeding lines in a number of vegetable crops like tomato (Kaemmer et al., 1995), beans (Hamann et al., 1995), pepper (Prince et al., 1995), and potato (Ford and Taylor, 1997; McGregor et al., 2000).

8.14 BREEDING LINES AND ACCESSION IDENTIFICATION

Several situations during a breeding program may require identification of breeding lines and accessions. Mislabeling is a common problem in breeding experiments due to the large number of lines that need to be handled. Breeding lines can get contaminated due to mixing of seed samples and cross contamination in field. Molecular markers are ideal for distinguishing closely related genotypes that differ in few morphological traits. Use of human minisatellite probe 33.15 and the M13 repeat sequences for their ability to distinguish sister lines of two F6 backcrosses were demonstrated by Stockton and Gepts (1994). A comparison of the utility of 33.15 and M13 probes with GACA and ribosomal DNA sequences with respect to the polymorphism detected was made. The GACA 4 repeat was observed to be least efficient in discriminating the closely related lines of beans. Kaemmer et al. (1995) fingerprinted tomato accessions using microsatellite probes. The authors reported the utility of the technique in purity testing of breeding lines and in F1 progeny testing. Using RAPD technique, Tivang et al. (1996) revealed variation among and within artichoke-breeding populations. Heterogeneity was observed within clonal cultivars. Roose and Stone (1996) reported the utility of RAPD and RFLP markers in distinguishing F_1 from F_2 seeds in asparagus and for evaluation of seed purity. Ten pairs of potential duplicate accessions in a total of 134 capsicum accessions were identified by Rodriguez et al. (1999) on the basis of RAPD markers. Further, misclassified and unclassified accessions were placed in the correct groups. Using microsatellite markers, Fischer and Bachmann (2000) distinguished 83 accessions of onion.

8.15 SEX IDENTIFICATION

Early identification of male and female plants can bring considerable efficiency in breeding programs of dioecious species. Jiang and Sink (1997) developed SCAR markers in asparagus which were linked to the sex locus at a distance of 1.6 cM. Codominants STS markers enabling the differentiation of XY from YY males in asparagus were developed by Buttner and Jung (2002).

8.16 MAP-BASED GENE CLONING

One of the most serious limitations to the advance of plant molecular biology and biotechnology is the difficulty in isolating genes responsible for specific characters, yield, disease resistance, insect resistance, and quality are just few of the important characters for which genetic variation exists within crop species, but for which the corresponding genes have not yet been cloned. The advent of genome mapping at the DNA level (especially RFLPs) has provided a method for localizing genes of economic importance to specific chromosomal positions. The ability to map any gene of economic importance to a defined chromosomal site opens the possibility of isolating genes via chromosome walking. This method is called map-based gene cloning. Map-based cloning consists of four major steps:

1. Development of a high-resolution molecular linkage map in the region of interest.
2. Identification of appropriate YAC or BAC clones for isolating putative clones harboring the gene of interest.
3. Verification through transformation that the target gene is isolated. In tomato, the availability of a high-density molecular map and a yeast artificial chromosome library potentially provides the foundation on which to initiate map-based gene cloning for genes underlying any trait that can be genetically mapped (Martin et al., 1992).

8.17 MARKER-ASSISTED SELECTION

In this technique, linkages are sought between DNA markers and agronomically important traits such as resistance to pathogens, insects and nematodes, tolerance to abiotic stresses, quality parameters, and quantitative traits. Instead of selecting for a trait, the breeder can select for a marker that can be detected very easily in the selection scheme. The essential requirements for marker-assisted selection in a plant-breeding program are as follows: DNA marker-based selection for disease resistant trait essentially requires following conditions: The identified DNA marker(s) should cosegregate or closely linked (1 cM or less) with the resistant trait. Alternatively, less tightly linked flanking markers should be available for the resistant gene(s). The availability of an efficient screening technique(s) for DNA markers, which can be practically feasible to handle large populations. The screening technique should have high reproducibility across laboratories. The screening technique should be cost effective with high reproducibility. A number of markers linked with monogenic disease resistance are available in vegetable crops (Kumar et al., 2014). Such mapping has been facilitated by the use of different kind of mapping populations like near isogenic lines developed by repeated back crossing, RILs developed by single seed decent or DH methods. Nowadays, bulk sergeant analysis is increasingly being used to map monogenic resistance, because it allows rapid mapping of genes.

8.18 GENETIC ENGINEERING

Vegetables play an important role in human nutrition and health. Vegetable crop productivity and quality are seriously affected by several biotic and abiotic stresses, which destabilize rural economies in many countries. Moreover, absence of proper postharvest storage and processing facilities leads to qualitative and quantitative losses. In the past four decades, conventional breeding has contributed significantly for the improvement of vegetable yields, quality, postharvest life, and resistance to biotic and abiotic stresses. However, there are many constraints in conventional breeding, which can only be overcome by advancements made in modern biology. In the last decade, various traits such as biotic stress resistance, quality, and storage life have been successfully engineered into vegetable crops and some of them have been commercialized. In recent years, significant progress has been made to manipulate vegetable crops for abiotic stress tolerance, quality improvement, and pharmaceutical and industrial applications. Although the progress in commercialization of transgenic vegetable crops has been relatively slow, transgenic vegetables engineered for nutraceutical and pharmaceutical use will contribute significantly to the value added agriculture in near future (Table 8.2) (Dalal et al., 2006).

Over the past 25 years, methods have been developed to selectively alter the genetic instructions in the DNA that direct the growth and development of living organisms. Collectively known as "genetic engineering" or rDNA technology, these methods allow scientists to identify, cut out, and then reconnect specific genes (or DNA segments) into a carrier DNA (or vector). This DNA segment can then be introduced into the same or a different organism, and when it executes its instructions (or is "expressed"), it will transfer the characteristic coded by the gene to the receiving organism. Because DNA is chemically identical among all organisms, the instructions on these cloned pieces of DNA can be readily exchanged and "understood" between organisms. Because of their simple genetic makeup, transferring DNA to or among bacteria is relatively easy. Thus, the first applications of rDNA technology were to introduce useful genes into bacteria in order to produce large amounts of specific products. Insulin, for example, is now produced by expressing the human insulin gene in bacteria. Approximately 70% of all cheese produced is now processed using a recombinant enzyme called chymosin produced in bacteria, rather than the very similar enzyme (rennet) isolated from the stomach lining of calves. Transferring DNA into higher organisms is somewhat more complex but has been achieved for most important agricultural plants and animals. Thus, in theory, any gene from any organism is potentially transferable to other organisms by rDNA techniques.

In plants, transfer of genes or transformation can be accomplished by several methods. One fascinating approach uses a bacterial pathogen, *A. tumefaciens* to transfer the desired DNA into the plant. The bacterium naturally transfers part of its DNA into the plant's chromosomes, where it then causes the production of compounds that the bacterium consumes. Scientists have learned how to "disarm" the pathogen so that it can no longer impose its own changes, but it retains the ability to transfer DNA into the host plant. Desired genes can be spliced into the bacterial DNA and then *Agrobacterium*, like a video editor, will transfer them into the plant without causing disease (Suslow and Bradford, 1999).

Phenylpropanoid pathway of secondary metabolism are involved in interactions with beneficial microorganisms (flavonoid inducers of the Rhizobium symbiosis), and in defense against pathogens (isoflavonoid phytoalexins). The phenyl propane polymer lignin is a major structural component of secondary vascular tissue and fibers in higher plants.The isolation of genes encoding key enzymes of the various phenylpropanoid branch pathways opens up the possibility of engineering important crop plants such (a) improved forage digestibility, by modification of lignin composition and/or content; (b) increased or broader-spectrum disease resistance, by introducing novel phytoalexins or structural variants of the naturally occurring phytoalexins, or by modifying expression of transcriptional regulators of phytoalexin pathways; and (c) enhanced nodulation efficiency, by engineering over-production of flavonoid nod gene inducers.

KEYWORDS

- **micropropagation**
- **somatic hybridization**
- **meristem & anther culture**
- **PCR**
- **RAPD**
- **RFLP**
- **AFLP**
- **MAP**
- **MAS**

REFERENCES

Aboshama, H. M. S. Direct Somatic Embryogenesis of Pepper (*Capsicum annuum* L.). *World J. Agric. Sci.* **2011,** *7* (6), 755–762.

Akhtar, S.; Mandal, A.; Sarker, K. K.; Alam, M. F. In Vitro Propagation of Eggplant through Meristem Culture. *Agric. Conspectus Sci. Cus*. **2008,** *73* (3), 149–155.

Alan, A.R.; Lim, W.; Mutschler, M.A. & Earle, E.D. Complementary Strategies for Ploidy Manipulations in Gynogenic Onion (*Allium cepa* L.). *Plant Sci.* **2007,** *173* (1), 25–31.

Ahlawat, A.; S. K. Pahuja, S. K.; Dhingra, H. R. Studies on Interspecific Hybridization in Cyamopsis Species. *Afr. J. Agric. Res.* **2013**, 8(27), 3590–3597.

Arifin, N. S.; Ozaki, Y.; Okubo, H. Genetic Diversity in Indonesian Shallot (*Allium cepa* var. *ascalonicum*) and *Allium* × *wakegi* Revealed by RAPD Markers and Origin of *A.* × *wakegi* Identified by RFLP Analyses of Amplified Chloroplast Genes. *Euphytica* **2000,** *111*, 23–31.

Ashkenazi, V.; Chani, E.; Lavi, U.; Levy, D.; Hillel, J.; Veilleux, R. E. Development of Microsatellite Markers in Potato and their Use in Phylogenetic and Fingerprinting Analyses. *Genome* **2001,** *44* (1), 50–62.

Bhojwani, S. S.; Razdan, M. K. *Plant Tissue Culture: Theory and Practice: Developments in Crop Science*. Elsevier: Amsterdam, 1983.

Boxus, P. Plant Biotechnology Applied To Horticultural Crops. In: *World Conference on Horticulture Research*, 1998; pp 17–20.

Bray, E. A.; Bailey-Serres, J.; Weretilnyk, E. Responses to Abiotic Stresses. In: *Biochemistry and Molecular Biology of Plants*; Gruissem, W.; Buchannan, B.; Jones, R., Eds.; ASPP: Rockville, MD, 2000; pp 1158–1249.

Bray, E. A. Abscisic Acid Regulation of Gene Expression during Water-Deficit Stress in the Era of the *Arabidopsis* Genome. *Plant Cell Environ.* **2002,** *25*, 153–161.

Capiati, D. A.; País, S. M.; Téllez-Iñón, M. T. Wounding Increases Salt Tolerance in Tomato Plants: Evidence on the Participation of Calmodulin-Like Activities in Cross-Tolerance Signaling. *J. Exp. Bot.* **2006,** *57*, 2391–2400.

Cansian, R. L.; Echeverrigaray, S. Discrimination among Cultivars of Cabbage Using Randomly Amplified Polymorphic DNA Markers. *Hortic. Sci.* **2000,** *35*, 1155–1158.

Chan, S. W. L. Chromosome Engineering: Power Tools for Plant Genetics. *Trends Biotechnol.* **2010,** *28* (12), 605–610.

Chakrabarty, D.; Yu, K. W.; Peak, K. Y. Detection of DNA Methylation Changes during Somatic Embryogenesis of Siberian Ginseng (*Eleuterococcus senticosus*). *Plant Sci.* **2003,** *165* (1), 61–68.

Dalal, M.; Dani, R. P.; Kumar, P. A. Current Trends in the Genetic Engineering of Vegetable Crops. *Sci. Hortic.* **2006,** *107*, 215–225.

Datta, S. Impact of Climate Change in Indian Horticulture—A Review. *Int. J. Sci., Environ. Technol.* **2013,** *2* (4), 661–671.

Dey, S. S.; Behera, T. K.; Munshi, A. D.; Sirohi, P. S. Studies on genetic divergence in bitter gourd (Momordica charantia L.). Indian J Hort. 2007, **64**, 53–57.

Morphological and molecular analyses define the genetic diversity of Asian bitter gourd (Momordica charantia L.) (PDF Download Available). Available from: https://www.researchgate.net/publication/266522543_Morphological_and_molecular_analyses_define_the_genetic_diversity_of_Asian_bitter_gourd_Momordica_charantia_L [accessed Jun 28, 2017].

Diao, W.P., Jia, Y.Y., Song, H., Zhang, X.Q., Lou, Q.F. and Chen, J.F. 2009. Efficient embryo induction in cucumber ovary culture and homozygous identification of the regenerate using SSR markers. *Scientific Horticulture* 119: 246 – 251.

Dirlewanger, E.; Isaac, P. G.; Ranade, S.; Belajouza, M.; Cousin, R.; Vienne, D. D. Restriction Fragment Length Polymorphism Analysis of Loci Associated with Disease Resistance Genes and Developmental Traits in *Pisum sativum* L. *Theor. Appl. Genet.* **1994,** *88* (1), 17–27.

De Guzman, E. V., Del Rosario, A. G.; Eusebio, E. C. The Growth and Development of Makapuno Coconut Embryo *in vitro*. III. Resumption of Root Growth in High Sugar Media. *Phil. Agric.* **1971,** *53*, 566–579.

Lamb, C. J.; Masoud, S.; Sewalt, V. J. H.; Paiva, N. L. Metabolic Engineering: Prospects for Crop Improvement through Genetic Manipulation of Phenyl Propanoid Biosynthesis and Defense Responses—A Review. *Gene* **1996,** *169*, 61–71.

Diwan, N.; Fluhr, R.; Eshed, Y.; Zamir, D.; Tanskley, S. D. Mapping of Ve in Tomato: A Gene Conferring Resistance to the Broad-Spectrum Pathogen, *Verticillium dahliae* Race 1. *Theor. Appl. Genet.* **1999,** *98* (2), 315–319.

Erickson, A. N.; Markhart, A. H. Flower Developmental Stage and Organ Sensitivity of Bell Pepper (*Capsicum annuum* L) to Elevated Temperature. *Plant Cell Environ.* **2002,** *25*, 123–130.

Fand, B. B.; Kamble, A. L.; Kumar, M. Will Climate Change Pose Serious Threat to Crop Pest Management: A Critical Review? *Int. J. Sci. Res. Publ.* **2012,** *2* (11), 15.

Fehér, A. Why Somatic Plant Cells Start to Form Embryos? In: *Somatic Embryogenesis. Plant Cell Monographs*; Mujid, A., Samaj, J. Eds.; Springer: Berlin/Heidelberg, 2005; Vol 2, pp 85–101.

Fischer, D.; Bachmann, K. Onion Microsatellites for Germplasm Analysis and their Use in Assessing Intra and Interspecific Relatedness within the Subgenus *Rhizirideum. Theor. Appl. Genet.* **2000,** *101*, 153–164.

Ford, R.; Taylor, P. W. J. The Application of RAPD Markers for Potato Cultivar Identification. *Austr. J. Agric. Res.* **1997,** *48*, 1213–1217.

Gemes, J.A., Balogh, P. and Ferenczy, A. 2002. Effect of optimal stage of female gametophyte and heat treatment on in vitro gynogenesis induction in cucumber (*Cucumis sativus* L.). *Plant Cell Replication*,**21**(2): 105–111.

Gwanama, C.; Labuschagne, M. T.; Botha, A. M. Analysis of Genetic Variation in *Cucurbita moschata* by Random Amplified Polymorphic DNA (RAPD) Markers. *Euphytica* **2000,** *113*, 19–24.

Groben, R.; Wrickle, G. Occurrence of Microsatellites in Spinach Sequences from Computer Databases and Development of Polymorphic SSR Markers. *Plant Breed.* **1998,** *117* (3), 271–274.

Hamann, A.; Zink, D.; Nagi, W. Microsatellite Fingerprinting in the Genus *Phaseolus. Genome* **1995,** *38*, 507–515.

Guha, S.; Maheshwari, S. C. In vitro production of embryos from anthers of Datura. Nature, 1964, 204, 4957, pp. 497.

Gürel, S.; Gürel, E. and Kaya, Z. Callus Development and Indirect Shoot Regeneration from Seedling Explants of Sugar Beet (*Beta bulgaris* L. ssp.) Cultured *in vitro. Turk. J. Botany* **2001,** *25*, 25–33.

Heras, L.; Vazquez, F. J.; Jimenez, J. M. C.; Vico, F. R. RAPD Fingerprinting of Pepper (*Capsicum annuum* L.) Breeding Lines. *Capsicum Eggplant Newslett.* **1996,** *15*, 37–40.

Hussain, A.; Qarshi, I. A.; Nazir, H.; Ullah, I. Plant Tissue Culture: Current Status and Opportunities. In: *Agricultural and Biological Sciences. Recent Advances in Plant In Vitro Culture*; Leva, A., Rinaldi, L. M. R., Eds.; InTech 2012, ISBN 978-953-51-0787-3.

Hill, M.; Witsenboer, H.; Zabeau, M.; Vos, P.; Kesseli, R.; Michelmore, R. PCR-Based Fingerprinting Using AFLPs as a Tool for Studying Genetic Relationships in *Lactuca* spp. *Theor. Appl. Genet.* **1996,** *93*, 1202–1210.

He, G.; Prakash, C. S.; Jarret, R. L. Analysis of Genetic Diversity in a Sweet Potato (*Ipomoea batatas*) Germplasm Collection Using DNA Amplification Fingerprinting. *Genome* **1995,** *38*, 938–945.

Hazra, P.; Samsu, H. A.; Sikder, D.; Peter, K. V. Breeding Tomato (*Lycopersicon esculentum* Mill.) Resistant to High Temperature Stress. *Int. J. Plant Breed.* **2007,** *1* (1), 31–40.

IPCC. In: *Climate Change—Impacts, Adaptation and Vulnerability*; Parry, M. L., Canziani, O. F., Palutikof, J. P., van der Linden, P. J., Hanson, C. E., Eds.; Cambridge University Press: Cambridge, UK, 2007; p 976.

Jahn, M.; Paran, I.; Hoffman, K.; Radwanski, E. R.; Livingsone, K. D.; Grube, R. C.; Aftergoot, E.; Lapidot, M.; Moyer, J. Genetic Mapping of the TSW Locus for Resistance to the Tospovirus Tomato Spotted Wilt Virus in *Capsicum* spp. and Its Relationship to the SW-5 Gene for Resistance to the Same Pathogen in Tomato. *Am. Phytopathol. Soc.* **2000,** *13* (6), 673–682.

Jiang, C.; Sink, K. C. RAPD and SCAR Markers Linked to the Sex Expression Locus M in Asparagus. *Euphytica* **1997,** *94*, 329–333.

Johanson, A.; Ives, C. L. An Inventory of the Agricultural Biotechnology for Eastern and Central Africa Region. Michigan State University, 2001; 62.

Javornik, B.; Bohanec, B.; Campion, B. Second Cycle Gynogenesis in Onion, *Allium cepa* L, and Genetic Analysis of the Plants. *Plant Breeding* **1998,** *117* (3), 275–278.

Khandka, D. K.; Nejidat, A.; Golan-Goldhirsh, A. Polymorphism and DNA Markers for *Asparagus* Cultivars Identified by Random Amplified Polymorphic DNA. *Euphytica* **1996,** *87*, 39–44.

Kharkongar, H. P.; Khanna, V. K.; Tyagi, W.; Rai, M.; Meetei, N. T. Wide Hybridization and Embryo-Rescue for Crop Improvement in Solanum. *Agrotechnology* **2012,** *S11*, 004. doi:10.4172/2168-9881.S11-004.

Kumar, V.; Rajvanshi, S. K.; Yadav, R. K. Potential Application of Molecular Markers in Improvement of Vegetable Crops. *Int. J. Adv. Biotechnol. Res.* **2014,** *15* (4), 690–707.

Huang, C. C.; Cui, Y. Y.; Weng, C. R.; Zabel, P.; Lindhout, P. Development of Diagnostic PCR Markers Closely Linked to the Tomato Powdery Mildew Resistance Gene ol-1 on Chromosome 6 of Tomato. *Theor. Appl. Genet.* **2000,** *100*, 918–924.

Kim, M.; Jang, I. C.; Kim, J. A.; Park, E. J, Yoon, M.; Lee, Y. Embryogenesis and Plant Regeneration of Hot Pepper (*Capsicum annuum* L.) through Isolated Microspore Culture. *Plant Cell Rep.* **2008,** *27*, 425–434.

Lantos, C.; Juhasz, A. G.; Somogyi, G.; Otvos, K.; Vagi, P.; Mihaly, R.; Kristof, Z.; Somogyi, N.; Pauk, J. Improvement of Isolated Microspore Culture of Pepper (*Capsicum annuum* L.) via Co-culture with Ovary Tissues of Pepper or Wheat. *Plant Cell Tissue Organ Cult.* **2009,** *97*, 285–290.

Karihaloo, J. L.; Brauner, S.; Gottleib, L. D. RAPD Variation in Eggplant (*Solanum melongena* L.) Solanaceae. *Theor. Appl. Genet.* **1995,** *90*, 767–770.

Kaemmer, D.; Weising, K.; Beyermann, B.; Borner, T.; Eggplen, J. T.; Kuhl, G. Oligonucleotide Fingerprinting of Tomato DNA. *Plant Breed.* **1995,** *114*, 12–17.

Kumar, V.; Naidu, M. M. Development in Coffee Biotechnology—In Vitro Plant Propagation and Crop Improvement. *Plant Cell Tissue Organ Cult.* **2006,** *87*, 49–65.

Leljak-Levanic, D.; Bauer, N.; Mihaljevic, S.; Jelaska, S. Changes in DNA Methylation During Somatic Embryogenesis in *Cucurbita pepo* L. *Plant Cell Rep.* **2004,** *23* (3), 120–127.

Litt, M.; Lutty, J. A. A Hypervariable Microsatellite Revealed by In Vitro Amplification of a Dinucleotide Repeat within the Cardiac Muscle Actin Gene. *Am. J. Human Genet.* **1989,** *44* (3), 397–401.

Lou, H.; Kako, S. Somatic Embryogenesis and Plant Regeneration in Cucumber. *Hortic. Sci.* **1994,** *29* (8), 906–909.

Martin, G. B.; Brommenschenkel, S. H.; Chunwongse, J.; Frary, A.; Ganal, M. W.; Spivey, R.; Wu, T.; Earle, E. D.; Tanksley, S. D. Map-Based Cloning of a Protein Kinase Gene Conferring Disease Resistance in Tomato. *Science* **1993,** *262*, 1432–1436.

Margale, E.; Herve, Y.; Hu, J.; Quiros, C. F. Determination of Genetic Variability by RAPD Markers in Cauliflower, Cabbage and Kale Local Cultivars from France. *Genet. Resour. Crop Evol.* **1995,** *42*, 281–289.

McGregor, C. E.; Lambert, C. A.; Greyling, M. M.; Louw, J. H.; Warnich, L. A Comparative Assessment of DNA Fingerprinting Techniques (RAPD, ISSR, AFLP and SSR) in Tetraploid Potato (*Solanum tuberosum* L.) Germplasm. *Euphytica* **2000,** *113*, 135–144.

Melotto, M.; Afanador, L.; Kelly, J. D. Development of a SCAR Marker Linked to the I Gene in Common Bean. *Genome* **1996,** *39*, 1216–1219.

Mourya, B.; Pflieger, S.; Blattes, A.; Lefebvre, V.; Palloix, A. A CAPS Marker to Assist Selection of Tomato Spotted Wilt Virus (TSWV) Resistance in Pepper. *Genome* **2000,** *43*, 137–142.

Noli, E.; Conti, S.; Maccaferri, M.; Sanguineti, M. C. Molecular Characterization of Tomato Cultivars. *Seed Sci. Technol.* **1999,** *27*, 1–10.

Prince, J. P.; Lackney, V. K.; Angeles, C.; Blauth. J. R.; Kyle, M. M. A Survey of DNA Polymorphism within the Genus *Capsicum* and the Fingerprinting of Pepper Cultivars. *Genome* **1995,** *38*, 224–231.

Paran, I.; Michelmore, R. W. Development of Reliable PCR-Based Markers Linked to Downy Mildew Resistance Genes in Lettuce. **1993,** *85* (8), 985–993.

Martinez, L.E.; Aguero, C.B.; Lopez,M.E.; Galmarini, C.R. Improvement of in vitro gynogenesis induction in onion (*Allium cepa* L.) using polyamines. *Plant Sci.* 2000,156:221–226.

Masterson, J. Stomatal Size in Fossil Plants: Evidence for Polyploidy in the Majority of Angiosperms. *Science* **1994,** *264*, 421–423.

Mayet, M. The New Green Revolution in Africa: Trojan Horse for GMO? In: A Paper Presented at a Workshop: "Can Africa feed itself"?—Poverty, Agriculture and Environment—Challenges for Africa, 6–9th June 2007, Oslo, Norway, Center for African Biosafety, 2007.

Metwally, E.I., Moustafa, S.A., El-Sawy, B.I. and Shalaby, T.A. Production of haploid plants from in vitro culture of unpollinated ovules of *Cucurbita pepo*. Plant Cell Tiss. Org. Cult. 1998,52: 117 – 121.

Mneney, E. E.; Mantel, S. H.; Mark, B. Use of Random Amplified Polymorphic DNA Markers to Reveal Genetic Diversity within and between Populations of Cashew (*Anacardium occidentale* L). *J. Hortic. Sci. Biotechnol.* **2001,** *77* (4), 375–383.

Morris, E. J. Modern Biotechnology: Potential Contribution and Challenges for Sustainable Food Production in Sub-Saharan Africa. *Sustainability* **2011,** *3*, 809–822.

Mtui, G. Y. S. Involvement of Biotechnology in Climate Change Adaptation and Mitigation: Improving Agricultural Yield and Food Security. *Int. J. Biotechnol. Mol. Biol. Res.* **2011,** *2* (13), 222–231.

Paran, I.; Aftergoot, E.; Shifriss, C. Variation in *Capsicum annuum* Revealed by RAPD and AFLP markers. *Euphytica* **1998,** *99*, 167–173.

Qi, L.; Friebe, B.; Zhang, P.; Gill, B. S. Homologous Recombination, Chromosome Engineering and Crop Improvement. *Chromosome Res.* **2007,** *15* (1), 3–19.

Narayan, R. Air pollution—A Threat in Vegetable Production. In: International Conference on Horticulture (ICH-2009) Horticulture for Livelihood Security and Economic Growth; Sulladmath, U. V., Swamy, K. R. M., Eds.; 2009, pp 158–159.

Buttner, R., Jung, S. M. C. AFLP Derived STS Markers for the Identification of Sex in *Asparagus officinalis* L. *Theor. Appl. Genet.* **2000,** *100*, 432–438.

Raghavan, V. Embryogenesis in angiosperms. Cambridge Univ. Press, Cambridge, U.K. **1986.**

Redenbaugh, K., Ed. *Synseeds: Applications of Synthetic Seeds to Crop Improvement.* CRC Press: Boca Raton, FL, 1993.

Roose, M. L.; Stone, N. K. Development of Genetic Markers to Identify Two *Asparagus* Cultivars. In: Proceedings of the VIII International Asparagus Symposium; Nichols M., Swain D., Eds.; 21 November 1993, Palmerston North, New Zealand, *Acta Hortic.* **1996,** *415*, 129–135.

Rodriguez, J. M.; Berke, T.; Engle, L.; Nienhuis, J. Variation among and within *Capsicum* Species Revealed by RAPD Markers. *Theor. Appl. Genet.* **1999,** *99*, 147–156.

Spooner, D. M.; Peralta, I. E.; Knapp, S. Comparison of AFLPs with Other Markers for Phylogenetic Inference in Wild Tomatoes [*Solanum* L. section *Lycopersicon* (Mill.) Wettst.]. *Taxon* **2005,** *54* (1), 43–61.

Santos, D.; Feveriro, P. Loss of DNA Methylation Affects Somatic Embryogenesis in *Medicago truncatula. Plant Cell, Tissue Organ Cult.* **2002,** *70* (2), 155–161.

Segui-Simarro, J. M.; Nuez, F. Embryogenesis Induction, Callogenesis, and Plant Regeneration by In Vitro Culture of Tomato Isolated Microspores and Whole Anthers. *J. Exp. Bot.* **2007,** *58*, 1119–1132.

Shim, S. L.; Jorgensen, R. B. Genetic Structure in Cultivated and Wild Carrots (*Daucus carota* L.) Revealed by AFLP Analysis. *Theor. Appl. Genet.* **2000,** *101*, 227–233.

Sharma, H. C.; Crouch, J. H.; Sharma, K. K.; Seetharama, N.; Hash, C. T. Applications of Biotechnology for Crop Improvement: Prospects and Constraints. *Plant Sci.* **2002,** *163* (3), 381–395.

Shalaby, T.A. Factors Affecting Haploid Induction Through in vitro Gynogenesis in Summer Squash (*Cucurbita pepo* L.) *Sci. Hort.* **2000,** *115* (115), 1–6.

Stockton, T.; Gepts, P. Identification of DNA Probes that Reveal Polymorphisms among Closely Related *Phaseolus vulgaris* Lines. *Euphytica* **1994,** *76*, 177–183.

Supena, E. D. J.; Suharsono, S.; Jacobsen, E.; Custers, J. B. M. Successful Development of a Shed-Microspore Culture Protocol for Doubled Haploid Production in Indonesian Hot Pepper (*Capsicum annuum* L.). *Plant Cell Rep.* **2006,** *25*, 1–10.

Suslow, T. V.; Bradford, K. J. Applications of Biotechnology in Vegetable Breeding, Production, Marketing, and Consumption. *Veg. Biotechnol.* **1999,** 1–18.

Sulistyaningsih, E.; Aoyagi, Y. and Tashiro, Y. Flower bud culture of shallot (*Allium cepa* L. Aggregatum group) with cytogenetic analysis of resulting gynogenic plants and somaclones. Plant Cell Tiss. Org. Cult. 2006, 86, 249–255.

Thomas, W. T. B.; Forster, B. P.; Gertsson, B. Doubled Haploids in Breeding. In: Maluszynski, M. et al. *Doubled Haploid Production in Crop Plants: A Manual.* Kluwer Academic Publishers: Dordrecht, The Netherlands, 2013, pp. 337–349.

Tivang, J.; Skroch, P. W.; Nienhuis, J.; De Vos, N. Randomly Amplified Polymorphic DNA (RAPD) Variation among and Within Artichoke (*Cynara scolymus* L.) Cultivars and Breeding Populations. *J. Am. Soc. Hortic. Sci.* **1996,** *121*, 783–788.

Tulecke, W. R. A Tissue Derived from the Pollen of *Ginkgo biloba. Science* 117, 599–600.

Touchell, D.; Smith, J.; Ranney, T. G. Novel Applications of Plant Tissue Culture. *Comb. Proc. Int. Plant Propagat. Soc.* **2008,** *22* (58), 22–25.

Treasury, H. M. Green Biotechnology and Climate Change. *Euro Biol.* 2009; p 12. Available online at http://www.docstoc.com/docs/15021072/Green-Biotechnology-and-Climate-Change.

Usman, M.; Hussain, Z.; Fatima, B. Somatic Embryogenesis and Shoot Regeneration Induced in Cucumber Leaves. *Pak. J. Bot.* **2011,** *43* (2), 1283–1293.

Vincent, C. M.; Martínez, F. X. The Potential Uses of Somatic Embryogenesis in Agroforestry Are Not Limited to Synthetic Seed Technology. *Rev. Bras. Fisiol. Veg.* **1998,** *10* (1), 1–12.

Vidyasagar, K. *National Conference on Plant Biotechnology*, Lady Doak College: Madurai, 2006.

Wang, P. J. Regeneration of Virus-Free Potato from Tissue Culture. In: *Plant Tissue Culture and Its Biotechnological Application*; Barz, W., Reinhard, E., Zenk, M. H., Springer Verlag: Berlin, Heidelberg, New York, 1977, pp 386–391.

Van Heusden, A. W.; Ooijen, J. W. V.; Ginkel, R. V. V.; Verbeek, W. H. J.; Wietsma, W. A.; Kik, C. A Genetic Map of an Interspecific Cross in *Allium* Based on Amplified Fragment Length Polymorphism (AFLPTM) Markers. *Theor. Appl. Genet.* **2000,** *100* (1), 118–126.

Villand, J.; Skroch, P. W.; Lai, T.; Hanson, P.; Kuo, C. G.; Nienhuis, J. Genetic Variation among Tomato Accessions from Primary and Secondary centres of Diversity. *Crop Sci.* **1998,** *38*, 1339–1347.

Vivek, B. S.; Simon, P. W. Linkage Relationships among Molecular Markers and Storage Root Traits of Carrot (*Daucus carota* L. spp. *sativus*). *Theor. Appl. Genet.* **1999,** *99*, 58–64.

Vos, P.; Hogers, R.; Bleeker, M.; Reijans, M.; Van, D. L. T.; Hornes, M.; Frijters, A.; Pot, J.; Peleman, J.; Kuiper, M. AFLP: New Technique for DNA Fingerprinting. *Nucleic Acids Res.* **1995,** *23*, 4407–4414.

Williams, J. G. K.; Kubelik, A. R.; Livak, K. J.; Rafalski, J. A.; Tingey, S. V. DNA Polymorphisms Amplified by Arbitrary Primers Are Useful as Genetic Markers. *Nucl. Acid Res.* **1990,** *18*, 6531–6535.

Yayeh, Z. (2005). The First Genetic Linkages Among Expressed Regions of the Garlic Genome. *J. American Soc. Hort. Sci.* **130**, *4*, 569–574.

Zimmermann, J. L. Somatic Embryogenesis: A Model for Early Development in Higher Plants. *Plant Cell* **1993,** *5*, 1411–1423.

Ziv, M. Morphogenesis of *Gladiolus* buds in bioreactors. Implication for Scaled-Up Propagation of Geophytes. In: Progress in Plant Cellular and Molecular Biology. Proceedings of the VIIth Internal. Congress on Plant Tissue and Cell Culture, Amsterdam, 27–29th June 1990; Nijkamp, H. J. J., Van der Plas, L. H. W., Van Aartrijk, J., Eds.; Kluwer Academic Publishers: Dordrecht, 1990, pp 119–124.

Zhang, R.; Xu, Y.; Yi, K.; Zhang, H.; Liu, L.; Gong, G. A. Genetic Linkage Map for Watermelon Derived from Recombinant Inbred Lines. *J. Am. Soc. Hortic. Sci.* **2004,** *129* (2), 237–243.

CHAPTER 9

BIOTECHNOLOGICAL APPROACHES FOR CROP IMPROVEMENT IN FLOWERS AND ORNAMENTAL PLANTS

CHANDRASHEKAR S. Y.[1*], HEMLA NAIK B.[2], and RASMI R.[3]

[1]*Department of Floriculture and Landscape Architecture, College of Horticulture, Mudigere, Karnataka, India*

[2]*Department of Horticulture, College of Agriculture, Shivamogga, Karnataka, India*

[3]*UHS, Bagalkot, University of Agriculture and Horticultural Sciences, Shivamogga, Karnataka, India*

[*]*Corresponding author: E-mail: chandrashekar.sy@gmail.com*

CONTENTS

ABSTRACT

The flowers and ornamental plants are grown for decoration, rather than food or raw materials, which comprises wide array of plants and are classified into several groups like cut flowers, ornamental grasses, lawn or turf grasses, potted and indoor plants, bedding plants, trees and shrubs, etc. The biotechnology has made tremendous impact both scientifically and economically in ornamental plants. It comprises a continuum of technologies, ranging from traditional biotechnology such as plant tissue culture to modern tools such as genetic engineering of plants. The key areas in which plant cell and tissue culture has direct application in ornamental horticulture are large-scale propagation of elite clones from a hybrid or specific parent lines, production of disease-free propagules etc. Genetic modification has been used for the development of varieties of numerous species. Molecular markers random amplified polymorphic DNA (RAPD), restriction fragment length polymorphism (RFLP), amplified fragment length polymorphism (AFLP), simple tandem repeats (STR), simple sequence repeats (SSR), sequence-tagged sites (STS), expressed tagged sites (ETS), etc. are used for the selection of commercially important characteristics such as length of the juvenile phase, chilling response, disease resistance, flower number of plant size. Genetic engineering is also applied to increase the vase life of flowers, by blocking the ethylene production of flowers. Ethylene triggers flower deterioration. The biochemical pathway of ethylene biosynthesis is well characterized and the crucial genes encoding 1-aminocyclopropane-l-carboxylate synthetase and 1-amninocyclo propane 1-carboxylate oxidase have been successfully sequenced.

9.1 INTRODUCTION

Flowers and ornamental plants are grown for decoration, rather than food or raw materials. They are most often intentionally planted for esthetic appeal. However, ornamental plants also have important uses such as for fragrance, for attracting wildlife pollination, and for checking the air pollution. The ornamental plants comprises wide array of plants and are classified into several groups like cut flowers, ornamental grasses, lawn or turf grasses, potted and indoor plants, bedding plants, trees and shrubs, *etc.*

Biotechnology is one of the rapidly developing areas of contemporary science. It can bring new ideas, improved tools, and novel approaches to the solution of some persistent, seemingly intractable problems in crop production. Biotechnology is the technique that uses living organisms, or substances from those organisms, to make or modify a product, to improve plants or animals, or to develop microorganisms for specific uses (Singh, 2013). It comprises a continuum of technologies, ranging from traditional biotechnology such as plant tissue culture to modern biotechnology such as genetic engineering of plants and represents the latest front in the ongoing scientific progress.

9.1.1 THE KEY AREAS OF ORNAMENTAL PLANTS HAVING BIOTECHNOLOGICAL APPLICATIONS

1. Multiplication of elite clones from hybrid or specific parental lines and rapid production of disease-free propagates through micropropagation or somatic embryogenesis.
2. Application of agricultural microbiology to produce microorganisms beneficial to cultivated crops including flowers.
3. Development of techniques based on the use of monoclonal antibodies and nucleic acid probes for the diagnosis of plant pests and diseases and the detection of foreign chemicals in final produce.
4. Application of genetic mapping techniques, based on the use of molecular markers, as an aid to conventional plant breeding programs.
5. Development of plant varieties through genetic engineering of plant species.

Ornamental plants are a group of plants where biotechnology has made tremendous impact both scientifically and economically. The contribution of the conventional breeding methods to the ornamental crop improvement had been very significant in building up the floriculture industry so far. Biotechnology has been applied to flowers for producing new flower colors and flower forms. It offers such potential for significant advances made in the improvement of ornamental crops. In ornamental crops, where novelty and originality imparts value addition for these market-oriented products, biotechnology has great potential. Improvement of crop characteristics and in turn plant production has major impact on floriculture business. The value addition may be in the form of changed architecture, promoting in vivo or in vitro propagation of recalcitrant genotypes, resistance to biotic and abiotic stresses, improved vase life or modification in flower color, shape, and period of blooming. Biotechnology can play a vital role in modifying these in terms of varietal development and multiplication and popularization of newly bred varieties.

9.1.2 ROLE OF BIOTECHNOLOGY IN CROP IMPROVEMENT OF ORNAMENTAL PLANTS

The fact that a whole plant can be regenerated from a single cell, explant, or organ makes tissue culture a valuable technique to proliferate genetically identical material and select interesting variants for commercial purposes. Totipotency, which states that cells are autonomic and, in principle, capable of regenerating to give a complete new plant, also allows a genetic change, made at the cellular level, to become an established traits of a whole plant. The newly introduced or selected trait can, subsequently, be passed on to future generations of the species by conventional breeding methods. Tissue culture is particularly important for vegetatively propagated crop species, since it reduces the labor associated with line maintenance and germplasm conservation. The micro propagation is also used for the exchange of disease-free plant material. Anther or pollen culture also facilitates in early achievement of homozygosity and is particularly useful to breed for recessive traits.

Techniques of biotechnology have been used by the floricultural industry, in both propagation and breeding. Meristem culture and micropropagation are used to generate virus-free, high-quality propagation stock by plant propagators. Breeders commonly use other tissue culture techniques to supplement breeding programs such as anther culture and embryo rescue. Breeders have also used marker-assisted breeding programs using restriction fragment length polymorphism (RFLP) analysis to generate gene linkage maps as an aid to conventional breeding techniques. Newer areas of biotechnology such as genomics, proteomics, and gene mapping have also been applied to floricultural plants. Genetic engineering of plants is entering a period of very rapid application, expanding the market for ornamental plants in the near future.

The key areas in which plant cell and tissue culture has direct application in ornamental horticulture are large-scale propagation of elite clones from a hybrid or specific parent lines through micropropagation and somatic embryogenesis, production of disease-free propagules, and meaningful development of plant varieties through cellular and molecular techniques in conjunction with the whole plant breeding.

9.2 IMPORTANT BIOTECHNOLOGICAL APPROACHES FOR CROP IMPROVEMENT OF ORNAMENTAL PLANTS

9.2.1 MICROPROPAGATION

Ornamental in vitro plant tissue culture for propagation was initially developed in England and France with orchids, chrysanthemum and carnations in the 1960s. One of the major aspects of plant biotechnology is the production of a large number of identical individuals via in vitro cloning. The plants produced through this technique are uniform and true to type with distinctive characteristics of increased vigor due to their higher health status. By the use of axillary shoot proliferation methods, micropropagation can be carried out successfully in chimeras, which are very important in ornamental plants. The techniques of meristem culture are now being used worldwide on commercial scale for micropropagation of almost all important genera of orchids, thus placing orchids within the reach of an average person. The process of adventitious plantlets formation is of prime importance for in vitro propagation of flower bulb crops. The best explants that can be used are bulb scales, stems, or buds, even though regeneration is possible from all parts of the bulbous plants (Tables 9.1 and 9.2). Adventitious shoot formation has been exploited in different cut flower species using a range of explants like leaf, internodes, petals, immature flower buds, root segments, petioles, flower stalks, etc.

TABLE 9.1 List of Some Important Flower Crops Propagated through In Vitro Techniques.

Name of the Crop	Source of Explant
Anthurium	Leaf segments, petiole, flower stalk segments, spathe, spadix
Foliage plants	Shoot tips
Chrysanthemum	Leaf segments, internodes, petals, petioles
Carnation	Leaf segments, internodes, petals
Gerbera	Leaf, petiole, flower stalk segment
Gladiolus	Inflorence stalk, leaf sections
Orchids	Epidermal peelings, leaf segments, root tips
Rose	Internodal segments, petals, leaf segments, immature embryos, root segments
Tuberose	Leaf segments, inflorescence, stalk segments, shoot apices
Lily and amaryllis	Segments of bulb scales

TABLE 9.2 Applications of Micropropagation Techniques.

Sl. No.	Crop(s)	Techniques Employed
1	Orchids	Meristem culture, seed culture
2	Bulbous ornamental plants	In vitro culture
3	Rose	In vitro culture
4	Pelargonium	Pollen culture
5	African violet	Adventitious bud
6	Spathiphyllum	Axillary or adventitious caulogenesis
7	Anthurium	Shoot tip and axillary branching
8	Chrysanthemum	Axillary or adventitious caulogenesis and meristem culture
9	*Ficus* spp.	Shoot tip and axillary shoot proliferation cultures
10	Bougainvillea	Shoot tip proliferation
12	Foliage plants	In vitro culture

9.2.2 PRODUCTION OF SPECIFIC PATHOGEN-FREE PLANTS

Tissue culture is being used to produce virus-free propagules and facilitate their mass propagation. Generally, ornamental plants are vegetatively propagated and the viruses and virus-like pathogen are transmitted mechanically. The detection of viruses in ornamental plants has become essential as no practical treatment exist to cure virus-infected material in the field. The use of virus-free propagated material is important in controlling virus diseases of ornamental crops. The most widely used techniques for detection of plant viruses are based either on the antibodies which recognize the surface structure of the viral protein or on complementary nucleic acid (DNA or RNA). The recent techniques based on immune probing like enzyme-linked immunosorbant assay and dot immune-binding assay permit quick detection of viral pathogens and mass screening of viruses. Currently, the technology for production of disease-free propagule is available for alstroemeria, carnation, chrysanthemum, dahlia, lilium, iris, freesia, gladiolus, hyacinth, etc.

9.2.3 IN VITRO POLLINATION AND EMBRYO RESCUE

The technique of in vitro pollination appears very promising for overcoming prefertilization barriers to incompatibility and raising new genotypes. The most critical step of in vitro pollination technique is the development of viable seeds from ovules and ovaries following fertilization. Ovule culture holds a great potential for raising hybrids which normally fail due to abortion of embryo at a rather early stage.

Interspecific and intergenic crosses are frequently carried out to transfer genes of interest from wild to cultivated species. Incompatibilities found in such crosses results in seeds with abortive embryos. Embryo recovery is effective in interspecific or intergeneric crosses. The objective of such crosses is to transfer alleles for disease resistance, environmental stress tolerance, high-yield potential or other desirable characteristics of species or genus to agronomically accepted cultivars. One of the objectives of this technique is to recover rare hybrids derived from incompatible crosses as well as to overcome seed dormancy by studying the nutritional and physiological aspects of embryo development and by testing seed viability. These rare hybrids can serve as a source of explants with high-totipotency tissues.

Although fertilization results in embryo formation in many interspecific or intergeneric crosses and in crosses between diploids and tetraploids, but due to bad endosperm formation, the embryo degenerates. Otherwise, these embryos can frequently grow and give rise to normal hybrid plants if a supplementary in vitro endosperm is provided. The embryo is recovered from the ovule some days after fertilization and cultivated in vitro. The development of a viable plant from an embryo depends on several factors such as genotype, embryo development stage at the moment of isolation, growth condition of the mother plant, culture medium composition, oxygen concentration, light, and temperature. The ideal time for embryo recovery varies from species to species or from one species cultivar to another. In case of lilium, alstroemeria, impatiens, helianthus, new varieties have been produced using the technique of in vitro pollination and embryo rescue to overcome the prezygotic and postzygotic compatibility, respectively.

9.2.4 SOMACLONAL VARIATION

There are many in vitro culture techniques, including somaclonal variation induced by mutagenic agents. Somaclonal variation is to designate all types of variation which occur in plants regenerated from plant-tissue culture. Mechanisms involved in somaclonal variation induction include gross karyotypic changes that accompany in vitro culture via callus formation, cryptic chromosome rearrangements, somatic permutation with changes of parts among sister chromatids, transposition of elements, genetic amplification or decrease, and several combination of these processes. In brief, somaclonal variation is the sum of the genetic variations (chromosome and genetic mutation) that are incorporated in the regenerated plants of a species. Part of such variation may exist prior to the in vivo culture and is produced in the in vitro culture. Somaclonal variants have been obtained in chrysanthemum, begonia, lisianthus, and daylily also.

9.2.5 SOMATIC EMBRYOGENESIS

Somatic embryogenesis is the formation of embryo from a cell other than a gamete or the product of gametic fusion. Somatic embryogenesis is a powerful tool for the improvement of ornamental plants, not only with regard to clonal propagation but also for other biotechnological applications as well. It has been successfully exploited in the improvement of crops like anthurium, alstroemeria, gladiolus, iris, lily, etc.

9.2.6 GENETIC ENGINEERING IN FLOWER CROPS

Genetic transformation is the transfer or introduction of a DNA sequence or, more specifically, of a gene to an organism without fertilization or crossing. The genetically transformed plants are called transgenic plants. Genetic induction is the controlled introduction of nucleic acids in a receiver genome without fertilization.

Different genetic transformation techniques have been established with the development of tissue culture techniques and genetic engineering which include the use of *Agrobacterium tumefaciens*, particle bombardment (Gene gun), polyethylene glycol (PEG)-mediated, electroporation, sonication, silica carbonate microparticles, microlaser, micro- and macro-injection and direct DNA application. All these plant genetic transformation techniques are classified into two groups, indirect and direct gene transfer.

Indirect transference is mediated by a vector such as *A. tumefaciens* or *A. rhizogenes*. Transformation via *A. tumefaciens* is the method most commonly used plant transformation.

Direct DNA transfer is based on physical or chemical methods, with PEG, electroporation, and the particle-acceleration methods which are largely used in gene transfer and expression studies. In Australia, transgenic carnation and violet plants have been developed with a half postharvest period. The following are the necessary steps used in genetic transformation:

1. isolation of a useful gene;
2. introduction of this into the plant cell;
3. integration of this gene in the plant genome;
4. fertile plant regeneration;
5. expression of the introduced gene in the regenerated plants; and
6. transmission of induced gene from generation to generation.

This method offers breeders some advantages, such as

1. it can change one characteristic without modifying the other;
2. it requires fewer generations and it is faster than backcrossing, and
3. it is more flexible, that is, it allows the introduction of new characteristics from other plant species and even animals and microorganisms.

In ornamental plants, characters like flower color, shape, longevity, plant morphology, and resistance to biotic and abiotic stress are modified through the use of molecular genetic methods. Gene isolation and combined with improvement of pigments biosynthesis have opened up new avenues to the generation of novel color varieties. Gene-transfer techniques are the prerequisite for the application of the recombinant techniques to flower breeding.

Among various approaches of gene transfer into the plant Agrobacterium-mediated gene transfer, protoplast-based direct gene transfer, and biolistic DNA transfer are the major techniques widely used for the transformation of floricultural crops. The best of the DNA delivery is the co-cultivation of regenerable explants with *Agrobacterium*. There are two strains—*A. tumefaciens and A. rhizogenes* which induce crown gall and hairy root diseases are two widely used gene transfer system in ornamental crops.

The success of *Agrobacterium*-mediated transformation depends upon

- cultivars selected;
- type of explants;
- *Agrobacterium* strain;
- condition of co-cultivation; and
- selection method and mode of regeneration of plantlets.

The global flower industry thrives on novelty. Genetic engineering is providing valuable means of expanding the floriculture gene pool so promoting the generation of new commercial varieties. There has been extensive research on the genetic transformation of different flowering plant species and many ornamental species have now been successfully transformed, including those which are most important commercially. To date, more than 30 ornamental species have been transformed, including anthurium, begonia, carnation, chrysanthemum, cyclamen, datura, daylily, gentian, gerbera, gladiolus, hyacinth, iris, lily, lisianthus, orchid, pelargonium, petunia, poinsettia, rose, snapdragon, and torenia (Deroles et al., 2002). New ornamental plant varieties are being created by breeders in response to consumer demand for new products. In general terms, engineered traits are valuable to either the consumer or the producer. At present, only consumer traits appear able to provide a return capable of supporting what is a still a relatively molecular breeding tool.

9.2.7 GENETIC MODIFICATION

Genetic modification (GM) has been used for the development of varieties of numerous important food species. Though not at the same scale, there are also research efforts in the field of ornamental plants for varietal development, especially for flowering ornamental plants. Development of these new varieties through hybridization or mutagenesis can be very difficult, lengthy, or improbable, if varieties are completely sterile, such as orchids. GM answers these constraints and provides a way for variety improvement. Table 9.3 provides a list of genes used in the development of GM ornamental plants. Biotechnology also shortens the duration of variety development in an industry where phenotypic novelty, such as flower color, is an attractive marketing factor.

Several traits of ornamental plants have already been modified including flower color, fragrance, flower shape, plant architecture, flowering time, postharvest life, and resistance for both biotic and abiotic stresses. Currently, at least 50 ornamental plants can now be transformed. Transgenic ornamental plants have been produced by several different techniques, the most common techniques being *Agrobacterium*-mediated transformation and particle bombardment.

Ornamental plant traits are classified according to their value in the market chain. There are traits with more value to the grower than to the consumer. These are traits related to ease of production and shipping such as disease

TABLE 9.3 Genes Used in the Development of GM Ornamental Plants.

Gene and Source(s)	Results	References
***F3'-5'h* gene** (petunia/pansy)	Overexpression produces blue flowers in combination with a silenced *dfr* gene in Carnation (Petunia) and Roses (Pansy)	Katsumoto et al. (2007)
CrtW (*Lotus japonicas*)	Overexpression changes petal color from light yellow to deep yellow or orange in Lotus	Suzuki et al. (2007)
CHS (*Gentian*)	Gene silencing produces white flowers in Gentian	Nishihara et al. (2006)
ANS (*Gentian*)	Gene silencing produces pale blue flowers in Gentian	Nakatsuka et al. (2008)
Ls (*Chrysanthemum*)	Less branching in Chrysanthemum	Han et al. (2007), Jiang et al. (2009)
Ipt (*Agrobacterium tumefaciens*)	Increased branching and reduced internode length in Chrysanthemum	Khodakovskaya et al. (2009)
RolC (*Agrobacterium rhizogenes*)	Dwarfed *Pelargoniums* and *Petunias*	Boase et al. (2004), Winefield et al. (1999)
MADS-Box (Orchid/Lily)	Ectopic expression changes the second round of petals into calyx in orchids and lilies	Thiruvengadam and Yang (2009)
Asl38/lbd41 (*Arabidopsis*)	Flowers turned into multiple column patterns in *Celosia cristata*	Meng et al. (2009)
Floral integrator genes (Arabidopsis)	Activate the floral identity genes; promotes flowering in appropriate conditions	Amasino and Michaels (2010), Jung and Muller (2009), Turck et al. (2008)
AP1 (*Chrysanthemum*)	Speeds up time to flowering in Chrysanthemum	Jiang et al. (2010), Shulga et al. (2010)
Cry1A (*Bacillus thuringiensis*)	Resistance to *Helicoverpa armigera* and *Spodoptera litura* in Chrysanthemum	Shinoyama and Mochizuki (2006), Soh et al. (2009)
CVB coat protein gene (*Chrysanthemum*)	Chrysanthemum Virus B (CVB) resistance	Skachkova et al. (2006)

TABLE 9.3 *(Continued)*

Gene and Source(s)	Results	References
Rdr1 (Rose)	Resistance to black spot in roses	Kaufmann et al. (2003)
Sarcotoxin gene (*Sarcofaga peregrine*)	Resistance against *Burkholderia caryophylli* in carnation	Yoshimura et al. (2007)
Rd29A:DREB1A (*Arabidopsis*)	Enhanced abiotic stress tolerance in *Chrysanthemum*	Hong et al. (2009, 2006a, 2006b)
ACO/ACS-coding genes (carnation/apple)	Increased vase life in carnation	Inokuma et al. (2008), Veres et al. (2004)
ERS1 (*Chrysanthemum*)	Mutated gene slows down yellowing of leaves in Chrysanthemum	Narumi et al. (2005)
Cp4, EPSPS (*Agrobacterium tumefaciens*)	Glyphosate herbicide tolerance in creeping bentgrass	Chai et al. (2003)

resistance and shelf life. Meanwhile, other traits have more value to the consumer such as novel colors, dwarfed plants, modified growth, improved fragrance, flower shapes, and flower sizes. A third category includes breeder traits such as traits that affect seed production such as male sterility.

Commercialization of genetically engineered flowers is currently confined to novel colored carnation. The production of novel color flower has been the first success story in floriculture genetic engineering. However, further products are expected given the level of activity in the field. Other traits that have received attention include floral scent, floral and plant morphology, senescence of flowers both on the plant and postharvest and disease resistance. To date, there are only a few ornamental genetically modified products (Table 9.4) in development and only one, a carnation genetically modified for flower color, in the market place. There are approximately 8 ha of transgenic carnation in production worldwide, largely in South America. The other breeding programs on color modification or alteration of plant architecture and height remain focused on rose, gerbera, and various pot plant species.

9.2.8 GENOME MAPPING AND MARKERS

Ornamental species such as ivy have been used to work out the basic information to build the framework. Molecular markers are used for mapping of cultivars and species. Molecular markers are used for the selection of commercially important characteristics such as length of the juvenile phase, chilling response, disease resistance, flower number of plant size. Some of the common markers include random amplified polymorphic DNA (RAPD), RFLP, amplified fragment length polymorphism (AFLP), simple tandem repeats (STR), simple sequence repeats (SSR), sequence-tagged sites (STS), expressed tagged sites (ETS), etc.

9.3 MOLECULAR BREEDING OF ORNAMENTAL CROPS FOR VARIOUS APPROACHES

9.3.1 FLOWER COLOR, FLAVOR, AND FRAGRANCE

In ornamental crops esthetic traits like plant morphology, novel color, etc. can be improved through different molecular approaches like introduction of foreign genes of bacterial, viral, and plant origin: overexpression or

TABLE 9.4 List of Transgenic ornamental plant species.

Sl. No.	Ornamental Plants	Transformation Method	References
1	*Alstromeria* sp.	*A. tumefaciens* and microprojectile	Kim et al. (2007), Lin et al. (2000)
2	*Anthurium andreanum*	*A. tumefaciens*	Chen and Kuehnle (1996)
3	*Anthurium majus*	*A. tumefaciens* and *A. rhizogenes*	Cui et al. (2001), Hoshino et al. (1998)
4	*Begonia* sp.	*A. tumefaciens*	Kiyokawa et al. (1996), Kishimoto et al. (2000)
5	*Cyclamen persicum*	*A. tumefaciens*	Aida et al. (1999), Sironi et al. (2000), Boase et al. (2000)
6	*Dahlia*	*A. tumefaciens*	Otani et al. (2013)
7	*Datura* sp.	*A. tumefaciens* and *A. rhizogenes*	Ducrocq et al. (1994)
8	*Delphinium sp*	*A. tumefaciens*	Hirose et al. (1998)
9	*Dendranthema* sp.	*A. tumefaciens* and microprojectile	Yepes et al. (1995, 1999), Takatsu et al. (1999)
10	*Dianthus caryophyllus*	*A. tumefaciens*, *A. rhizogenes*, and microprojectile	Miroshimchenko and Dolgov (2000), Nakano et al. (1999), Zuker et al. (1995)
11	*Eustoma grandiflorum*	*A. tumefaciens* and microprojectile	Hecht et al. (1997), Giovanni et al. (1996), Takahashi et al. (1998)
12	*Euphorbia pulcherrima*	Electroporation	Vik et al. (2000)
13	*Forsythia intermedia*	*A. tumefaciens*	Rosati et al. (1996)
14	*Gentiana* sp.	*A. rhizogenes* and microprojectile	Vinterhalter et al. (1999), Hosokawa et al. (2000)
15	*Gerbera hybrid*	*A. tumefaciens*	Nagaraju et al. (1998), Reynoird et al. (2000)
16	*Gladiolus grandiflorus*	*A. tumefaciens* and microprojectile	Bablu and Chawala (2000), Kamo et al. (1995a, 1995b)
17	*Hemerocallis* sp.	microprojectile	Ling et al. (1998)
18	*Hibiscus syriacus*	*A. tumefaciens*	Yang et al. (1996)
19	*Hyacinth*	microprojectile	Langeveld et al. (2000)
20	*Iris germanica*	*A. tumefaciens*	Jeknic et al. (1999)

TABLE 9.4 *(Continued)*

Sl. No.	Ornamental Plants	Transformation Method	References
21	*Linum* sp.	*A. tumefaciens*	Ling and Binding (1997)
22	*Lilium longiflorum*	microprojectile	
23	*Orchid* sp.	*A. tumefaciens* and microprojectile	Belarmino and Mii (2000), Yang et al. (1999), Knapp et al. (2000)
24	*Ornithogalum* sp.	microprojectile	De Villiers et al. (2000)
25	*Osteospermum* sp.	*A. tumefaciens*	Allavena et al. (2000)
26	*Pelargonium* sp.	*A. tumefaciens* and *A. rhizogens*	Krishnaraj et al. (1997), Boase et al. (1998), Pellegrineschi and Davoliomariani (1996)
27	*Petunia hybrida*	*A. tumefaciens* and vacuum infiltration	Van der Meer (1999), Tjokrokusumo et al. (2000)
28	*Rhododendron* sp.	*A. tumefaciens* and microprojectile	Pavingevora et al. (1997), Hsia and Korban (1998)
29	*Rosa* sp.	*A. tumefaciens*	Souq et al. (1996), Van der Salm et al. (1997)

suppression of native or introduced genes. Major traits targeted in floricultural crops include flower color modification, enhanced postharvest attributes, insect and disease resistance, altered flower and plant morphology. For modification of phenotypes, three genes, for example, structural genes, bacterial genes, and regulatory genes are involved. Transformation of plant cells to modify flower color requires:

- a thorough knowledge of the biosynthetic pathway;
- characterization of each gene in these intricate steps; and
- isolation, cloning, and expression in an alien environment.

Methods

1) Selection of plant which is incapable of synthesizing a DFR which acts on DHK or produces reduced level of compared to a wild-type plant or produces a DFR with a reduced substrate specificity for DHK compared to DHQ or DHM followed in carnation, rose, chrysanthemum, and gerbera for blue, lilac, violet, or purple.
2) Wherein the nucleotide sequence encoding a DFR is from Petunia.
3) Wherein the nucleotide sequence encoding an F 3′ 5′ H is from petunia, pansy, China aster, anemone, iris, hyacinth, bell flower.

Improvement of flower color with genetic manipulation through modification of certain enzymes in metabolic pathways is given later (Tables 9.5 and 9.6).

The first successful application of genetic engineering for flower color modification was petunia to produce crimson colored pelargonidin pigments by transferring Al gene from *Zea mays* which codes for a specific protein dihydroquercetin-4-reductase. The first antisense technology has been used genetically engineered petunia to incorporate antisense Chs gene (Chalcone Synthase gene) to alter flower color.

Recently, color modification genes reported in various flower crops, namely, petunia (*An2*, *An4*, *Fl*, *Hfl*, *Po*, *Rt*), ipomoea (*Tpn 2*, *Tpn 1*), torenia (*ANS* gene), rose (*CHS c-DNA*), etc. Blue carnation is developed through characterization of anthocyanin and use of antisense suppression to block the expression of a gene encoding flavonone-3-hydrogenase. In antirrhinum, novel yellow color has been tried with GM of chalcone and aurone flavanoid biosynthesis.

TABLE 9.5 Molecular Targets for Modification of Flower Color.

Crop	Flower Trait	Target	References
Petunia hybrida	White to red and pattern	Mutant maize gene	Aswath and Hanur (2009)
	Purple to white and pattern	Chalcone synthase	
	White to red	Maize LC regulatory gene	
	Sepal color genes to dark purple	Maize LC regulatory gene	
	White to pale yellow	Flavonoid biosynthesis	
	White to pink	Flavonol synthase and DHFR genes	
	Red to deep purple sectors	F 3′ 5′ H gene	
Rose	Red to pink	Anthocyanin biosynthesis	
	Red to light red/magenta red	Anthocyanin biosynthesis	
	Red to blue	Delphinidin accumulation	
Dianthus	White to mauve, violet	Flavonoid biosynthesis	
	Orange to cream	F 3H gene	
Gerbera	Red to pink, cream	Chalcone synthase	
Dendranthemum	Pink to white	Flavonoid biosynthesis	
Torentia	Blue to white	Anthocyanin biosynthesis	
	Blue to red	Cytochrome P450 gene	

The major world players engaged in genetic engineering of cut flowers are *Florigene*, *Calgene Pacific*, and *DNA Plant Technology* (DNAP). *Florigene* (The Netherlands) was founded in 1989 as a joint venture between *DNAP*, *Znadunie* (which is now called S&G Seeds, belonging to *Sandoz*, Switzerland), and the *Rabobank Biotech Venture Fund*. Florigene was one of the first companies to obtain an alteration of color by genetic engineering. Working together with chrysanthemum breeder *Fides*, *Florigene* has transformed the pink chrysanthemum variety "Moneymaker" into a white flower, by blocking the *Chs* gene responsible for pigment synthesis. Since there are many white chrysanthemum varieties, the newly created variety, called *Floriant*, will not gain market share. The development of *Floriant* was meant as a test case for genetic engineering of flowers and the approval procedure of the Dutch government. *Calgene Pacific* (*Cl'*) was established in Melbourne, Australia, in 1986. Shareholders include *DNAP* (USA), *Fides* (The Netherlands), and *Suntory* (Japan). Last year coat protein (CP) acquired its Dutch competitor *Florigene.*

TABLE 9.6 List of Important Crops Where Flower Modification Was Successful.

Crop	Gene Incorporated	Strategy	References
Chrysanthemum	*Chs*	Sense orientation	Aswath and Hanur (2009)
		Antisense orientation	
Gerbera	*Gchs*	Sense orientation	
	I/gdfr	Antisense orientation	
Lisianthus	*Chs*	Antisense orientation	
Rose cv. Royality	*Chs*	Antisense orientation	
		Sense orientation	
Carnation	*Anti-f3h*	Antisense orientation	

Identification, isolation, and transferring of genes responsible for colors are the main focus of research at CP. In 1991, it isolated the key genes responsible for the colors blue and red. CP's main research project is the development of blue flowers, particularly a blue rose. Of the 10 most popular flowers only the Freesia has blue varieties would command a market share close to that of red, if they were freely available. Blue transgenic petunia was reported in 1992. Blue carnations and chrysanthemum are being tested this year, whereas the blue rose is expected in it. *DNAP* in Oakland, USA, together with researchers at the *University* of *California* at Davis, has also developed a transgenic chrysanthemum with altered flower color. Just recently, DNAP has been the first to report the development of a transgenic rose. Friable embryogenic tissues of rose have been transformed and reproduced into flowering plants. Although the transferred marker genes are of no direct commercial interest, the procedure facilitates the introduction of desirable genes, especially those controlling flower colors, into commercial cultivars of rose.

9.3.2 ENHANCED POSTHARVEST ATTRIBUTES

Genetic engineering is also applied to increase the vase life of flowers, by blocking the ethylene production of flowers. Ethylene triggers flower deterioration. The biochemical pathway of ethylene biosynthesis is well characterized and the crucial genes encoding 1-aminocyclopropane-l-carboxylate synthetase and 1-amninocyclo propane 1-carboxylate oxidase have been successfully sequenced clone and incorporated into flower crops like petunia and carnation in antisense orientation.

9.3.2.1 INSECT–DISEASE RESISTANCE

Isolation, characterization, and sequencing of specific gene coding resistance to virus, fungi, insects, and bacteria paved the way their transfer to flower crops. Such efforts have been made in gerbera, chrysanthemum, and carnation. For viral resistance, TSWV-N gene and stem explants are used for chrysanthemum, dahlia, gerbera, etc. Transgenic *Dendrobium* orchid transformed with cymbidium mosaic virus have been developed. Transgenic plants that express the viral CP genes become partially resistant to infection. The technique involves the transfer of a DNA copy of virus satellite RNA into plants. Transformed plants produce large amounts of satellite RNA when exposed to satellite RNA host virus decreases in virus replication and reduces development of disease symptoms. *Calgene Pacific (CP)* is now applying this technology to carnation which is susceptible to *Fusarium* attack.

9.3.3 ALTERATION IN FLOWER AND PLANT MORPHOLOGY

Recently, a number of regulatory genes have been identified, isolated, and characterized that govern the determination of floral meristem organ primordial which are termed as MADS box genes and ABC genes. Several such genes regulating plant morphology and development in flower crops have been described previously (Table 9.7).

TABLE 9.7 List of Genes Regulating Plant Morphology and Development in Flower crops.

Gene(s)	Function	Reference
Clavata, Wuschel	Establishment and maintenance of shoot apical meristem	Aswath and Hanur (2009)
GA insensitive (gai)	Stem elongation and plant height	
Brassinosteroid gene	Plant height (dwarfing)	
MAX	More axillary branching	
Lazy, TAC 1	Branching angles of tillers	
Phytochrome	Shading response and harvest index	
Rol C	Regulation on plant branching and architecture	

9.4 DNA FINGER PRINTING

DNA finger printing is a technique used to distinguish between individuals of the same species using only samples of their DNA. DNA profiling exploits highly variable repeat sequences called variable number tandem repeats. These loci are variable enough that two unrelated humans are unlikely to have the same alleles.

Basically there are two types of DNA finger printing:

9.4.1 CLASSICAL HYBRIDIZATION-BASED FINGER PRINTING

It is practiced by cutting of genomic DNA with a restriction enzyme. In this method, DNA is digested with the restriction enzymes and DNA fragments are separated according to their size by electrophoresis on a gel. This gel is southern blotted into a membrane and specific fragments are made visible by hybridization with labeled probe.

1. RFLP analysis is found to be useful for estimating genetic diversity (Tanksley et al., 1992), to assist in the conservation of endangered species and plant genetic resources. It is also used for plant genome mapping.

9.4.2 POLYMERASE CHAIN REACTION (PCR)-BASED FINGER PRINTING

It amplifies the amounts of a specific region of DNA using oligonucleotide primers and a thermostable DNA polymerase. The amplified products are separated by electrophoresis on agarose gels and detected with ethidium bromide. This method is useful for estimating genetic diversity, identification of species or cultivars, genome mapping, population genetics, etc.

9.4.2.1 RANDOMLY AMPLIFIED POLYMORPHIC DNA

This is the efficient method for genome mapping and characterization of genetic resources. It is based on repeated amplification of DNA sequences using arbitrary primers to provide DNA fingerprints (Williams et al., 1990).

9.4.2.2 SINGLE SSR PRIMER AMPLIFICATION REACTION

In this case, exponential amplification occurs from the single primer reaction with a particular SSR and polymorphism is SSR based. So, multiple loci are detected from a genome using single PCR Reaction.

9.4.2.3 DAF (DNA AMPLIFICATION FINGERPRINTING)

It exploits single arbitrary primers for amplification of DNA based on PCR. This method is effective for genetic typing and mapping.

9.4.2.4 ARBITRARY PRIMED POLYMERASE CHAIN REACTION

This is one type of RAPD in which discrete amplification patterns are generated through the employment of single primers of 10–50 bases in length.

9.4.2.5 RANDOMLY AMPLIFIED MICROSATELLITE POLYMORPHISMS

The method consists of amplification of genomic DNA using arbitrary (RAPD) primers followed by separation with electrophoresis and hybridization of dried gel with microsatellite oligonucleotide probes. It is used in genetic finger printing of closely related species.

9.4.2.6 AMPLIFIED FRAGMENT LENGTH POLYMORPHISM

This marker is based on PCR and used for rapid screening of genetic diversity. An AFLP technique generates hundreds of highly replicable markers from DNA of any organism and allows high-resolution genotyping of finger printing quality (Vos et al., 1995). It has broad application in systemics, population genetics and mapping of quantitative trait loci.

9.4.2.7 INTERSIMPLE SEQUENCE REPEAT MARKERS

In this case, microsatellite-based primers are used to amplify inter-SSR DNA sequences. Here, a number of microsatellite anchored at the 3′ end are used for amplifying genomic DNA which increases their specificity (Zietkiewicz et al., 1994). An unlimited number of primers are synthesized for various combinations of di-, tri-, tetra-, and pentanucleotides with an anchor made up of a few bases and have broad range applications in plant species.

9.4.2.7.1 Mitochondrial Analysis: mt DNA Is Useful in Determining the Unclear Identities

It is one of the important molecular techniques used in studying the extent and distribution of variation in gene pools maintained in various ornamental crops. The marker developed from different molecular technique can be combined to obtain DNA finger prints in important ornamental corps. DNA finger printing is useful in protecting our indigenous wealth of ornamental plants or varieties developed in our country (Table 9.8). It protects intellectual property protection rights of the breeders.

Uses of this technique

- To identify unable plants in the trade.
- Genetic mapping of ornamental plants.
- To assist in cultivar identification, breeding program, and evolutionary research in commercial ornamental plants.

TABLE 9.8 List of Molecular Markers Used in Ornamental Plants.

Sl. No.	Flower Crops	Markers	Uses	References
1	*Alstromeria* sp.	RAPD	Genetic variation and Genotypic identification	Beyermann et al. (1992), Anatassopoulos and Keil (1996), Han et al. (1999, 2000)
		AFLP	Genetic diversity and construction of linkage map	
2	*Anthurium* spp.	RAPD	Genetic relationship of cultivars	Ranamukhaarachchi et al. (2001)
3	*Caladium*	AFLP	Genetic relationship	Loh et al. (2001)
4	*Camelia* spp.	Microsatellite	Development of population genetic characterization	Kaundun and Matsumoto (2002)
5	*Cephalotaxus*	AFLP	Analysis of genetic diversity	Zhang et al. (1999)
6	*Cymbidium*	RAPD	Identification of cultivars and analysis of genetic diversity	Obara-Okeyo and Kako (1998)
7	*Cynodon*	DAF	Genetic relationship	Caetano-Anolles et al. (1995)
8	*Dahlia*	AFLP	Detection of genetic diversity and cultivar analysis	Wegner and Debener (2008)
9	*Dendranthema* sp.	RAPD/DAF	Genetic variability	Wolff and Peters-Van Rijn (1993), Wolff et al. (1995), Scott et al. (1996)
10	*Dianthus caryophyllus*	RAPD	Selection for resistance to Fusarium oxysporum, Genetic relationship among cultivars	Scovel et al. (2001), Tejaswini et al. (2008)
11	*Dianthus* spp.	RFLP	Selection of disease resistance	Scovel et al. (2001), Vainstein et al. (1991)
		Microsatellite	Identification of *Dianthus* varieties	
12	*Euphorbia* (*Poinsettia*)	RAPD/DAF	Identification of genetic relationship	Ling et al. (1997), Starman et al. (1999)
13	*Gladiolus* spp.	RAPD, SRAP	Genetic relationship, Genetic diversity	Takatsu et al. (2001), Geeta et al. (2014)
14	*Geranium*	DAF	Genetic relationship	Starman and Abbitt (1997)
15	*Gerbera*	RFLPs	Genetic fidelity	Tzuri et al. (1991)
16	*Heliconia*	RAPD	Genetic relationship	Marouelli et al. (2010)

TABLE 9.8 *(Continued)*

Sl. No.	Flower Crops	Markers	Uses	References
17	*Hemerocallis fulva*	AFLP	Genetic similarity	Tomkins et al. (2001)
18	*Hibiscus*	AFLP, RAPD	Genetic relationship	Cheng et al. (2004), Barik et al. (2006)
19	*Lilium* spp.	RAPD		Langeveld et al. (2000)
20	*Iris setosa, I. laevigata* *I. uniflora*	RAPD	Genetic identification	Arnold et al. (1999)
21	*Ixora coccinea, I. javanica*	RAPD	Genetic diversity among cultivars	Rajaseger et al. (1997)
22	*Lilium* spp.	RAPD RFLP	Genetic relationship, genetic variation, Identification of parentage	Yamashingi (1995), Haruki et al. (1998)
23	*Lupinus* spp.	RAPD/AFLP	Phylogenetic relationship, linkage map	Talhinhas et al. (2003)
24	*Ophrys aranaeola*	Microsatellite	Characterization of microsatellite and cross species amplification	Soliva et al. (2000)
25	*Orchids*	RFLP	Analysis of maternal linkage	Xue et al. (2010)
26	*Osteospermum*	RAPD/RFLP	Genetic variation	Faccioli et al. (2000)
27	*Ozothammus*	RAPD	Genetic diversity	Ko et al. (1996)
28	*Paeonia*	RAPD	Classification	Suo et al. (2005)
29	*Petunia*	RAPD/DAF AFLP, RFLP	Linkage map, genetic relationship, Identification of clones containing Rf gene, Gene mapping	De Wet et al. (2008), Cerny et al. (1996), Jianhua et al. (1997), Bentolila and Hanson (2001), Stronuner et al. (2000)
30	*Rosa* *R. multiflora*	RAPD Microsatellite	Linkage map, parentage analysis, genetical variability, classification of genus, germplasm analysis, Construction of genetic linkage	Tzuri et al. (1991), Hubbard et al. (1992), Rajapakse et al. (1992), Millan et al. (1996), Debener et al. (1996), Hibrand-Saint Oyant et al. (2008)

- RAPD markers are utilized to determine the genetic relationship of different cultivars of ornamental crops.
- To analyze rank correlation of results of contributing characters as reported in carnation.
- To assist in intra- and interspecific breeding of new cultivars of ornamental plants through RAPD markers.

9.4 FUTURE ROLE OF BIOTECHNOLOGY

In the near future, efforts should be devoted to obtain transformed plants from apparently recalcitrant species. As noted above, there is a need to increase the number of isolated genes, particularly those conferring resistance to pests, diseases, and abiotic stresses, and for quality improvement. Attention should be given to genes coding for the reproductive process, such as genes for self incompatibility and for male sterility and to genes which influence the interaction between host and symbionts in nitrogen fixation. The development of an appropriate methodology for producing artificial seeds will open the way to considerable progress in agriculture through better establishment and uniformity of crops. Mapping the genomes of the most important crops should be one of the priorities of current and future genetic research. This will result in the production of new crop plants with desirable traits and in a better understanding of plant physiological processes.

One of the greatest concerns about biotechnology is patent legislation relating to genes and their use in plant breeding. With the high cost of biotechnological research, coupled with the large investments by private companies operating in this sector, patent legislation is necessary to protect new genetic products. However, there is a risk that while new biotechnology will certainly improve knowledge of crop genetics, patents could severely limit its application in crop improvement. This problem should not be underestimated; a solution must be found that takes account of both private interests and agricultural progress.

However, biotechnologies renew every day, and researchers in ornamental horticulture should take advantages of these technologies to use with their own specifications and finally promote the development of research and industry. Hence, it can be expected that more GM ornamental products will be released in the future. As more GM cutflower varieties are released, public awareness will increase. Certain traits of ornamental horticulture may also be compatible to the production of secondary metabolites, including pharmaceuticals.

KEYWORDS

- **flowers**
- **ornamental plants**
- **micropropagation**
- **somaclonal variation**
- **embryo rescue**
- **DNA finger printing**
- **arbitrary primed polymerase chain reaction (AP-PCR)**
- **randomly amplified microsatellite polymorphisms (RAMPO)**

REFERENCES

Aida, R.; Hirose, Y.; Kishimoto, S.; Shibata, M. *Agrobacterium tumefaciens*-Mediated Transformation of *Cyclamen persicum* Mill. *Plant Sci.* **1999,** *148*, 1–7.

Allavena, A.; Giovannini, A.; Spena, A.; Zottini, M.; Accotto, G. P.; Vaira, A. M.; Berio, T. Genetic engineering of Osteospermum Spp: A case story, *Acta Hort.* **2000**, 508, 129–133.

Amasino, R. M.; Michaels, S. D. The Timing of Flowering. *Plant Physiol.* **2010,** *154*, 516–520.

Anatassopoulos, E.; Keil, M. Assessment of Natural and Induced Genetic Variation in *Alstroemeria* Using Random Amplified Polymorphic DNA (RAPD) Markers. *Euphytica* **1996,** *90*, 235–244.

Arnold, M. L.; Bennett, B. D.; Zimmer, E. A. Natural Hybridization between *Iris fulva* and *I. hexagona*: Pattern of Ribosomal DNA Variation. *Evolution* **1990,** *44*, 1512–1521.

Aswath, C.; Hanur, V. S. Genes to Fulfil Consumers Demand. *Indian Hortic.* **2009,** *54* (10), 30–33.

Bablu, P.; Chawla, H. S. *Agrobacterium*-Mediated Transformation in Gladiolus. *J. Hortic. Sci. Biotechnol.* **2000,** *75*, 400–404.

Barik, S.; Senapati, S. K.; Aparajita, A.; Mohapatra, A.; Rout, G. R. Identification and Genetic Variation among *Hibiscus* Species (Malvaceae) Using RAPD Markers. *Z. Naturforsch.* **2006,** *61*, 123–128.

Belarmino, M. M.; Mii, M. *Agrobacterium*-Mediated Genetic Transformation of a *Phalaenopsis* Orchid. *Plant Cell Rep.* **2000,** *19*, 435–442.

Bentolila, S.; Hanson, M. Identification of a BIBAC Clone that Co-segregates with the Petunia Restorer of Fertility *Rp* Gene. *Mol. Gene. Genom.* **2001,** *266*(2), 223–230.

Beyermann, B.; Numberg, P.; Weihe, A.; Meixner, M.; Epolen, J. T.; Bomer, T. Fingerprinting Genomes with Oligonucleotide Probes Specific for Simple Repetitive DNA Sequences. *Theor. Appl. Genet.* **1992,** *83*, 691–694.

Boase, M. R.; Bradley, J. M.; Borst, N. K. An Improved Method for Transformation of Regal Pelargonium (*Pelargonium × domesticum* Dubonnet) by *Agrobacterium tumefaciens. Plant Sci.* **1998,** *139*, 59–69.

Boase, M. R.; Spiller, G. B.; Peters, T. A. Transgenic *Cyclamen persicum* Mill. Produced from Etiolated Hypocotyls, Stably Express the *gus A* Reporter Gene in Petals, Scapes, Leaf Laminae, Petioles and Corms 27 Months after Infection with *Agrobacterium tumefaciens*. *SIVB Congr. In Vitro Biol.* **2000,** *36*, 1035.

Boase, M.; Winefield, C.; Lill, T.; Bendall, M. Transgenic Regal Pelargoniums That Express the rolC Gene from *Agrobacterium rhizogenes* Exhibit a Dwarf Floral and Vegetative Phenotype. *In Vitro Cell Dev. Biol.* **2004,** *40*, 46–50.

Caetano-Anolles, G.; Callahan, L. M.; Williams, P. E.; Weaver, K. R.; Gresshoff, P. M. DNA Amplification Fingerprinting Analysis of Bermudagrass (*Cynodon*): Genetic Relationships between Species and Interspecific Crosses. *Theor. Appl. Genet.* **1995,** *91*(2), 228–235.

Cerny, T. A.; Caetnotano- Anolles, G, Trigiano, R. N.; Starman, T. W. Molecular Phylogeny and DNA Amplification Fingerprinting of *Petunia* Taxa. *Theor. Appl. Genet.* **1996,** *92*, 1009–1016.

Chai, M. L.; Wang, B. L.; Kim, J. Y.; Lee, J. M.; Kim, D. H. *Agrobacterium*-Mediated Transformation of Herbicide Resistance in Creeping Bentgrass and Colonial Bentgrass. *J. Zhejiang Univ. Sci.* **2003,** *4* (3), 346–351.

Chen, F. C.; Kuehnle, A. R. Obtaining Transgenic Anthurium through *Agrobacterium*-mediated Transformation of Etiolated Intemodes. *J. Am. Soc. Hortic. Sci.* **1996,** *121*, 47–51.

Cheng, Z.; Lu, B. R.; Sameshima, K.; Fu, D. X.; Chen, J. K. Identification and Genetic Relationships of Kenaf (*Hibiscus cannabinus* L.) Germplasm Revealed by AFLP Analysis. *Genet. Resour. Crop Evol.* **2004,** *51* (4), 393–401.

Cui, M. L.; Takayanagi, K.; Kamada, H.; Nishimura, S.; Handa, T. Efficient Shoot Regeneration from Hairy Roots of *Antirrhinum majus* L. Transformed by *rol* Type MAT Vector System. *Plant Cell Rep.* **2001,** *20*, 55–59.

De Villiers, S. M.; Kamo, K.; Thomson, J. A.; Bomman, C. H.; Berger, D. K. Biolistic Transformation of Chincherinchee (*Ornithogalum*) and Regeneration of Transgenic Plants. *Physiol. Plant* **2000,** *109*, 450–455.

De Wet, L. A. R.; Barker, N. P.; Peter, C. I. The Long and the Short of Gene Flow and Reproductive Isolation: Inter-Simple Sequence Repeat (ISSR) Markers Support the Recognition of Two Floral Forms in *Pelargonium reniforme* (Geraniaceae). *Biochem. Syst. Ecol.* **2008,** *36* (9), 684–690.

Debener, T, Bartels, C.; Mattiesch, L. RAPD Analysis of Genetic Variation between a Group of Rose Cultivars and Selected Wild Rose Species. *Mol. Breed.* **1996,** *2*, 321–327.

Deroles, S. C.; Boase, M. R.; Lee, C. E. Peters, T. A. Gene Transfer to Plants. In: *Breeding for Ornamental plants: Classical and Molecular Approaches*; Vainstein, A., Ed.; Kluwer Academic Publishers: Dordrecht, 2002; pp 155–196.

Ducrocq, C.; Sangwan, R. S.; Sangwannorreel, B. S. Production of *Agrobacterium*-Mediated Transgenic Fertile Plants by Direct Somatic Embryogenesis from Immature Zygotic Embryos of *Datura innoxia*. *Plant Mol. Biol.* **1994,** *25*, 995–1009.

Faccioli, P.; Terzi, V.; Pecchionoi, L.; Berio, T.; Giovanni, A.; Allavena, A. Genetic Diversity in Cultivated *Osteospermum* as Revealed by Random Amplified Polymorphic DNA Analysis. *Plant Breed.* **2000,** *119*, 351–355.

Geeta, S. V.; Shirol, A. M.; Nishani, S.; Shiragur, M.; Varuna, K. J. Assessing Genetic Diversity of Gladiolus Varieties Using SRAPs Markers. *Res. J. Agric. Sci.* **2014,** *5* (4), 658–661.

Giovanni, A.; Pecchioni, N.; Allavena, A. Genetic Transformation of Lisianthus (*Eustoma grandiflorum* Griseb) by *Agrobacterium rhizogenes. J. Genet. Breed.* **1996,** *50*, 35–39.

Han, T. H.; De Jeu, M.; Van Eck, H.; Jacobsen, E. Genetic Diversity of Chilean and Brazilian *Alstroemeria* Species Assessed by AFLP Analysis. *Heredity* **2000,** *84* (5), 564–569.

Han, T. H.; van Eck, H. J.; De Jeu, M. J.; Jacobsen, E. Optimisation of AFLP Fingerprinting of Organisms with a Large-Sized Genome: a Study on *Alstroemeria* sp. *Theor. Appl. Genet.* **1999,** *98*, 465–471.

Han, B.; Suh, E.; Lee, S.; Shin, H.; Lim, Y. Selection of Non-branching Lines Induced by Introducing Ls-like cDNA into Chrysanthemum (*Dendranthema grandiflorum* (Ramat.) Kitamura) "Shuho-no-chikara". *Sci. Hortic.* **2007,** *115*, 70–75.

Haruki, K.; Hasoki, T.; Nako, Y. Tracing the Parentages of Some Oriental Hybrid Lily Cultivars by PCR-RLFP Analysis. *J. Jpn. Soc. Hortic. Sci.* **1998,** *67*, 352–359.

Hecht, M.; Ecker, R.; Stay, R.; Watad, A. A.; Hirchberg, J.; Mann, V.; Altman, A.; Ziv, M. *Agrobacterium*-Mediated Transformation of Lisianthus (*Eustoma grandiflorum*). In: Proc. 3rd International ISHS Symposium on In Vitro Culture and Horticultural Breeding, Jerusalem, Israel. *Acta Hortic.* **1997,** *447*, 339–340.

Hibrand-Saint Oyant, L.; Crespel, L.; Rajapakse, S.; Zhang, L.; Foucher, F. Genetic Linkage Maps of Rose Constructed with New Microsatellite Markers and Locating QTL Controlling Flowering Traits. *Tree Genet. Genomes* **2008,** *4* (1), 11–23.

Hirose, Y.; Aida, R.; Shibata, M. *Agrobacterium tumefaciens*-Mediated Transformation of Delphinium (*Delphinium* spp.) *Breed. Sci.* **1998,** *48*, 117.

Hong, B.; Tong, Z.; Ma, N.; Kasuga, M.; Yamaguchi-Shinozaki, K.; Gao, J. Expression of the Arabidopsis DREB1A Gene in Transgenic Chrysanthemum Enhances Tolerance to Low Temperature. *J Hortic. Sci. Biotechnol.* **2006a,** *81*, 1002–1008.

Hong, B.; Tong, Z.; Ma, N.; Li, J.; Kasuga, M.; Yamaguchi-Shinozaki, K.; Gao, J. Heterologous Expression of the AtDREB1A Gene in Chrysanthemum Increases Drought and Salt Stress Tolerance. *Sci. China C: Life Sci.* **2006b,** *49*, 436–445.

Hong, B.; Ma, C.; Yang, Y.; Wang, T.; Yamaguchi-Shinozaki, K.; Gao, J. Overexpression of AtDREB1A in 690 Chrysanthemum Enhances Tolerance to Heat Stress. *Plant Mol. Biol.* **2009,** *70*, 231–240.

Hoshino, Y.; Turkan, I.; Mii, M. Transgenic Bialaphos-Resistant Snapdragon (*Antirrhinum majus* L.) Produced by *Agrobacterium rhizogenes* Transformation. *Sci. Hortic.* **1998,** *76*, 37–57.

Hosokawa, K.; Matsuki, R.; Oikawa, Y.; Yammura, S. Production of Transgenic Gentian Plants by Particle Bombardment of Suspension-Culture Cells. *Plant Cell Rep.* **2000,** *19*, 454–458.

Hsia, C. N.; Korban, S. S. Micro-Projectile-Mediated Genetic Transformation of Rhododendron Hybrids. *Am. Rhododendron Soc. J.* **1998,** *52*, 187–191.

Hubbard, M.; Kelly, J.; Rajapakse, S.; Abbott, A.; Ballard, R. Restriction Fragment Length Polymorphisms in Rose and their Use for Cultivar Identification. *Hortic. Sci.* **1992,** *27* (2), 172–173.

Inokuma, T.; Kinouchi, T.; Satoh, S. Reduced Ethylene Production in Transgenic Carnations Transformed with ACC Oxidase cDNA in Sense Orientation. *J. Appl. Hortic.* **2008,** *10*, 3–7.

Jiang, B.; Miao, H.; Chen, S.; Zhang, S.; Chen, F.; Fang, W. The Lateral Suppressor-Like Gene, DgLsL, Alternated the Axillary Branching in Transgenic Chrysanthemum (*Chrysanthemum morifolium*) by Modulating IAA and GA Content. *Plant Mol. Biol. Rep.* **2009,** *28*, 144–151.

Jiang, D.; Liang, J.; Chen, X.; Hong, B.; Jia, W.; Zhao, L. Transformation of Arabidopsis Flowering Gene FT to from Cut Chrysanthemum 'Jinba' by Agrobacterium Mediate. *Acta Hortic.* **2010,** *37*, 441–448.

Jeknic, Z.; Lee, S. P.; Davis, J.; Ernst, R. C.; Chen, T. H. H. Genetic Transformation of *Iris germanica* Mediated by *Agrobacterium tumefaciens*. *J. Am. Soc. Hortic. Sci.* **1999,** *124*, 575–580.

Jianhua, Z.; McDonald, M. B.; Sweeny, P. M. Testing for Genetic Purity in *Petunia* and *Cyclamene* Seed Using Random Amplified Polymorphic DNA. *Hortic. Sci.* **1997,** *32*, 246–247.

Jung, C.; Muller, A. E. Flowering Time Control and Applications in Plant Breeding. *Trends Plant Sci.* **2009,** *14*, 563–573.

Kamo, K.; Blowers, A.; Smith, F.; Vaneck, J. Stable Transformation of Gladiolus by Particle Gun Bombardment of Cormels. *Plant Sci.* **1995a,** *110*, 105–111.

Kamo, K.; Blowers, A.; Smith, F.; Vaneck, J.; Lawson, R. Stable Transformation of Gladiolus Using Suspension Cells and Callus. *J. Am. Soc. Hortic. Sci.* **1995b,** *120*, 347–352.

Katsumoto, Y.; Mizutani, Holton, M. F.; Nakamura, T. A.; Tanaka, Y. Engineering of the Rose Flavonoid Biosynthetic Pathway Successfully Generated Blue-Hued Flowers Accumulating Delphinidin. *Plant Cell Physiol.* **2007,** *48* (11), 1589–1600.

Kaufmann, H.; Mattiesch, L.; Lorz, T. D. Construction of a BAC Library of *Rosa rugosa* Thunb., Assembly of a Contig Spanning Rdr1, a Gene that Confers Resistance to Blackspot. *Mol. Genet. Genomics* **2003,** *268*, 666–674.

Kaundun, S. S.; Matsumoto, S. Heterologous Nuclear and Chloroplast Microsatellite Amplification and Variation in Tea, *Camellia sinensis. Genome* **2002,** *45* (6), 1041–1048.

Khodakovskaya, M.; Vakov, R.; Malbeck, J.; Li, A.; Li, Y.; McAvoy, R. Enhancement of Flowering and Branching Phenotype in *Chrysanthemum* by Expression of ipt under the Control of a 0.821 kb Fragment of the LEACO1 Gene Promoter. *Plant Cell Rep.* **2009,** *28*, 1351–1362.

Kim, I. B.; Raemakers, C. J. J. M.; Jacobsen, E.; Visser, R. G. F. Efficient Production of Transgenic *Alstroemeria* Plants by Using *Agrobacterium tumefaciens*. *Ann. Appl. Biol.* **2007,** *151* (3), 401–412.

Kishimoto, S.; Aida, R.; Shibata, M. *Agrobacterium tumefaciens*-Mediated Transformation of *Begonia* × *hiemalis*. *J. Jpn. Soc. Hortic. Sci.* **2000,** *69*, 424.

Kiyokawa, S.; Kikuchi, Y.; Kamda, H.; Harada, H. Genetic Transformation of *Begonia tuber hybrida* by *Rirolgenes*. *Plant Cell Rep.* **1996,** *15*, 606–609.

Knapp, J. E.; Kausch, A. P.; Chandlee, J. M. Transformation of Three Genera of Orchid Using the *bar* Gene as a Selectable Marker. *Plant Cell Rep.* **2000,** *19*, 893–898.

Ko, H. L.; Henry, R. J.; Beal, P. R.; Moisander, J. A.; Fisher, K. A. Distinction of *Ozothamnus diosmifolius* (Vent.) DC Genotypes Using RAPD. *Hortic. Sci.* **1996,** *31*, 858–861.

Krishnaraj, S.; Bi, Y. M.; Saxena, P. K. Somatic Embryogenesis and *Agrobacterium*-mediated Transformation System for Scented Geraniums (*Pelargonium* sp. Frensham). *Planta* **1997,** *201*, 434–440.

Langeveld, S. A.; Pham, K.; van Schadewijk-Nieuwstad, M.; Burma, D.; Langens-Gerrits, M.; Bach, A.; Derks, T. F. L. M. Genetic Transformation of Bulbous Crops. In: VIIIth Symposium on flower Bulbs. *Acta Hortic.* **2000,** *62*.

Lin, H. S.; van der Toom, C. J. G.; Raemakers, K. J.; Visser, R. G. F.; De Jue, M. J.; Jacobsen, E. Genetic Transformation of *Alstroemeria* Using Particle Bombardment. *Mol. Breed.* **2000,** *6*, 369–377.

Ling, H. Q.; Binding, H. Transformation in Protoplast Cultures of *Linum usitatissimum* and *L. suffruticosum* Mediated with Peg and with *Agrobacterium tumefaciens*. *J. Plant Physiol.* **1997,** *151*, 479–488.

Ling, J. T.; Huang, J.; Sauve, R. Genetic Transformation and Regeneration of Transgenic 'Stella de Oro'daylily. *In Vitro Cell. Dev. Biol.—Plant* **1998,** *34*, 1053.

Ling, J. T.; Sauve, R.; Gawel, N. Identification of *Poinsettia* Cultivars Using RFLP Marker, *Hortic. Sci.* **1997,** *32*, 122–124.

Loh, J. P.; Kiew, R.; Keet, A.; Gan, L. H.; Gan, Y. Y. Amplified Fragment Length Polymorphism (AFLP) Provides Molecular Markers for the Identification of *Caladium bicolor* Cultivars. *Ann. Bot.* **2001,** *84*, 155–161.

Marouelli, L. P.; Inglis, P. W. Ferreira, M. A.; Buso, G. S. C. Genetic Relationships among *Heliconia* (Heliconiaceae) Species Based on RAPD Markers. *Genet. Mol. Res.* **2010,** *9* (3), 1377–1387.

Meng, L.; Sun, X.; Li, F.; Liu, H.; Feng, Z.; Zhu, J. Modification of Flowers and Leaves in Cockscomb (*Celosia cristata*) Ectopically Expressing Arabidopsis ASYMMERTIC LEAVES2-LIKE38 (ASL38/LBD41) Gene. *Acta Physiol. Plant.* **2009,** *32*, 315–324.

Millan, F.; Osuma, F.; Cobos, S.; Torres, A.; Cubero, J. I. Using RAPDs to Study Phylogenetic Relationships in *Rosa. Theor. Appl. Genet.* **1996,** *92*, 273–277.

Miroshimchenko, D. V.; Dolgov, S. N. Production of Transgenic Hygromycin Resistant Carnation (*Dianthus caryophylluus* L.) after Cocultivation with *Agrobacterium tumefaciens*. In: The 19th International Symposium on Improvement of Ornamental Plants. *Acta Hortic.* **2000,** *508*, 163–114.

Nagaraju, V.; Srinivas, G. S. L.; Sita, G. L. *Agrobacterium*-Mediated Genetic Transformation in Gebera Hybrid. *Curr. Sci.* **1998,** *74*, 630–634.

Nakano, M.; Koike, Y.; Watanabe, Y.; Suzuki, S. Production of Transgenic Plantlets of Carnation via *Agrobacterium*-mediated Transformation of Petal Explants. *Bull. Fac. Agric., Niigata Univ.* **1999,** *51* (2), 105–114.

Nakatsuka, T.; Mishibaa, K. I.; Abe, Y.; Kubota, A.; Kakizaki, Y.; Yamamura, S.; Nishihara, M. Flower Color Modification of Gentian Plants by RNAi-Mediated Gene Silencing. *Plant Biotechnol.* **2008,** *25*, 61–68.

Narumi, T.; Aida, R.; Ohmiya, A.; Satoh, S. Transformation of Chrysanthemum with Mutated Ethylene Receptor Genes: mDG-ERS1 Transgenes Conferring Reduced Ethylene Sensitivity and Characterization of the Transformants. *Postharv. Biol. Technol.* **2005,** *37*, 101–110.

Nishihara, M.; Nakatsuka, T.; Hosokawa, K.; Yokoi, T.; Abe, Y.; Mishiba, K. I.; Yamamura, S. Dominant Inheritance of White-Flowered and Herbicide-Resistant Traits in Transgenic Gentian Plants. *Plant Biotechnol.* **2006,** *23*, 25–31.

Obara-Okeyo, P.; Kako, S. Genetic Diversity and Identification of *Cymbidium* Cultivars as Measured by Random Amplified Polymorphic DNA (RAPD) Markers. *Euphytica* **1998,** *99*, 95–101.

Otani, Y.; Chin, D. P.; Masahiro, M. Establishment of *Agrobacterium*-Mediated Genetic Transformation System in *Dahlia. Plant Biotechnol.* **2013,** *30*, 135–139.

Pavingevora, D.; Briza, J.; Koytek, K.; Nicedermeierova, H. Transformation of *Rhododendmn* spp. Using *Agrobacterium tumefaciens* with a GUS-Intron Chimeric Gene. *Plant Sci.* **1997,** *122*, 165–171.

Pellegrineschi, A.; Davoliomariani, O. *Agrobacterium rhizogenes*-Mediated Transformation of Scented Geranium. *Plant Cell Tissue Org. Cult.* **1996,** *47*, 79–86.

Rajapakse, S.; Hubbard, M.; Kelly, J. W.; Abbotr, A. G.; Ballard, R. E. Identification of Rose Cultivars by Restriction Fragments Length Polymorphism. *Sci. Hortic.* **1992,** *52*, 237–245.

Rajaseger, G.; Tan, H. T. W.; Turner, I. M.; Kumar, P. P. Analysis of Genetic Diversity among *Ixora* cultivars (Rubiaceae) Using Random Amplified Polymorphic DNA. *Ann. Bot.* **1997,** *80* (3), 355–361.

Ranamukhaarachchi, D. G.; Henny, R. J.; Guy, C. L.; Li, Q. B. DNA Fingerprinting to Identify Nine *Anthurium* Pot Plant Cultivars and Examine their Genetic Relationship. *Hortic. Sci.* **2001,** *36*(4), 758–760.

Reynoird, J. P.; Dewitte, W.; Prinsen, E.; Noin, M.; Chriqui, D.; Van Onckelen, H. Shooty Tumours Induced on Gerbera Hybrid Leaf Explants by *A. tumefaciens* Strain 82319 Are Characterized by High Endogenous Cytokinin Levels. In: The 19th International Symposium on Improvement or Ornamental Plants. *Acta Hortic.* **2000,** *508*, 45–48.

Rosati, C.; Cadic, A.; Renou, C.; Duron, M. Regeneration and *Agrobacterium*-mediated Transformation of *Forsythia* × *intermedia* Spring Glory. *Plant Cell Rep.* **1996,** *16*, 114–117.

Scott, M. C.; Caetan-Anolles, G.; Trigiano, R. N. DNA Amplification Fingerprinting Identifies Closely Related Chrysanthemum Cultivars. *J. Am. Soc. Hortic. Sci.* **1996,** *121*, 1043–1048.

Scovel, G, Ovadis, M.; Vainstein, A.; Reuven, M.; Ben-Yephet, Y. Marker Assisted Selection for Resistance to *Fusarium oxysporum* in the Greenhouse Carnation. *Acta Hortic.* **2001,** *552*, 151–156.

Shinoyama, H.; Mochizuki, A. Insect Resistant Transgenic *Chrysanthemum dendranthema* × *grandiflorum* (Ramat.) Kitamura. In: Proceedings of the 22nd International Eucarpia Symposium Section Ornamental plants: Breeding for Beauty, 2006; pp 177–183.

Shulga, O.; Mitiouchkina, T.; Shennikova, A.; Skryabin, K.; Dolgov, S. Over Expression of AP1-Like Genes from Asteraceae Induces Early Flowering in Transgenic Chrysanthemum Plants. *In Vitro Cell. Dev. Biol. Anim.* **2010,** *46*, 109–110.

Singh, P. *Essentials of Plant Breeding*; Kalyani Publisher: Kolkata, India, 2013; pp 207–219.

Sironi, F.; Rufforli, B.; Giovanni, A.; Allavena, A. Morphogenesis a Genetic Transformation of *Cyclamen persicum* Mill. In: 4th International Symposium on In Vitro Horticultural breeding, 2000; p 80.

Skachkova, T. S.; Mitiouchkina, T. Y.; Taran, S. A.; Dolgov, S. V. Molecular Biology Approach for Improving Chrysanthemum Resistance to Virus B. In: Proceedings of the 22nd International Eucarpia Symposium Section Ornamental plants: Breeding for Beauty, 2006; pp 185–192.

Soh, H.; Han, Y.; Lee, G.; Lim, J.; Yi, B.; Lee, Y.; Choi, G.; Park, Y. Transformation of *Chrysanthemum morifolium* with Insecticidal Gene (Cry1Ac) to Develop Pest Resistance. *Hortic., Environ. Biotechnol.* **2009,** *50*, 57–62.

Soliva, M.; Gautschi, B.; Salzmann, C.; Tenzer, I.; Widmer, A. Isolation and Characterization of Microsatellite Loci in the Orchid *Ophrys araneola* (Orchidaceae) and a Test of Cross-Species Amplification. *Mol. Ecol.* **2000,** *9* (12), 2178–2179.

Souq, F.; Coutus Thevenot, P.; Yean, H.; Delbard, G.; Maizere, Y.; Barbe, J. P.; Boulay, M.; Morisot, A.; Ricci, P. Genetic Transformation of Roses, 2 Examples: One on Morphogenesis, the Other on Anthocyanin Biosynthetic Pathway. In: Second International Symposium on Roses, Antibes, France. *Acta Hortic.* **1996,** *424*, 381–338.

Starman, T. W.; Abbit, S. Evaluating Genetic Relationships of *Geranium* Using Arbitrary Signatures from Amplification Profiles. *Hortic. Sci.* **1997,** *31*, 729–741.

Starman, T. W.; Duan, X.; Abbit, S. Nucleic Acid Scanning Technique Distinguished Closely Related Cultivars of *Poinsettia. Hortic. Sci.* **1999,** *34*, 1119–1122.

Strommer, J.; Gerats, A. G. M.; Sanago, M.; Molnar, S. J. A Gene-Based RFLP Map of Petunia. *Theor. Appl. Genet.* **2000,** *100*, 899–905.

Suo, Z.; Li, W.; Yao, J.; Zhang, H.; Zhang, Z.; Zhao, D. Applicability of Leaf Morphology and Intersimple Sequence Repeat Markers in Classification of Tree Peony (Paeoniaceae) Cultivars. *Hortic. Sci.* **2005,** *40* (2), 329–334.

Suzuki, S.; Nishihara, M.; Nakatsuka, T.; Misawa, N.; Ogiwara, I.; Yamamura, S. Flower Color Alteration in *Lotus japonicus* by Modification of the Carotenoid Biosynthetic Pathway. *Plant Cell Rep.* **2007,** *26*, 951–959.

Tanksley, S. D.; Ganal, M. W.; Prince, J. P.; de Vicente, M. C.; Bonierbale, M. W.; Broun, P.; Fulton, T. M.; Giovannoni, J. J.; Grandillo, S.; Martin, G. B.; Messeguer, R.; Miller, J. C.; Miller, L.; Paterson, A. H.; Pineda, O.; Riider, M. S.; Wing, R. A.; Wu, W.; Young, N. D. High Density Molecular Linkage Maps of the Tomato and Potato Genomes. *Genetics* **1992,** *132* (4) 1141–1160.

Takatsu, Y, Miyamoto, M.; Inoue, E.; Yamada, T.; Manabe, T.; Kasumi, M.; Hayashi, M.; Sakuma, F.; Marubashi, W.; Niwa, M. Interspecific Hybridization among Wild *Gladiolus* species of Southern Africa Based on Randomly Amplified Polymorphic DNA Markers. *Sci. Hortic.* **2001,** *91* (3–4), 339–348.

Takatsu, Y.; Nishizawa, Y.; Hibi, T.; Akutsu, K. Transgenic Chrysanthemum (*Dendranthema grandiflorum* (Ramat.) Kitamura) Expressing a Rice Chitinase Gene Shows Enhanced Resistance to Gray Mold *(Botrytis cinerea). Sci. Hortic.* **1999,** *82*, 113–123.

Talhinhas, P.; Neves-Martins, J.; Leitao, J. AFLP, ISSR and RAPD Markers Reveal High Levels of Genetic Diversity among *Lupinus* spp. *Plant Breed.* **2003,** *122* (6), 507–510.

Tejaswini, H. P.; Sreedhara, S. A.; Anand, L. Molecular Markers for Working Out Genetic Relationship among Genotypes of Carnation (*Dianthus caryophyllus* L.). *Indian J. Gen. Plant Breed.* **2008,** *68*, 93–95.

Thiruvengadam, M.; Yang, C. H. Ectopic Expression of Two MADS Box Genes from Orchid (*Oncidium gower* Ramsey) and Lily (*Lilium longiflorum*) Alters Flower Transition and Formation in *Eustoma grandiflorum*. *Plant Cell Rep.* **2009,** *28*, 1463–1473.

Tjokrokusumo, D.; Heinrich, T.; Wylie, S.; Potter, R.; McComb, J. Vacuum Infiltration of *Petunia hybrid* Pollen with *Agrobacterium tumefaciens* to Achieve Plant Transformation. *Plant Cell Rep.* **2000,** *19*, 792–797.

Tomkins, J. P.; Wood, T. C.; Barnes, L. S.; Westman, A.; Wing, A. R. A. Evaluation of Genetic Variation in the Daylily *(Hemerocallis* spp.) Using AFLP Markers. *Theor. Appl. Genet.* **2001,** *102*(4), 489–496.

Turck, F.; Fornara, F.; Coupland, G. Regulation and Identity of Florigen: FLOWERING LOCUS T Moves Center Stage. *Ann. Rev. Plant Biol.* **2008,** *59*, 573–594.

Tzuri, G, Hillel, J.; Lavi, V.; Haberfeld, A.; Vainstein, A. DNA Fingerprint Analysis of Ornamental Plants. *Plant Sci.* **1991,** *76*, 91–97.

Vainstein, A.; Hillel, J.; Lavi, U.; Tzuri, G. Assessment of Genetic Relatedness in *Carnation* by DNA Fingerprint Analysis. *Euphytica* **1991,** 56, 225–229.

Van der Meer, I. M. *Agrobacterium*-Mediated Transformation of *Petunia* Leaf Disks. *Plant Cell Cult. Protocols* **1999,** *111*, 327–334.

Van der Salm, T. P. M.; van der Toorn, C. J. G.; Bouwer, R.; Tencate, C. H. H.; van der Kreiken, W. M.; Dons, H. J. M. Production of *ml* Genes Transformed Plants of *Rosa hybrid* L., Characterization of their Rooting Ability. *Mol. Breed.* **1997,** *3*, 39–47.

Veres, A.; Kiss, E.; Toth, E.; Toth, A.; Heszky, L. Effect of an Apple Derived Antisense ACC-Synthase cDNA on the Ethylene Production and the Vase Life of Carnation (*Dianthus caryophyllus* L.), Genetic Variation for Plant Breeding. In: Proceedings of the 17th EUCARPIA General Congress, Tulln, Austria, Vol 11, 2004; pp 267–271.

Vik, N. I.; Gjerde, H.; Bakke, K.; Hvoslef-Eide, A. K. Stable Transformation of Poinsettia via DNA Electrophoresis. *Acta Hortic.* **2001,** *560*, 101–103.

Vinterhalter, B.; Orbovic, V.; Vinterhalter, D. Transgenic Root Culture of *Gentiana punctate* I. *Acta Soc. Bot. Polon.* **1999,** *68*, 275–280.

Vos, P.; Hogers, R.; Bleeker, M.; Reijans, M.; van de Lee, T.; Hornes, M.; Freijters, A.; Pot, J.; Peleman, J.; Kuiper, M.; Zabeau, M. AFLP: a New Concept for DNA Fingerprinting. *Nucleic Acids Res.* **1995,** *21*, 4407–4414.

Wegner, H.; Debener, T. Novel Breeding Strategies for Ornamental Dahlias: Molecular Analyses of Genetic Distances between Dahlia Cultivars and Wild Species. *Eur. J. Hortic. Sci.* **2008,** *73*(3), 97–103.

Williams, J. G. K.; Kubelik, A. R.; Livak, K. J.; Rafalski, J. A.; Tingey, S. V. DNA polymorphisms amplified by arbitrary primers are useful as genetic markers. Nucleic Acids Res., **1990**, 18, 6531–6535.

Winefield, C.; Lewis, D.; Arathoon, S.; Deroles, S. Alteration of Petunia Plant Form through the Introduction of the rolC Gene from *Agrobacterium rhizogenes*. *Mol. Breed.* **1999,** *5*, 543–551.

Wolff, K.; Zietkiewitz; and Hofstra, H. Identification of *Chrysanthemum* cultivars and Stability of DNA Fingerprints, DNA Fingerprint Pattern. *Theor. Appl. Genet.* **1995,** *91*, 439–447.

Wolff, K.; Peters-Van Rijn, J. Rapid Detection of Genetic Variability in *Chrysanthemum* (*Dendranthema grandiflora* tzvelio) Using Random Primers. *Heredity* **1993,** *71*, 335–341.

Xue, D.; Feng, S.; Zhao, H.; Jiang, H.; Shen, B.; Shi, N.; Lu, J. The Linkage Maps of *Dendrobium* Species Based on RAPD and SRAP Markers. *J. Genet. Genomics* **2010,** *37*, 197–204.

Yamashingi, M. Detection of Section Specific Random Amplified Polymorphic DNA (RAPD) Markers in *Lilium*. *Theor. Appl. Genet.* **1995,** *91*, 830–835.

Yang, J.; Lee, H. J.; Shin, D. H.; Oh, S. K.; Seon, J. H.; Paek, K. Y.; Han, K. H. Genetic Transformation of *Cymbidium* Orchid by Particle Bombardment. *Plant Cell Rep.* **1999,** *18*, 978–984.

Yang, L. Y.; Vazquwz-Tello, A.; Hidaka, M.; Masaki, H.; Uozumi, T. *Agrobacterium*-Mediated Transformation of *Hibiscus syriacus* and Regeneration of Transgenic Plants. *Plant Tissue Cult. Lett.* **1996,** *13* (2), 161–167.

Yepes, L. M.; Mittak, V.; Slightom, J. L. *Agrobacterium tumefaciens* versus Biolistic Mediated Transformation of *Chrysanthemum* cvs. Polaris and Golden Polaris with Nucleocapsid Protein Genes of Three Tospovirus. In: International Symposium on Cut Flowers of Tropic Species. *Acta Hortic.* **1999,** *482*, 209–218.

Yepes, L. M.; Mittak, V.; Pang, S. Z.; Gonsalves, C.; Slightom, J. L.; Gonsalves, D. Biolistic Transformation of *Chrysanthemum* with the Nucleocapsid Gene of Tomato Spotted Wilt Virus. *Plant Cell Rep.* **1995,** *14*, 694–698.

Yoshimura, Y. Ohishi, K. Narita, R. Ohya, T. Production of Transgenic Carnation Introduced Sarcotoxin Gene. *Res. Bull. Aichi-ken Agric. Res. Center* **2007,** *39*, 1–6.

Zhang, D.; Din, M. A.; Prince, R. A. Discrimination and Genetic Diversity of *Cephalotaxus* Accessions Using ALFP Markers. *J. Am. Soc. Hortic. Sci.* **1999,** *125*, 404–412.

Zietkiewicz, E.; Rafalski, A.; Labuda, D. Genome fingerprinting by simple sequence repeat (SSR)-anchored polymerase chain reaction amplification. *Genomics*, **1994**, 20, pp. 176–183.

Zuker, A.; Chang, P. L. F.; Ahroni, A.; Cheah, K.; Woodson, W. R.; Bressan, R. A.; Watad, A. A.; Hasegawa, P. M.; Vainstein, A. Transformation of Bombardment. *Sci. Hortic.* **1995,** *64*, 177–185.

CHAPTER 10

RECENT ADVANCES IN THE DEVELOPMENT OF TRANSGENIC CROP PLANTS, BIOSAFETY ASPECTS, AND FUTURE PERSPECTIVES

SUBHANKAR ROY-BARMAN*, RAVINDRA A. RAUT, ATRAYEE SARKAR, NAZMIARA SABNAM, SANJUKTA CHAKRABORTY, and PALLABI SAHA

Department of Biotechnology, National Institute of Technology, Durgapur 713209, India

Corresponding author. E-mail: subhankarroy.barman@bt.nitdgp.ac.in

CONTENTS

ABSTRACT

Genetic improvement of crop plants is not new; we have been modifying plant genomes for thousands of years for our well-being. Development of transgenic crop plants is an outcome of increasing human population and incidence of biotic/abiotic stress determinants. The cost-effective approach of genetic engineering allows for a relatively fast cross-species gene transfer. A number of crop plants have been genetically engineered for resistance to insect pests, fungal and viral pathogens, nematodes etc. using a variety of approaches. Genetically modified crop plants have been developed for tolerance to various abiotic stress conditions such as osmosis, salt, drought, temperature, environmental pollutants and so on. Plants have been engineered for better nutrient utilization as well as enhancement of nutrition quality in food. Crop plants have been engineered for molecular farming in order to generate sufficient antigenic vaccines, antibodies, netraceutical and therapeutic proteins. Recently, plant genomes have also been modified for enhancement in the production of biofuel. It is natural to think about the biosafety aspects of transgenic crop plants especially with respect to health and ecological issues. The cause of concern arises due to the phenomena of various types of gene flow in nature. The selectable markers can be removed from the genetically engineered plants using approaches such as co-transformation, multi-autotransformation, site-specific recombination, Cre/lox recombination system etc. Recently, genome editing technology, which allows plant breeding without introducing a transgene, is expected to generate many new crop varieties with traits that can satisfy various kinds of demands for commercialization genetically improved crop plants.

10.1 INTRODUCTION

The human species emerged on this earth about 300,000 years ago. Since then, we have been working for our well-being and improvement of food quality, quantity, shelter, etc., without thinking of the nature and the mother earth, only to satisfy our needs. Human activities have caused enormous changes in physical, chemical, geological, biological, and atmospheric domains of our planet. Genetic improvement of crop plants is not new; we have been modifying plant genomes for thousands of years. The major crop species were domesticated about 5000–10,000 years ago. This has connection to human civilization of which development is intricately linked to agricultural growth. The biggest challenge even today is to produce sufficient amount of food for the exploding human population on earth. In 1800, the world population was 1 billion people, whereas in 1900, at the beginning of the 20th century, the population increased to 1.65 billion people, and by 2000, this number increased to over 6 billion (http://www.plantcell.org/site/teachingtools/teaching.xhtml). The world human population is expected to reach 9 billion by the year 2050, and we need to at least double crop production, especially rice, by this time (Sheehy and Mitchell, 2011). Crop plants are under continuous threat by various biotic and abiotic stresses. Therefore, a continuous effort is to be exercised to ensure tomorrow's food security, which is not endorsed by today's food sufficiency. We are still caught up with limited success of cultural practices, environmentally unhealthy use of pesticides, and decreasing arable land area.

To feed the several billion people living on this planet, the production of high-quality food must increase with reduced inputs, but this accomplishment will be particularly challenging in the face of global environmental change. Plant breeders need to focus on traits with the greatest potential to increase yield. Hence, new technologies must be developed to accelerate breeding through improving genotyping and phenotyping methods and by increasing the available genetic diversity in breeding germplasm. Most of the gain will come from delivering these technologies in developing countries, but the technologies will have to be economically viable and readily disseminated. Crop improvement through breeding brings immense value relative to investment and offers an effective approach to improving food security. However, to meet the recent Declaration of the World Summit on Food Security (FAO, 2009) for production of 70% more food by 2050, an average annual increase in production of 44 million metric tons per year is required. Particularly challenging for society will be the changes in weather patterns that will require alterations in farming practices and infrastructure, for example, water storage and transport networks. The likely impacts on global food production are many because one-third of the world's food is produced on irrigated land. Along with agronomic- and management-based approaches to improving food production, improvements in a crop's ability to maintain yields with lower water supply and quality will be critical (Tester and Langridge, 2010). By and large, we need to increase the tolerance of crops to biotic and abiotic stress conditions by several folds.

Modern tools of plant biotechnology can complement conventional plant breeding in an economically useful way to genetically improve crop plants. In genetically modified (GM) crop plants, their genome is engineered using tools of genetic engineering such as recombinant DNA technology, which is complemented by our knowledge of molecular biology. In this approach, different DNA fragments from various useful sources are put together to create a new molecule that is introduced into the plant genome for desired purposes. Thus, essentially transgenic plants are those plants containing DNA from other organisms. Remarkably, while developing transgenic plants, the genetic engineer enjoys advantage of cross-species gene transfer and considerable reduction in time toward generating an improved transgenic line for a specific crop plant. In the distant and recent past, we have relied on domestication of crop plants, development of hybrid seeds, and experienced "green revolution" through advances in plant breeding technologies. In recent years, we have been witnessing a "gene revolution" that is making remarkable advance in the field of plant biotechnology. Genetic engineering involves cloning of desired genes, development of designer gene constructs, and transfer of transgenes to the organism concerned. Specific changes are introduced in the genome of crop plants using the tools of genetic engineering. Over last three decades, a large number of transgenic plants have been developed across different classes of crops with various improved agronomic characteristics. The main focus has been development of transgenic crop plants for enhanced resistance to bacterial diseases, fungal diseases, virus, nematode, insect pests, etc. and tolerance to drought, salinity, flooding, heavy metals, etc. However, although many GM crop plants have been developed, only a few of them have made their way to the field. On the contrary, the land area under GM crop cultivation has increased steadily over last decade though it has mainly remained restricted to the countries such as the United States, Argentina, Brazil, Canada, India, China, etc. (Table 10.1). Some of the GM crop plants that are being grown in the field are cotton, corn, soybean, canola, sugarbeet, papaya, alfalfa, brinjal, etc.

The improvement of agricultural production and productivity as well as the future versatility of agricultural production are bound to be dependent on the rational utilization of modern plant biotechnology. We stand at the convergence of an unbelievable plethora of new technologies, such as recombinant DNA technology, information technology, and high-throughput genomics, to enhance our understanding of the structure and function of the genomes and to apply this information for improvement of plants and animals. Products arising from modern biotechnology such as GM or transgenic crops are providing new opportunities to achieve sustainable productivity gains in agriculture.

However, ever since GM crop plants were generated, there has been lot of hot debates on application of GM crops over conventional breeding and recently organic farming. Development of GM crop varieties has raised a wide range of ethical, environmental, economic, social, and political

TABLE 10.1 Global Area of Biotech Crops in 2015 and 2016: by Country (Million Hectares**) (reproduced from ISAAA, 2016)

	Country	2015	2016	% increase in 2016 over 2015
1.	USA*	70.9	72.9	3
2.	Brazil*	44.2	49.1	11
3.	Argentina*	24.5	23.8	-3
4.	Canada*	11.0	11.6	5
5.	India*	11.6	10.8	-7
6.	Paraguay	3.6	3.6	0
7.	Pakistan*	2.9	2.9	0
8.	China*	3.7	2.8	-24
9.	South Africa*	2.3	2.7	17
10.	Uruguay*	1.4	1.3	-7
11.	Bolivia*	1.1	1,2	9
12.	Australia*	0.7	0.9	29
13.	Philippines*	0.7	0.8	14
14.	Myanmar*	0.3	0.3	0
15.	Spain*	0.1	0.1	0
16.	Sudan*	0.1	0.1	0
17.	Mexico*	0.1	0.1	0
18.	Colombia*	0.1	0.1	<0.1
19.	Vietnam	<0.1	<0.1	<0.1
20.	Honduras	<0.1	<0.1	<0.1
21.	Chile	<0.1	<0.1	<0.1
22.	Portugal	<0.1	<0.1	<0.1
23.	Bangladesh	<0.1	<0.1	<0.1
24.	Costa Rica	<0.1	<0.1	<0.1
25.	Slovakia	<0.1	<0.1	<0.1
26.	Czech Republic	<0.1	<0.1	<0.1
27.	Burkina Faso	0.5	--	--
28.	Romania	<0.1	--	--
	Total	179.7	185.1	3.0

*Biotech mega-countries growing 50,000 hectares or more
**Rounded-off to the nearest hundred thousand or more

concerns. However, no debates have finally resulted in any unanimously agreed policy. Rather, it has run into a deadlock between various stakeholders such as scientists, farmers, politicians, bureaucrats, and administrators. In the meanwhile, the common man is left in the sidelines only. GM technologies have the potential toward ensuring food security through development of enhanced resistance to biotic stresses, increased tolerance to abiotic stresses, nutritional quality improvement, and use of lesser and lesser pesticides and fertilizers at the same time. On the other hand, organic farming may not ensure increased productivity due to higher incidence of pests and diseases, although it utilizes farmers' knowledge of growing crops while maintaining diversity of crops (Azadi and Ho, 2010). Our choice of going with the GM crops should be determined by the following alarming situations: (1) availability of limited and/or gradually reducing land area for agriculture due to urbanization, (2) exploding human population, (3) constantly changing climatic conditions, (4) limited water availability, (5) increasing incidence of pests and diseases resistant to various pesticides that have been in use so far, (6) biosafety issues, etc.

However, environmental stresses, gradual development of pest resistance to pesticides, alarming increase in population explosion, and food shortage are major concerns of mankind on this globe. Limited natural resources cannot fulfill the food demand of every individual. Thus, we keep experiencing numerous cases of malnutrition especially in the developing and underdeveloped countries. Producing crops with improved quality and quantity is imperative for growing food demand through sustainable agriculture that could be attained using conventional selection and breeding coupled with genetic engineering (Ashraf and Akram, 2009). The application and development of biotechnology have led to newer opportunities and possibilities to enhance qualitative and quantitative enhancement of crop plants (Sun, 2008; Yamaguchi and Blumwald, 2005). Biotechnology for genetic improvement has become a sustainable strategy to combat deficiencies in food by enhancing proteins, carbohydrates, lipids, vitamins, and micronutrient composition (Sun, 2008; Zimmermann and Hurrell, 2002). Major emphasis of agricultural biotechnology has been on traits for improvement in crops related to insect and herbicide resistance, nutritional quality, virus resistance, shelf life, and biofuel production since the 1990s. Thus, to ensure future food security, it is advisable to carefully embrace GM crop plants, especially when genetic engineering offers (1) feasibility of cross-species gene transfer, (2) time saving approach of generating new cultivars, (3) cost-effective strategy toward quality and quantity improvement of crop plants.

In this chapter, we are presenting the work on development of GM crops carried out mostly during the last decade, the major biosafety concerns relevant to the use of transgenic crops, possible precautionary measures for commercial use of GM crops, and future perspectives on generation of "designer biotech crops." However, it must be admitted that the discussion may not still be very exhaustive, and just in case we have missed to acknowledge anybody's work or reference, it is purely unintentional and we sincerely apologize for such lacunae.

10.2 GENETIC ENGINEERING OF RESISTANCE TO BIOTIC STRESS

Some of the limiting factors in crop production are various pests, diseases, and weeds, which are considered as biotic stresses. The limitations associated with chemical methods (mainly, the environmental hazard) and other conventional breeding methods of control necessitated the development of alternative methods for developing new cultivars with higher resistance to biotic stresses. Genetic engineering approach has proven worth to select for the resistance sources from across the species and introduce the agronomical useful genes into the desired plants to provide resistance against different biotic stresses. Numerous strategies have been taken up for last more than three decades for enhancement of resistance against insect pests, nematodes, fungi, bacteria, and virus, which are injurious to crop plants and also cause diseases.

10.3 ENGINEERED RESISTANCE TO INSECT PESTS

A number of transgenic crop plants have been developed for increased resistance to variety of insect pests using variety of strategies (Table 10.2).

10.3.1 USING BT-TOXIN

One of the most skyrocketing achievements in plant biotechnology is development of insect resistant crops expressing crystal proteins from *Bacillus thuringiensis* (*Bt*). *Bt* is a Gram-positive bacterium that produces proteinaceous crystalline (Cry) inclusion bodies during sporulation. It also produces cytotoxins that synergize the activity of Cry toxins (Tohidfar and Khosravi, 2015). It is known that the *Bt* crystal proteins (δ-endotoxin) are toxic to lepidopterans, dipterans, and coleopterans, and at the same time, it is nontoxic to humans and animals (Ahmad et al., 2012).

TABLE 10.2 Recent Developments of Transgenic Crop Plants Resistant to Insect Pests.

Sl. No.	Transgene (Gene Name)	Source of Transgene	Crop and Cultivar	Resistance to	Method of Transformation	Reference
1.	*amiR-24*	*Bacillus thuringiensis*	Tobacco	Cotton bollworm (*Helicoverpa armigera*)	ATMT	Agrawal et al. (2015)
2.	*Bt* (δ-endotoxin gene)	*Bacillus thuringiensis*	Rice	Striped stem borer (*Chilo suppressalis*)	ATMT	Gao et al. (2015)
3.	*HaAK*	*Arabidopsis* sp.	Arabidopsis	Cotton bollworm (*Helicoverpa armigera*)	ATMT	Liu et al. (2015b)
4.	*cry3A*	*Bacillus thuringiensis*	Potato	Colorado potato beetle (*Leptinotarsa decemlineata Say*)	ATMT	Guo et al. (2014)
5.	*cry1Ia8*	*Bacillus thuringiensis* (*Btc008*)	Cabbage	Diamondback moth *Plutella xylostella* (*Linnaeus*)	ATMT	Yi et al. (2C13)
6.	*HaHR3*	Cotton bollworm (*Helicoverpa armigera*)	Tobacco	Cotton bollworm (*Helicoverpa armigera*)	ATMT	Xiong et al. (2013)
7.	*EcR*	Cotton bollworm (*Helicoverpaarmigera*)	Tobacco	Cotton bollworm (*Helicoverpa armigera*)	ATMT	Zhu et al. (2012)
8.	*Cry1Ab*	*Bacillus thuringiensis*	Rice	Lepidopteron Leaf folder and stem borer	ATMT	Qi et al. (2012)
9.	*cry1Ab* and *cry1Ac*	*Bacillus thuringiensis*	Chickpea	Cotton bollworm (*Helicoverpa armigera*)	ATMT	Mehrotra et al. (2011)
10.	*Nlsid-1* and *Nlaub*	*Nilaparvata lugens*	Rice	Rice brown plant hopper	ATMT	Zha et al. (2011)
11.	*CYP6AE14* *P450 gene*	Cotton bollworm (*Helicoverpa armigera*)	Cotton	Cotton bollworm (*Helicoverpa armigera*)	ATMT	Mao et al. (2011)
12.	*m-Cry1Ac*	*Bacillus thuringiensis*	Sugarcane	Sugarcane stem borer	Microprojectile bombardment	Weng et al. (2011)

TABLE 10.2 *(Continued)*

Sl. No.	Transgene (Gene Name)	Source of Transgene	Crop and Cultivar	Resistance to	Method of Transformation	Reference
13.	*CpTI* (cowpea trypsin inhibitor) and *Bt/CpTI* (Bt transgene linked to CpTI)	Cowpea and *Bacillus thuringiensis*	Rice	Rice stem borers (*Scirpophaga incertulas*, *Chilo suppressalis*, and *Sesamia inferens*) and rice leaf-folder (*Cnaphalocrocis medinalis*)	ATMT	Yang et al. (2011b)
14.	*Cry1Ab*, *Cry1Ac*, and *Cry2A*, *cry1C*	*Bacillus thuringiensis Berliner* (*Bt*)	Rice	Striped stem borer (*Chilo suppressalis Walker*)	ATMT	Yang et al. (2011a)
15.	*aadA*		Eggplant	Fruit and shoot borer (*Leucinodes orbonalis*)	Microprojectile bombardment	Singh et al. (2010)
16.	*Bt cry1Ab*	*Bacillus thuringiensis*	Tobacco	Lepidopterans	Chloroplast transformation strategy via bombardment	Jabeen et al. (2010)
17.	*cry2Aa*	*Bacillus thuringiensis* (*Btc008*)	Chickpea	Pod borer (*Helicoverpa armigera*)	ATMT	Acharjee et al. (2010)
18.	*NaPI* and *StPin1A*	Potato (*Solanum tuberosum*)	Cotton	*Helicoverpa punctigera* and *Helicoverpa armigera*	ATMT	Dunse et al. (2010)
19.	*SPI2c* (serine protease inhibitors)	*Solanum nigrum*	*Solanum nigrum*	Generalist insect herbivores	ATMT	Hartl et al. (2010)
20.	*Sporamin* (trypsin inhibitor) and *CeCPI* (phytocystatin)	Sweet potato and taro	Tobacco	Cotton bollworm (*Helicoverpa armigera*)	ATMT	Senthilkumar et al. (2010)
21.	*ASAL* (*Allium sativum* leaf agglutinin)	Garlic (*Allium sativum*)	Rice	GLH (*Nephotettix virescens*) and BPH (*Nilaparvata lugens*)	ATMT	Sengupta et al. (2010)

*A*TMT, Agrobacterium tumefaciens-mediated transformation, GLH, green leafhoppers; BPH, brown plant hopper.

A ***cry2Aa*** gene with a sequence-modified open-reading frame encoding an insecticidal crystal protein from *Bt* was introduced into chickpea (*Cicer arietinum* L.) to confer resistance to *Helicoverpa armigera* (Acharjee et al., 2010). Maize has been transformed with either *Bt* ***cry1Ab***, ***cry1Ac***, or ***cry9C*** to protect it against *Ostrinia nubilalis* and *Sesamia nonagriodes*, or with ***cry1F*** to protect it against *Spodoptera frugiperda*, and with ***cry3Bb***, ***cry34Ab***, and ***cry35Ab*** to protect it against the rootworms of the genus *Diabrotica* (James, 2012). *Bt* toxin genes ***cryIA***, ***cryIAc***, and ***cry3A*** have been expressed in soybean (Macrae et al., 2005), chickpea (Sanyal et al., 2005), and alfalfa (Tohidfar et al., 2013), respectively, for insect resistance. Cotton plants were engineerd using *Bt* toxin gene ***cryIAb*** for protection against cotton bollworm (Tohidfar et al., 2008). Transgenic cruciferous vegetables have been developed for use against *Plutella xylostella* (James, 2012). The *Bt* toxins have been introduced in soybean using either one or two cry genes among ***cry1A*b**, ***cry1Ac***, ***cry1F*** (James, 2013). In field trials, transgenic sugarcane plants expressing high levels of modified ***cry1Ac*** have been shown to provide effective control against stem borers (Weng et al., 2011).

10.3.1.1 MODE OF ACTION OF BT TOXINS

Cry proteins once ingested by the insect are solubilized in the midgut and are then cleaved there by digestive proteases. Some of the resulting polypetides bind to midgut epithelial cell receptors resulting in cell lysis and finally insect death (Gahan et al., 2010).

10.3.2 USING LECTINS

Lectins are carbohydrate-binding peptides or proteins that occur abundantly in seeds and storage tissues of different plants. One of the most important direct defense responses in plants against the attack by phytophagous insects is the production of these peptides or proteins. Lectins have been found to be useful to protect the plants against insect pests, especially the sap-sucking insects (Joshi et al., 2010). The lectins from snowdrop or garlic were found to be injurious to insects but not to mammals (Fitches et al., 2010). The most important protein examined is the lectin from snowdrop (*Galanthus nivalis* agglutinin, GNA). **GNA** has been reported to have the capability to affect the metabolic activity of brown plant hopper (BPH), white-backed plant hopper and green leafhopper pests of rice (Nagadhara et al., 2003). GM rice plant expressing snowdrop lectin gene [*Galanthus nivalis agglutinin* (GNA)] demonstrated reduced survival and fecundity of insects, impaired insect development, and an inhibitory effect on BPH feeding (Brar et al., 2009; Nagadhara et al., 2004; Tang et al., 2001). Transgenic rice with GNA (snow drop) has shown resistance to BPH (*Nilaparvata lugens*) (Li et al., 2005). Transgenic potato expressing *gna* gene showed reduced damage to leaves (Bell et al., 2001). It has been observed that *Allium sativum* leaf agglutinin, the garlic lectin gene, possesses the insecticidal activity against BPH and green leaf hopper (Saha et al., 2006) as has been observed in rice cv. IR64-induced hopper resistance. *Allium cepa* agglutinin has been reported to show insecticidal property to control sap-sucking insects (Hossain et al., 2006).

10.3.2.1 MODE OF ACTION OF LECTINS

The most likely mechanisms underlying the entomotoxic activity of lectins involve interactions with different glycoproteins or glycan structures in insects, which may interfere with a number of physiological processes in these organisms. Since lectins possess at least one carbohydrate-binding domain and different sugar specificities, and considering the variety of glycan structures in the bodies of insects, possible targets for lectin binding are numerous.

10.3.3 USING INHIBITORS AGAINST PROTEASES

Some cultivars including cotton expressing Cowpea trypsin inhibitor (CpTI) have been commercially released in China in 2000. Oryzacystatin 1 (OC1) isolated from rice seeds has been successfully introduced into various crops like rice (Duan et al., 1996), wheat (Altpeter et al., 1999), oilseed rape (Rahbé et al., 2003), and eggplant (Ribeiro et al., 2006). It protects these plant species against beetle attacks and, in some cases, aphids (Sharma et al., 2004). In a remarkable multigene approach, a *Bt*-corn called *Bt*-Xtra containing three genes including *cry1Ac* from *Bt*, *bar* from *Streptomyces higroscopicus*, and potato proteinase inhibitor (*pinII*) has been produced, where the only inhibitor gene was *pinII*. Potato type I and II serine protease inhibitors (PIs) are produced by solanaceous plants as a defense mechanism against insects and microbes. Co-expression of potato type I and II proteinase inhibitors conferred cotton plants protection against a major insect pest, *Helicoverpa punctigera* (Dunse et al., 2010).

10.3.3.1 MODE OF ACTION OF PROTEASE INHIBITORS

Plant PIs are able to protect plants against insect attacks by interfering with the proteolytic activity of insects' digestive gut. Among the proteinaceous PIs, serine and cysteine PIs are abundant in plant seeds and storage tissues (Reeck et al., 1997) and may contribute to their natural defense system against insect predation. Proteinase inhibitors have been found to affect growth and development of many insects.

10.3.4 USING INHIBITORS AGAINST α-AMYLASES

One potential class of inhibitors is α-amylase inhibitors as they can control seed weevils, which are highly dependent on starch as energy source. The bean (*Phaseolus vulgaris*) amylase inhibitor gene was expressed in seeds of transgenic garden pea (*Pisum sativum*), and other grain legumes and seeds from these transgenics were resistant to stored product pests such as larvae of bruchid beetles and field pests such as larvae of the pea weevil *Bruchus pisorum* (Morton et al., 2000). The α-amylase inhibitor gene from *P. vulgaris* was introduced to chickpea and the transformed plants showed a significant resistance to bruchid weevil (Ignacimuthu and Prakash, 2006). When the same gene was expressed in *Coffea arabica*, the seed extracts from resultant transgenics were had an inhibiting amylolytic enzyme activity up to 88% (Barbosa et al., 2010).

10.3.5 OTHER INSECTICIDAL PROTEINS

Other insecticidal proteins such as antibodies, wasp and spider toxins, microbial insecticides, and insect peptide hormones have also been used to generate various transgenic plants. Some bacterial species like *Bt* has become the source of insecticidal activities during vegetative growth. They produce **Vip3A** protein against lepidopteran insects. Unlike *Bt* toxins, Vips do not need to be solubilized in the insect gut. They bind to receptors in the insect gut different from those targeted by Cry proteins (Lee et al., 2006a). *Vip3Aa20*, the modified form of *vip3Aa1* gene, showed insecticidal effects against a wide host range including the corn earworm, the black cutworm, the fall armyworm, and the Western bean cutworm (Tohidfar and Khosravi, 2015).

10.4 ENGINEERED RESISTANCE TO FUNGAL PATHOGENS

Various transgenic crop plants have been developed for enhanced resistance to number of fungal pathogens using variety of strategies (Table 10.3).

10.4.1 USING GENES FOR CHITINASES AND GLUCANASES

In recent years, several laboratories have transformed plants with genes encoding β-1,3-glucanase and chitinase in order to develop transgenic crops with enhanced resistance to fungal diseases. Chitinase appears to have been used probably most frequently to obtain transgenics in various crops for effective control of fungal pathogens. The genes for **chitinase** from varied sources have been used to generate transgenics in grapevine (Yamamoto et al., 2000), rice (Datta et al., 2001; Kim et al., 2003; Kumar et al., 2003; Mei et al., 2004; Takakura et al., 2000), peanut (Rohini and Rao, 2001), cucumber (Kishimoto et al., 2002), tobacco (Carstens et al., 2003), potato (Chye et al., 2005; Moravčíková et al., 2004, 2007), cotton (Tohidfar et al., 2005), trifoliate orange (Mitani et al., 2006), strawberry (Vellicce et al., 2006), oilseed rape (Melander et al., 2006), taro (He et al., 2008a), pea (Hassan et al., 2009), finger millet (Ignacimuthu and Caesar, 2012), tomato (Girhepuje and Shinde, 2011), etc. Conversely, the gene for **glucanase** has been used to generate transgenics in tobacco (Cheong et al., 2000), flax (Wróbel-Kwiatkowska et al., 2004), rice (Akiyama et al., 2004), Indian mustard (Mondal et al., 2007), etc.

10.4.1.1 MODE OF ACTION OF CHITINASES AND GLUCANASES

Chitin constitutes one of the major components of the cell walls of many fungal pathogens such as *Rhizoctonia solani*, and it can be hydrolyzed by chitinase. β-1,3-Glucanase is known to degrade glucans which are also present in the fungal cell walls.

10.4.2 USING OTHER ANTIFUNGAL GENES

Apart from chitinases and glucanases, many other antifungal proteins and peptides such as thaumatin-like protein (TLP), ribosome-inactivating protein (RIP), *A. cepa* Antimicrobial protein (*Ace*-AMP1b), *Raphanus*

TABLE 10.3 Recent Developments of Transgenic Crop Plants Resistant to Fungal Pathogens.

Sl. No.	Transgene (Gene Name/Notation)	Source of Transgene	Crop and Cultivar	Resistance to	Method of Transformation	Reference
1.	*Chitinase*	*Streptomyces griseus* HUT6037	*Brassica juncea* (RAYA ANMOL)	Wide range of Fungal pathogens	ATMT	Ahmad et al. (2015)
2.	*HaGLP1* (Germin-like proteins)	Sunflower (*Helianthus annuus*)	*Arabidopsis thaliana* plants ecotype Columbia (Col-0, accession CS1092)	*Sclerotinia sclerotiorum* and *Rhizoctonia solani*	ATMT	Beracochea et al. (2015)
3.	*Chi11*	Rice	Finger millet (*Eleusine coracana* (L.) Gaertn. GPU45)	Leaf blast *Pyricularia grisea*	ATMT	Ignacimuthu et al. (2012)
4.	*KP4*	Totivirus (UMV4)	Maize (H99, B73)	*Ustilago maydis*	ATMT	Allen et al. (2011)
5.	*RsAFP2*	Radish (*Raphanus sativus*)	Chinese wheat variety Yangmai 12	*F. graminearum* and *R. cerealis*	Particle bombardment	Li et al. (2011a)
6.	*GbTLP1* (*Gossypium barbadense* thaumatin-like protein gene)	Cotton (*Gossypium barbadense* L.)	Tobacco (*Nicotiana tabacum*)	*Verticillium dahlia*, *Fusarium oxysporum*	ATMT	Munis et al. (2010)
7.	*MsDef*1 (*Medicago sativa* defensin gene)	Alfalfa (*Medicago sativa*)	Tomato (Castle Rock)	*Fusararium oxysporum*	ATMT	Abdallah et al. (2010)
8.	*ThEn42* (endochitinase); *StSy* (stilbene synthase)	*Trichoderma harzianum*; Grape	Banana *(Musa cavendish*, AAA, cv. Grand Nain)	*Mycosphaerella fijiensis Botrytis cinerea*	Particle bombardment	Vishnevetsky et al. (2010)

TABLE 10.3 *(Continued)*

Sl. No.	Transgene (Gene Name/Notation)	Source of Transgene	Crop and Cultivar	Resistance to	Method of Transformation	Reference
9.	*RCC2* (rice chitinase gene)	Rice	Grape (*Vitis vinifera* L. cv. Neo Muscut)	*Uncinula necator* *Elisinoe ampelina*	ATMT	Yamamoto et al. (2010)
10.	*NPR1*	*Arabidopsis thaliana*	Cotton	*Verticillium dahliae* isolate TS2 *Fusarium oxysporum* f. sp. *vasinfectum*, *Rhizoctonia solani*, and *Alternaria alternata*	ATMT	Parkhi et al. (2010)
11.	*Pi-d2*	Rice (variety Digu)	Rice variety Lijiangxintuanheu Taipei 309, Nipponbare, and Zhonghua 9	Blast strain ZB15	ATMT	Chen et al. (2010)
12.	*Rs-AFP2* defensin gene	*Raphanus sativus*	(*Oryza sativa* L. cv. Pusa basmati 1)	*Magnaporthe oryzae* *Rhizoctonia solani*	ATMT	Jha et al. (2010)
14.	*Lc* (leaf color)	Maize (*Zea mays*)	Apple (*Malus* × *domestica* cv. "Holsteiner Cox")	*Venturia inaequalis*	–	Flachowsky et al. (2010)
15.	Wasabi defensin gene	*Wasabia japonica*	Rice (*Oryza sativa* cv. Sasanishiki)	*Magnaporthe grisea*	ATMT	Kanzaki et al. (2002)
16.	*Ech*42 (endochitinase)	*Trichoderma atroviride*	Apple (Marshall McIntosh)	*Venturia inaequalis*	ATMT	Bolar et al. (2001)
17.	p*PGIP* (pear fruit polygalacturonase inhibitor protein)	Pear	Tomato	*Botrytis cinerea*	ATMT	Powell et al. (2000)

ATMT, *Agrobacterium tumefaciens*-mediated transformation.

sativus antifungal protein (*Rs*-AFP2), *Dahlia merckii* antimicrobial protein (*Dm*-AMP1), *Mirabilis jalapa* antimicrobial protein (*Mj*-AMP2), etc. have been very useful in conferring fungal disease resistance in transgenic plants. Wheat plants have been successfully engineered to express ***Ace*-AMP1** to confer resistance against powdery mildew and Karnal bunt diseases (Roy-Barman et al., 2006). Rice plants have also been genetically engineered (GE) using the same gene to have a wide-spectrum increased resistance against both bacterial and fungal pathogens (Patkar and Chattoo, 2006). Transgenic *indica* rice expressing ***Mj*-AMP2** showed enhanced resistance to the rice blast fungus (Prasad et al., 2008). Enhanced resistance to rice blast and sheath blight was achieved in transgenic rice overexpressing ***Rs*-AFP2** (Jha and Chattoo, 2009) and ***Dm*-AMP1** (Jha et al., 2009). Transgenic maize plants expressing the Totivirus antifungal protein, **KP4**, is highly resistant to corn smut fungus *Ustilago maydis* (Allen et al., 2011). Sunflower germin-like protein ***Ha*GLP1** promotes ROS accumulation and enhances protection against fungal pathogens such as *Sclerotinia sclerotiorum* and *R. solani* in transgenic *Arabidopsis thaliana* (Beracochea et al., 2015). Transgenic apple plants overexpressing the ***Lc*** gene of maize showed increased resistance to apple scab caused by *Venturia inaequalis* and fire blight caused by *Erwinia amylovora*) but, also had some altered growth habit (Flachowsky et al., 2010). High resistance to *S. sclerotiorum* in transgenic soybean plants was achieved by expressing ***OXDC*** (oxalate decarboxylase) gene (Cunha et al., 2010). Expression of **defensin** gene from radish in transgenic wheat conferred increased resistance to *Fusarium graminearum* and *Rhizoctonia cerealis* (Li et al., 2011a). Plant defensins are cysteine-rich proteins that play an important role in defense against fungal pathogens. They have a strong potential to be used for engineering disease resistance in crops because of their potent antifungal activity.

10.4.2.1 MODES OF ACTION OF ACE-AMP1, RS-AFP2, DM-AMP1, MJ-AMP2, RIPS, AND TLPS

***Ace*-AMP1** is a lipid-transfer protein with sequence homology and structural analogies to plant nonspecific lipid-transfer proteins (ns-LTPs). In contrast to ns-LTPs isolated from radish and maize, Ace-AMP1 is unable to transfer phospholipids from liposomes to mitochondria due to the presence of aromatic residues in the domain corresponding to a lipid-binding pocket found in true lipid transfer proteins. However, the underlying mechanism of action is not very clear.

***Mj*-AMPs** have been identified in the seeds of *M. jalapa* and their structural and biological properties resemble those of defensins, a class of antimicrobial peptides. The Mj-AMPs exhibit a broad spectrum of antifungal activity since they are active against number of plant pathogens.

***Rs*-AFP2**, a plant defensin from the seeds of *R. sativus*, interacts with glucopyranosylceramide (GlcCer) present in the plasma membrane of fungal hyphae, leads to increased K^+ efflux and Ca^{2+} influx, membrane potential changes, and exerts antifungal activity against a broad spectrum of plant pathogenic filamentous fungi by causing hyperbranching and growth reduction of the hyphal tips.

***Dm*-AMP1**, also a plant defensing, interacts with mannosylated sphingolipids occurring in the outer plasma membrane and displays a broad-spectrum antifungal activity.

RIPs exhibit RNA *N*-glycosidase activity and depurinate the 28S rRNA of the eukaryotic 60S ribosomal subunit. This results in failure of binding of elongation factor-2 and cessation of protein synthesis by the altered ribosome.

TLPs are involved in the acquired systemic resistance and in response to biotic stress, causing the inhibition of hyphal growth and reduction of spore germination, probably by a membrane permeabilization mechanism and/or by interaction with pathogen receptors.

10.4.3 USING GENES FOR RESISTANCE SOURCES

Several *R* genes (resistance) associated with innate immunity of plants have been identified and isolated from various sources (Ballvora et al., 2002; Pel et al., 2009). The ***LpiO*** gene, one of the tested effectors from *Solanum* species, when co-expressed along with ***Rpi-blb1*** (as resistance gene) in *Nicotiana benthamiana*, it led to rapid identification of *Rpi-sto1* and *Rpi-pta1* as resistance genes to late blight (Vleeshouwers et al., 2008). Stacking of three broad-spectrum potato *R* genes (*Rpi*), ***Rpi-sto1*** (*Solanum stoloniferum*), ***Rpi-vnt1.1*** (*Solanum venturii*), and ***Rpi-blb3*** (*Solanum bulbocastanum*) in potato showed HR against pathogenic effects of *Phytophtora* (Zhu et al., 2012). Activating phytoalexins in plants against disease is another strategy for protection against pathogens. Genetic transformation of rice with **stilbene synthase gene** (*STS*) of Vst1, a key enzyme in synthesis of phytoalexin in grape improved resistance to *Piricularia orizae* (Coutos-Thévenot et al., 2001). Similarly, transgenic barley has been developed to resist powdery mildew (Liang et al., 2000).

Ectopic expression of *OsCDR1*, encoding a predicted aspartate protease, in Arabidopsis and rice conferred enhanced resistance against bacterial and fungal pathogens (Prasad et al., 2009). More recently, the role of mitogen-activated protein kinase (MAPK) cascade in the regulation of genes responsible for phytoalexin synthesis in rice in response to UV and blast infestation was reported (Wankhede et al., 2013). MAPK kinase is a key component of MAPK cascade. It was found that expression of phytoalexin in rice increased specifically under UV radiation. Subsequently, generation of transgenic rice lines expressing *OsMKK6* gene was shown to overproduce of phytoalexins. Resistance to several fungal and bacterial diseases has been obtained by overexpressing the nonexpressor of pathogenesis-related genes-1 (NPR1) in various plant species with apparently minimal or no pleiotropic effects. Resistance against various fungal pathogens and reniform nematode in transgenic cotton plants has been achieved by expressing Arabidopsis *NPR1* (Parkhi et al., 2010). Expression of this gene in transgenic cotton plants also enhanced resistance against *Thielaviopsis basicola*. These plants exhibited stronger and faster induction of most of these defense-related genes, particularly *PR1*, *thaumatin*, *glucanase*, *LOX1*, and *chitinase* (Kumar et al., 2013).

10.5 ENGINEERED RESISTANCE TO BACTERIAL PATHOGENS

A variety transgenic crop plants have been developed for improved resistance to number of bacterial pathogens using different strategies (Table 10.4).

10.5.1 USING ERFS

The expression of cotton ethylene responsive transcription factors (ERF) in tobacco showed exhibition of greater level of resistance to *Xanthomonas* (Champion et al., 2009). It is to be noted that bacterial blight is a destructive disease of domesticated rice (*Oryza sativa*) caused by the pathogen *Xanthomonas oryzae* pv. *oryzae*.

10.5.1.1 MODE OF ACTION OF ERF

The ERF have been demonstrated to have a role in controlling the expression of pathogenesis-related (PR) genes (Grennan, 2008).

TABLE 10.4 Recent Developments of Transgenic Crop Plants Resistant to Bacterial Pathogens.

Sl. No.	Transgene (Gene Name/Notation)	Source of Transgene	Crop and Cultivar	Resistance to	Method of transformation	References
1.	*hRPN*	*Erwinia amylovora*	Pears (*Pyrus communis cv.* "*Passe Crassane*")	*Erwinia amylovora*	ATMT	Malnoy et al. (2005)
2.	*FALL39* (precursor for the antimicrobial peptide LL-37)	*Homo sapiens*	Chinese cabbage (*Brassica rapa* cv. Osome)	*Psanthomonas carotovorum*	ATMT	Jung et al. (2012), Gudmundsson et al. (1995)
3.	*Bs2*	Pepper	Tomato (*Solanum lycopersicum)*	*Xanthomonas*	ATMT	Horvath et al. (2012), Tai et al. (1999)
4.	*PR1*	Grape vine (*Vitis* interspecific hybrid)	Tobacco (*N. tabacum* "Samsun")	*Pseudomonas syringae*	ATMT	Li et al. (2011c)
5.	*Pflp*	Sweet pepper (*Capsicum annuum*)	Banana cultivars "SukaliNdiizi" and "Nakinyika"	*Xanthomonas campestris*	ATMT	Namukwaya et al. (2011)
6.	*Xa21*	Rice (conserved)	Tomato (*Lycopersocin esculentum* cultivars Roma, Rio Grande, Pusa Ruby, Pant Bahr, and Avinash) Sweet orange (cvs. Hamlin, Natal, Pera, and Valencia)	*Pseudomonas solanacearum* *Xanthomonas axonopodis*	ATMT	Afroz et al. (2010), Mendes et al. (2010)
7.	*Indolicidin*	Cow (*Bovis* neutrophils)	Tobacco (*Nicotiana tabacum* var. Xanthi)	*Erwinia carotovora*	ATMT	Bhargava et al. (2007), Collinge et al. (2010)
8.	*CB* (Cecropin B)	*Hylophora cecropia*	Tomato (*S. lycopersicum* cv. Microtom	*Ralstonia solanacearum* *Xanthomonas campestris*	ATMT	Jan et al. (2010)

TABLE 10.4 *(Continued)*

Sl. No.	Transgene (Gene Name/Notation)	Source of Transgene	Crop and Cultivar	Resistance to	Method of transformation	References
9.	*EFR*	*Arabidopsis thaliana*	Tobacco (*N. benthamiana*) Tomato (*S. Lycopersicum* var. Moneymaker)	Broad spectrum bacterial resistance	ATMT	Lacombe et al. (2010), Fillatti et al. (1987), Horsch et al. (1985)
10.	*Rxo1*	Maize R gene	Rice	*Xanthomonas oryzae*	ATMT	Wally and Punja (2010), Zhao et al. (2005)
11.	*AtNPR1*	*Arabidopsis thaliana*	Duncan grapefruit Hamlin sweet orange Tomato (*Lycopersicon esculentum*)	*Xanthomonas citri* *Xanthomonas* sp.	ATMT	Zhang et al. (2010b), Lin et al. (2004)
12.	*βhth*(β-hordothionin gene)	Barley	Tobacco	*Pseudomonas solanaceaurm*	ATMT	Collinge et al. (2010), Charity et al. (2005)
13.	*Lc*	Maize (*Zea mays*)	Apple (*Malus domestica* cv. "Holsteiner Cox")	*Erwinia amylovora*	ATMT	Li et al. (2007a, 2007b)
14.	*SN1*	*S. chacoense*	Potato (*S. tuberosum* subsp. *tuberosum* cv. Kennebec)	*Erwinia arotovora*	ATMT	Almasia et al. (2008)
15.	*Chit33*, *chit42* (chitinase encoding genes)	*Trichoderma harzianum* strain CECT2431	Tobacco (*Nicotiana tabacum* var. Xhanti)	*Pseudomonas syringae*	ATMT	Mercedes et al. (2006)
16.	*GOX* (glucose oxidase)	*Aspergillus niger*	Rice (*Oryza sativa* L. ssp. Japonica cv. Taipei 309)	*Xanthomonas oryzae*	Particle bombardment	Kachroo et al. (2003)
17.	*D4E1*	Synthetic (plasmid ubi7-D4E1)	Poplar hybrid (*Populus tremula* × *P. alba* clone 717 IB 4)	*Agrobacterium tumefaciens* *Xanthomonas populi*	ATMT	Mentag et al. (2003)

TABLE 10.4 *(Continued)*

Sl. No.	Transgene (Gene Name/Notation)	Source of Transgene	Crop and Cultivar	Resistance to	Method of transformation	References
18.	*Asth1* (thionin)	Oat (*Avena sativa* cv. Zenshin	Rice seed (*Oryza sativa* cvs. Nipponbare and Chiyohonami)	*Burkholderia glumae Burkholderia plantarii*	ATMT	Iwai et al. (2002)
19.	*PPO-P1*	Potato (*Solanum tuberosum* cv. Katahdin)	Tomato (*Lycopersicon esculentum* Mill. cv. Money Maker)	*Pseudomonas syringae*	ATMT	Li and Steffens (2002)
20.	*Tsi1*	Tobacco (*Nicotiana tabacum* cv. Samsun NN and cv. Xanthi)	Hot pepper (*C. annuum* cv. Nockwang)	*Xanthomonas campestris*	ATMT	Shin et al. (2002), Park et al. (2001)
21.	*Msi-99* (magainin-2 analog)	Frog (*Xenopus laevis*)	Tobacco (*Nicotiana tabacum* var. Petit Havana)	*Pseudomonas syringae*	Particle bombardment	DeGray et al. (2001)
22.	*myp30* (magainin analog)	Frog (*Xenopus laevis*)	Tobacco (*Nicotiana tabacum* L. cv. Kentucky 14)	*Erwinia arotovora*	ATMT	Li et al. (2001), Zasloff (1987)
23.	*exp1* (*N*-oxoacyle-homoserine biosynthesis)	*E. carotovora*	Tobacco (*Nicotiana tabacum* cv. Samsun)	*Erwinia arotovora*	ATMT	Mäe et al. (2001)
24.	*Msarco* (Sarcotoxin IA gene)	*Sarcophega peregrina*	Tobacco (*Nicotiana tabacum* cv. Sumsun NN)	*Erwinia arotovora Pseudomonas syringae*	ATMT	Mitsuhara et al. (2000)
25.	*cecB* (cecropin B gene)	*Bombyx mori*	Rice (*Oryza sativa* L. Japonica cv. Nipponbare)	*Xanthomonas oryzae*	ATMT	Hiei et al. (1994)

ATMT, *Agrobacterium tumefaciens*-mediated transformation.

10.5.2 USING HAIRPIN GENES

The **harpin** (*hrp*) genes encode type III secretory pathways and are required by many phytopathogenic bacteria for pathogenesis on susceptible hosts and to elicit a hypersensitive response (HR) on nonhost or resistant host plants. Several studies indicated that enhanced **HrpNEa** levels in transgenic plants have effectively increased resistance to bacteria (Malnoy et al., 2005). Harpin NEa (HrpNEa) is encoded by the gene hrpN located on the chromosome of *Erwinia* causing the fire blight disease of apple. HrpNEa is a known inducer of systemic acquired resistance (SAR) in plants. Transgenic plants resistant to bacterial pathogens have been produced making use of this property.

10.5.2.1 MODE OF ACTION OF HAIRPIN PROTEINS

When hrp genes are secreted to the plant cells from bacterial pathogens, localized cell death happens through series of reactions like involving accumulation of reactive oxygen species (ROS).

10.5.3 USING TOXIN DETOXIFYING GENE FROM THE PATHOGEN

Another approach for engineering of plant resistance against bacterial disease is based on the transformation with a gene encoding a toxin-detoxifying enzyme from the pathogen itself. *Pseudomonas syringae* pv. *tabaci* produces the toxin called tabtoxin. In plants, tabtoxin is converted to tabtoxinine-β-lactam, which inhibits glutamine synthase leading to an accumulation of cytotoxic ammonia. The pathogen protects itself against the toxin by expression of the tabtoxin resistance gene (*ttr*), which is able to protect *P. syringae* by acetylating tabtoxin to an inactive form. The transgenic tobacco, expressing *ttr* gene, displayed a reduction in disease symptoms (Batchvarova et al., 1998).

Recently, a plant ferrodoxin-like protein (PFLP) was transferred to *Arabidopsis*. Expression of PFLP enhanced resistance to bacterial disease. PFLP is a photosynthetic type ferredoxin with an N-terminal signal peptide for chloroplast localization. Presence of PFLP in transgenic plants conferred resistance against bacterial disease (Lin et al., 2010). Expression of this gene in transgenic banana also enhanced resistance to wilt disease caused by *Xanthomonas* sp. (Namukwaya et al., 2012). Expression of a synthesized gene encoding cationic peptide Cecropin B in transgenic tomato plants enhanced resistance against bacterial diseases (Jan et al., 2010). Resistance in the susceptibility to *Xanthomonas axonopodis* pv. *citri* was achieved in transgenic *Citrus sinensis* plants expressing rice ***Xa21*** (Mendes et al., 2010).

10.6 ENGINEERED RESISTANCE TO VIRAL PATHOGENS

A lot of transgenic crop plants have been developed for enhanced resistance to various viral pathogens using different strategies (Table 10.5).

Hundreds plant viruses have been identified till date, which cause various diseases and significant crop losses. Viral diseases are conventionally controlled using certified virus free planting material, eradicating infected plants and spraying chemicals against virus vectors. Additionally, coat protein-mediated resistance to viruses has been one of the successes of plant genetic engineering. Several major crop plants have been engineered using this approach, to resist important viral pathogens. The resistant cultivars that have been commercialized include potato event HLMT15-15, which is tolerant to PYV (Potato Y Virus) or potato event RBMT21-350, which is resistant to PLRV (Potato Leaf Roll Virus) (James, 2013). Transgenic tobacco expressing defective cucumber mosaic virus (CMV) replicase-derived dsRNA was produced to achieve high level of resistance (Ntui et al., 2014). The ability of the sense and antisense RNA for the replication-associated protein encoded by *AC1* (African cassava mosaic virus replication-associated) or *C1* gene of Gemini viruses was also assessed to protect plants against viral infection (Zhang et al., 2005). It was also reported that presence of defective movement proteins in the transgenic plants conferred resistance to viruses, as they are associated with their growth and development in planta (Hallwass et al., 2014; Peiró et al., 2014).

10.7 ENGINEERED RESISTANCE TO NEMATODES

Nematodes, which are not readily controlled by pesticides or other control options, cause an estimated $118b annual loss to world crops (McCarter, 2009). Although natural resistance genes are unavailable for many crops to plant breeders, transgenic plants can provide significant amount of nematode resistance for such crops. Approaches, such as limiting use of dietary protein uptake by nematodes from the crops or by preventing root invasion without a direct lethality or use of RNA interference (RNAi) can take control over wide range of nematodes. A variety of transgenic crop plants have been

TABLE 10.5 Recent Developments of Transgenic Crop Plants Resistant to Viral Pathogens.

Sl. No.	Transgene (Gene Name/Notation)	Source of Transgene	Crop and Cultivar	Resistance to	Method of transformation	Reference
1.	Coat protein gene, V2 gene and replication-associated gene	*Tomato yellow leaf curl virus*-Oman (TYLCV-OM)	*Tomato (Solanum lycopersicum* L.)	TYLCV-OM	ATMT	Ammara et al. (2015)
2.	*pC5*, *pC6*	RGSV	Japonica rice (Nipponbare)	Rice grassy stunt virus (RGSV)	ATMT	Shimizu et al. (2012)
3.	*CP*	Tobacco Streak Virus	Sunflower (*Helianthus annuus* L.)	Tobacco Streak Virus	ATMT	Pradeep et al. (2012)
4.	*Rep* (Replication initiation protein)	*Banana bunchy top virus* (BBTV)	Banana (*Musa* spp.)	*Banana bunchy top virus* (BBTV)	ATMT	Shekhawat et al. (2012)
5.	Capsid protein	AMV BPMV SMV	Soybean (Throne)	*Alfalfa mosaic virus* (AMV), *Bean pod mosaic virus* (BPMV), and *Soybean mosaic virus* (SMV)	ATMT	Zhang et al. (2011b)
6.	FL-CP (coat protein of *Cassava Brown Streak Uganda* virus)	*Cassava Brown Streak Uganda* virus	Cassava (*Manihot esculenta* Crantz)	*Cassava Brown Streak Uganda* virus	ATMT	Yadav et al. (2011)
7.	*CP*	*Soyabean Dwarf virus*	Soybean	Soybean dwarf virus	Particle bombardment	Tougou et al. (2007)
8.	*TOGT* (glucosyl transferase)	Tobacco	Tobacco plants (*N. tabacum* cv. Samsun NN)	Potato virus Y	ATMT	Matros and Mock (2004)
9.	*CP*	*Citrus tristeza virus strain T-305*	Mexican lime (*Citrus aurantifolia* Swing.)	*Citrus tristeza* virus	ATMT	Domínguez et al. (2000)

ATMT, *Agrobacterium tumefaciens*-mediated transformation; CP, coat protein.

developed for improved resistance to various nematode pests using diverse kind of strategies (Table 10.6).

10.8 TRANSGENIC DEFENSE BASED ON PEPTIDES AND PROTEINS

In the GE plants, the feeding of nematodes is targeted and it involves overexpression of cysteine proteinase inhibitors (cystatins) that interfere with intestinal digestion of their dietary protein taken in from the plant. **Cystatins** have a proven wide value against a range of nematodes with differing modes of parasitism (Fuller et al., 2008). A cystatin from the tropical root crop, taro, when expressed in tomato conferred resistance against *Meloidogyne* (Chan et al., 2010). The acetylcholinesterase-inhibiting peptide when expressed in *A. thaliana* suppressed the number of female *Heterodera schachtii* (beet cyst nematode) by more than 80%, while in transgenic potato plants, its expression resulted in almost 95% resistance to *Globodera pallida* (Lilley et al., 2011). When nicotinic acetylcholine receptors (nAChR)-binding peptide was expressed in transgenic potato plants that secreted the peptide from their root tips, it resulted in an effective resistance up to 77% against potato cyst nematode in both containment glasshouse and field trials (Atkinson et al., 2012).

10.8.1 MODE OF ACTION nAChRs-BINDING PEPTIDE

The nAChR-binding peptide is taken up from the environment by certain chemosensory sensilla within the anterior amphidial pouches and it undergoes retrograde transport along some chemoreceptive neurons to their cell bodies and a limited number of interneurons. Chemoreception was only impaired when that transport had been completed.

10.9 TRANSGENIC DEFENSE BASED ON RNAi

In the RNAi process, double-stranded RNA (dsRNA) triggers silencing of specific target genes through mRNA degradation. RNAi in *A. thaliana* plants expressing dsRNA from hairpin and/or inverted repeat constructs reduced transcript abundance of targeted **parasitism genes** in *H. schachtii* (Patel et al., 2008, 2010; Sindhu et al., 2009). This led to a significant reduction in female members (between 23% and 64%) with considerable variation

TABLE 10.6 Recent Developments of Transgenic Crop Plants Resistant to Nematode Pests.

Sl. No.	Transgene (Gene Name/ Notation)	Source of Transgene	Crop and Cultivar	Resistant to	Method of Transformation	References
1.	*16D10*	Conserved root-knot nematode (RKN) gene *16D10*	Wine grape (*V. vinifera* cv. Chardonnay)	Root-knot nematode	ATMT	Yang et al. (2013)
2.	CCII (cystatin)	Maize kernel	Plantain (*Musa AAB* cv. Gonja Manjaya)	*Radopholus similis*, *Helicotylenchus multicinctus*	ATMT	Roderick et al. (2012)
3.	*OC-I* (*Oryza* cystatin I)	Rice Nihonbare (*Oryza sativa* L. *japonica*)	Sweet potato (cv. Xushu 18 and cv. Lizixiang) Alfalfa (*Medicago sativa)* hybrid Regen-SY	*Pratylenchus penetrans*	ATMT	Gao et al. (2011d), Abe et al. (1987)
4.	*NPR1*	Arabidopsis	Tobacco Cotton (*Gossypium hirsutum* cv. Coker 312)	*Meloidogyne incognita*	ATMT	Priya et al. (2011), Parkhi et al. (2010), Sunilkumar and Rathore (2001)
5.	*CeCP1*	Taro (*Colocasia esculenta*) Kaosiang No. 1	Tomato (*Solanum lycopersicum* Mill.) cultivar CLN2468D	*Meloidogyne incognita*	ATMT	Chan et al. (2010)
6.	*Cry5* B	Plant codon-modified from *Bacillus thuringiensis Cry5B*	Tomato (*Lycopersicon esculentum* Mill. var. Rutgers select)	*Meloidogyne incognita*	ATMT	Li et al. (2008)
7.	*CaMi*	Pepper (*Capsium annum* L. (line PR205)	Tomato	Root-knot nematode	ATMT	Chen et al. (2007)

TABLE 10.6 *(Continued)*

Sl. No.	Transgene (Gene Name/ Notation)	Source of Transgene	Crop and Cultivar	Resistant to	Method of Transformation	References
8.	*Cry6 A*	Plant codon-modified from *Bacillus thuringiensis Cry6A*	Tomato	*Meloidogyne incognita*	ATMT	Li et al. (2007a, 2007b)
9.	*Mi-1.2*	Wild-type tomato (*L. peruvianum*)	Eggplant (*S. melongena* cv. HP83) Tomato (*L. esculentum* cv. Moneymaker)	*Meloidogyne javanica*	ATMT	Goggin et al. (2006), Milligan et al. (1998)
10.	*PIN2* (protease inhibitor)	Potato	Wheat (*T. durum* PDW215)	*Heterodera avenae*	ATMT	Vishnudasan et al. (2005)
11.	*Hero A*	Tomato	Tomato line LA1792	Potato cyst nematode resistance		Sobczak et al. (2004), Ernst et al. (2002)
12.	*OC-II* (*Oryza* cystatin II)	Rice Nihonbare (*Oryza sativa* L. *japonica*)	Alfalfa (*Medicago sativa*) hybrid Regen-SY	*Pratylenchus penetrans*	ATMT	Samac and Smigocki (2003), Abe et al. (1987)
13.	*GAD (Glutamate decarboxylase)*	Chimeric or mutant version of tobacco *GAD*	Tomato (*Nicotiana tabacum* L. cvs. Delgold and Samsun NN	*Meloidogyne incognita*	ATMT	McLean et al. (2003)

ATMT, *Agrobacterium tumefaciens*-mediated transformation.

between lines. In chimeric soybean, RNAi for **fibrilin** gene of *H. glycines* resulted in variable and nonsignificant effects (Li et al., 2010). Soybean composite plants derived from hairy root cultures engineered to silence either of **two ribosomal proteins, a spliceosomal protein or synaptobrevin**, of *H. glycines* by RNAi resulted in 81–93% reduction female members in the roots of transgenic plants (Klink et al., 2009). Similarly, by targeting mRNA splicing factor **prp-17** or an uncharacterized gene ***cpn-1***, high reduction in egg production was achieved (Li et al., 2010). A high level of resistance to root-knot nematode was also achieved by targeting a **parasitism gene** expressed in the subventral gland cells of *Meloidogyne incognita*. When dsRNA complementary to the ***16D10*** gene was expressed in transgenic *A. thaliana*, the resulting lines displayed a significant reduction (63–90%) in the number of galls and their size with a corresponding reduction in total egg production. A broad spectrum of resistance against *M. incognita*, *Meloidogyne javanica*, *Meloidogyne arenaria*, and *Meloidogyne hapla* was achieved since there is a high level of homology between the 16D10 sequences of different *Meloidogyne* species. Significant reduction in egg masses were achieved in transgenic Arabidopsis (~60%) and tobacco (~70%) expressing siRNA against the secreted peptide (16D10) of *Meloidogyne chitwoodi* (Dinh et al., 2014). Almost complete resistance to *Meloidogyne* infection was reported in tobacco plants expressing dsRNA corresponding to **splicing factor or integrase** (Yadav et al., 2006) and of four genes targeted in transgenic soybean roots with reduction of gall number by more than 90% (Ibrahim et al., 2011). However, all host-delivered RNAi targeting of *Meloidogyne* genes did not result in a resistance phenotype. Partial silencing of ***MjTis11***, a putative transcription factor of *M. javanica* did not significantly affect either nematode development or fecundity (Fairbairn et al., 2007). Crossing transgenic lines expressing more than a single line of engineered defense provided higher levels of resistance to *M. incognita* than either parent plants (Charlton et al., 2010). Such additive effect may raise the efficacy and durability of RNAi-based defenses. RNAi against nematode effector protein gene (***NULG1a***) from *M. javanica* in Arabidopsis reduced nematode population in the roots by 80% (Lin et al., 2013).

10.10 GENETIC ENGINEERING FOR TOLERANCE TO ABIOTIC STRESS

Plant growth and final yield are often affected due to abiotic stresses such as salt, drought, flooding, extreme temperature, and oxidative stresses. One of the principal causes of crop failure worldwide is abiotic stress causing a reduction in average yields of most crops by more than 50% (Bray et al., 2000). It is predicted that more than half of the arable land will be affected with salts by 2050. The condition is terrifically alarming and necessitates increasing crop productivity. Conversely, responses of plants under abiotic stresses are complex, involving multiple genes with additive effects (Fig. 10.1) In the field of plant genetic engineering, major emphasis has been given to introduce genes encoding compatible organic osmolytes, heat shock proteins, plant growth regulators, late embryogenesis abundant (LEA) proteins, and transcription factors responsible for activation of a subset of gene expressions. Agricultural productivity could be increased significantly if designer crops can be produced to cope up with environmental stresses. It is also remarkable that with changing environmental conditions in the recent past, most of the efforts have been diverted toward generation of abiotic stress-tolerant transgenic plants.

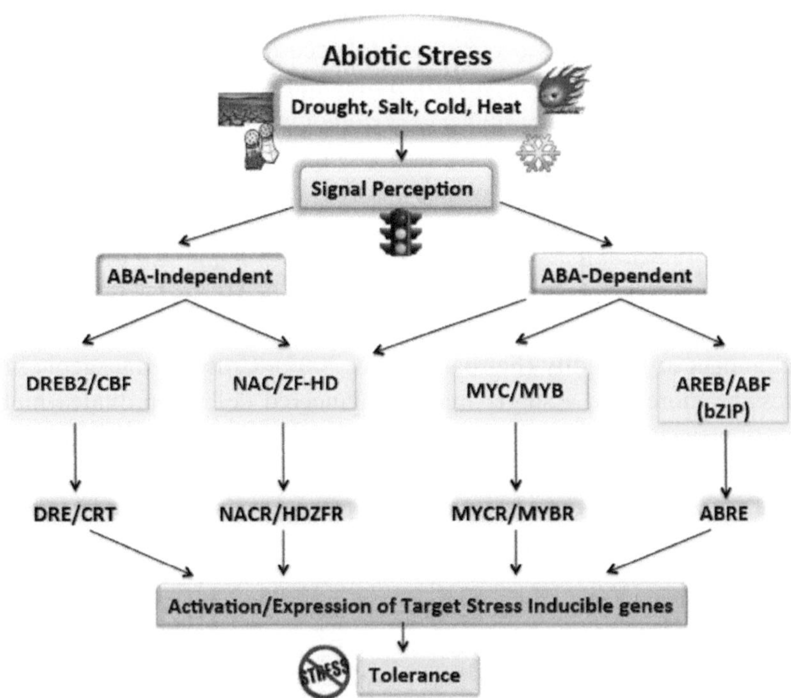

FIGURE 10.1 A schematic explanation of signal transduction pathways and its components involved in gene expression under abiotic stress. Under stress, ABA biosynthesis activates two regulatory ABA-dependent gene expressions: MYC/MYB and bZIP/ABRE. ABA-independent signal transduction pathway involves ERF family of transcription factors.

10.10.1 USING GENES FOR SYNTHESIS OF OSMOTIC PROTECTANTS

It is well known that some organic solutes play an important role in induction of drought tolerance (Ashraf and Foolad, 2007). A number of genes play an important role in the synthesis of osmoprotectants in stress-tolerant plant like proline, glycinebetaine, polyamines, mannitol, trehalose, and galacinol, which are known to accumulate during osmotic adjustment. Some of the genes required for synthesis of such osmotic protectants have been used to engineer crop plants for improved tolerance to various abiotic stress conditions.

10.10.2 USING GENES FOR SYNTHESIS OF GLYCINE BETAINE

Introduction of a gene encoding **choline oxidase** (*codA*) in *Brassica juncea* (Parsad et al., 2000) and rice (Mohanty et al., 2002) resulted in increased tolerance to salt stress due to enhanced levels of glycine betaine. Similarly, increased accumulation of chloroplastic glycine betaine in tomato engineered using the same gene raised the level of stress tolerance (Park et al., 2007). When the *CMO* gene encoding **choline monooxygenase** was expressed in tobacco plants, it resulted in improved tolerance to drought (Shen et al., 2002), and in transgenic rice, it resulted in enhanced tolerance to salt and temperature stress (Shirasawa et al., 2006). Similarly, transgenic cotton (*Gossypium hirsutum*) plants expressing ***AhCMO*** accumulated 26–131% more glycine betaine and showed tolerance to salinity (Zhang et al., 2009). **Choline dehydrogenase** encoding gene *betA* when expressed in maize conferred the plants with higher drought tolerance (Quan et al., 2004), and when expressed in cotton, it resulted in enhanced tolerance to chilling conditions (Zhang et al., 2012). Expression of **choline oxidase** gene *COX* in rice resulted in improved tolerance to saline conditions (Su et al., 2006), whereas in potato, it conferred higher tolerance to oxidative, drought, and salt stress conditions (Ahmad et al., 2008).

10.10.3 USING GENES FOR SYNTHESIS OF PROLINE

Soybean (De Ronde et al., 2004) and petunia (Yamada et al., 2005a) have been GE to produce proline and the transgenics were found to demonstrate enhanced tolerance to heat and drought. Transgenic tobacco plants expressing *P5CR* encoding **pyrroline-5-carboxylate synthase** showed tolerance to drough.

10.10.4 USING GENES FOR SYNTHESIS OF MANNITOL AND TREHALOSE

Expression of *TPS1* encoding **trehalose-6-phosphate synthase** in tobacco (Karim et al., 2007) and rice (Jang et al., 2003) resulted in increased drought tolerance, whereas in tomato (Cortina and Culiáñez-Macià, 2005), it caused higher tolerance to both, oxidative, drought, and salinity stress. Engineering other trehalose biosynthesis genes such as ***otsA*** and ***otsB*** also improved drought tolerance in transgenic plants (Garg et al., 2002). Transgenic expression of an ***mtlD*** involved in the biosynthesis of manitol developed higher tolerant lines against oxidative stress, drought, and salinity stress (Abebe et al., 2003).

10.10.5 USING GENES ENCODING LATE EMBRYOGENESIS ABUNDANT PROTEINS

LEA proteins get accumulated in plants under stress and help them to maintain structure of cellular membranes, ionic balance, water binding, and they also seem to act as molecular chaperons under drought stress conditions. Thus, they are also believed to have vital role in stress tolerance of plants (Babu et al., 2004; Gosal et al., 2009).

When LEA gene ***HVA1*** was transformed into rice and bread wheat, it increased tolerance to drought in both the cases (Sivamani et al., 2000). Expression of LEA gene ***ME-lea n4*** in transgenic *Lactuca sativa* (Park et al., 2005a) and *Brassica camesptris* enhanced drought tolerance in either case (Park et al., 2005b). Transgenic expression of ***PMA1959*** and ***PMA80*** LEA in rice resulted in enhanced dehydration tolerance (Cheng et al., 2002). Overexpression of ***OsLEA 3-1*** in transgenic rice also caused increased tolerance to drought stress under field conditions (Xiao et al., 2007).

10.10.6 USING GENES ENCODING TRANSCRIPTION FACTORS

Transcription factors are DNA-binding proteins required to transcribe and regulate genes. Researchers have been continuously putting their efforts to identify, characterize, clone, and use the know-how to engineer crop plants to protect them against different stress conditions. Overexpression of ***ZmDREB2A*** in maize (Qin et al., 2007) and groundnut (Bhatnagar-Mathur et al., 2009) promoted stress tolerance. Likewise, through overexpression of ***AtDREB*** in wheat (Pellegrineschi et al., 2004), rice (Kim and Kim, 2009), and groundnut (Bhatnagar-Mathur et al., 2014), enhanced drought tolerance was achieved. Transgenic expression of ***OsDREB*** in rice was found to increase activity of genes involved in the tolerance of drought, high salt, and cold response, whereas overexpression of ***GhDREB1*** in tobacco showed significant chilling tolerance only (Shan et al., 2007). Similarly, when *Arabidopsis* **DREB1B** was constitutively expressed in transgenic potato, it enhanced drought and freezing tolerance (Movahedi et al., 2012). Stress-inducible expression of ***GmDREB1*** conferred salt tolerance in transgenic alfalfa (Jin et al., 2010). Drought and salt tolerance was improved in transgenic Arabidopsis expressing ***NAC*** (NAM, ATAF, and CUC) transcriptional factor from *Arachis hypogea* (Liu et al., 2011). Transgenic rice plants expressing ***OsNAC*** in the root system improved drought tolerance under field conditions (Jeong et al., 2010).

10.10.7 USING METAL TOLERANCE

Enhancement of ***TaALMT1*** expression helped increasing Al^{3+} resistance of wheat (Pereira et al., 2010). This was the first report of a major food crop being stably transformed for greater Al^{3+} resistance. Transgenic overexpression of *CcMT1* gene in *A. thaliana* has shown increased plant biomass and chlorophyll content as well as low content of copper and cadmium metals in shoots and roots compared with wild-type plants under copper and cadmium metal stress (Sekhar et al., 2011). Transgenic rice plants expressing cadmium tolerance gene **yeast cadmium factor (*YCF1*)** has been developed (Islam and Khalekuzzaman, 2015). This transgenic rice plants have the ability to uptake cadmium from soil, and it is stored into cell vacuoles and protects rice grain from cadmium. This way soil also will be free from cadmium through the process of phytoremediation. Overexpression of the same gene caused enhancement of heavy metal tolerance in *B. juncea* (Bhuiyan et al., 2011). Overexpression of the *Tamarix hispida* **ThMT3** gene not only increased copper tolerance but also the induction of adventitious root in *Salix matsudana* (Yang et al., 2015). The transgenic tobacco plants expressing a *Trichoderma virens* **GST** are more tolerant to cadmium, but it did not enhance accumulation of the metal in the plant biomass (Dixit et al., 2011).

Enhanced heavy metal tolerance was achieved and accumulation was also demonstrated in transgenic sugar-beet plants expressing *Streptococcus thermophilus* ***StGCS-GS*** in presence of cadmium, zinc, and copper (Liu et al., 2015a). Some of transgenic crop plants developed for improved tolerance to metal ions are listed in Table 10.7.

10.11 ENGINEERING TOLERANCE TO SALT STRESS

Soil salinity is one of the major constraints in today's agriculture, affecting an estimated 45 million hectares of irrigated land and is expected to increase due to global climate changes and as a consequence of various agricultural practices (Munns and Tester, 2008). The deleterious effects of salt stress include slower growth rates, reduced tillering, and abnormal reproductive development, which, in turn, affect crop yield. Various mechanisms of salinity tolerance of crops such as ion exclusion, osmotic tolerance, and tissue tolerance can be genetically improved.

Osmotic tolerance is regulated by long-distance signals that reduce shoot growth and is triggered before Na^+ accumulation in the shoots. Thus, when ***TmHKT1;5-A*** was introgressed from *Triticum monococcum* into a durum wheat, it resulted in a significant improvement in grain yield under high salt stress by increasing its ion exclusion (James et al., 2012; Munns et al., 2012). Na^+ and Cl^- transport processes in roots reduce the accumulation of toxic concentrations of Na^+ and Cl^- within leaves during ion exclusion. Both, high affinity **potassium transporter (*HKT*)** gene family and the **salt overly sensitive (*SOS*)** pathway have been implicated in having a crucial role in regulating Na^+ transport within a plant system. Genetic engineering of expression of these genes has been frequently reported to alter accumulation of Na^+ in the shoot. However, transgenic approaches to improve salinity tolerance using ***HKT1***s have not been so successful. *HKT2* has been reported to increase salinity tolerance, although not through Na^+ exclusion (Mian et al., 2011). Overexpression of genes in the SOS pathway has been reported to result in increased salt tolerance in transgenic Arabidopsis. Constitutive expression of **CaXTH3, a hot pepper xyloglucan endotransglucosylase/hydrolase** enhanced tolerance to salt in transgenic tomato plants (Choi et al., 2011). Additionally, it also increased drought tolerance in these plants. Overexpression of **osmotin gene** in tomato conferred tolerance to salt and drought. The transgenic plants showed significantly higher relative water content, chlorophyll content, proline content, and leaf expansion than the wild-type plants under stress conditions (Goel et al., 2010). Ectopic

TABLE 10.7 Recent Developments of Transgenic Crop Plants Tolerant to Metal Ions.

Sl. No.	Transgene (Gene Name/Notation)	Source of Transgene	Crop and Cultivar	Tolerance to	Method of transformation	Reference
1.	*YCF1*	*S. cerevisiae*	Rice, BRRI dhan29	Cadmium	ATMT	Islam and Khalekuzzaman (2015)
2.	*ThMT3* (Type III metallothionein gene)	*Tamarix hispida*	*S. matsudana Koidz.* var. *matsudana*	Copper	ATMT	Yang et al. (2015)
3.	*StGCS-GS* (γ-glutamylcysteinesynthetase-glutathione synthetase)	*Streptococcus thermophilus*	*Beta vulgaris* L. US-8916	Copper, zinc, cadmium	ATMT	Liu et al. (2015a)
4.	*YCF1* (yeast cadmium factor 1)	*S. cerevisiae*	*Brassica juncea* cv. *Rai-5*	Cadmium	ATMT	Bhuiyan et al. (2011)
5.	*LMT1* (aluminum metallothionein *TaA*)	Wheat	*Triticum aestivum* Bob White 26	Aluminum	Particle bombardment	Pereira et al. (2010)
6.	*TvGST* (glutathione transferase gene)	*Trichoderma virens*	*Nicotiana tabacum*	Cadmium	ATMT	Dixit et al. (2011)
7.	*MdSPDS1* (spermidine synthase)	*Malus sylvestris*	*Pyrus communis* L. "Ballad"		ATMT	Wen et al. (2008)
8.	*merA18* (bacterial mercury detoxification gene)	*Arabidopsis thaliana*	*Liriodendron tulipifera*	Mercury	Microprojectile bombardment	Rugh et al. (1998)

ATMT, *Agrobacterium tumefaciens*-mediated transformation.

expression of the same gene led to enhanced salt tolerance in transgenic chilli pepper (*Capsicum annum* L.) (Subramanyam et al., 2011). Overexpression of ***TaNHX2*** enhanced salt tolerance of "composite" and whole transgenic soybean (Cao et al., 2011) and tomato (Yarra et al., 2012) plants. Similarly, transgenic sweet potato plants expressing ***LOS5*** gene were developed to tolerate salt stress (Gao et al., 2011a). Stress-inducible transgenic expression of *GmGSTU4* shaped the metabolome of transgenic tobacco plants toward increased salinity tolerance (Kissoudis et al., 2015). Transgenic overexpression of mutagenized version of **Δ^1-pyrroline-5-carboxylate synthetase (*P5CS*)** in transgenic *indica* rice resulted in enhanced proline accumulation and salt stress tolerance (Kumar et al., 2010). ***SUV3*** overexpressing transgenic rice plants are reported not only known to be salt tolerant, but it also conserved physicochemical properties and microbial communities of rhizosphere (Sahoo et al., 2015). Some of transgenic crop plants developed for improved tolerance to salt stress are listed in Table 10.8.

Overexpression of the **ethylene-responsive factor gene *BrERF4*** from *Brassica rapa* increased tolerance to salt and drought in Arabidopsis plants, and it also affected the growth and development significantly (Seo et al., 2010). Similarly, overexpression of ***GsGST*** encoding glutathione-S-transferase, from wild soybean (*Glycine soja*) enhanced drought and salt tolerance in transgenic tobacco (Ji et al., 2010). Expressing a ***BADH*** gene from *Atriplex micrantha* enhanced salinity tolerance in transgenic maize (Di et al., 2015). Conversely, constitutive and stress-inducible overexpression of a native **aquaporin gene (*MusaPIP2;6*)** in transgenic banana plants demonstrated its pivotal role in salt tolerance (Sreedharan et al., 2015).

In case of ***tissue tolerance***, salt is compartmentalized at the cellular and intracellular level under highly saline condition. The mechanisms contributing to tissue tolerance include synthesis of compatible solutes, accumulation of Na^+ in the vacuole, and production of enzymes catalyzing detoxification of ROS. Increasing the abundance of proteins involved in the synthesis of compatible solutes (such as proline and glycinebetaine), vacuolar Na^+/H^+ antiporters (NHX), vacuolar H^+ pyrophosphatases (e.g., AVP1), and enzymes responsible for the detoxification of ROS have had differing levels of success in improving tolerance of crop plants to salinity (Roy et al., 2014). Enhanced salt tolerance in transgenic wheat expressing a **vacuolar Na^+/H^+ antiporter** gene was observed. However, often reports do come about under performance of transgenic plants and low salt stress conditions. Such kind of effects may probably be regulated by use of stress-inducible promoters.

Ca^{2+} mediates many aspects of plant growth and development. Ca^{2+} signaling cascade is activated upon perception of environmental cues on the

TABLE 10.8 Recent Developments of Transgenic Crop Plants Tolerant to Salt Stress.

Sl. No.	Transgene (Gene Name/Notation)	Source of Transgene	Crop and Cultivar	Method of Transformation	Reference
1.	*BADH*, betaine aldehyde dehydrogenase gene	*Atriplex micrantha*	Maize elite inbred lines, Zheng58 and Qi319	ATMT	Di et al. (2015)
2.	*AcPIP2*, plasma membrane Aquaporin gene	*Atriplex canescens*	*Nicotiana benthamiana*, *Arabidopsis thaliana Col-1*	ATMT	Li et al. (2015)
3.	*OCPI2*, chymotrypsin protease inhibitor	*Oryza sativa PB-1*	*Arabidopsis thaliana Columbia*	ATMT	Tiwari et al. (2015)
4.	WT-*PhyA S599A-PhyA*	*Avena sativa*	Zoysia grass (*Zoysia Japonica* Steud.) Creeping bentgrass (*Agrostis stolonifera* L.)	ATMT	Gururani et al. (2015)
5.	*MusaPIP2;6*, aquaporin gene	Banana cv. *Karibale Monthan*	Banana cv. *Karibale Monthan*	ATMT	Sreedharan et al. (2015)
6.	*SOS2*, salt overly sensitive gene	*Populus trichocarpa*	Aspen hybrid clone Shanxin Yang (*Populus davidiana* × *Populus bolleana*)	ATMT	Yang et al. (2015)
7.	*LCY-ε*, lycopene ε-cyclase	*Ipomoea batatas* cv. *Yulmi* wild type	*Ipomoea batatas*	ATMT	Kim et al. (2012)
8.	*AtNHX1*, Na^+/H^+ antiporter gene	*Arabidopsis thaliana*	*Zea mays*	ATMT	Li et al. (2013)
9.	*JcDREB*, stress responsive DNA binding transcription factor	*Jatropha curcas*	*Arabidopsis thaliana*	ATMT	Yu et al. (2013)
10.	(*DREB1B*), dehydration-responsive element-binding factor 1	Arabidopsis	*Solanum tuberosum* cv. Desiree	ATMT	Movahedi et al. (2012)

TABLE 10.8 *(Continued)*

Sl. No.	Transgene (Gene Name/Notation)	Source of Transgene	Crop and Cultivar	Method of Transformation	Reference
11.	*TaWRKY 19 TaWRKY2,* stress-responsive *WRKY* gene	*Triticicum aestivum* L. cultivar Xifeng 20	*Arabidopsis thaliana* Col-0	ATMT	Niu et al. (2012)
12.	Osmotin gene	*Nicotiana tabaccum* cv. Wisconsin 38	*Capsicum annuum*	ATMT	Subramanyam et al. (2011)
13.	*CaXTH3*, bacterial mercury detoxification gene	Hot pepper	*Solanum lycopersicum* cv. Dotaerang	ATMT	Choi et al. (2011)
14.	*AhNAC*, NAC gene	*Arachis hypogea*	*Arabidopsis thaliana* (Col-1)	ATMT	Liu et al. (2010)
15.	*GST*, Glutathione transferase gene	*Suadea salsa*	Arabidopsis	ATMT	Qi et al. (2010)
16.	*ADC1 ADC2*, arginine decarboxylase	*Tritordeum*	*Oryza sativa*	Particle bombardment	Liu et al. (2007)

ATMT, *Agrobacterium tumefaciens*-mediated transformation.

cell membrane, resulting in the regulation of gene expression and protein activities (Batistič and Kudla, 2012). In crop plants, such as rice, apple, barley, tobacco, and tomato, overexpression of genes encoding proteins in Ca^{2+} signaling pathways have been shown to improve the growth of the plants during salt stress.

A gene for **monohydroascorbate reductase** (MDAR) has been isolated from halophytic mangrove *Avicennia marina* and expressed under CaMV 35S promoter in tobacco plants following ATMT (Kavitha et al., 2010). Overexpression of *Am-MDAR* was found to increase salt tolerance in transgenic tobacco compared to untransformed control plants. The protein was localized in the chloroplast of transgenic tobacco as presence of a transit peptide at the N terminus of *Am*-MDAR already suggested. Upregulation of *Am-MDAR* under stress conditions such as salt stress, H_2O_2, high light intensity, and iron load and its localization in the chloroplast point toward a crucial role for this protein in the stress tolerance of *A. marina.*

Various stress factors produce ROS, which can cause damage to plants. Enzymes such as superoxide dismutase, catalase, and peroxidase have the capacity to act as antioxidants and neutralize the effect of ROS (Ahmad et al., 2010). ***OsMT1a*** overexpressing transgenic rice plants, which had enhanced ascorbate peroxidase (APX) activity, showed enhanced tolerance to water limited conditions (Yang et al., 2009). Conversely, chilling tolerance at the booting stage has been increased in rice by transgenic overexpression of the **APX gene, *OsAPXa*** (Sato et al., 2011). Cytosolic APX has been found to help plants acclimatize better under conditions heat and drought stress (Koussevitzky et al., 2008). The increased production of **glutathione reductase (GSH)** can be triggered by the stimulation of pathways involved in the metabolism of sulfur and cysteine. Manipulation of improvement of tolerance to oxidative stress was observed with engineering of GSH biosynthesis pathway (Sirko et al., 2004). Increased salinity tolerance and better growth were reported in transgenic tobacco plants by overexpressing glyoxalate pathway enzymes. In this case, increased GSH content maintained higher reduced to oxidized GSH ratio (GSH:GSSG) and minimized lipid peroxidation (Yadav et al., 2005). Overexpression of ***TaEXPB23***, a wheat expansin gene, improved oxidative stress tolerance in transgenic tobacco plants (Han et al., 2015).

10.12 ENGINEERING DROUGHT TOLERANCE

Drought is one of the prime abiotic stresses in the world. Crop yield losses due to drought stress are considerable. A variety of approaches have been used to alleviate the problem of drought. Conventional plant breeding or genetic engineering seems to be an efficient and economic means of tailoring crops to enable them to grow successfully in drought-prone environments. It has been observed that in all above approaches discussed in order of achieving tolerance to salt stress and osmotic protection, the resultant transgenics are also found to be tolerant to drought conditions. *AcPIP2* encoding a plasma membrane intrinsic protein from halophyte *Atriplex canescens*, enhanced plant growth rate and abiotic stress tolerance when overexpressed in *A. thaliana* (Li et al., 2015). ***OsSDIR1*** (*O. sativa* SALT-AND DROUGHT-INDUCED RING FINGER 1) overexpression greatly improved drought tolerance in transgenic rice (Gao et al., 2011d). Overexpression of ***TsCBF1*** gene conferred improved drought tolerance in transgenic maize (Zhang et al., 2010a). Both, drought and salinity tolerance were enhanced in transgenic sweet potato (Fan et al., 2012) expressing ***BADH*** from spinach, and in transgenic groundnut expressing ***AtNHX1*** (Asif et al., 2011). When a wheat ***TaMYB30-B*** encoding R2R3-MYB protein was engineered into Arabidopsis, the resulting transgenic plants showed improved drought stress tolerance (Zhang et al., 2012). Remarkably, expression of ***Arabidopsis enhanced drought Tolerance1/HOMEODOMAIN GLABROUS11*** conferred drought tolerance in transgenic rice without compromising the yield factor (Yu et al., 2013). Transgenic tobacco (*Nicotiana tabacum* cv. *Xanthi-nc*) overexpressing Arabidopsis ***LOS5/ABA3*** also resulted in enhanced drought tolerance (Yue et al., 2011). Some of transgenic crop plants developed for improved tolerance to drought stress are listed in Table 10.9.

10.13 ENGINEERING TOLERANCE TO HIGHER OR LOWER TEMPERATURE

The increase in global mean surface temperature is projected to be in the range of 1.5–4°C by the end of the 21st century, which is due to global warming. We have been experiencing a lot of seasonal variations for last several years now due to climate change. The plants in the field are also experiencing increased levels of heat stress. The major world food crops are already underperforming with heat as one of the stress factors (Lobell and Gourdji, 2012; Teixeira et al., 2013). Critical reproductive stages are under threat due to this reason. Extreme temperature regimes in temperate and subtropical agricultural zones cause significant yield loss (Teixeira et al., 2013). Some species and cultivars are more sensitive to heat stress (Lobell and Gourdji, 2012), and in the due course of time, they may somewhat adapt

TABLE 10.9 Recent Developments of Transgenic Crop Plants Tolerant to Drought Stress.

Sl. No.	Transgene (Gene Name/Notation)	Source of Transgene	Crop and Cultivar	Method of Transformation	Reference
1.	*BdWRKY36* (*WRKY* transcription factor)	*Brachypodium distachyon*	*Nicotiana tabaccum*	ATMT	Sun et al. (2015)
2.	*AtEDT1/HDG11* (Homodomain-leucine zipper transcription factor enhanced drought tolerance/HOMEODOMAIN GLABROUS11)	*Arabidopsis thaliana Col-0*	*Oryza sativa japonica*	ATMT	Yu et al. (2013)
3.	*TaMYB30-B* (*MYB* type gene)	Wheat of different ploidy levels	*Arabidopsis thaliana*	ATMT	Zhang et al. (2012)
4.	*OtsA, OtsB* (trehalose-6-P-synthase, trehalose-6-P-phosphatase)	*E. coli*	Rice Pusa Basmati-1 (PB-1)	ATMT	Ahmad et al. (2012)
5.	*TaWRKY19* (WRKY-type transcription factor)	*Triticum aestivum*	*Arabidopsis thaliana*	ATMT	Niu et al. (2012)
6.	*TaWRKY2* (stress-responsive *WRKY* gene)	*Triticicum aestivum* L. cultivar Xifeng 20	*Arabidopsis thaliana* Col-0	Vaccum infiltration method via ATMT	Niu et al. (2012)
7.	*ATHB-7* (homeodomain-leucine zipper (HD-Zip) transcription factor gene)	*Arabidopsis thaliana*	*Lycopersicum lycopersicon DTL 20*	ATMT	Mishra et al. (2012)
8.	*SoBADH* (betain aldehyde dehydrogenas gene)	*Spinacia oleracia*	*Ipomoea batatas* cv. Sushu-2	ATMT	Fan et al. (2012)
9.	*AVP1* (Vacuolar H^+-Pyrophosphatase gene)	*Arabidopsis thaliana*	*Gossypium hirsutum* cv. *Coker 312*	ATMT	Pasapula et al. (2011)
10.	*LOS5* (molybdenum-cofactor sulfurase)	*Arabidopsis thaliana*	*Nicotiana tabacum* cv. *Xanthi*	ATMT	Yue et al. (2011)

TABLE 10.9 *(Continued)*

Sl. No.	Transgene (Gene Name/Notation)	Source of Transgene	Crop and Cultivar	Method of Transformation	Reference
11.	*BADH* (betaine aldehyde dehydrogenase gene)	Spinach	Potato cv. gannongshu	ATMT	Zhang et al. (2011b)
12.	*Osmotin*	Tobacco	*S. lycopersicum* cv. Pusa Ruby	ATMT	Goel et al. (2010)
13.	*TsCBF1* (abiotic stress responsive transcription factor)	*Thellungiella halophila*	Maize	Particle bombardment	Zhang et al. (2010a)
14.	*OsNAC10*	*Oryza sativa*	*Oryza sativa* cv. Nipponbare	ATMT	Jeong et al. (2010)
15.	*TPS* and *TPP* (trehalose synthases)	Yeast	*Arabidopsis thaliana*	ATMT	Miranda et al. (2007)
16.	*SAMDC* (*S*-adenosyl methioninedecarboxy)	Human	*Nicotiana tabacum* var. xanthi	ATMT	Waie and Rajam (2003)
17.	*AtOAT* (Ornithine amino transferase)	*Arabidopsis thaliana*	*Nicotiana tabaccum*	ATMT	Roosens et al. (2002)

ATMT, *Agrobacterium tumefaciens*-mediated transformation.

to heat stress naturally. In some heat tolerant crops, specific thermoprotective genes are constitutively expressed at higher levels (Bita et al., 2011). However, the capacity of plants to evolve naturally against temperature fluctuations, like any other adaptations, will also be a slow process. The conventional plant breeding methods has not been very successful against abiotic stresses, especially higher temperature, because of the complexity of the phenomenon itself. Therefore, genetic engineering of plants for enhanced heat tolerate could be a way to combat the effects of global rise in temperature on crop productivity.

There are at least three approaches that have been used for engineering heat tolerance. A number of proteins associated with diverse cellular metabolic activities have been overexpressed in transgenic experiments with the view of enhancing heat tolerance. These include proteins found to be involved in metabolism of amino acids and their derivatives, protein biosynthesis, photosynthetic activity, redox homeostasis and hormonal regulation, etc. Higher heat tolerance in transgenic plants were achieved through overexpression of **L-aspartate-α-decarboxylase** from *Escherichia coli* in tobacco, gene for **arginine decarboxylase** enzyme (involved in polyamine biosynthesis) from *Avena sativa* in *Solanum melongena*, *Saccharomyces cerevisiae* gene encoding for **S-adenosyl-L-methionine decarboxylase** (SAMDC) enzyme (involved in polyamine biosynthesis) in *Solanum lycopersicum*, **spermine synthase** gene in *A. thaliana*, *Rosa chinesis* gene encoding for **translation initiation factor** in *A. thaliana*, *Zea mays* gene encoding for **elongation factor** in *Triticum aestivum*, ***AtFKBP62*** gene in *A. thaliana*, *Cajanas cajan* gene encoding for **cyclophilin** chaperone in *A. thaliana*, *A. thaliana* gene for **thioredoxin-like protein** (a foldase and holdase chaperone) in *A. thaliana* (Grover et al., 2013), etc. Some of transgenic crop plants developed for improved tolerance to higher temperature stress are listed in Table 10.10.

ROS scavenging pathways also help plants face stress responses. Thus, heat tolerance was improved when *A. thaliana* gene encoding for **nucleotide diphosphate kinase** was overexpressed in *Solanum tuberosum*, *S. lycopersicum* gene for **GDP-mannose pyrophosphorylase** was overexpressed in *N. tabacum*, *O. sativa* gene for **chloroplast protein**-enhancing stress tolerance overexpressed in *A. thaliana* (Grover et al., 2013), and *A. thaliana* gene for **cytokinin oxidase/dehydrogenase** ectopically expressed in *N. tabacum* (Macková et al., 2013).

Sometimes, general stress-related proteins have been ectopically overexpressed. Thus, higher heat tolerance was achieved when *Xerophyta viscosa* gene encoding for **stress-associated protein 1** (SAP1; a cell membrane-binding protein) was overexpressed in *A. thaliana*, *Populus tremula* gene

TABLE 10.10 Recent Developments of Transgenic Crop Plants Tolerant to Heat Stress.

Sl. No.	Transgene (Gene Name and Notation)	Source of Transgene	Crop and Cultivar	Method of Transformation	Reference
1.	*CSD1*, *CSD2* (copper/zinc superoxide dismutase), *CCS*	*Arabidopsis thaliana*	*Arabidopsis thaliana* ecotype Columbia	ATMT	Guan et al. (2013)
2.	*TPS1* (trehalose synthesis)	*Saccharomyces cerevisiae*	*Medicago sativa* L. cv. Regen SY27x	ATMT	Suárez et al. (2009)
3.	*TPS* and *TPP* (trehalose synthesis)	*Saccharomyces cerevisiae* strain W303-1A	*Arabidopsis thaliana* Col-0 ecotype	ATMT	Miranda et al. (2007)

ATMT: *Agrobacterium tumefaciens*-mediated transformation.

encoding for a stable protein overexpressed in *A. thaliana*, and when *A. thaliana* gene for SAP5 was overexpressed in *G. hirsutum* (Grover et al., 2013). *A. thaliana* SAP5 positively regulates salt and osmotic stress tolerance through its E3 ubiquitin ligase activity.

Additionally, a diverse kind of proteins has been employed in development of transgenics for improved heat tolerance, such as overexpression of *Agrobacterium rhizogenes* gene encoding for **β-glucosidase** in *Rubia cordifolia*, *C. cajan* gene for hybrid proline rich in *A. thaliana*, *Malus domestica* gene encoding for vacuolar proton translocating inorganic **pyrophosphatase** in *M. domestica*, *Z. mays* gene encoding for **acetyl cholinesterase** in *N. tabacum*, and *A. thaliana* gene for **CYP710A1** in *A. thaliana* (Grover et al., 2013; Senthil-Kumar et al., 2013), **Annexin protein** from *N. nucifera* in transgenic *A. thaliana* (Chu et al., 2012). On the other hand, overexpression of **SlCZFP1**, a novel TFIIIA-type zinc finger protein from tomato conferred enhanced cold tolerance in transgenic Arabidopsis and rice (Zhang et al., 2011a). Some of transgenic crop plants developed for improved tolerance to lower temperature stress are listed in Table 10.11.

10.14 ENGINEERING FOR REMOVAL OF ENVIRONMENTAL POLLUTANTS

Human activities and industrial development generate large amounts of chemicals that often contaminate soil and water. Prevalent contaminants include petroleum hydrocarbons, polycyclic aromatic hydrocarbons, halogenated hydrocarbons, pesticides, solvents, metals, and salts. Among these, halogenated hydrocarbons, such as polychlorinatedbiphenyls (PCBs) and chlorophenols, are persistent environmental pollutants (Wang et al., 2015). In general, cleaning up environmental pollutants using wild-type plants leads to the accumulation of PCBs in shoots and roots of plants that may be released to the soil or get again into the atmosphere (Akin et al., 2009; Xia et al., 2009). However, it appears like GE plants can handle the situation better. Phytoremediation is now emerging as a promising strategy and attracting much attention due to its advantages of being less expensive, environmentally sustainable, and esthetically acceptable compared to physical and chemical methods (Krämer, 2005). A lot of the studies have shown the removal rate of PCBs or 2,4-DCP using conventional plants is inadequate and slow (Zeeb et al., 2006). The primary reason is that plants lack the necessary enzymatic machinery involved in bacteria or mammals for efficient cleavage of aromatic structure. Alternatively, there is increasing opportunity in using

TABLE 10.11 Recent Developments of Transgenic Crop Plants Tolerant to Cold Stress.

Sl. No.	Transgene (Gene Name and Notation)	Source of Transgene	Crop and Cultivar	Method of Transformation	Reference
1.	*CsTK* (transketolase)	cDNA library	*Cucumis sativa* L. cv. Jinyou 3	ATMT	Bi et al. (2015)
2.	*DREB1B* (dehydration-responsive element-binding factor 1)	*Arabidopsis* sp.	*Solanum tuberosum* L.	ATMT	Movahedi et al. (2012)
3.	*TaWRKY19*	*Triticum aestivum* L. cv. Xifeng 20	*Arabidopsis* ecotype Columbia plants (Col-0)	ATMT	Niu et al. (2012)
4.	*GmbZIP1*	*Glycine max* L.	*Nicotiana tabacum* W38	ATMT	Gao et al. (2011c)
5.	*SlCZFP1* (*Solanum lycopersicum* cold zinc finger protein 1)	*Solanum lycopersicum* var. D. Huang	*Arabidopsis thaliana* (L.) Heynh. (ecotype Wassilewskija, Ws-2), *Oryza sativa* L. cultivar Kita-ake	ATMT	Gao et al. (2011d)
6.	*ZmMKK4* (mitogen-activated protein kinase kinase)	*Z. mays* L. cv. Zhengdan 958	*Arabidopsis thaliana* (ecotype Columbia, Col-0	ATMT	Kong et al. (2011)
7.	*OsAPXa* (ascorbate peroxidase)	*OsAPXa* cDNA library	*L.* cv. Oborozuki	ATMT	Sato et al. (2011)
8.	*At-CBF1* (Arabidopsis C-repeat-binding factor 1)	*Agrobacterium tumefaciens*	*Solanum lycopersicum* var. Shalima	ATMT	Singh et al. (2011)
9.	*JcDREB*	*Jatropha curcas*	*Arabidopsis thaliana* CK	ATMT	Tang et al. (2011)
10.	*mtlD* (mannitol-1-phosphate dehydrogenase)	*Escherichia coli*	*Lycopersicon esculentum* M. cv. Pusa Uphar	ATMT	Khare et al. (2010)
11.	*TERF2/LeERF2* (ethylene responsive factor)	*Solanum lycopersicum*	*Solanum lycopersicum* cv. Lichun and *Nicotiana tabacum* cv. NC89	ATMT	Zhang and Huang (2010c)
12.	*ThCAP* (cold acclimation protein)	*Tamarix hispida*	*Populus davidiana, P. bolleana*	ATMT	Guo et al. (2009)
13.	*TPP1* (trehalose synthase)	cDNA library of Nona Bokra	*Oryza sativa* L. ssp. *indica* pv. Nona, *Oryza sativa* L. ssp. *japonica*	ATMT	Ge et al. (2008)

TABLE 10.11 *(Continued)*

Sl. No.	Transgene (Gene Name and Notation)	Source of Transgene	Crop and Cultivar	Method of Transformation	Reference
14.	*Osmyb4* (cold-induced transcription factor)	*Oryza sativa*	*Malus pumila* Mill. Cv. Greensleeves	ATMT	Pasquali et al. (2008)
15.	*WCOR15* (cold-induced gene)	*Triticum aestivum* L. Mironovskaya 808	*Nicotiana tabacum* cv. "Petit Havana"	ATMT	Shimamura et al. (2006)
16.	*CBF3/DREB1A* and *ABF3*	*Arabidopsis thaliana*	*Oryza sativa* cv. Nakdong	ATMT	Oh et al. (2005)
17.	*OsSAP1* (*Oryza sativa* subspecies indica stress-associated protein gene)	*Oryza sativa* subsp. indica var. Pusa Basmati-1	*Nicotiana tabacum* var. Xanthi	ATMT	Mukhopadhyay et al. (2004)
18.	*betA* (Choline dehydrogenase)	*Escherichia coli*	*Nicotiana tabacum* L. cv. samsun	ATMT	Holmström et al. (2000)
19.	*Nt 107* (Glutathione-S-transferase/ glutathione peroxidase)	*Nicotiana tabacum*	*Nicotiana tabacum* L. cv. Xanthi NN	Self-pollination	Roxas et al. (2000)

ATMT, *Agrobacterium tumefaciens*-mediated transformation.

phytoremediation, which will be greatly enhanced by using transgenic plants bearing bacterial genes involved in xenobiotic metabolism, leading to a wider application in the field (Abhilash et al., 2009).

The **2,3-dihydroxybiphenyl-1,2-dioxygenase (BphC.B)**, a key enzyme of aerobic catabolism of a variety of aromatic compounds, was cloned from a soil metagenomic library, then was expressed in alfalfa driven by CaMV 35S promoter using *Agrobacterium*-mediated transformation. The tolerance capability of transgenic line BB11 toward complex contaminants of PCBs/2,4-DCP significantly increased compared with nontransgenic plants (Wang et al., 2015). Strong dissipation of PCBs and high removal efficiency of 2,4-DCP were exhibited in a short time. It was confirmed that expressing BphC.B would be a feasible strategy to help achieving phytoremediation in mixed contaminated soils with PCBs and 2,4-DCP.

10.14.1 MODE OF ACTION OF 2,3-DIHYDROXYBIPHENYL-1,2-DIOXYGENASE

BphC found in a range of Gram-negative and Gram-positive bacteria that aerobically assimilate biphenyl could utilize nonheme ferrous iron to cleave the aromatic nucleus of catechols meta (adjacent) to the yellow substance. BphC is involved in aerobic catabolism of a variety of aromatic compounds including phenol, naphthalene, and polychlorinated biphenyls.

10.15 GENETIC ENGINEERING OF TOLERANCE TO HERBICIDES

The herbicides used in the earlier days have been very destructive for most plants and their use is undesirable for the environment. Among newer herbicides, glyphosate has been widely used for it can be degraded by soil microorganisms. However, with the development of herbicide tolerant crop plants, herbicides can now be applied over the top of crops during the growing season to control weed population more effectively (Ahmad et al., 2012). The glyphosate-tolerant maize, soybean, canola, and cotton are the most abundant lines among those crops (Tohidfar and Khosravi, 2015).

10.15.1 HOW DOES GLYPHOSATE WORK?

Glyphosate, the active component of Roundup®, is used across in the field as nonselective postemergence herbicide. Glyphosate works as an analog of enolpuruvate by binding to and inhibiting the enzyme **5-enolpyruvylshikimate-3-phosphate synthase (EPSPS)**, which is an active component in the shikimate pathway, leading to the synthesis of chorismate-derived metabolites such as the aromatic amino acids. Thus, inactivation of this enzyme by glyphosate means killing of the plant due to the absence of aromatic amino acids for complete renewal of proteins (Tohidfar and Khosravi, 2015).

10.15.2 STRATEGIES FOR GENERATING HERBICIDE TOLERANT CROPS

Quite a few transgenic plants have been developed for tolerance to various herbicides in crops such as soybean, corn, cotton, and canola more than a decade ago. By and large, there are two approaches that can be used to create herbicide tolerant crops: one-way is to modify the degree of sensitivity of the target enzymes so that the sensitivity of plant to the herbicide is reduced or eliminated. Examples of the first approach include glyphosate and acifluorfen tolerance. Transgenic plants tolerant to the herbicide acifluorfen have been produced through overexpression of the target enzyme involved in chlorophyll biosynthesis (Lermontova and Grimm, 2000). This herbicide inhibits chlorophyll biosynthesis. Herbicide resistant *Amaranthus palmeri* has been developed recently by expressing glyphosate-insensitive herbicide target site gene, ***EPSPS*** involved in the shikimate cycle wherein it catalyzes the reversible addition of the enolpyruvyl moiety of phosphoenolpyruvate to shikimate 3-phosphate (Gaines et al., 2010). A highly glyphosate insensitive EPSPS was created by DNA shuffling in the gene from *Vitisi vinifera* and transgenic introduction of such a gene in rice and Arabidopsis improved tolerance to glyphosate (Tian et al., 2015).

The other approach is to engineer the herbicide detoxification pathway into the plant. Resistance to glufosinate and bromoxynil is based on the second approach. In this approach, introducing a gene in the plant system metabolizes the herbicide concerned. For example, in the case of herbicide Ignite/Basta, the ***bar*** resistance gene from *Streptomyces hygroscopicus* was used to detoxify the herbicide. The expression of *bar* gene responsible for resistance to herbicides was demonstrated in sweet potato (Zang et al., 2009). Previously, various transgenic plants expressing the *bar* gene were developed in sugarbeet, popular plants, aspen, oilseed rape, tomato, potato, alfalfa, and tobacco. Imidazolinone resistance (IR) ***XA17*** gene was incorporated into some maize lines for resistance to imazaquin and nicosulfuron herbicides (Menkir et al., 2010). Transgenic tobacco expressing a tau class GST isoenzyme ***GmGSTU4*** from soybean is active as GSH-dependent peroxidase (GPOX) and shows catalytic activity for diphenyl ether herbicide fluorodifen/alachlor (Benekos et al., 2010). The gene encoding **glyphosate *N*-acetyltransferase** (*Gat*) from *Bacillus licheniformis* into the plant it deactivates glyphosate into a nontoxic *N*-acetylglyphosate (Siehl et al., 2007). The soybean and corn plants expressing *GAT* gene were tolerant to glyphosate (Castle et al., 2004).

10.15.3 MODE OF ACTION OF PHOSPHINOTHRICIN ACETYL TRANSFERASE

The protein phosphinothricin-*N*-acetyl transferase (PAT) is produced in GE plants by genes isolated from *Streptomyces viridochromogenes* (pat gene) or *S. hygroscopicus* (bar gene). PAT is used against selection agents, such as phosphinothricin, bialaphos, and glufosinate ammonium in GM crops. These agents interefere with the functioning of glutamine synthetase/glutamate synthase cycle and the conversion of glutamate and ammonia to glutamine is blocked. The pathway again turns functional only when PAT detoxifies the selection agent by acetylation.

10.15.4 CROP TOLERANCE TO BROADLEAF AND GRASS HERBICIDES

Substrate preferences of bacterial **aryloxyalkanoate dioxygenase enzymes (AADs)** that can effectively degrade 2,4-D were investigated and in addition to their activity on 2,4-D, some members of this class can act on other widely used herbicides. ***AAD-1*** cleaves the aryloxyphenoxy propionate family of grass-active herbicides, and ***AAD-12*** acts on pyridyloxyacetate auxin herbicides such as triclopyr and fluroxypyr. Maize plants transformed with an *AAD-1* gene showed robust crop resistance to aryloxyphenoxy propionate herbicides over four generations and were also not injured by 2,4-D applications at any growth stage. Arabidopsis plants expressing *AAD-12* were resistant to 2,4-D as well as triclopyr and fluroxypyr, and transgenic soybean plants expressing *AAD-12* maintained field resistance to 2,4-D over five generations. These results showed that single *AAD* transgenes can provide simultaneous resistance to a broad group of agronomically important classes of herbicides, including 2,4-D, with utility in both monocot and dicot crops (Wright et al., 2010).

Some of transgenic crop plants developed for improved tolerance to environmental pollutants and herbicides are listed in Table 10.12.

TABLE 10.12 Recent Developments of Transgenic Crop Plants Tolerant to Pollutants and Herbicides.

Sl. No.	Transgene (Gene Name and Notation)	Pollutants	Source of Transgene	Crop and Cultivar	Method of Transformation	Reference
1.	*CuZnSOD* (CuZn superoxide dismutase) and *APX* (ascorbate peroxidase	Sulfur dioxide	oxidative stress-inducible SWPA2 promoter (SSA plants)	*Ipomoea batatas*	Particle bombardment	Kim et al. (2015)
2.	*VvEPSPS* (5-enolpyruvyl shikimate-3-phosphate synthase)	Glyphosate	*Vitis vinifera*	*Oryza sativa* L. ssp. *japonica*, *Arabidopsis* sp.	ATMT, DNA shuffling	Tian et al. (2015)
3.	*BphC.B* (2,3-dihydroxybiphenyl-1,2-dioxygenase)	PCBs and 2,4-DCP	Soil metagenomic library	*Medicago sativa* L. cv. Gongnong No. 1	ATMT	Wang et al. (2014)
4.	*TaEXPB23* (−expansin gene)	Methyl viologen	*Triticum aestivum* L.	*Nicotiana tabacum* L. cv. NC89	ATMT	Han et al. (2015)
5.	*TaALMT1*	Aluminum	*Triticum aestivum*	*Triticum aestivum* Bob White 26 “SH9826” line (BW26)	Particle bombardment	Pereira et al. (2010)
6.	*GmGSTU4*	Diphenyl ether and chloroacetanilide	*Glycine max*	*Nicotiana tabacum* L. cultivar Basmas	ATMT	Kostantinos et al. (2010)
7.	*aad-1*, *aad-12* (aryl oxyalkanoate dioxygenase)	Aryl oxyphenoxypropionate, triclopyr, and fluroxypyr	*Ralstonia eutropha*	*Arabidopsis thaliana*, *Glycine max*, *Zea mays*	ATMT	Wright et al. (2010)

ATMT, *Agrobacterium tumefaciens*-mediated transformation.

10.16 GENETIC ENGINEERING OF PHOSPHORUS UTILIZATION EFFICIENCY IN PLANTS

Phosphorus is one of the three major nutrient requirements of plants. Its low availability, mobility, and high fixation in soils make it a constraint worldwide for crop productivity. However, molecular biology provides great opportunities to improve phosphorus use efficiency in plants. It is also to be noted that phosphorus mainly comes from nonrenewable resource and, therefore, "smart" crop plants have to be developed for better phosphorus use efficiency. Plants have multiple adaptation systems evolved for efficient utilization of phosphorus from soil (Tian et al., 2012). Therefore, it is important to understand, identify, and use the genes involved in various adaptation processes.

Among Pi transporters (PT), Pht1 mainly function in Pi acquisition from soils and translocation from roots to other parts of plants. Biomass and yield of transgenic rice plants were not coincidently improved. However, overexpression of ***NtPht1;1***, ***OsPht1;2***, or ***OsPht1;8*** facilitated Pi acquisition (Jia et al., 2011; Liu et al., 2010; Park et al., 2010). This was due to toxicity of excess phosphorus. Thus, overexpression of Pht1 in crops should be integrated with soil/farm management in order to improve crop phosphorus use efficiency. Similarly, overexpressing a transcription factor, **phosphate starvation response 2** (*OsPHR2*), a major component in phosphorus signaling pathways in rice, resulted in increased phosphorus concentration but inhibited plant growth, which might have been caused due to excessive amounts of P in leaves (Zhou et al., 2008). Similar results were also observed in modifying **SPX** and **miR399**. Suppressing *OsSPX1* in rice and overexpressing ath-miR399d from Arabidopsis in tomato led to excessive phosphorus accumulation in leaves and subsequently inhibited plant growth (Gao et al., 2010; Wang et al., 2009). Conversely, overexpression of a transcription factor, **PTF1** (Pi starvation induced transcription factor 1), enhanced phosphorus use efficiency in both, rice and maize (Li et al., 2011b; Yi et al., 2005). Therefore, improving phosphorus use efficiency through transgenic technology of introducing the critical genes in phosphorus signaling networks requires more insights into the physiological and molecular connections between components.

10.17 GENETIC ENGINEERING OF OILSEED CROPS FOR FISH OIL

Fatty acids (FAs) with 20 carbons or more in length containing three or more cis-double bonds, that is, very long-chain polyunsaturated fatty acids

(VLC-PUFAs) are essential components of human nutrition. These FAs are the major constituents of mammalian retinal, brain and testis membrane phospholipids and play important roles in cellular and tissue metabolism regulating membrane fluidity and thermal adaptation (Sayanova and Napier, 2011). VLC-PUFAs, especially eicosapentaenoic acid (EPA) and docosahexaenoic acid (DHA) play critical roles in human health and development. Mainly, the fishes, some fungi, marine bacteria, and microalgae are the sources of VLC-PUFAs. Dietary sources of VLC-PUFAs are predominantly met up from marine fish and seafood. However, the mankind has been putting enormous pressure on marine ecosystems due to increasing demand for fish and fish oils and United Nations' Food and Agriculture Organization estimated that more than 70% of world's fish stocks are either exploited or depleted (Sayanova and Napier, 2011). At the same time, commercial cultivation of marine microorganisms and aquaculture are not sustainable and cannot compensate for the shortage in fish supply. Therefore, there is an obvious requirement for an alternative and sustainable source for VLC-PUFAs. Plant oils are relatively inexpensive and are commonly considered to be healthier than animal fats, as they contain relatively high amounts of unsaturated FAs. Plant oils are rich in C18 FA, including the essential FA linoleic acid (LA) and α-linolenic acid (ALA), but are devoid of LC-PUFAs, such as arachidonic acid (ARA), EPA, and DHA, which typically only enter the human diet as oily fish (Ruiz-López et al., 2015). Marine fishes are rich in these beneficial FAs.

These days, it is possible to produce seed oils with a desirable FA composition using latest genetic engineering techniques, which is impossible to achieve by traditional breeding techniques. The health benefits of consumption of oily fish and the ω-3 long-chain polyunsaturated fatty acids (LC-PUFA) such as reducing the risk of cardiovascular disease and related metabolic conditions are now widely recognized. Metabolic engineering demonstrated the feasibility of making EPA and DHA in the seed oils of transgenic *Camelina sativa* plants (Usher et al., 2015). Generation of LC-PUFAs in transgenic plants was demonstrated almost two decades ago in transgenic tobacco and Arabidopsis. The transgenic plants mainly accumulated ω-6 γ-linolenic acid (GLA) and stearidonic acid (SDA) in the leaves. GLA and SDA as high as 70% were achieved using seed-specific promoters (Hong et al., 2002; Qiu et al., 2002; Sato et al., 2004). Similar level of GLA accumulation was achieved in transgenic safflower expressing D6-desaturase from *Saprolegnia diclina* (Nykiforuk et al., 2012). The ALA-specific Δ6-desaturase from *P. vialii* was cloned under a seed-specific promoter and introduced into Arabidopsis and linseed (Ruiz-López et al., 2009). It has been some years since the successful reconstitution of the ω-3 LC-PUFA biosynthetic pathway in plants using multiple desaturases and elongases was achieved. Genetic transformation of different oilseeds crops with multiple genes is not the technical barrier today. The feasibility of making EPA in a transgenic plant was successfully demonstrated by expressing algal components of the alternative pathway in the leaves of Arabidopsis (Qi et al., 2004). Intriguingly, it generated C20 ω-6 LC-PUFA ARA in addition to moderate amounts of EPA. Seed-specific accumulation of EPA in linseed was achieved where genes of the conventional δ-6-pathway was expressed (Abbadi et al,. 2004). Resultant transgenic seeds contained low levels of EPA, and very high levels of C18 δ6-desaturation products. The authors thus hypothesized that this unwanted build-up of a biosynthetic intermediate was as a consequence of poor acyl exchange between different metabolic pools, and the concept has been defined as "substrate dichotomy" (Napier, 2007). These findings formed the basis for further attempts to increase the levels of target FAs (EPA, DHA) and reduce the levels of undesired biosynthetic intermediates (such as the δ6-desaturation product GLA). Several studies have confirmed the ability to make significant levels of EPA, with minimum levels of GLA (Cheng et al., 2010; Wu et al., 2005). It is now technically possible to accumulate fish oil-like levels of ω-3 LC-PUFAs in the seed oils of transgenic plants similar to that found in fish oils, in which EPA and DHA accumulate up to 20% of total FAs (Napier et al., 2015). Recently, seed oils have been engineered to produce EPA and/or DHA at levels similar to fish oils (Ruiz-López et al., 2014). However, successful conversion of native plant FAs such as LA and ALA to LC-PUFAs such as EPA and DHA in seeds requires a coordinated expression of multiple genes. Recent advances in engineering of oilseed crops has led to the accumulation of ω-3 LC-PUFAs at fish oil levels, demonstrating the efficacy of acyl-CoA desaturases over previously used lipid-linked desaturases, resolving the substrate dichotomy problem (Ruiz-López et al., 2015).

10.18 GENETIC ENGINEERING FOR QUALITY NUTRITION AND HEALTH

Mineral nutrients are found in very poor quantities in staple food crops. Resultant effect, the poorest, especially those surviving on same kind of staple food, is also most vulnerable to mineral deficiency diseases. Malnutrition is more prevalent in the developing world because it is found that the nutritious food is often not reaching the poor and needy. This is also attributed by poverty, which often occurs due to ill health and an inability to work, the typical consequences of malnutrition. Therefore, poverty, malnutrition, and poor health form a triangle which the poor and needy find difficult to escape (Farre et al., 2011). The mankind relies on food not only to fill the stomach and get enough energy, but also for essential nutrients required to maintain a good state of health and active immune system. Thus, food security is one of the main pillars of health and well-being of the society as a whole. Adequate nutrition is required to ensure lower morbidity and mortality from both infectious and noninfectious diseases. It is particularly important in children and pregnant women where the lack of essential nutrients can lead to irreversible physical and mental damage during development (Hoddinott et al., 2008).

Various strategies have been proposed to deal with micronutrient deficiencies including the provision of mineral supplements, the fortification of processed food, the biofortification of crop plants at source with mineral-rich fertilizers and the implementation of breeding programs and genetic engineering approaches to generate mineral-rich varieties of staple crops (Gómez-Galera et al., 2010). Biofortification focuses on enhancing the qualities of essential mineral nutrients in the edible part of staple crops. Agronomic intervention, plant breeding, or genetic engineering can achieve incorporation of mineral nutrients in crops, whereas plant breeding and genetic engineering can command bioavailability of minerals as well (Gómez-Galera et al., 2010). Some of transgenic crop plants developed for improved nutritional quality, oil production, etc. are listed in Table 10.13.

10.18.1 GENETIC ENGINEERING FOR ENHANCEMENT OF MINERAL MICRONUTRIENTS

Plants take up inorganic nutrients from the environment and metabolically synthesize organic nutrients. We can focus on genetic engineering strategies such as increasing the solubility of these nutrients in the rhizosphere, mobilizing them in the plants, transporting them to storage organs, increasing the storage capacity of the plant, and maximizing bioavailability for enhancement of mineral micronutrients in food crops (Gómez-Galera et al., 2010). Unlike most other minerals, deficiency of nutrients such as iron, zinc, selenium, and calcium have serious implications. Thus, there is a need of designer crops with enhancement of necessary micronutrients (Naqvi et al., 2009).

TABLE 10.13 Recent Developments of Transgenic Crop Plants with Improved Nutritional Quality, Oil Production, etc.

Sl. No.	Transgene (Gene Name and Notation)	Function	Source of Transgene	Crop and Cultivar	Method of Transformation	Reference
1.	*dgat1-1*	Acetyl glyceride oil production	*Arabidopsis thaliana*	*Camelina sativa*	ATMT	Liu et al. (2015c)
2.	*gus* and *nptII*	High seed production	*E. coli*	*Camellia sinensis* L.O. Kuntze	Biolistic mediated	Sandal et al. (2015)
3.	*FAD2*	Oleic acid production	*Linum usitatissimum* L. cDNA library	*Linum usitatissimum* L. cv. Glenelg	ATMT	Chen et al. (2015)
4.	*OtΔ6* (Δ6-desaturase gene), PSE1 *(TcΔ5)* Δ5-desaturase gene, *PsΔ12* (*Δ*12-desaturase gene), Pi-x3 (x3-desaturase)	Omega-3 LC-PUFA production	*Ostrococcus tauri*, *Physcomitrella patens*, *Thraustochytrium* sp. (*Phytophthora sojae*, *Phytophthora infestans*)	*Camelina sativa*	ATMT	Ruiz-López et al. (2014)
5.	*FAD2* (*Δ*12-desaturase)	Oleic acid production	*J. curcas* seed cDNA library	*Jatropha curcas* (Jc-MD)	ATMT	Qu et al. (2012)
6.	*Cg1* (Corngrass1)	Starch content increase	*Zea mays*	*Panicum virgatum*	ATMT	Chuck et al. (2012)
7.	*Δ6-desaturase*, *Δ12-/Δ6-desaturases*	c-Linolenic acid production	*Saprolegnia diclina*, *Mortierella alpina*	*Carthamus tinctorius*	ATMT	Nykiforuk et al. (2012)
8.	*OASS* (*O*-acetylserine sulfhydrylase)	Enhanced levels of cysteine and Bowman–Birk protease inhibitor in seeds	*cDNA library*	*Glycine max* L. cv. Maverick	ATMT	Kim et al. (2012)

TABLE 10.13 *(Continued)*

Sl. No.	Transgene (Gene Name and Notation)	Function	Source of Transgene	Crop and Cultivar	Method of Transformation	Reference
9.	*acyl-CoA D6-desaturase*	Oil production	*Micromonas pusilla*	*S. cerevisiae* strain INVSc1, *Nicotiana benthamiana*, *Arabidopsis thaliana* (Col 0)	ATMT	Petrie et al. (2010a)
10.	*ZmLEC1* (LEAFY COTYLEDON1), *ZmWRI1* (WRINKLED1)	Oil production	*Zea mays*	*Zea mays*	ATMT	Shen et al. (2010)
11.	*LEC2* (LEAFY COTYLEDON2)	DHA biosynthesis	*Arabidopsis thaliana*	*Nicotiana benthamiana*	ATMT	Petrie et al. (2010b)
12.	*OsSPL14*, *SPL16*	Grain size, shape, and quality	*Oryza sativa* Nipponbare	*Oryza sativa* indica Ri22, SNJ	ATMT	Jiao et al. (2010)
13.	*idi*, *crtE*, *crtB*, *crtI*, *crtY*, *crtZ*, and *crtW*	Carotenoid synthesis	*Pantoea ananatis*, *Brevundimonas* sp. strain SD212, and *Paracoccus* sp. strain N81106	*Brassica napus* L. cultivar Westar	ATMT	Fujisawa et al. (2009)
14.	*CrtI* (carotene desaturase), *psy* (phytoene synthase)	Vitamin A synthesis	*Pantoea ananatis*	*Oryza sativa*, GR	Biolistic methods and ATMT	Tang et al. (2009)
15.	*Zmpsy1* (*Zea mays* phytoene synthase 1), *PacrtI* (*Pantoea ananatis* phytoene desaturase), *Gllycb* (*Gentiana lutea* lycopene β-cyclase), *Glbch* (*G. lutea*	Carotenoid synthesis	*Pantoea ananatis*, *Gentiana lutea*, *Paracoccus* sp.	*Zea mays* L. (cv. M37W	Combinatorial nuclear transformation	Zhu et al. (2008)

TABLE 10.13 *(Continued)*

Sl. No.	Transgene (Gene Name and Notation)	Function	Source of Transgene	Crop and Cultivar	Method of Transformation	Reference
	β-carotene hydroxylase), *Paracrt W* (*Paracoccus* β-carotene ketolase)					
16.	*CordapA*, *LKR/SDH* (lysine-ketoglutarate reductase/saccharophine dehydrogenase)	Lysine biosynthesis	Corynebacterium	*Zea mays*	ATMT	Frizzi et al. (2008)
17.	*AVP1* (*Arabidopsis* vacuolar pyrophosphatase)	Enhanced growth under phosphorus limitation	*Arabidopsis thaliana Col-0 ecotype*	*Oryza sativa* var. japonica "Taipei 309," *Lycopersicon esculentum* Mill. cultivar Money Maker	ATMT	Yang et al. (2007)
18.	*Δ6 elongases*	Very long-chain polyunsaturated fatty acids (VLCPUFAs)	*Thraustochytrium* sp.	*Brassica juncea* BJ5	ATMT	Guohai et al. (2005)
19.	*tyrA*, *pds1*, *hpt1*, and *ggh*	Vitamin E synthesis	*Synechocystis* sp. PCC 6803	*Arabidopsis thaliana* var. Columbia *Brassica napus* L. cultivar ebony and *Glycine max* cultivar A3244	ATMT	Karunanandaa et al. (2005)
20.	*Δ6-desaturase gene*	Oil production	*Jatropha curcas*	*Arabidopsis thaliana* D1	ATMT	Reddy and Thomas (1996)

ATMT, *Agrobacterium tumefaciens*-mediated transformation; GR, Golden Rice.

10.18.1.1 IRON

Background: A major challenge with iron is that only the ferrous form (Fe^{2+}) is soluble and available to plants for uptake, whereas the ferric form (Fe^{3+}) is sequestered into insoluble complexes with soil particles (Gómez-Galera et al., 2012). Plants have evolved two counter strategies, first, by secreting reductases into the soil converting ferric iron into the ferrous form, and second, by releasing chelating agents known as phytosiderophores (PS) that can be reabsorbed by the roots as PS-Fe^{3+}.

Strategies for improvement of iron levels in plants include increasing the export of both reductases and PS, overexpression of iron transporter proteins, overexpression of ferritin, which stores large amounts of iron in a bioavailable form and the expression of phytase, which breaks down phytate and makes the stored iron easier to be absorbed in the human digestive system. For example, overexpressing the enzymes nicotianamine synthase (NAS) and/or nicotianamine aminotransferase (NAAT) in transgenic rice significantly increased the iron content (Johnson et al., 2011; Zheng et al., 2010).

10.18.1.2 ZINC

Background: Zinc deficiency affects more than 2 billion people worldwide manifesting as a spectrum of symptoms including hair loss, skin lesions, fluid imbalance (inducing diarrhea), and eventually wasting of body tissues (Hambidge and Krebs, 2007).

Strategies to increase the zinc content of plants have concentrated on transport and accumulation (Palmgren et al., 2008). The expression of NAS/NAAT and transporters such as *Os*ysl15 and *Os*irt1 in rice can increase the levels of both zinc and iron since many PS and transporters can interact with them (Lee et al., 2012).

10.18.1.3 SELENIUM

Background: Selenium is a component of enzymes and other proteins containing the amino acids selenocysteine and selenomethionine, required for the interconversion of thyroid hormones; therefore selenium and iodine deficiency can have similar symptoms (Khalili et al., 2008).

Genetic engineering strategies have focused on storage and accumulation of selenium to increase its levels. Expression of Arabidopsis ATP sulfurylase in mustard increased the selenium content in shoots and roots (Pilion-Smits et al., 1999).

10.18.1.4 CALCIUM

Background: Soluble calcium is an electrolyte and signaling molecule, but most of the calcium in the human body is present in its mineralized form as a component of bones and teeth. The replenishment of serum calcium by bone resorption is slow, so dietary calcium deficiency in the short term can lead to electrolyte imbalance and over the long term can cause osteoporosis.

Genetic engineering strategies to increase the calcium content of plants include the expression of calcium transporters such as *At*CAX1, which increased the calcium content of carrots and potatoes by up to threefold (Connolly, 2008; Park et al., 2005c).

10.18.2 NUTRIENT ENHANCERS AND ANTINUTRIENTS

Background: Mineral bioavailability can be increased by promoting the accumulation of enhancers or eliminating antinutrients that regulate the absorption of plant minerals by the human digestive system (Gibson, 2007). Some key nutrients doubly act as enhancers, like ascorbate and β-carotene, promoting iron uptake by chelating and/or reducing Fe^{3+} and prevent interactions with phytate and polyphenols (García-Casal, 2000). Phytic acid is a key antinutrient abundant in cereals, legumes, and oil seeds where it binds all the principal mineral nutrients and sequesters them into stable complexes that cannot be absorbed (López et al., 2002).

The amount of phytic acid in seeds can be reduced by silencing genes involved in its biosynthesis, such as myo-inositol-1-phosphate synthase (Nunes et al., 2006) or 1D-myo-inositol 3-phosphate synthase (Kuwano et al., 2009). Expression of a thermostable recombinant fungal phytase increased iron bioavailability in wheat (Brinch-Pedersen et al., 2006) and maize (Chen et al., 2008).

10.18.3 GENETIC ENGINEERING FOR ENHANCEMENT OF ORGANIC NUTRIENTS

Human can synthesize almost all the organic compounds needed for normal physiological activity, except the essential nutrients such as some of the amino acids, FAs, and vitamins.

10.18.3.1 ESSENTIAL AMINO ACIDS

Nine amino acids are constitutive essential nutrients because they cannot be synthesized de novo by human, and others are essential under certain specific cases, like child development or metabolic disorders. The most relevant examples are lysine, threonine, tryptophan, methionine, and cysteine. Staple cereals are poor sources of lysine and threonine, and staple legumes are poor sources of tryptophan, methionine, and/or cysteine (Zhu et al., 2007a).

Two strategies to tackle amino acid deficiency are engineering plants to produce proteins containing essential amino acids; and engineering amino acid metabolism to increase the availability of essential amino acids in the product. Lysine was the first target, its content increased up to 4.2% in transgenic rice and wheat (Sindhu et al., 1997; Stöger et al., 2001). In a significant development, 12 and 8 residues of lysine were added to endogenous cereal storage proteins barley hordothionine to produce HT12 and high lysine protein to produce HL8 (Jung and Carl, 2000). Further achievements in improving lysine yield include increment by 55% in maize seeds by the expression of the lysine-rich storage protein (sb401) (Yu et al., 2004), 47% in maize by the expression of lysine-rich animal protein such as porcine α-lactalbumin (Bicar et al., 2008), and 26% in maize seeds by expressing a heterotypical Arabidopsis lysyl tRNA synthetase that inserts lysine residues in place of other amino acids during the synthesis of seed storage proteins (Wu et al., 2007). The lysine content of maize has also been increased by using RNAi silencing one of the zein storage protein genes allowing the protein complement to be filled with lysine-rich storage proteins (Segal et al., 2003).

Amaranthus hypochondriacus seed storage protein is rich in all the essential amino acids and has a composition almost ideal for human consumption. Transgenic maize seeds expressing the AH protein contained up to 32% more protein than wild-type seeds containing higher levels of lysine, tryptophan, and isoleucine (Rascón-Cruz et al., 2004). Similarly, transgenic potato tubers expressing AH contained 45% more protein than normal (Chakraborty et al., 2000), while transgenic wheat seeds contained nearly 2.5% AH as a proportion of total seed protein, increasing the levels of lysine to 6.4% and tyrosine to 3.8% (Tamás et al., 2009).

Expression of feedback-insensitive dihydrodipicolinate synthase in maize increased lysine levels from to 30%, with concomitant increase in threonine (Frizzi et al., 2008). The key rate-limiting enzyme in tryptophan synthesis—anthranilate synthase—catalyzes the conversion of chorismate to anthranilate. Thus, expressing a feedback-insensitive version, the tryptophan level was increased by 400-fold in rice (Wakasa et al., 2006), 30-fold in potato tubers (Yamada et al., 2005b), and 20-fold in soybean seeds (Ishimoto et al., 2010).

10.18.3.2 ESSENTIAL FATTY ACIDS

The health-promoting ω-3 and ω-6 polyunsaturated fatty acids (PUFAs) need to be obtained from diet (Djoussé et al., 2011). Once acquired, simple ω-3 PUFAs such as ALA can be converted into more complex VLC-PUFAs like ARA, which can be converted back to the simpler species.

The VLC-PUFAs have been biosynthesized by expressing microbial desaturases and elongases in linseed, soybean, and mustard (Abbadi et al., 2004; Kinney et al., 2004; Wu et al., 2005).

10.18.3.2.1 Vitamin A

The reduced form of vitamin A (retinal) is required for the production of rhodopsin, essential for eyesight. Vitamin A deficiency affects more than 4 million children each year, up to 500,000 of who become partially or totally blind (Harrison, 2005).

The overexpression of DXP synthase in tomato produced a carotenoid precursor that increased the pathway flux enhancing the total carotenoid content (Enfissi et al., 2005). Cassava roots expressing the bacterial ***CrtB*** gene accumulated 34 times normal carotenoid level (Welsch et al., 2010). The replacement of the daffodil gene with its maize ortholog in Golden Rice 2 produced significant amounts of β-carotene (Paine et al., 2005). The same genes when expressed in maize yielded kernels with much higher amounts of β-carotene (Naqvi et al., 2010; Zhu et al., 2008). Further, expression of three

Erwinia genes encoding phytoene synthase (CrtB), phytoene desaturase (CrtI), and lycopene beta-cyclase (CrtY) in golden potato, causing diversion of carotenoid synthesis from the α- to the β-branch (Diretto et al., 2007), and expression of the cauliflower ***Or*** gene in tubers increasing the storage capacity for carotenoids have also been observed (López et al., 2008).

10.18.3.2.2 Vitamin C

Ascorbate (vitamin C) is an antioxidant and also cofactor of several enzymes, including those required for the synthesis of collagen, carnitine, cholesterol, and certain amino acid hormones. Vitamin C deficiency causes the ulceration disease scurvy, resulting in the breakdown of connective tissues (Bartholomew, 2002).

Overexpression of L-gulonolactone oxidase (***GLOase***) in lettuce caused a sevenfold improvement in ascorbate fresh weight (Jain and Nessler, 2000). Similarly, a twofold increase by expressing the same gene in potato tubers and a six times increase by expressing the rice dhar gene from the ascorbate recycling pathway in multivitamin maize (Naqvi et al., 2010) were achieved. Co-expression of stylo 9-cis-epoxycarotenoid dioxygenase and yeast D-arabinono-1,4-lactone oxidase improved not only vitamin C level but also the tolerance to drought and chilling in transgenic tobacco and stylo plants (Bao et al., 2016).

10.18.3.2.3 Vitamin B9

Folate (vitamin B9) is the source of tetrahydrofolate essential for DNA synthesis and many other core metabolic reactions. Folate deficiency causes macrocytic anemia and elevated levels of homocysteine, but in pregnant women, it can lead to the neural tube defect—spinal bifida in the fetus (Scholl and Johnson, 2000).

Two transgenic tomato lines, one expressing GCH1 enhancing the cytosolic (pterin) branch and the other ADCS1 enhancing the PABA branch, were crossed (de la Garza et al., 2007). The resultant single line released their individual bottlenecks of only double the enhancement from normal folate level to achieve a 25-fold increase in folate levels. This strategy in rice endosperm resulted in a 100-fold increase in folate levels, indicating its powerful potential in developing-country settings where rice is the staple diet (Storozhenko et al., 2007).

10.18.3.2.4 Vitamin E

Vitamin E comprises eight related molecules known as tocochromanols—powerful antioxidants—protecting FAs, low-density lipoproteins and other components of cell membranes from oxidative stress.

The α/γ tocopherol ratio in transgenic lettuce plants was increased by expressing the Arabidopsis γ-tocopherol methyltransferase (γ-TMT), achieving near complete conversion to α-tocopherol in the best-performing ones (Cho et al., 2005). Similarly, a 10.4-fold increase in α-tocopherol levels and a 14.9-fold increase in β-tocopherol levels in soybean seeds expressing Perilla frutescens γ-TMT was achieved (Tavya et al., 2007). The constitutive expression of two Arabidopsis cDNA clones encoding ρ-hydroxyphenylpyruvate dioxygenase (HPPD) and 2-methyl-6-phytylplastoquinol methyltransferase (MPBQ MT) increased the tocopherol content by threefold in transgenic maize (Naqvi et al., 2011).

10.19 GENETIC ENGINEERING FOR MOLECULAR FARMING

The production of recombinant proteins (including pharmaceuticals and industrial proteins) and other secondary metabolites in plants is known as plant molecular farming. It has long been considered as a promising strategy not only for agriculture and industry but also to produce valuable recombinant proteins for human and veterinary medicine. Some of the products are now commercially available. The process involves the growing, harvesting, transport, storage, and downstream processing of extraction and purification of the protein (De Wilde et al., 2002). It has been proven over the years that plants have the ability to produce even more complex functional mammalian proteins with therapeutic activity, such as human serum proteins and growth regulators, antibodies, vaccines, hormones, cytokines, enzymes, and antibodies (Liénard et al., 2007). Various plant expression platforms such as plant cell suspensions, plant tissues, whole plants; aquatic plants, etc. can be used for production of recombinant proteins (Fig. 10.2). A number plant types and systems have been used for expression of vaccine antigens (Rybicki, 2009). Initially, the systems that were edible by humans and animals, or had "Generally Regarded As Safe" (GRAS) status, were considered with the assumption that the vaccines would be eaten without further processing (Rybicki, 2010). The systems mainly include *Nicotiana* spp., *A. thaliana*, alfalfa, spinach, potatoes, duck-weed, strawberries, carrots, tomatoes, aloe, and single-celled algae. Proteins have also been expressed in seeds of maize,

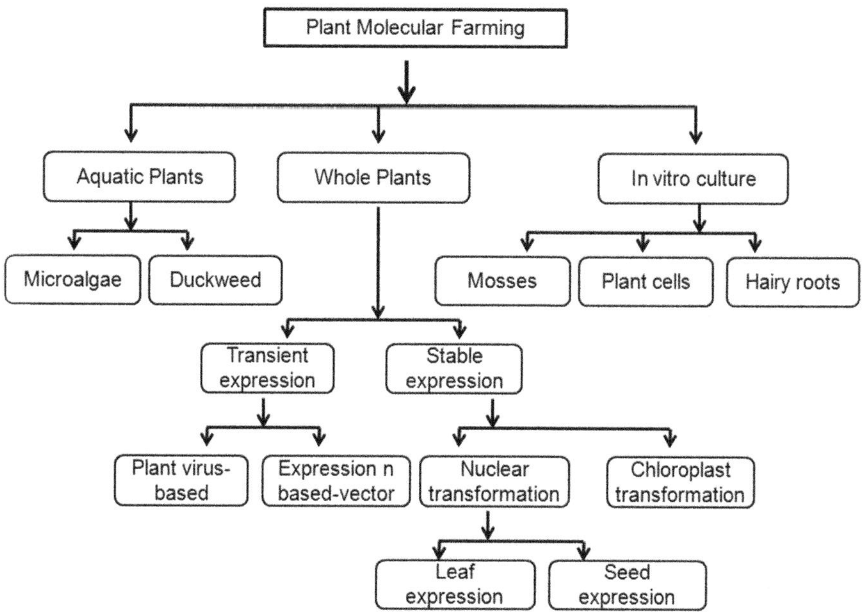

FIGURE 10.2 Plant expression platforms for production of various recombinant proteins. (Adapted from Xu et al., 2012).

rice, beans, and tobacco; in potatoes, tomatoes, and strawberries; in suspension cell cultures of tobacco and maize; in hairy root cultures; and in transformed chloroplasts of a variety of plant species. Human growth hormone, the first recombinant plant-derived pharmaceutical protein and the first recombinant antibody were produced in transgenic plants in 1986 and 1989, respectively (Barta et al., 1986; Hiattet al., 1989). However, avidin, the first recombinant protein was expressed in transgenic maize for commercial purpose in 1997 only (Hood et al., 1997). These trials really demonstrated that plants could be turned into biofactories for the large-scale production of recombinant proteins. This has been possible due to their ability to perform posttranslational modifications that make these recombinant proteins fold properly and maintain their structural and functional integrity. Transgenic plants are gaining more attention as the new generation bioreactors due to increasing demand for biopharmaceuticals, coupled with the high costs and inefficiency of the existing production systems (Knäblein, 2005), which include yeast, microbes, animals cells (Jones et al., 2003), and transgenic animals (Harvey et al., 2002). Thus, transgenic plants are now being seen as suitable alternatives as warehouse of molecular farming. Conventional hosts have some peculiar limitations. On the contrary, use of plant systems as production units for various vaccines came out with number of potential advantages over the existing systems such as ability for large-scale production, reduction in cost, feasibility of oral delivery, and being useful units for production of glycosylated vaccines (Obeme et al., 2011).

10.19.1 ANTIGENIC VACCINES DEVELOPED IN PLANTS

Largely, vaccine is an antigenic preparation that activates immune system against a given disease. Quite a few such vaccines have been produced in the plant system. These include the hepatitis B surface antigen that has been expressed in transgenic potatoes (Richter et al., 2000), tomato (He et al., 2008b), banana (Kumar et al., 2005), and in tobacco cell suspension culture (Sojikul et al., 2003). Heat labile enterotoxin B subunit of *E. coli* has been expressed in potato (Lauterslager et al., 2001), maize (Chikwamba et al., 2002), tobacco (Rosales-Mendoza et al., 2009), and soybean (Moravec et al., 2007). Similarly, vaccine against gastroenteritis has been raised in maize against corona virus (Tuboly et al., 2000). Several crops, such as tobacco, tomato, and rice, have been used to express the cholera toxin B subunit of *Vibrio cholerae* (Daniell et al., 2001a; Mishra et al., 2006; Nochi et al., 2007). Transgenic tomato expressing RSV (respiratory syncytial virus) fusion (F) protein has been developed to be used as edible vaccine against RSV (Sandhu et al., 2000). Plants such as different leafy crops, cereals, legumes, oilseeds, fruits, vegetables, cell cultures, algae, etc. have been used for the production of biopharmaceutical proteins (Fischer et al., 2004). Some vegetables such as potato, tomato, and carrot have been reported to express vaccines (Walmsley and Arntzen, 2000). Potato has been used as a model plant for the production of oral vaccines (Polkinghorne et al., 2005). Tomato is the new system used as such an expression system. Proplastids of cultured carrot cells have been shown to express recombinant proteins (Daniell et al., 2005). Lettuce, celery cabbage, and cauliflower are among other plants that are being used as production system for the vaccines (Koprowski, 2005). Several plant-made vaccines for veterinary purposes, including avian influenza, Newcastle disease, foot-and-mouth disease, and enterotoxigenic *E. coli*, have been expressed in the plant (Lentz et al., 2010; Ling et al., 2010). Pigeon pea and peanut have been used to express the hemagglutinin protein of rinder pest (Satyavathi et al., 2003). The vaccine has already been commercialized (World Health Organization, 2007) against human papilloma virus (HPV), which is the causal organism of cervical cancer in women.

HPV virus and L1 proteins were generated in transgenic potato and tobacco plants (Santi et al., 2006). Others such as the L1 protein of human papillomavirus types 11 and 16 (Giorgi et al., 2010; Maclean et al., 2007), the Norwalk virus capsid protein (Mason et al., 1996), and the H5N1 pandemic vaccine candidate (D'Aoust et al., 2010) have been expressed in one or two of the plants like tobacco, potato, and carrots. Among fruit crops, expression of foreign proteins (vaccines) in banana with the help of promoter has been demonstrated (Trivedi and Nath, 2004). Vaccine production in papaya has also been demonstrated through expression of novel synthetic vaccine SPvac (Carter and Langridge, 2002; Sciutto et al., 2002). It is known that human immunodeficiency virus type 1 (HIV-1) is a dreadful disease worldwide. Effective vaccination will be much useful to control this virus. The expression of HIV-1 antigens expression in plants has been reported by a number of scientists (Bogers et al., 2004) and the production of HIV-1 subtype G Gag-derived proteins in *Nicotiana* spp. has also been demonstrated (Meyers et al., 2008). Thus, it is possible that vaccine production is done using all these systems on a large scale for systemic and oral immunization. Antigenic F1-V fusion protein from *Yersinia pestis*, the causal organism of bubonic and pneumonic plague, was expressed in lettuce for plant-based vaccine production (Rosales-Mendoza et al., 2010).

10.19.2 ANTIBODIES PRODUCED IN PLANTS

Recombinant antibodies were expressed for the first time in plants (Hiatt et al., 1989). Then on, different moieties ranging from single chain Fv fragments (ScFvs, which contain the variable regions of the heavy and light chains joined by a flexible peptide linker) to Fab fragments (assembled light chains and shortened heavy chains), small immune proteins (SIP), IgGs, and chimeric secretory IgA and single-domain antibodies have been expressed in plants (Ismaili et al., 2007; Xu et al., 2007). There are two main approaches that are being employed to produce biologically active whole antibodies in plants. One approach is crosspollination of individually transformed plants expressing light or heavy chains, resulting in high yield which reaches 1–5% of total plant protein (Hiatt et al., 1989; Ma et al., 1994). The other one involves cotransformation of the heavy and light chain genes on a single- (Düring et al., 1990), two- (Villani et al., 2009), or more expression cassettes (Nicholson et al., 2005). One of such plant derived antibodies; a secretory antibody against a surface antigen of *Streptococcus mutans* was actually found to be as effective as the original mouse IgG, in protecting against

S. mutans colonization on teeth (Ma et al., 1998). Recently, a HIV-specific monoclonal antibody produced in maize seeds was found to be as active as its Chinese hamster ovary-derived counterpart (Ramessar et al., 2008a, 2008b). Similar to antigenic vaccines, there are several different plant produced antibodies that are being tested in the clinical trials.

Plants do not produce antibodies naturally. However, plants can correctly assemble functional antibody molecules encoded by mammalian antibody genes. Pathogen toxins cause many plant diseases. One such disease is the soybean sudden death syndrome (SDS) caused by the fungal pathogen *Fusarium virguliforme*. It has so far not been possible to isolate the pathogen from diseased foliar tissues. One or more toxins produced by the pathogen have been considered to cause this foliar SDS. One of these possible toxins, FvTox1, was recently identified. Expression of anti-FvTox1-1 in stable transgenic soybean plants resulted in enhanced foliar SDS resistance compared to that in nontransgenic control plants (Brar and Bhattacharya, 2012).

10.19.3 NEUTRACEUTICAL AND THERAPEUTIC PROTEINS PRODUCED IN PLANTS

Antimicrobial nutraceuticals such as human lactoferrin and lysozymes have been successfully produced in several crops (Huang et al., 2008; Stefanova et al., 2008) and are also commercially available. There are other neutraceuticals that are under clinical trials.

A human growth hormone was the first therapeutic human protein to be expressed in plants (Barta et al., 1986). Human serumal bumin, which is normally isolated from blood, was produced in transgenic tobacco and potato (Sijmons et al., 1990). Since then, several human proteins have been expressed in the plants, which include epidermal growth factor (Bai et al., 2007; Wirth et al., 2004), α-, β-, and γ-interferons, which are used in treating hepatitis B and C (Arlen et al., 2007), erythropoietin, which promote red blood cell production (Musa et al., 2009; Weise et al., 2007), interleukin used in treating Crohn's disease (Elías-López et al., 2008; Fujiwara et al., 2010), insulin used for treating diabetes (Nykiforuk et al., 2006), human glucocerebrosidase used for the treatment of Gaucher's disease in GE carrot cells (Shaaltiel et al., 2007) and some other plants. Some of these therapeutics are at different stages of clinical trials or at the verge of commercialization.

10.19.4 NONPHARMACEUTICAL PROTEIN DERIVED FROM PLANTS

The nonpharmaceutical plant-derived proteins or industrial proteins such as avidin, trypsin, aprotinin, β-glucuronidase (GUS), peroxidase, laccase, cellulase are available in the market. Molecular farming of cell-wall degrading enzymes such as cellulases, hemicellulases, xylanases, and ligninases are of great importance for the biofuel industry required for production of cellulosic ethanol (Chatterjee et al., 2010; Mei et al., 2009; Sticklen, 2008). Other nonhydrolytic proteins such as cell wall disintegrating carbohydrate-binding modules of cell wall degrading enzymes, and the cell wall loosening proteins like the expansins that are useful in enhancing the efficiency of cell wall degradation by disrupting the different polysaccharide networks and thereby allowing increase accessibility of the hydrolytic enzymes to the substrate are potential candidates for molecular farming (Obeme et al., 2011).

Other potential nonpharmaceutical plant-derived technical proteins that are being explored and optimized for production include polyhydroxyalkanoate (PHA) copolymers, and poly(3-hydroxybutyrate) (PHB), which are biodegradable plastic-like compounds (Conrad, 2005; Matsumoto et al., 2009). However, it should be noted that only a few plant-derived pharmaceuticals have been approved for molecular farming so far (Obeme et al., 2011).

10.20 GENETIC IMPROVEMENT OF BIOFUEL CROP PLANTS

Major obstacles in biofuel production include lack of biofuel crop domestication; low oil yields of relevant crop plants as well as recalcitrance of lignocellulose to chemical and enzymatic breakdown. Research and development efforts for biofuel production are targeted at obtaining renewable liquid fuels from plant biomass. Researchers have gathered quite some knowledge on the genetic and genomic resources available for improvement biofuel crops. Biofuel production from various crop plants has already been demonstrated. This knowledge will be used to produce the next generation of biofuel crops by increasing lipid content with respect to some specific FAs and by optimizing the hydrolysis of plant cell walls to release fermentable sugars.

Commercially, bioethanol is derived from corn and biodiesel is obtained from plants with a high content in FAs such as soybean, canola, and sunflower. However, an alternative to these crops is required because corn and soybean are some of the major food crops and the yields of starch and plant oil in these crops are too low to cover the huge demand of transportation fuels. This has prompted the development of alternative biofuel production based on lignocellulosic biomass (Schubert, 2006; Sticklen, 2008; Tilman et al., 2009). Lignocellulose, composed of the polysaccharides cellulose and hemicellulose, and lignin, a phenolic polymer, are some of the most abundant biomaterials on earth (Pauly and Keegstra, 2008). Biofuel crops are to be grown in a strategic manner. Biofuel crops should not take away arable land area, at least, for major crops, and at the same time, it is advisable to avoid application of any fertilizer or pesticide. Methods for efficient genetic transformation of switchgrass, Jatropha, poplar and *Brachypodium* using *Agrobacterium* have been developed. Some of the other achievements in this field also includes production of PHB in transgenic switchgrass (Somleva et al., 2008) and development of tissue culture techniques for the propagation of *Miscanthus* and Jatropha explants (Sujatha et al., 2008).

10.20.1 ENGINEERING OF PLANT OIL METABOLISM

Increasing seed oil production is a major goal for global agriculture to meet the high demand for oil consumption by humans and for biodiesel production. Overexpression of *ZmLEC1* (maize LEAFY COTYLEDON1) increased seed oil by as much as 48%, but it resulted in reduction in seed germination and leaf growth in maize, whereas overexpression of *ZmWRI1* (maize WRINKLED1) resulted in an oil increase similar to overexpression of *ZmLEC1* without affecting germination, seedling growth, or grain yield (Shen et al., 2010). Triacylglycerols (TAGs) from plant seed storage oils are excellent sources for the generation of biodiesel (Durett et al., 2008; Dyer et al., 2008). Trans-esterification of plant TAGs with methanol is done in the presence of acid or alkali to produce fatty-acid methyl esters (FAMEs). Biofuel crops such as soybean and Jatropha have either low or unpredictable oil yields. Redirecting the biosynthesis of specific types of FAs are needed to achieve optimal biodiesel production increasing oil content in plants. The strategies for optimal FAME production includes lowering of the levels of both saturated and polyunsaturated FAs, while increasing the amount of monounsaturated FAs, such as palmitoleate (C16:1) or oleate (C18:1) (Durett et al., 2008). Downreglation of FATB, an acyl-ACP thioesterase, in soybean caused accumulation of oleic acid up to 85% from 18% in the wild type and reduction in the levels of the saturated FA palmitate (Buhr et al., 2002). Expression levels of enzymes involved in synthesis of TAG have been engineered to increase oil content in seeds. Overexpression of a fungal diacylglycerol acyltransferase (DAGT2) enzyme led to increase in oil content in soybean seeds and Arabidopsis (Lardizabal et al., 2008). Activation of the FA biosynthetic pathway has been an alternative means to increase seed oil content in plants. For example, total FA and lipid seed content increased in transgenic Arabidopsis overexpressing of two soybean transcription factors **Dof4** and **Dof11** (Wang et al., 2007). **Dof4** and **Dof11** activated lipid biosynthesis in Arabidopsis through activity of acetyl CoA carboxylase and long-acyl-CoA synthase, respectively. Similarly, the lipid content in transgenic canola seeds was increased by 40% by overexpression of the yeast glycerol 3-phosphate dehydrogenase (ghpd1) (Vigeolas et al., 2007). It is noteworthy that engineering of oil accumulation in vegetative tissues, such as leaves, is an attractive approach to increase overall yield of oils for biodiesel production.

10.20.2 ENGINEERING PLANT LIGNOCELLULOSE

One of the main areas of research and development is the study of synthesis of plant cell wall components and their degradation. The plant cell walls are composed of cellulose, hemicellulose, and lignin. The highly complex nature of lignocellulose requires costly and harsh pretreatments to gain access to monosaccharides. The ultimate goal is develop improved lignocellulosic characteristics for easier and more efficient breakdown. The lignin component has repeatedly been pointed out as the major factor contributing to cell wall recalcitrance for access to monosaccharides (Akin, 2007; Weng et al., 2008). Significant improvement in fermentable sugar release from lignocellulose was achieved by downregulating certain monolignol biosynthetic enzymes in transgenic alfalfa. Enhanced enzymatic cell wall hydrolysis was correlated with lower lignin amounts in alfalfa lines silenced for cinnamate 4-hydroxylase, hydroxycinnamoyl CoA:shikimate hydroxycinnamoyl transferase, and coumaroyl shikimate 3-hydroxylase (Chen and Dixon, 2007). The enzymes such as cinnamoyl CoA reductase (CCR) and cinnamyl alcohol dehydrogenase (CAD) function at later stages in monolignol biosynthesis. Downregulaiton of CCR and CAD in alfalfa lines led to significant improvements in enzymatic saccharification efficiency. This plasticity of lignin can be exploited for engineering of lignin compositions for improved lignin extraction from a given plant biomass. For example, maize cell walls with coniferyl ferulate as an additional component had improved enzymatic hydrolysis and sugar release (Grabber et al., 2008).

Still, the improvement of plant biomass characteristics for biofuel production is at infancy. Biofuel crops have been identified and are at various levels of domestication and cultivar selection, whereas genetic and genomic resources for these species, including draft genome sequences and transformation protocols, are constantly being developed. Major breakthroughs on the understanding of lipid metabolism and plant cell wall biosynthesis and structure are still needed to overcome low oil yields and the recalcitrance of lignocellulose, respectively, for efficient and cost-competitive conversion to biofuels (Vega-Sánchez and Ronald, 2010).

10.21 ENGINEERING MULTIPLE GENES IN PLANTS

Recently, it is seen that increasing number of researchers are transferring multiple genes in order to generate plants with ambitious phenotypes. This allows researchers to achieve objectives that were once thought almost impossible. The potentiality of this approach now appears limitless. The transformation methods used earlier for plants were developed with the implicit intention to introduce one or two genes, which mainly included a primary transgene and a selectable or screenable marker and corresponding protocols had been optimized accordingly (Twyman et al., 2002). Technical bottlenecks limiting the number of genes to be transferred to plants had introduced a serious limitation to the progress of plant biotechnology in past decades (Carpell and Christou, 2004; Dafny-Yelin and Tzfira, 2007; Halpin, 2005). Current ability to transfer multiple genes into plants enables researchers to study and manipulate entire metabolic pathways, express multimeric proteins or protein complexes, and study complex genetic and regulatory networks. However, various multiple gene transfer methods certainly still have some limitations. The methods may involve conventional gene stacking method such as crossing transgenic lines, sequential transformation of desired cultivars using either more than a single vector each containing a single transgene or a single vector consisting of more than a single transgene of interest, in addition to the necessary selectable or screenable markers. The more the number of transgenes, the lesser will be the chance that all of them get integrated in the genome and expressed. It would thus require larger populations of plants to be screened to identify rare transgenic lines with the most sought-after genotype. Essentially all these methods aim to achieve the creation of a SMART locus, that is, a locus containing stable multiple arrays of transgenes (Naqvi et al., 2009).

The major application of multiple gene transfer to plants has been in the analysis and modification of metabolic pathways, which requires a number of genes. For example (1) three genes were engineered each into potato and rice for carotenoid pathway, linseed, tobacco, and Arabidopsis for PUFA synthesis and canola for vitamin E synthesis; (2) up to four genes were engineered into Arabidopsis, canola, and soybean for vitamin E synthesis; (3) four genes were engineered into maize for carotenoid, ascorbate, and folate pathways; (4) up to five genes were engineered into maize for carotenoid pathway, (5) five genes were engineered into soybean for PUFA synthesis; (6) seven genes were engineered into canola for carotenoid pathway; and (7) up to nine genes were engineered into mustard for PUFA synthesis.

10.21.1 ENGINEERING FOR VITAMIN E SYNTHESIS

Arabidopsis ***pds1***, ***hpt1***, and ***vte1*** genes were introduced in canola using an *Agrobacterium*-linked cotransformation strategy. The tocochromanol content was doubled in best performing lines (Raclaru et al., 2006). The subsets of ***tyrA***, ***pds1***, ***hpt1***, and ***ggh*** genes were introduced into Arabidopsis, canola, and soybean using *Agrobacterium*-linked cotransformation strategy to increase tocochromanol levels. Best result obtained was 15-fold tocochromanol increase in soybean (94% tocotrienols) (Karunanandaa et al., 2005).

10.21.2 ENGINEERING FOR PUFA SYNTHESIS

Arabidopsis was transformed with three different combinations of two desaturases and one elongase using an *Agrobacterium*-linked cotransformation strategy. It resulted in increased EPA and ARA content (Hoffmann et al., 2009). Two desaturases and one elongase were again separately engineered into Arabidopsis using an *Agrobacterium*-linked cotransformation strategy. In this case, EPA and ARA content increased at the expense of ALA (Qi et al., 2004). Similarly, when linseed and tobacco were genetically transformed with paired combinations of six different FA desaturases and elongases increase in GLA and SDA levels were achieved (Abbadi et al., 2004). Introduction of five genes encoding FA desaturases and elongases from two microbial species and Arabidopsis into soybean using an *Agrobacterium*-linked cotransformation strategy achieved a 20% increase in EPA (Kinney et al., 2004). When nine genes encoding FA desaturases and elongases from five microbial species were engineered into mustard using

Agrobacterium-linked cotransformation strategy, a 25% increase in ARA and a 15% increase in EPA was achieved (Wu et al., 2005).

10.21.3 ENGINEERING FOR CAROTENOID PATHWAY

Introduction of the genes such as ***crtI***, ***crtB***, and ***crtY*** from two species into rice using a combo-linked/unlinked strategy (multiple genes on two T-DNAs), a 23-fold increase in β-carotene levels was achieved in rice grains (Ye et al., 2000). When the same set of genes was engineered into potato there was a 20-fold increase in carotenoid levels (Ravanello et al., 2003). Introduction of ***psy1***, ***crtI***, ***lycb***, ***bch***, and ***crtW*** genes from four species using an unlinked direct transfer cotransformation strategy in maize recovered transgenic plants with a range of phenotypes reflecting different carotenoid contents (Zhu et al., 2008). The genes ***idi***, ***crtE***, ***crtB***, ***crtI***, ***crtY***, ***crtZ***, and ***crtW*** from three species were introduced into canola using *Agrobacterium*-linked cotransformation strategy with the aim to increase carotenoid levels, particularly ketocarotenoids and resultant effect a tremendous enhancement in the level of carotenoids was observed (Fujisawa et al., 2009).

10.21.4 ENGINEERING CAROTENOID, ASCORBATE, AND FOLATE PATHWAYS

The genes such as ***Zmpsy1*** and ***PacrtI*** for carotenoid pathway, ***Dhar*** for ascorbate pathway, and ***folE*** for folate pathway were engineered into maize using an unlinked direct transfer cotransformation strategy to increase levels of β-carotene, folate, and ascorbate in the endosperm. Significant increases in all three nutrients were achieved producing "super-nutritious" cereals (Naqvi et al., 2009).

10.21.5 ENGINEERING ABIOTIC STRESS TOLERANCE

Abiotic stress tolerance has mainly been achieved in plants by the transfer of a single gene (Muthurajan and Balasubramanian, 2009). Since abiotic stress tolerance of plants is a very complex trait and involves multiple physiological and biochemical processes, it is thought that the improvement of plant stress tolerance should involve pyramiding of multiple genes. Therefore, the generation of transgenic plants by introducing two or more foreign genes has become one of the important goals of plant genetic engineers to combat abiotic stresses (Gouiaa et al., 2012). A novel cultivar of maize expressing ***betA*** and **TsVP** (encoding V^-H^+-PPase from *Thellungiella halophila*) was developed using conventional cross hybridization technique (Wei et al., 2011). Development of GE maize plants expressing two genes ***ApGSMT2*** and ***ApDMT2*** from the bacterium *Aphanothece halophytica* with an enhanced ability to synthesize glycine betaine was reported (He et al., 2013). Effectiveness of co-expression of two heterologous abiotic stress tolerance genes ***HVA1*** and ***mtlD*** in maize (*Z. mays*) was demonstrated to confer drought and salt tolerance (Nguyen et al., 2013).

10.21.6 ENGINEERING BIOTIC STRESS TOLERANCE

In a few cases, the genes for chitinases and glucanases have been expressed together in a given host to attain even a higher degree of fungal disease resistance. Sometimes they have been also used together with some other antifungal genes. For example, transgenic potato-expressing chitinase and glucanase (Chang et al., 2002); rice-expressing chitinase and RIP (Kim et al., 2003); rice-expressing chitinase and glucanase (Mei et al., 2004); soybean-expressing chitinase and RIP (Li et al., 2004); tomato-expressing glucanase and an antifungal protein (***alf*AFP**) (Chen et al., 2006); rice-expressing chitinase, glucanase, and RIP (Zhu et al., 2007b); barley-expressing chitinase and TLP (Tobias et al., 2007); and carrot-expressing chitinase, glucanase, and a cationic peroxidase (Wally et al., 2009) for enhanced fungal disease resistance were developed. When modified ***cry1Ab*** and ***cry1Ac*** genes from *Bt* were pyramided in transgenic chickpea (*C. arietinum* L.), it improved resistance to pod borer insect *H. armigera* (Mehrotra et al., 2011).

10.22 BIOSAFETY CONCERNS AND EXPERIMENTAL STRATEGIES

10.22.1 BIOSAFETY

Biosafety is all about ensuring safety and security of both, ecology and human health, so that biological integrity is maintained. It is about minimizing the perceived risks to environment and human health from the handling of genetically modified organisms (GMOs) developed through modern biotechnology. It is remarkable that convention on biological diversity addresses the conservation and sustainable use of biodiversity. Governments from 130 countries agreed the Cartagena Protocol on Biosafety in Montreal in January 2000. It sets out rules for risk assessment, risk management, Advance Informed Agreement (AIA), technology transfer, and capacity building. AIA procedures will take care of the transgenic plants introduced into the environment intentionally, which may threaten biodiversity.

10.22.2 BIOSAFETY CONCERNS

Transgenesis has been in use for over 20 years for genetic improvement of crop plants. Transgenic crops generally carry foreign genes inserted randomly in the genome, and their commercialization is frequently prevented by public concern over *health and environmental safety issues*. Transgenic crop products are the most highly regulated items in the world. In recent years, there have been calls in the United States to relax some of the rules for their oversight. But, controversies over the safety of transgenic food products continue to resonate, particularly in Europe, Africa, and recently in the Far East. Numerous national and international scientific panels have concluded that food derived through transgenic approaches is as safe as food produced otherwise. In fact, the foodborne pathogens pose a much greater threat to human health. However, scary stories continue to appear in the media and questions continue to be asked about the adequacy of current regulatory systems to determine the safety of our food, transgenic or otherwise (DeFrancesco, 2013). It is thought people would show more preference for GM foods if they were eco-friendly.

The great success of GM crops has had an enormous impact on world crop production and cultivation pattern of agricultural species (James, 2006). The extensive environmental release and cultivation of GM crop varieties have aroused tremendous biosafety concerns and debates worldwide (Stewart et al., 2000). Biosafety issue has already become a crucial factor in constraining the further development of transgenic biotechnology and wider application of GM products in agriculture. There are a number of ***biosafety-related concerns*** in general, but the most important ones envisaged as ***ecological risks*** can be summarized as follows:

1. Direct and indirect effects of toxic transgenes (e.g., the *Bt*-insect-resistance gene) to nontarget organisms (O'Callaghan et al., 2005; Oliveira et al., 2007); insect pests may develop resistance to crops with *Bt* toxin.

2. Influences of transgenes and GM plants on biodiversity, ecosystem functions, and soil microbes (Oliveira et al., 2007); it may lead to monoculture and threaten crop genetic diversity with a possible genetic erosion over a period of time.
3. Transgene escape to crop landraces and wild relatives through gene flow and its potential ecological consequences (Lu and Snow, 2005; Mercer et al., 2007); potential transfer of genes from herbicide-resistant crops to wild or weedy relatives thus creating "superweeds."
4. Potential risks associated with the development of resistance to biotic-resistance transgenes in the target organisms (Li et al., 2007a, 2007b; Wu, 2006).

Among the above environmental biosafety issues, transgene escape from a GM crop variety to its non-GM crop counterparts or wild relatives has aroused tremendous debates worldwide Lu and Snow, 2005). This is because transgene escape can easily occur via gene flow that may result in potential ecological consequences, if significant amounts of transgenes constantly move to non-GM crops and wild relative species. Despite the potential benefits of transgenic crops, there are also concerns regarding the possible environmental and agronomic impacts if the transgenes escape and get established in natural or agricultural ecosystems. From an agronomic point of view, the transfer of novel genes from one crop to another may have many implications, including depletion in the quality of seeds leading to a change in their performance and marketability. Concerns over the ecological impacts of transgenic crops largely depend upon whether or not a crop has wild relatives and the ability to cross pollinate them. If crops hybridize with wild relatives and gene introgression occurs, wild populations could incorporate transgenes that change their behavior and they could present a serious threat as weeds or competitors in natural communities. This is particularly true when these transgenes can bring evolutionary selective advantages or disadvantages to crop varieties or wild populations. It is therefore essential to properly address the most relevant questions relating to the transgene outflow and its potential environmental consequences on a science-based altitude. Some of these concerns have been discussed in detail in the following section.

10.22.3 GENE FLOW IN THE NATURE

Because transgene technology has profound effect on management of biotic stress, transgenic plants may have substantial impact in the coming years.

The only risk of transgenic crops to the environment that might be permanent is ***gene flow*** from the crop to the close relative. Darwin considered mutations to be the basis of species evolution, but the mutations need not have been in the genes of the species in question. Instead, mutations can also come from a related species, or even further afield. The genome of any species or even cultivar is constantly changing, and thus the term "genetic purity" of varieties is inappropriate, as varieties continually change through selection, breeding, or genetic engineering. However, genes regularly move within species, to and from crops, as well as to their con-specific progenitors, feral and weedy forms. This gene movement between sexually compatible individuals, known as "vertical gene flow," can occur between varieties and strains and among some readily interbreeding species. "Horizontal gene flow" can occur between the kingdoms or distantly related species. It is far more common in prokaryotic organisms, for example, phages move from plasmids with antibiotic resistance among bacteria. A third type of gene introgression called "diagonal gene flow" occurs between the crops and distantly related, hardly sexually interbreeding relatives, within a genus, or among closely related genera. The risks are quite different from genes flowing to natural ecosystems versus ruderal and agro-ecosystems. Transgenic herbicide resistance posses a major threat if introgressed into weedy relatives, whereas disease and insect resistance pose less so (Gressel, 2014).

Naturally, various marker traits such as AFLP, RAPD, SSLP, RFLP, chloroplast, etc. move from crops to weeds. Incorporation of crop genes into wild and weedy relative populations (i.e., introgression) has long been of interest to ecologists and weed scientists. Potential negative outcomes that result from crop transgene introgression (e.g., extinction of native wild relative populations; invasive spread by wild or weedy hosts) have not been documented, and few examples of transgene introgression exist. However, molecular evidence of introgression from nontransgenic crops to their relatives continues to emerge, even for crops considered as low-risk candidates for transgene introgression. Recently, there are reports of gene flow from crops to relatives via pollen for traits such as resistance to imidazolinone, chlorotoluron, difenzoquat, glufosinate, glyphosate, disease, etc.

10.22.4 IS THE TRANSGENE FLOW DELETERIOUS TO LANDRACES, WEEDS, AND IN GENERAL?

Some argue for preserving "genetic purity" of landraces, although others counter-argue that transgene flow of crop protection traits into landraces will facilitate their continued cultivation. Farmers are less likely to abandon the landraces for higher yielding cultivars or hybrids if the landraces are resistant to most insects and pathogens, or if weeds can be cost-effectively controlled without hand labor. Thus, if the intention really is to preserve the cultivation of landraces, and not just have their presence in gene banks, then such gene flow is obligatory (Gressel, 2014). Crop protection traits have already introgressed from crops to related weeds. Here, the answer for crop protection traits is usually affirmative for herbicide resistance, but not so much for other traits such as disease or insect resistance, where data are more ambiguous owing to the sporadic nature of disease and insect incidences affecting weeds. Such genes would clearly increase weed fitness when microbial or insect biocontrol measures are used against the weed (Gressel, 2014). It is a basic assumption of many technology detractors that transgene flow to wild species is deleterious, and that transgenes will "takeover" local genes and "contaminate" natural populations, cause a loss of "genetic purity" and lead to a loss of biodiversity. Swamping is usually doubtful, as pollen must get from the crop to the distant wild ecosystem. Pollen loses vitality with time, and there will be a distance between the two ecosystems, and distance equals time. Whether the transgene becomes established in an ecosystem then depends on the nature of the transgene (Gressel, 2014). There are two potential risks following transgene introgression from crops to their wild or weedy relatives as depicted.

10.22.5 HOW TO DEAL WITH "TRANSGENE FLOW" WITHIN THE ECOSYSTEM?

Technologies have been proposed to contain genes within crops (chloroplast transformation, male sterility) that imperfectly prevent gene flow by pollen to the wild. Pollen that carries a transgene is required in almost all transgene introgression models. Hence, transgene introgression could be completely prevented if pollen does not develop, and multiple methods have been used to decrease pollen fertility via *genetic male sterility* or *cytoplasmic male sterility* (CMS). Conversely, since pollens contain no cytoplasm, it is probably safe if we can contain the selectable markers within the chloroplast for generation of transgenic plants. Chloroplasts do not enter into the male gamete and thus, the possibility of transgene escape via pollens is negated. On the other hand, containment does not prevent related weeds from pollinating crops. Repeated backcrossing with weeds as pollen parents results in gene establishment in the weeds. Transgenic mitigation relies on coupling crop protection traits in a tandem construct with traits that lower the fitness of the related weeds. Mitigation traits can be morphological (dwarfing, no seed shatter) or chemical (sensitivity to a chemical used later in a rotation). Tandem mitigation traits are genetically linked and will move together. Mitigation traits can also be spread by inserting them in multicopy transposons, which disperse faster than the crop protection genes in related weeds. Thus, there are gene-flow risks mainly to weeds from some crop protection traits, and these risks will have to be dealt with (Gressel, 2014).

10.22.6 WHAT IS REQUIRED FOR SUCCESSFUL TRANSFER OF A TRANSGENE FROM PLANT TO EITHER A MICROBE OR MAMMALIAN CELL?

Transfer of plant DNA into microbial or mammalian cells under normal conditions of dietary exposure would require all of the following events to occur: (1) removal of the relevant gene from the plant genome, probably as linear fragments; (2) protection of the gene from nuclease degradation in the plant as well as animal gastrointestinal tract; (3) uptake of the gene with dietary DNA; (4) transformation of bacteria or competent mammalian cells; (5) insertion of the gene into the host DNA by rare repair or recombination events into a transcribable unit; and finally (6) continuous stabilization of the inserted gene (FAO/WHO, 2000).

10.22.7 REMOVAL OF SELECTABLE MARKERS FROM GM CROPS

During the efficient genetic transformation of plants with the gene of interest, some selectable marker genes are also used in order to identify the transgenic plant cells or tissues. Usually, antibiotic- or herbicide-selective agents and their corresponding resistance genes are used to introduce economically valuable genes into crop plants. From the biosafety authority and consumer viewpoints, the presence of selectable marker genes in transgenic crops released may be transferred to weeds or pathogenic microorganisms in the gastrointestinal tract or soil, making them resistant to treatment with herbicides or antibiotics, respectively. Sexual crossing also raises the problem of transgene expression because redundancy of transgenes in the genome may trigger homology-dependent gene silencing. The future potential of transgenesis technologies for crop improvement depends greatly on our abilities to engineer stable expression of multiple transgenic traits in a predictable fashion and to prevent the transfer of undesirable transgenic material to nontransgenic crops and related species. Therefore, it is now essential to develop an efficient marker-free transgenesis system. These considerations underline the development of various approaches designed to facilitate timely elimination of transgenes when their function is no longer needed. Due to the limited availability of suitable selectable marker genes, the stacking of transgenes will be increasingly desirable in future. The production of marker-free transgenic plants is now a critical requisite for their commercial deployment and also for engineering multiple and complex trait. Here we describe the current technologies to eliminate the selectable marker genes in order to develop marker-free transgenic plants and also discuss the regulation and biosafety concern of GM crops.

The genetic markers developed for use for genetic transformation of plants have been derived from either bacterial or plant sources and can be divided into two types: selectable and screenable markers. *Selectable markers* are those that allow the selection of transformed cells, or tissue explants, by their ability to grow in the presence of an antibiotic such as hygromycin, and kanamycin or a herbicide like glyphosate. In addition to selecting for transformants, such markers can be used to follow the inheritance of a foreign gene in a segregating population of plants. The co-introduction of selectable marker genes, especially antibiotic-resistance genes, is required for the initial selection of plant cells that are complemented with a new trait. ***Screenable markers*** encode gene products whose enzyme activity can be easily assayed, allowing not only the detection of transformants but also an estimation of the levels of foreign gene expression in transgenic tissue. Markers such as GUS, luciferase, or β-galactosidase allow screening for enzyme activity by histochemical staining or fluorimetric assay of individual cells and can be used to study cell-specific as well as developmentally regulated gene expression.

A number of selectable marker genes, mostly conferring resistance to antibiotics or herbicides, have been used previously for plant transformation studies. However, the most commonly used selectable markers are (1) *nptII* and *hpt* genes (for resistance to the aminoglycoside antibiotics, kanamycin and hygromycin, respectively) and (2) *bar* gene (for resistance to herbicide phosphinothricin).

10.22.7.1 THE CAUSE OF CONCERN

The successful use of antibiotics in medicine has now become a problem. The presence of selectable marker genes, especially those which include genes coding for antibiotic resistance and which are essential for the initial selection of transgenic plants, is considered undesirable. This is because the transgenes integrate at random positions in the genome leading to possible unwanted mutations and unpredictable expression patterns. The drawbacks of traditional markers are well felt even in practical research, which includes the following.

1. There are only a few selectable markers available for each crop species. But, different marker gene systems are required for the retransformation of plants that have already been GM.
2. If several marker genes left over from various developmental phases accumulate in a plant, it is possible that the stability of the GE trait can be impaired.
3. The probability of unforeseen effects (pleiotropic effect) occurring in the plants increases with the number of transferred genes and marker genes because one gene may affect the functionality of the other.
4. In addition, there is a potential risk of *horizontal gene transfer* and *vertical gene transfer* that could create environmental problems.

However, the most confident way to overcome all the concerns is just to remove the cause of concern, that is, the selectable marker gene itself. Therefore, there is a need for the development of techniques for the efficient production of "clean" marker-free transgenic plants. Thus, the development of efficient techniques for the removal of selectable markers, as well as the directed integration of transgenes at safe locations in the genome, is of great interest to biotech companies. Furthermore, the removal of selectable marker genes will also have a technical advantage, since the number of available selectable marker genes is limiting, and stacking of transgenes will become more and more desirable in the near future. In the next generation of transgenic plants, antibiotic-resistance markers will be the exception rather than the rule. However, there is still a long way to go before sufficient new procedures and strategies are designed, optimized, and become available with the scientific community.

Selectable marker gene-free transgenic rice harboring the garlic leaf lectin gene exhibited resistance to sap-sucking planthoppers (Sengupta et al., 2010) and another set of marker-free transgenic plants had enhanced seed tocopherol content (Woo et al., 2015). Chilling tolerance was improved in marker-free transgenic tomato plants through induced transgenic expression of ***At-CBF1*** (Singh et al., 2011). Selectable marker-free transgenic potato plants expressing ***cry3A*** against the Colorado potato beetle (*Leptinotarsa decemlineata* Say) were also developed (Guo et al., 2015).

10.23 TOOLS OF MODERN GENETIC ENGINEERING

10.23.1 METHODS OF ELIMINATION OF SELECTABLE MARKERS

There are several strategies to exclude the selectable genes from transgenic plants, such as cotransformation, site-specific recombination, multi-autotransformation vector, transposition system, and homologous recombination (HR).

10.23.1.1 COTRANSFORMATION

The cotransformation method is a very simple method to eliminate the marker gene from the nuclear genome. Cotransformation involves transformation with two plasmids that target insertion at two different plant genome loci. One plasmid carries a selective marker gene and the other carries the GOI. The following three methods are used in the cotransformation system: (1) Two different vectors carried by different *Agrobacterium* strains followed by ATMT (De Neve et al., 1997) or two plasmids are introduced in the same tissue by means of particle bombardment (Kumar et al., 2010); (2) two different vectors introduced into the same *Agrobacterium* cell for plant transformation (Sripriya et al., 2008); and (3) two T-DNAs can be introduced a single binary vector (two T-DNA system) for genetic transformation of plants (Miller et al., 2002). In these cotransformation systems, selectable marker genes and target genes are not placed between the same pair of T-DNA borders. Instead, they are placed into separate T-DNAs, which are expected to segregate independently in a Mendelian fashion. In this method, the selectable marker can be eliminated (Fig. 10.3) from the plant genome at the time of segregation and recombination that occurs during sexual reproduction by selecting on the transgene of interest and not the SMG in progeny.

The advantages of cotransformation methods include the high adaptability of conventional, unmodified *Agrobacterium*-mediated gene transfer methods and easier handling of the binary vectors because the two T-DNA are separated and, hence, target gene T-DNA can be handled independently of selectable marker T-DNA.

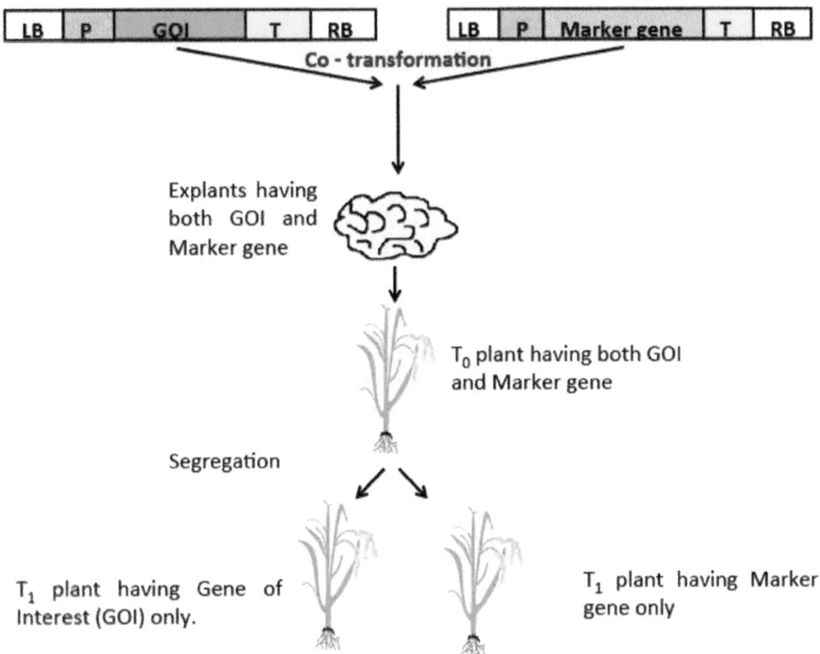

FIGURE 10.3 Diagram showing various steps of cotransformation method for generation of marker-free transgenic plants. (Adapted from Tuteja et al., 2012).

The limitations of the methods described above are very time consuming and compatible only for fertile plants. The tight linkage between co-integrated DNAs limits the efficiency of cotransformation. Indeed, integration of selectable marker and the transgene is at indiscriminate event: both the selectable marker and transgene may integrate in the same loci and that is not feasible for cotransformation.

10.23.1.2 MULTI-AUTOTRANSFORMATION

The multi-autotransformation (MAT) vector system represents a highly sophisticated approach for the removal of nuclear marker genes (Ebinuma et al., 1997). It is a unique transformation system that uses morphological changes caused by oncogene isopentenyltransferase (*ipt*) or rhizogene (the *rol* gene) of *A. tumefaciens* which control the endogenous levels of plant hormones and the cell responses to plant growth regulators as the selection marker. Expression of the *ipt* gene causes abnormal shoot morphology called extreme shooty phenotype (ESP), which subsequently reverts into normal shoots due to the excision of *ipt* gene by the function of "hit-and-run" cassette system (Ebinuma and Komamine, 2001). In this MAT system, a chosen GOI is placed adjacent to a multigenic element flanked by RS recombination sites (Fig. 10.4). A copy of the selectable *ipt* gene from *A. tumefaciens* is inserted between these recombinase sites, together with the yeast *R* recombinase gene and this entire assembly is situated within a T-DNA element for the *Agrobacterium*-mediated transformation of plant tissues. In this plant transformation system, neither antibiotic- nor herbicide-resistance genes are necessary as a selection marker. In addition, this system of transformation allows for repeated transformation of genes of interest in a plant (Sugita et al., 2000). The MAT vector system is a positive selection system that gives the advantage of regeneration to the transgenic cells without killing the nontransgenic cells.

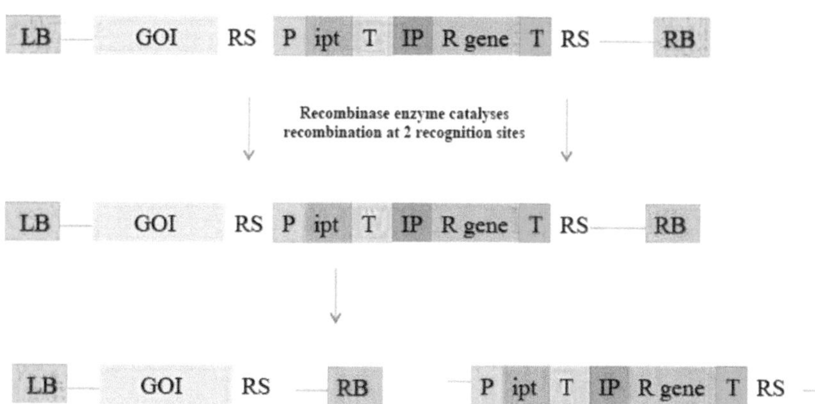

FIGURE 10.4 Multi-autotransformation (MAT). Oncogene (*ipt*) for selection of transgenic plants and a site-specific recombination system (R/Rs) are used in the principle of MAT. Recombinase (R) gene expression is under the chemically inducible promoter (IP) in order to avoid early removal of ipt gene. R catalyses recombination between two directly oriented recognition sites (RS) and removes a "hit-and-run" cassette from the plant genome (abbreviations: P, promoter; T, terminator; GOI, gene of interest; LB, left border; RB, right border). (Adapted from Tuteja et al., 2012).

10.23.1.3 SITE-SPECIFIC RECOMBINATION

Recombination is very clear phenomenon in biological systems: it occurs between two homologous DNA molecules. In bacteriophage, site-specific recombination takes place between defined excision sites in the phage and in the bacterial chromosome. In site-specific recombination, DNA-strand exchange takes place between segments possessing only a limited degree of sequence homology (Coates et al., 2005). The site-specific recombination methods in plants have been developed to delete selection markers to produce marker-free transgenic plants or to integrate the transgene into a predetermined genomic location to produce site-specific transgenic plants (Nanto and Ebinuma, 2008). Basically, three site-specific recombination systems are well known and are described in the following sections for the elimination of selectable marker.

10.23.1.4 CRE/LOX RECOMBINATION SYSTEM

The Cre/loxP system consists of two components: (a) two loxP sites each consisting of 34-bp inverted repeats cloned in direct orientation flanking a DNA sequence and (b) the *cre* gene encoding a 38-kDa recombinase protein that specifically binds to the loxP sites and excises the intervening sequence along with one of the loxP sites. The Cre/loxP system has been tested in several plants including Arabidopsis (Zuo et al., 2001), *Nicotiana* (Gleave et al., 1999), *Z. mays* (Zhang et al., 2003) and *O. sativa* (Sreekala et al., 2005). *One of the greatest advantages* of the Cre/lox system is the specificity of the enzyme for its 34-bp recognition sequence. With a few exceptions, it is difficult to insert and excise genes with precision in the plant genome without a site-specific recombination system. One of the major limitations of this system is that marker gene removal from transgenic plants using the Cre/lox recombination system of bacteriophage P1 requires retransformation and out-crossing approaches that are laborious and time-consuming.

10.23.1.5 FLP/FRT RECOMBINATION SYSTEM

In the FLP/FRT site specific system of the 2-μm plasmid of *S. cerevisiae*, the FLP enzyme efficiently catalyses recombination between two directly repeated FLP recombination target (frt) sites, eliminating the intervening sequence. By controlled expression of the FLP recombinase and specific placement of the frt sites within transgenic constructs, the system can be applied to eliminate the marker genes following selection (Cho, 2009). It is possible to make an inducible FLP/frt site-specific recombinase system. However, one of the limitations of the process is it requires the process of retransformation to get both FLP and frt in the same system. A heat-inducible strategy for the elimination of selection marker genes was also reported in vegetatively propagated plants like potato (Cuellar et al., 2006) (Fig. 10.5).

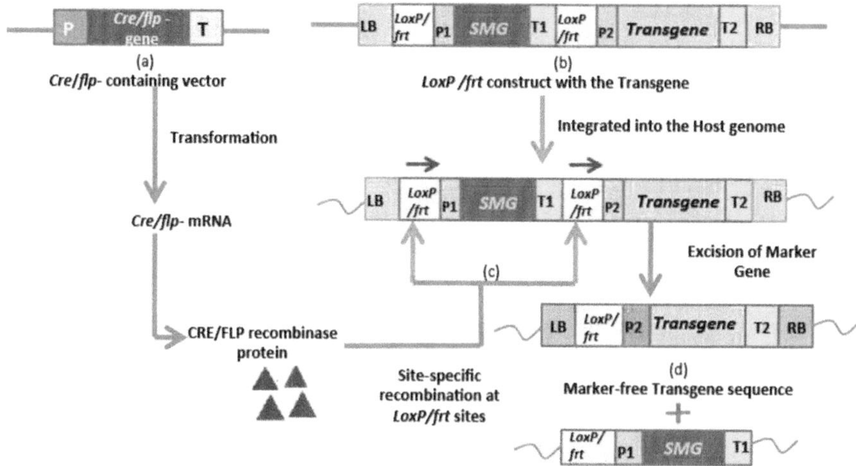

FIGURE 10.5 Development of selectable-marker-free transgenic plants using Cre/lox and/or FLP/FRT recombination system: (a) *cre/flp* encoding CRE-a Tyr recombinase protein or FLP is expressed under a constitutive or inducible promoter (chemical/heat shock). (b) *loxP/frt* flank the selectable marker gene on the T-DNA but not the transgene. These constructs can be introduced into the plant system via sequential transformation or cotransformation. Alternatively, genetic crossing can be used to bring both the constructs in the same plant. Following transformation, CRE/FLP gets access to *LoxP/frt* and causes site-specific recombination (c), thereby resulting into generation of selectable marker-free transgene sequence (d). Abbreviations: P, promoter; T, terminator; P1, promoter for marker gene; T1, terminator for marker gene; P2, promoter for transgene; T2, terminator for transgene; SMG, selectable marker gene; LB, left border; RB, right border sequence.

10.23.1.6 TRANSPOSITION-BASED METHODS

In general, all *Activator (Ac)* elements are identical, 4563 bp in length from maize. Transposase are the proteins that stimulate the movement of Ac. Deletions of Ac elements created *Dissociator (Ds)* elements in which all or part of this transposase was eliminated. This lack of transposase activity accounts for the inability of Ds elements to move in the absence of Ac. The transposase that is encoded by Ac elements can move throughout the cell and excise any Ds or Ac element. Thus, the Ac/Ds transposase is said to be transacting ability. Two transposon-mediated strategies have been developed to generate marker-free transgenic plants. The Ac/Ds elements can be introduced into the plant genomes and can be very useful in removing the selectable marker gene as depicted (Fig. 10.6).

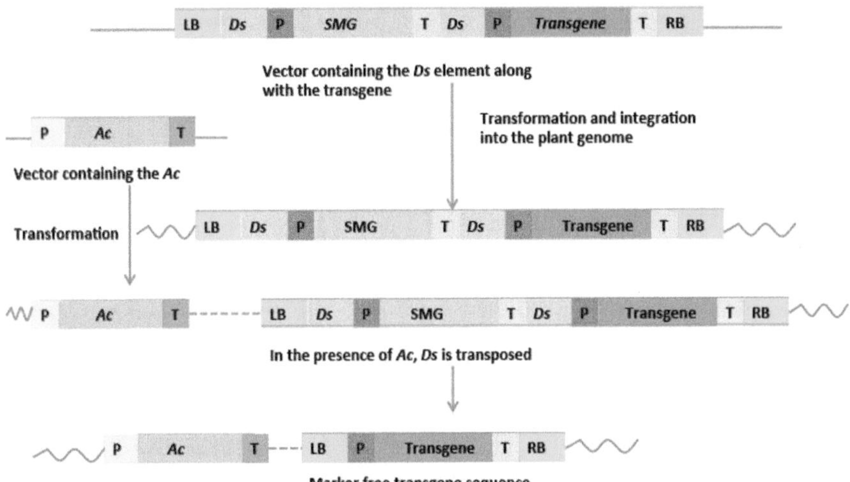

FIGURE 10.6 Development of selectable-marker-free transgenic plants using of Ac/Ds transposition system: Individual sequential transformation or cotransformation or genetic crossing can be used to introduce the constructs consisting of activator (*Ac*) and dissociator (*Ds*) elements in plants. An inducible promoter (chemical/heat shock) should regulate the expression of *Ac* gene in case of cotransformation or when both the expression cassettes are placed in a single vector backbone. *Ds* gets transposed (in the absence of *Ac*, *Ds* cannot get transposed) when *Ac* integrates into the genome and gets expressed, along with the selectable marker gene resulting in a marker-free transgene sequence. Abbreviations: P, promoter; T, terminator; SMG, selectable-marker gene; *Ac*, activator element; *Ds*, dissociation element; LB, left border; RB, right border.

10.23.1.7 POSITIVE SELECTION METHOD

In positive selection, GM cells are identified and selected without causing any injury or death to the nontransformed cell population (negative selection). In this case, the selectable marker gives the transformed cell the capacity to metabolize some compounds that are not usually metabolized. This fact will give the transformed cells an advantage over the nontransformed ones. The addition of this new compound in the culture medium, as nutrient source during the regeneration process, allows normal growth and differentiation of transformed cells. However, the nontransformed cells will not be able to grow and regenerate de novo plants.

10.23.1.7.1 The Gus Gene

The *gus* gene from *E. coli* codes for the GUS enzyme is widely used as a reporter gene in transgenic plants. In this system, GUS enzyme produced in the transformed cells hydrolyses benzyladenine *N*-3-glucuronide, glucuronide derivative of benzyladenine, which is an inactive form of the plant hormone cytokinin and releases benzyladenine, which is active cytokinin, in the medium. This cytokinin thus generated stimulates the transformed cell to regenerate, whereas the development of nontransformed cell is arrested. This marker system has been used for effective recovery of some transgenic plants (Okkels et al., 1997).

10.23.1.7.2 The manA Gene

The man gene from *E. coli* codes for the phosphomannose isomerase (PMI) enzyme. Mannose is converted into mannose-6-phosphate by endogenous hexokinase. Thus, when mannose is added to the culture medium, plant growth in the nontransfomed tissue may be minimized due to mannose-6-phosphate accumulation. PMI converts mannose-6-phosphate into fructose-6-phosphate, which in turn, is immediately channelized to glycolysis can be used as the sole carbohydrate source for the transformed cells. The mannose-6-phosphate cannot be metabolized the nontransformed cell and toxicity in plant cells was shown to be responsible for apoptosis, or programed cell death, through induction of an endonuclease, responsible for DNA laddering (Stein and Hansen, 1999). Mannose-6-phosphate accumulation also causes phosphate and ATP starvation, and thus, the critical functions such as cell division and elongation are retarded, giving rise to growth inhibition. Therefore, mannose turns out to be a very useful selection agent.

10.23.1.7.3 The xylA and DOGR1 Genes

Another positive selection system similar to PMI is the *xylA* encoding xylose isomerase isolated from *Thermoanaerobacterium thermosulfurogenes* or from *Streptomyces rubiginosus*. Transgenic plants of potato, tobacco, and tomato were successfully selected in xylose-containing media. Another gene *DOGR1*, from yeast, encoding 2-deoxyglucose-6-phosphate phosphatase (2-DOG-6-P) was has also been developed as a positive selection system. This marker confers resistance to 2-deoxyglucose (2-DOG) when overexpressed in transgenic plants. This system has been used to develop transgenic tobacco and potato plants (Kunze et al., 2001).

10.24 CISGENICS AND INTRAGENICS VERSUS TRANSGENICS

Although scientists add genes to crops via crop breeding, the breeding progeny is not considered as a GM crop because the introgressed genes and the regulatory sequences belong to the same host crop genus or in rare cases to the host's cross-breedable crop. One of the major concerns of the general public about transgenic crops relates to the mixing of genetic materials between species that cannot hybridize by natural means. To meet this concern, the two transformation concepts ***cisgenesis*** and ***intragenesis*** were developed as alternatives to transgenesis. Both concepts imply that plants must only be transformed with genetic material derived from the species itself or from closely related species capable of sexual hybridization. Furthermore, foreign sequences such as selection genes and vector-backbone sequences should be absent. If the donor gene and all of transgene's regulatory sequences belong to the same crop species or belong to the host's cross-breedable species, the resulting crop is called ***cisgenic***. In the ***cisgenic technology***, the cisgenic must be an identical copy of the host's native gene cassette, including its regulatory sequences integrated in the host plant in the normal sense orientation. In the ***intragenic technology***, gene cassettes containing specific gene sequences from crops are inserted into the crop that belongs to the same breedable gene pool. In this case, the promoters and terminators of different genes can regulate the gene-coding sequences. Cisgene and intragene constructs are depicted (Fig. 10.7). Cisgensis has been applied for improved baking quality of durum wheat using *1Dy10* (Gadaleta et al., 2008), late blight resistance in potato using *R* genes (Haverkort et al., 2009), scab resistance in apple using *HcrVf2* (Vanblaere et al., 2011), fungal disease resistance in grapevine using *VVTL-1* and *NtpII* (Dhekney et al., 2011), and improved grain phytase activity in barley using *HvPAPHY_a* (Holme et al., 2012). Intragenesis has been applied for high amylopectin content in potato using GBSS (de Vetten et al., 2003), scab resistance in apple using *HcrVf2* (Joshi et al., 2011), preventing black spot bruise in potato using *Ppo*, *R1*, and *PhL* (Rommens et al., 2006), gray mold resistance in strawberry using *PGIP*, limiting level of acrylamide in French fries from potato using *StAs1*, *StAs2* (Chawla et al., 2012), reducing lignin level in alfalfa using *Comt* (Weeks et al., 2008) and improving drought tolerance in perennial ryegrass using *Lpvp1* (Bajaj et al., 2008). Several surveys show higher public acceptance of intragenic/cisgenic crops compared to transgenic crops. The sexually compatible gene pool carries a high potential for generating plants with environmental, economic, and health benefits that may be essential for meeting the global need for a more efficient and sustainable crop production.

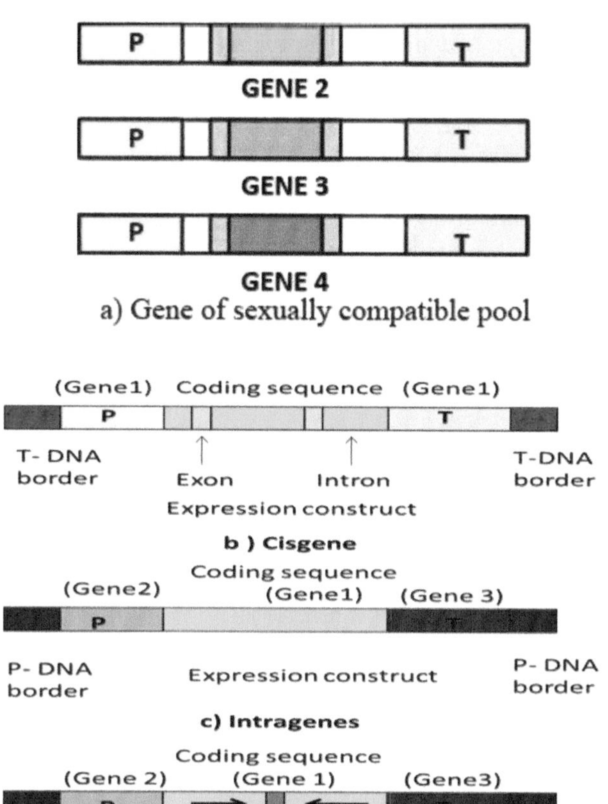

FIGURE 10.7 The cisgene is an identical copy of a gene from the sexually compatible pool including promoter, introns, and terminator. (a, b) The cisgene is inserted within *Agrobacterium*-derived T-DNA borders when following *Agrobacterium*-mediated transformation. Intragenesis allows in vitro recombination of elements isolated from different genes within the sexually compatible gene pool (a, c). (Adapted from Holme et al., 2013).

10.25 PLASTID GENETIC ENGINEERING

Plastid genetic engineering, with several unique advantages including transgene containment, has made significant progress in the last two decades in various biotechnology applications including development of crops with high levels of resistance to insects, bacterial, fungal, and viral diseases, different types of herbicides, drought, salt and cold tolerance, CMS, metabolic engineering, phytoremediation of toxic metals and production of many vaccine antigens, biopharmaceuticals, and biofuels. However, useful traits should be engineered via chloroplast genomes of several major crops (Clarke and Daniell, 2011).

Plastid transformation was initially developed in *Chlamydomonas* and tobacco, but, it is now feasible in a broad range of species. It now is widely used in basic research and for biotechnological applications. Selection of transgenic lines where all copies of the polyploid plastid genome are transformed requires efficient markers. A number of traits have been used for selection such as photoautotrophy, resistance to antibiotics, and tolerance to herbicides or to other metabolic inhibitors. The most successful and widely used markers are derived from bacterial genes that inactivate antibiotics, such as *aadA* that confers resistance to spectinomycin and streptomycin, although the presence of a selectable marker that confers antibiotic resistance is not desirable for many biotechnological applications.

Selectable markers for plastid transformation routinely involve those for (1) photosynthesis such as *petA*, *ycf3*, and *rpoA* in tobacco (Klaus et al., 2003), and *rbcL* in tobacco (Kode et al., 2006); (2) antibiotic resistance such as *aphA-6* for kanamycin in tobacco (Huang et al., 2002) and cotton (Kumar et al., 2004) and *rrnS* for spectinomycin and spectromycin in tomato (Nugent et al., 2005); (3) herbicide resistance such as *bar* for phosphinothricin in tobacco, EPSP for glyphosate in tobacco (Ye et al., 2003), and *HPPD* for diketonitrile in tobacco (Dufourmantel et al., 2007), (4) metabolism such as *BADH* for betaine aldehyde in tobacco (Daniell et al., 2001b) and *ASA2* for Trp analogues in tobacco (Barone et al., 2009). One of the highly remarkable markers, *aadA*-encoding resistance to spectinomycin and spectromycin, has been routinely used over time for transformation of number of crop plants such as rice (Lee et al., 2006b), tomato (Ruf et al., 2001), oilseed rape (Cheng et al., 2005; Hou et al., 2003), carrot (Kumar et al., 2004), soybean (Dufourmantel et al., 2004), lettuce (Lelivelt et al., 2005; Ruhlman et al., 2010), cauliflower (Nugent et al., 2006), cabbage (Liu et al., 2007), sugarbeet (De Marchis et al., 2009), eggplant (Singh et al., 2010), etc.

10.26 SMALL RNA ENGINEERING

MicroRNA-based genetic modification technology (miRNA-based GM tech) can be used for increasing crop yields and quality. It is one of the most promising solutions that contribute to agricultural productivity directly by developing superior crop cultivars with enhanced biotic and abiotic stress tolerance and increased biomass yields. Manipulating miRNAs and their targets in transgenic plants including constitutive, stress-induced, or tissue-specific expression of miRNAs or their targets, RNAi, expressing miRNA-resistant target genes, artificial target mimic, and artificial miRNAs are some of the useful strategies. In general, miRNAs and their targets not only provide an invaluable source of novel transgenes but also inspire the development of several new GM strategies, allowing advances in breeding novel crop cultivars with agronomically useful characteristics. Applications of microRNA-based gene regulation for crop improvement (Fig. 10.8) and strategies for developing miRNA-based GM crops (Fig. 10.9) are depicted. Specifically, RNA silencing has been a powerful tool that has been used to engineer various crop plants in last two decades. Based on the siRNAs-mediated RNA silencing (RNAi) mechanism, transgenic plants were designed to trigger RNA silencing by targeting pathogen genomes. Diverse targeting approaches have been developed based on the difference in precursor RNA for siRNA production, including sense/antisense RNA, small/long hairpin RNA and artificial miRNA precursors (Prins et al., 2008; Simón-Mateo and Garcia, 2011). Approaches to induce RNAi (Fig. 10.10) include (1) sense or antisense

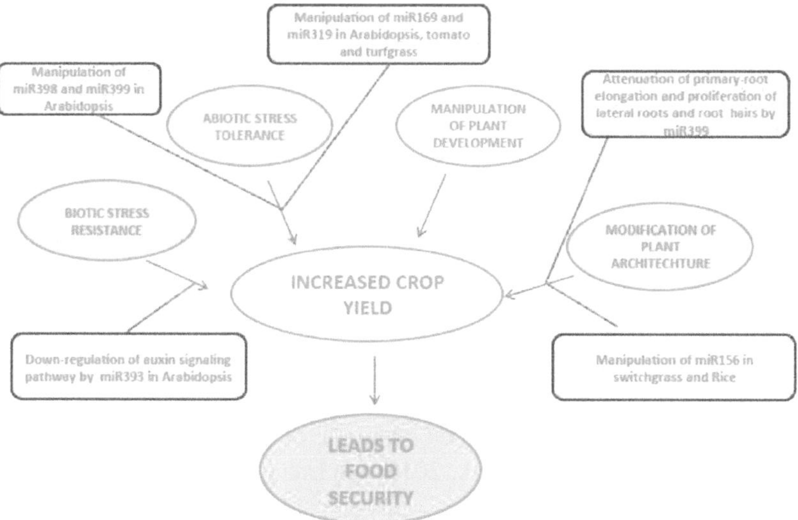

FIGURE 10.8 Applications of miRNA-based gene regulation in crop improvement. MicroRNA-based GM technology can help address food insecurity by either enhancing crop adaptations to extrinsic environmental stresses or increasing intrinsic yield potential in plants. (Adapted from Zhou and Luo 2013)

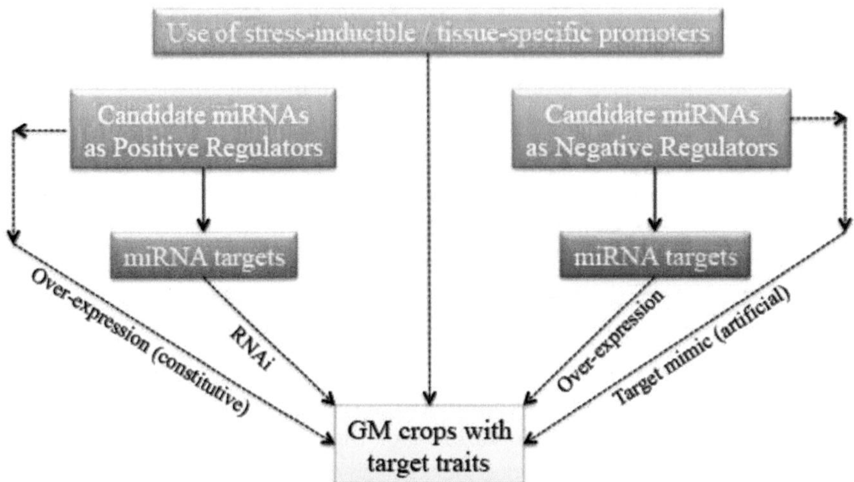

FIGURE 10.9 Strategies for development of genetically modified crop plants using miRNA approach. (Adapted from Zhou and Luo 2013).

viral sequences in transgene-mediated resistance; (2) virus-derived hpRNA transgene-mediated resistance; (3) artificial microRNA-mediated resistance, etc. RNAi of *JcFAD2-1* in transgenic Jatropha increased the proportion of oleic acid versus linoleic through genetic engineering, enhancing the quality of its oil (Qu et al., 2012). High-level resistance to banana bunchy top virus infection has been achieved in transgenic banana plants expressing small-interfering RNAs targeted against viral replication initiation (Shekhawat et al., 2012). RNAi-based resistance has been demonstrated in transgenic tomato plants against Tomato yellow leaf curl virus-Oman (Ammara et al., 2015). When transketolase activity was decreased by means of antisense technology in cucumber, it reduced the photosynthetic rate, seed germination, growth yield and tolerance to low temperature, and weak light stress (Bi et al., 2015). Silencing of both *FAD2* genes in stable transformants of flax, which was high in LA, led to high level of oleic acid (Chen et al., 2015). Cotton plants expressing CYP6AE14 dsRNA showed enhanced resistance to bollworms (Mao et al., 2011). RNAi-mediated silencing of *HaHR3* gene (Xiong et al., 2013) and *HaAK* gene (Liu et al., 2015b) in transgenic cotton also disrupted development of this insect pest. Transgenic plants overexpressing insect-specific microRNA, which is an effective alternative to *Bt*-toxin, acquired insecticidal activity against *H. armigera*. RNAi-mediated knockdown of midgut genes in transgenic rice has been a valuable tool to control the hemipteran insect *N. lugens* (Zha et al., 2011). Pest resistance was also increased

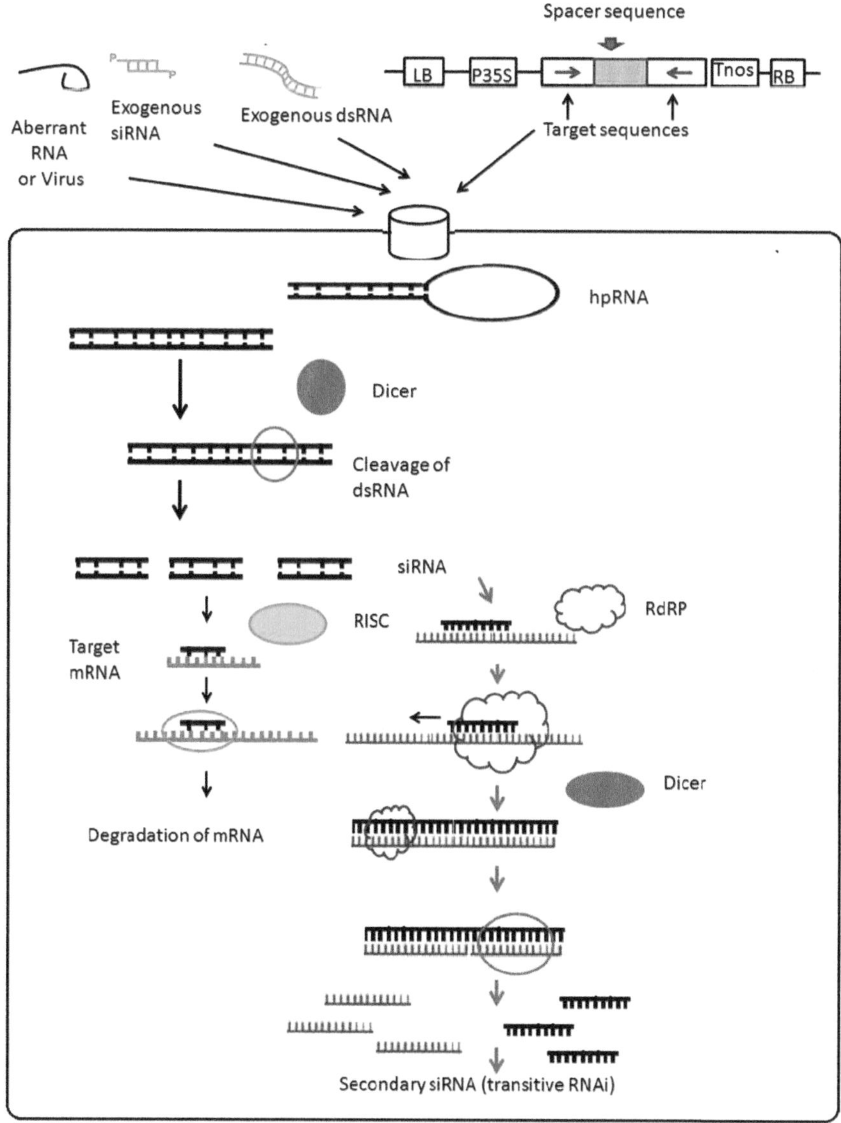

FIGURE 10.10 The RNA-mediated gene silencing in plants: Double-stranded RNA (dsRNA) generated through aberrant gene expression from a foreign gene, virus infection or tandem-repeat sequence due to insertion of a transposon/retrotransposon is digested into 21–25 nucleotide-long short interfering RNA (siRNA), by *Dicer* (an RNaseIII-like RNase), which functions as a template for the targeted degradation of mRNA in RISC (RNA-induced silencing complex) and also acts as the primer for RNA-dependent RNA polymerase (RdRp) to amplify the secondary dsRNA, although some differences between endogenous and foreign genes have been found for secondary RNAi.

in transgenic tobacco plants expressing dsRNA of an insect-associated gene ***EcR*** (Zhu et al., 2012). In addition, the transgenic tobacco plants expressing *H. armigera EcR* dsRNA were also resistant to another lepidopteran pest, the beet armyworm, *Spodoptera exigua*, due to the high similarity in the nucleotide sequences of their *EcR* genes. Transgenic maize plants with improved salt tolerance have been made free from selectable marker too (Li et al., 2010). Tomato plants resistant to Gemini viruses have been developed using artificial transacting small siRNA (Singh et al., 2015). RNAi-mediated resistance in transgenic cassava exhibited resistance to cassava brown streak Uganda virus (Yadav et al., 2011).

10.27 FUTURE PERSPECTIVES

Research on transgenic crops is expected to increase dramatically, given the whole world scenario. It is expected that there will be development and commercial release of several new abiotic and biotic stress-tolerant transgenic crop lines and biofuel plant platforms, coupled with vivid discussion at the public, academic, and government interface on biosafety of transgenic crops. These aspects will be the foci of future long-term monitoring programs because they have greater potential to alter plant fitness and to increase weedy or invasive tendencies, compared with traits in current commercial transgenic crops. Novel molecular strategies for monitoring and strategies for containment will also be foci of future studies. Monitoring approaches that survey transgenic crops and wild or weedy populations at crucial steps along the introgression process could also provide empirical data for enhancement, evaluation, and utilization of population models of transgene introgression (Kwit et al., 2011).

The growing demand for food is one of the major challenges to humankind. We have to safeguard both biodiversity and arable land for future agricultural food production, and we need to protect genetic diversity to safeguard ecosystem resilience. We must produce more food with less input, while deploying every effort to minimize risk. Agricultural sustainability is no longer optional but mandatory (Jacobsen et al., 2013). The traditional techniques are no longer sufficiently powerful to satisfy current and future needs for the three targets mentioned above. A combination of approaches will likely be needed to significantly improve the stress tolerance of crops in the field. These will include mechanistic understanding and subsequent utilization of stress response and stress acclimation networks, with careful attention to field growth conditions, extensive testing in the laboratory, greenhouse, and the field; the use of innovative approaches that take into consideration the genetic background and physiology of different crops; the use of enzymes and proteins from other organisms; and the integration of QTL mapping and other genetic and breeding tools (Mittler and Blumwald, 2010). Understanding of genomics paradigms has advanced considerably in the past decade. This resulted in a more integrative and deeper comprehension of how genetic and epigenetic processes regulate plant growth and development and response to the environment. The era of omics, including genomics, transcriptomics, epigenomics, proteomics, and metabolomics, is poised to facilitate biotechnological improvement of crops, particularly for physiological phenotypes that are controlled by complex genetic and epigenetic mechanisms (Moshelion and Altman, 2015). Further advances in plant biotechnology and agriculture depend on the efficient combination and application of diverse scientific inputs.

10.27.1 GENOME EDITING

In the genome-editing era, the dissemination of plants developed by advanced genetic engineering is not hampered by technological aspects but by the understanding and acceptance of such technologies in society. Researchers, the public, and regulatory bodies should proactively discuss the socially acceptable integration of genome-editing crops, if they recognize that the agricultural use of genome-editing can satisfy the needs of breeders and consumers alike and improve global food security (Araki and Ishii, 2015). Genome editing tool such as sequence-specific nucleases (SSNs) harness DNA editing repair pathways. SSNs enable precise genome editing by introducing DNA double-strand breaks (DSBs) that subsequently trigger DNA repair by either nonhomologous end joining (NHEJ) or HR (Fig. 10.11). The NHEJ is error-prone and frequently introduces small deletions and insertions at the junction of the newly rejoined chromosome, some of which cause gene knockouts by generating frameshift mutations. In genome editing by HR, DNA templates bearing sequence similarity to the break site are used to introduce sequence changes at the target locus. HR can be used to change single amino acids or small stretches of amino acids in proteins, or single base pairs or groups of base pairs in control elements. Thus, DNA repair by HR is a precise gene-targeting method.

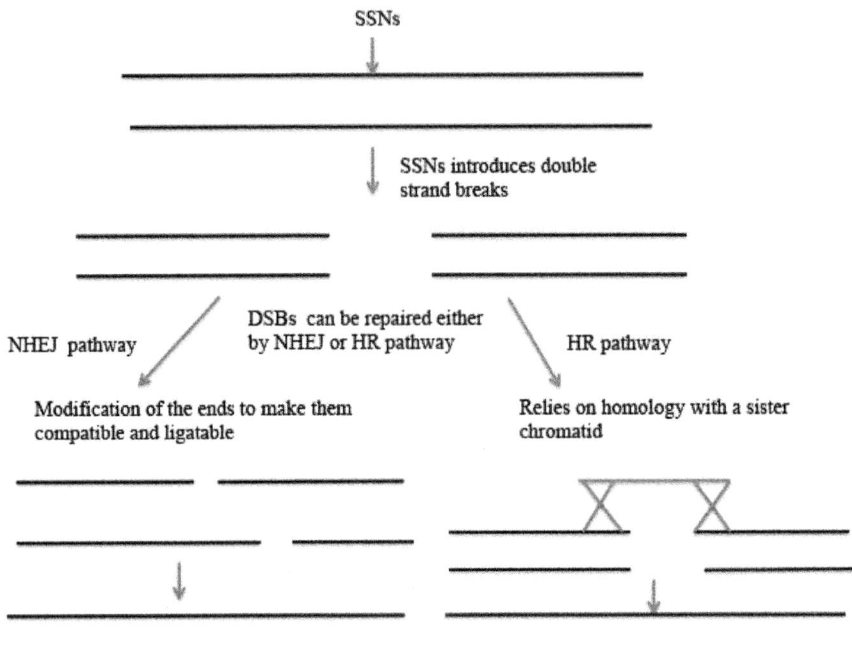

FIGURE 10.11 Genome repairing pathways after DSBs are induced by SSNs such as ZFNs, TALENs, and CRISPR/Cas system. (Adptered from Gao C, 2015).

Considering the regulatory and social hurdles associated with transgenic crops, novel and latest biotechnological tools like SSNs, namely, zinc finger nucleases (ZFNs), transcription activator-like effector nucleases (TALENs), and clustered regulatory interspaced short palindromic repeat (CRISPR)/ Cas-based RNA-guided DNA endonucleases have emerged. These tools allow precise insertion of specific genes for modification or replacement of genes at their specific genomic location without involving any other source of DNA. Genome editing is an advanced genetic engineering tool that can more directly modify a gene within a plant genome. The absence of foreign DNA, most notably selectable markers in the final product, and introduction of genes derived from the same plant species, should help to increase consumer acceptance of novel GM plant products developed with these technologies. Thus, with the emergence of such technologies, the time is right to visit the benefits of genetic modification and to begin the development of novel, consumer-acceptable products. These tools can be used for precision genome engineering and agriculture. These novel biotechnological tools have been successfully demonstrated in Arabidopsis, tobacco, rice, sorghum, and *Brachypodium* (Jiang et al., 2013; Townsend et al., 2009; Shan et al., 2013a, 2013b). Genome editing allows plant breeding without introducing a transgene, and this has led to new challenges for the regulation and social acceptance of GMOs. Genome editing is, thus, expected to generate many new crop varieties with traits that can satisfy the various demands for commercialization, by utilizing plant genomic information. This modern genome editing technology can produce novel plants that are similar or identical to plants generated by conventional breeding techniques, and therefore, creating distinct boundaries with regards to GMO regulations (Araki et al., 2014; Camacho et al., 2014; Hartung and Schiemann, 2014; Kanchiswamy et al., 2015; Voytas and Gao, 2014). Therefore, an appropriate regulatory response is urgently required toward the social acceptance of genome-edited crops. Recent reports regarding genome editing of major crops, including barley (*Hordeum vulgare*), maize (*Z. mays*), rice (*O. sativa*), soybean (*Glycine max*), sweet orange (*C. sinensis*), tomato (*S. lycopersicum*), and wheat (*Triticum*), have demonstrated a high efficiency of indels. Most notably, three homeoalleles of *TaMLO* were simultaneously edited in hexaploid bread wheat, resulting in heritable resistance to powdery mildew (Wang et al., 2014). Moreover, maize, which has indels in *ZmIPK1* is expected to have improved nutritional value as a result of decreased phosphorus content (Liang et al., 2014; Shukla et al., 2009). Furthermore, rice with indels in *OsBADH2* (Jiang et al., 2013; Shan et al., 2013) may appeal to consumers in view of its improved fragrance (Bradbury et al., 2008; Chen et al., 2008). Such results show that genome editing dramatically simplifies plant breeding even in major crops, with potential impact on the future of agriculture and human nutrition. However, most of these reports did not address potential off-target mutations. The occurrence of off-target mutations is one of the crucial issues in the agricultural use of genome editing. Some off-target mutations are likely to result in silent or loss-of-function mutations, others might lead to immunogenicity or toxicity in the food products by changing amino acids within a protein. However, there is no documented instance of any adverse effect resulting from foods produced from GM plants (Goodman and Tetteh, 2011).

Genome-editing tools are expected to become a method of choice, in addition to other novel technologies, for allelic modifications, gene replacement, structural characterization of the proteome, and posttranslational modifications. Multinational research is already taking into account the biology–agriculture crosstalk, paving the way to more effective and productive development of new cultivars. Recent studies have identified a large number of genetic and molecular networks underlying plant adaptation to adverse environmental growth conditions. All of these studies emphasize the complexity of the various traits and their polygenic nature. All biotechnological applications should be scrutinized with respect to global food security, economic, sociological, legal, and ethical considerations, aiming at public acceptance.

Genome editing based on SSNs is one of the most promising novel plant breeding technologies for crop improvement. Gene knockouts are valuable for generating new genetic variants, and genome editing can be used to make knockout collections for agronomically important crop plants such as rice and maize. The plants created by SSN mutagenesis do not appear to have any foreign DNA in their genomes, and are often indistinguishable from natural variants or those produced by conventional mutagenesis. They may therefore fall outside the existing regulations affecting GM crops. Additionally, because SSNs can be used to introduce single nucleotides or long stretches of DNA at predefined genomic sites, the types of insertion they produce may avoid the position effects associated with random insertion by traditional transgenesis. Further, if multiple transgenes are inserted at the same site, such a gene stack will be inherited as a single Mendelian locus, allowing introduction of several different transgenes into the genome. We believe that progress in genome editing in plants promises to open exciting new avenues for crop improvement. Scientists must not slow down on advancing these promising technologies because such advancements may well lead to yet more powerful technologies in favor of public.

10.28 PRESENT AREAS OF EMPHASIS

1. Gene mining and genome editing, followed by integration of transgenic approach to conventional plant breeding will be very useful in developing "biotech crops."
2. Utilization of "clean gene" or "marker-free" transgene technologies should be one of the main essences of the new transgene technologies.
3. Toxicity and allergenicity tests should be done on a case-by-case basis to assess the perceived risk of the transgenic food products.

4. Thorough implementation and supervision of biosafety guidelines via development of a network between agricultural universities/ institutes, other relevant laboratories, and biosafety committees is a must.
5. A comprehensive survey needs to be conducted for the degree of crossability between crop species and their wild relatives, existing wild and weedy relatives in an environment into which the transgenic crops are intended to be released so that we do not compromise with the health of the environment.
6. There is an urgent need for balanced risk assessment procedures with better models of monitoring system, on step-by-step and case-to-case basis, for studying the deleterious effects of herbicide, insect and disease resistant crops.
7. Sensible and realistic decisions should be taken by the policy-makers not to release these crops at the centers of origin, delicate ecological zones and the pockets rich in biodiversity, considering the potential impact of transgenic crops on genetic diversity.
8. Ecologists should be commissioned to comprehend the effect on biodiversity in the long run, after commercialization of transgenic crops, on regular basis. Additionally, monitoring and mapping of the biodiversity of hot spots should be regularly conducted.
9. There is an urgent need of an in-depth study to address the effects of transgenic plants on nontarget animals, plants and other organisms, etc., since not much scientific information is available in this area.
10. Public awareness program is a must regarding not only the benefits offered by transgenic crops, but also on their perceived risks and the need to protect valuable genetic resources. Literacy programs in schools/colleges/universities for basic understanding of modern genetics, molecular biology, etc. to make safe and responsible use of transgenic products is mandatory.
11. Government coordinated public–private partnership program in transgenic research and development will be very useful.

KEYWORDS

- genetic engineering
- *Bt*-toxin
- protease inhibitors
- chitinases
- glucanases
- phosphinothricin acetyl transferase (PAT)
- RNAi
- Cre/lox recombination system
- biosafety

REFERENCES

Abbadi, A.; Domergue, F.; Bauer, J.; Napier, J. A.; Welti, R.; Zähringer, U.; Cirpus, P.; Heinz, E. Biosynthesis of Very Long-Chain Polyunsaturated Fatty Acids in Transgenic Oilseeds: Constraints on their Accumulation. *Plant Cell* **2004,** *16* (10), 2734–2748.

Abdallah, N. A.; Shah, D.; Abbas, D.; Madkour, M. Stable Integration and Expression of a Plant Defensin in Tomato Confers Resistance to Fusarium Wilt. *GM Crops* **2010,** *1* (5), 344–350.

Abe, K.; Emori, Y.; Kondo, H.; Suzuki, K.; Arai, S. Molecular Cloning of a Cysteine Proteinase Inhibitor of Rice (*Oryza cystatin*). Homology with Animal Cystatins and Transient Expression in the Ripening Process of Rice Seeds. *J. Biol. Chem.* **1987,** *262* (35), 16793–16797.

Abebe, T.; Guenzi, A. C.; Martin, B.; Cushman, J. C. Tolerance of Mannitol-Accumulating Transgenic Wheat to Water Stress and Salinity. *Plant Physiol.* **2003,** *131* (4), 1748–1755.

Abhilash, P. C.; Jamil, S.; Singh, N. Transgenic Plants for Enhanced Biodegradation and Phytoremediation of Organic Xenobiotics. *Biotechnol. Adv.* **2009,** *27* (4), 474–488.

Acharjee, S.; Sarmah, B. K.; Ananda Kumar, P.; Olsen, K.; Mahon, R.; Moar, W. J.; Moore, A.; Higgins, T. J. V. Transgenic Chickpeas (*Cicer arietinum* L.) Expressing a Sequence-Modified Cry2Aa Gene. *Plant Sci.* **2010,** *178* (3), 333–339.

Afroz, A.; Chaudhry, Z.; Rashid, U.; Ali, G. M.; Nazir, F.; Iqbal, J.; Khan, M. R. Enhanced Resistance against Bacterial Wilt in Transgenic Tomato (*Lycopersicon esculentum*) Lines Expressing the Xa21 Gene. *Plant Cell, Tissue Organ Cult.* **2011,** *104* (2), 227–237.

Agrawal, A.; Rajamani, V.; Reddy, V. S.; Mukherjee, S. K.; Bhatnagar, R. K. Transgenic Plants over-Expressing Insect-Specific MicroRNA Acquire Insecticidal Activity against *Helicoverpa armigera*: An Alternative to Bt-Toxin Technology. *Transgen. Res.* **2015,** 1–11.

Ahmad, B.; Khan, M. S.; Ambreen, A. H.; Khan, I.; Haider, A.; Khan, I. Agrobacterium-Mediated Transformation of *Brassica juncea* (L.) Czern. with Chinitase Gene Confererring Resistance against Fungal Infection. *Pak. J. Bot.* **2015,** *47* (1), 211–216.

Ahmad, P.; Ashraf, M.; Younis, M.; Hu, X.; Kumar, A.; Akram, N. A.; Al-Qurainy, F. Role of Transgenic Plants in Agriculture and Biopharming. *Biotechnol. Adv.* **2012,** *30* (3), 524–540.

Ahmad, R.; Kim, M. D.; Back, K.-H.; Kim, H.-S.; Lee, H.-S.; Kwon, S.-Y.; Murata, N.; Chung, W.-I.; Kwak, S.-S. Stress-Induced Expression of Choline Oxidase in Potato Plant Chloroplasts Confers Enhanced Tolerance to Oxidative, Salt, and Drought Stresses. *Plant Cell Rep.* **2008,** *27* (4), 687–698.

Ahmad, R.; Kim, Y.-H.; Kim, M.-D.; Kwon, S.-Y.; Cho, K.; Lee, H.-S.; Kwak, S.-S. Simultaneous Expression of Choline Oxidase, Superoxide Dismutase and Ascorbate Peroxidase in Potato Plant Chloroplasts Provides Synergistically Enhanced Protection against Various Abiotic Stresses. *Physiol. Plant.* **2010,** *138* (4), 520–533.

Akin, D. E. Grass Lignocellulose. *Applied Biochemistry and Biotecnology*; Humana Press: New York, 2007; pp 3–15.

Akiyama, T.; Arumugam Pillai, M.; Sentoku, N. Cloning, Characterization and Expression of OsGLN2, a Rice Endo-1,3-β-Glucanase Gene Regulated Developmentally in Flowers and Hormonally in Germinating Seeds. *Planta* **2004,** *220* (1), 129–139.

Allen, A.; Islamovic, E.; Kaur, J.; Gold, S.; Shah, D.; Smith, T. J. Transgenic Maize Plants Expressing the Totivirus Antifungal Protein, KP4, Are Highly Resistant to Corn Smut. *Plant Biotechnol. J.* **2011,** *9* (8), 857–864.

Almasia, N. I.; Bazzini, A. A.; Esteban Hopp, H.; Vazque-Rovere, C. Overexpression of Snakin-1 Gene Enhances Resistance to *Rhizoctonia solani* and *Erwinia carotovora* in Transgenic Potato Plants. *Mol. Plant Pathol.* **2008,** *9* (3), 329–338.

Altpeter, F.; Diaz, I.; McAuslane, H.; Gaddour, K.; Carbonero, P.; Vasil, I. K. Increased Insect Resistance in Transgenic Wheat Stably Expressing Trypsin Inhibitor CMe. *Mol. Breed.* **1999,** *5* (1), 53–63.

Ammara, U.; Mansoor, S.; Saeed, M.; Amin, I.; Briddon, R. W.; Al-Sadi, A. M. RNA Interference-Based Resistance in Transgenic Tomato Plants against Tomato Yellow Leaf Curl Virus-Oman (TYLCV-OM) and its Associated Betasatellite. *Virol. J.* **2015,** *12* (1), 38.

Araki, M.; Ishii, T. Towards Social Acceptance of Plant Breeding by Genome Editing. *Trends Plant Sci.* **2015,** *20* (3), 145–149.

Araki, M.; Nojima, K.; Ishii, T. Caution Required for Handling Genome Editing Technology. *Trends Biotechnol.* **2014,** *32* (5), 234–237.

Arlen, P. A.; Falconer, R.; Cherukumilli, S.; Cole, A.; Cole, A. M.; Oishi, K. K.; Daniell, H. Field Production and Functional Evaluation of Chloroplast-Derived Interferon-α2b. *Plant Biotechnol. J.* **2007,** *5* (4), 511–525.

Ashraf, M.; Foolad, M. R. V. Roles of Glycine Betaine and Proline in Improving Plant Abiotic Stress Resistance. *Environ. Exp. Bot.* **2007,** *59* (2), 206–216.

Ashraf, M.; Akram, N. A. Improving Salinity Tolerance of Plants through Conventional Breeding and Genetic Engineering: An Analytical Comparison. *Biotechnol. Adv.* **2009,** *27* (6), 744–752.

Asif, M. A.; Zafar, Y.; Iqbal, J.; Iqbal, M. M.; Rashid, U.; Ali, G. M.; Arif, A.; Nazir, F. Enhanced Expression of AtNHX1, in Transgenic Groundnut (*Arachis hypogaea* L.) Improves Salt and Drought Tolerence. *Mol. Biotechnol.* **2011,** *49* (3), 250–256.

Atkinson, H. J.; Lilley, C. J.; Urwin, P. E. Strategies for Transgenic Nematode Control in Developed and Developing World Crops. *Curr. Opin. Biotechnol.* **2012,** *23* (2), 251–256.

Azadi, H.; Ho, P. Genetically Modified and Organic Crops in Developing Countries: A Review of Options for Food Security. *Biotechnol. Adv.* **2010,** *28* (1), 160–168.

Babu, R. C.; Zhang, J.; Blum, A.; David Ho T-H.; Wu, R.; Nguyen, H. T. HVA1, a LEA Gene from Barley Confers Dehydration Tolerance in Transgenic Rice (*Oryza sativa* L.) via Cell Membrane Protection. *Plant Sci.* **2004,** *166* (4), 855–862.

Bai, J.-Y.; Zeng, L.; Hu, Y.-L.; Li, Y.-F.; Lin, Z.-P.; Shang, S.-C.; Shi, Y.-S. Expression and Characteristic of Synthetic Human Epidermal Growth Factor (hEGF) in Transgenic Tobacco Plants. *Biotechnol. Lett.* **2007,** *29* (12), 2007–2012.

Bajaj, S.; Puthigae, S.; Templeton, K.; Bryant, C.; Gill, G.; Lomba, P.; Zhang, H.; Altpeter, F.; Hanley, Z. Towards Engineering Drought Tolerance in Perennial Ryegrass Using its Own Genome. In 6th Canadian Plant Genomics Workshop, 2008; p 62.

Ballvora, A.; Ercolano, M. R.; Weiß, J.; Meksem, K.; Bormann, C. A.; Oberhagemann, P.; Salamini, F.; Gebhardt, C. The R1 Gene for Potato Resistance to Late Blight (*Phytophthora infestans*) Belongs to the Leucine Zipper/NBS/LRR Class of Plant Resistance Genes. *Plant J.* **2002,** *30* (3), 361–371.

Bao, G.; Zhuo, C.; Qian, C.; Xiao, T.; Guo, Z.; Lu, S. Co-expression of NCED and ALO Improves Vitamin C Level and Tolerance to Drought and Chilling in Transgenic Tobacco and Stylo Plants. *Plant Biotechnol. J.* **2016,** *14* (1), 206–214.

Barbosa, A. E. A. D.; Albuquerque, É. V. S.; Silva, M. C. M.; Souza, D. S. L.; Oliveira-Neto, O. B.; Valencia, A.; Rocha, T. L.; Grossi-de-Sa, M. F. α-Amylase Inhibitor-1 Gene from *Phaseolus vulgaris* Expressed in *Coffea arabica* Plants Inhibits α-Amylases from the Coffee Berry Borer Pest. *BMC Biotechnol.* **2010,** *10* (1), 44.

Barone, P.; Zhang, X.-H.; Widholm, J. M. Tobacco Plastid Transformation Using the Feedback-Insensitive Anthranilate Synthase [α]-Subunit of Tobacco (ASA2) as a New Selectable Marker. *J. Exp. Bot.* **2009,** *60* (11), 3195–3202.

Barta, A.; Sommergruber, K.; Thompson, D.; Hartmuth, K.; Matzke, M. A.; Matzke, A. J. M. The Expression of a Nopaline Synthase—Human Growth Hormone Chimaeric Gene in Transformed Tobacco and Sunflower Callus Tissue. *Plant Mol. Biol.* **1986,** *6* (5), 347–357.

Bartholomew, M. James Lind's Treatise of the Scurvy (1753). *Postgrad. Med. J.* **2002,** *78* (925), 695–696.

Batchvarova, R.; Nikolaeva, V.; Slavov, S.; Bossolova, S.; Valkov, V.; Atanassova, S.; Guelemerov, S.; Atanassov, A.; Anzai, H. Transgenic Tobacco Cultivars Resistant to *Pseudomonas syringae* pv. *tabaci*. *Theor. Appl. Genet.* **1998,** *97* (5–6), 986–989.

Batistič, O.; Kudla, J. Analysis of Calcium Signaling Pathways in Plants. *Biochim. Biophys. Acta (BBA)—Gen. Subj.* **2012,** *1820* (8), 1283–1293.

Bell, H. A.; Fitches, E. C.; Marris, G. C.; Bell, J.; Edwards, J. P.; Gatehouse, J. A.; Gatehouse, A. M. R. Transgenic GNA Expressing Potato Plants Augment the Beneficial Biocontrol of *Lacanobia oleracea* (Lepidoptera: Noctuidae) by the Parasitoid *Eulophus pennicornis* (Hymenoptera; Eulophidae). *Transgen. Res.* **2001,** *10* (1), 35–42.

Benekos, K.; Kissoudis, C.; Nianiou-Obeidat, I.; Labrou, N.; Madesis, P.; Kalamaki, M.; Makris, A.; Tsaftaris, A. Overexpression of a Specific Soybean GmGSTU4 Isoenzyme Improves Diphenyl Ether and Chloroacetanilide Herbicide Tolerance of Transgenic Tobacco Plants. *J. Biotechnol.* **2010,** *150* (1), 195–201.

Beracochea, V. C.; Almasia, N. I.; Peluffo, L.; Nahirñak, V.; Hopp, E. H.; Paniego, N.; Heinz, R. A.; Vazquez-Rovere, C.; Lia, V. V. Sunflower Germin-Like Protein HaGLP1 Promotes ROS Accumulation and Enhances Protection against Fungal Pathogens in Transgenic *Arabidopsis thaliana*. *Plant Cell Rep.* **2015,** *34* (10), 1717–1733.

Bhargava, A.; Osusky, M.; Hancock, R. E.; Forward, B. S.; Kay, W. W.; Misra, S. Antiviral Indolicidin Variant Peptides: Evaluation for Broad-Spectrum Disease Resistance in Transgenic *Nicotiana tabacum*. *Plant Sci.* **2007,** *172* (3), 515–523.

Bhatnagar-Mathur, P.; Rao, J. S.; Vadez, V.; Dumbala, S. R.; Rathore, A.; Yamaguchi-Shinozaki, K.; Sharma, K. K. Transgenic Peanut Overexpressing the DREB1A Transcription Factor Has Higher Yields under Drought Stress. *Mol. Breed.* **2014,** *33* (2), 327–340.

Bhatnagar-Mathur, P.; Jyostna Devi, M.; Vadez, V.; Sharma, K. K. Differential Antioxidative Responses in Transgenic Peanut Bear No Relationship to their Superior Transpiration Efficiency under Drought Stress. *J. Plant Physiol.* **2009,** *166* (11), 1207–1217.

Bhuiyan, M. S. U.; Min, S. R.; Jeong, W. J.; Sultana, S.; Choi, K. S.; Song, W. Y.; Lee, Y.; Lim, Y. P.; Liu, J. R. Overexpression of a Yeast Cadmium Factor 1 (YCF1) Enhances Heavy Metal Tolerance and Accumulation in *Brassica juncea*. *Plant Cell, Tissue Organ Cult.* **2011,** *105* (1), 85–91.

Bi, H.; Dong, X.; Wu, G.; Wang, M.; Ai, X. Decreased TK Activity Alters Growth, Yield and Tolerance to Low Temperature and Low Light Intensity in Transgenic Cucumber Plants. *Plant Cell Rep.* **2015,** *34* (2), 345–354.

Bicar, E. H.; Woodman-Clikeman, W.; Sangtong, V.; Peterson, J. M.; Yang, S. S.; Lee, M.; Scott, M. P. Transgenic Maize Endosperm Containing a Milk Protein Has Improved Amino Acid Balance. *Transgen. Res.* **2008,** *17* (1), 59–71.

Bita, C. E.; Zenoni, S.; Vriezen, W. H.; Mariani, C.; Pezzotti, M.; Gerats, T. Temperature Stress Differentially Modulates Transcription in Meiotic Anthers of Heat-Tolerant and Heat-Sensitive Tomato Plants. *BMC Genomics* **2011,** *12* (1), 384.

Bogers, W. M. J. M.; Bergmeier, L. A.; Ma, J.; Oostermeijer, H.; Wang, Y.; Kelly, C. G.; Ten Haaft, P.; Singh, M.; Heeney, J. L.; Lehner, T. A Novel HIV-CCR5 Receptor Vaccine Strategy in the Control of Mucosal SIV/HIV Infection. *AIDS* **2004,** *18* (1), 25–36.

Bolar, J. P.; Norelli, J. L.; Harman, G. E.; Brown, S. K.; Aldwinckle, H. S. Synergistic Activity of Endochitinase and Exochitinase from *Trichoderma atroviride* (*T. harzianum*) against the Pathogenic Fungus (*Venturia inaequalis*) in Transgenic Apple Plants. *Transgen. Res.* **2001,** *10* (6), 533–543.

Bradbury, L. M. T.; Gillies, S. A.; Brushett, D. J.; Waters, D. L. E.; Henry, R. J. Inactivation of an Aminoaldehyde Dehydrogenase Is Responsible for Fragrance in Rice. *Plant Mol. Biol.* **2008,** *68* (4–5), 439–449.

Brar, D. S.; Virk, P. S.; Jena, K. K.; Khush, G. S. Breeding for Resistance to Planthoppers in Rice. In: *Planthoppers: New Threats to the Sustainability of Intensive Rice Production Systems in Asia* 2009; pp 401–409.

Brar, H. K.; Bhattacharyya, M. K. Expression of a Single-Chain Variable-Fragment Antibody against a *Fusarium virguliforme* Toxin Peptide Enhances Tolerance to Sudden Death Syndrome in Transgenic Soybean Plants. *Mol. Plant–Microbe Interact.* **2012,** *25* (6), 817–824.

Bray, E. A.; Bailey-Serres, J.; Weretilnyk, E. Responses to Abiotic Stresses. *Biochem. Mol. Biol. Plants* **2000,** *1158*, e1203.

Brinch-Pedersen, H.; Hatzack, F.; Stöger, E.; Arcalis, E.; Pontopidan, K.; Holm, P. B. Heat-stable Phytases in Transgenic Wheat (*Triticum aestivum* L.): Deposition Pattern, Thermostability, and Phytate Hydrolysis. *J. Agric. Food Chem.* **2006,** *54* (13), 4624–4632.

Buhr, T.; Sato, S.; Ebrahim, F.; Xing, A.; Zhou, Y.; Mathiesen, M.; Schweiger, B.; Kinney, A.; Staswick, P. Ribozyme Termination of RNA Transcripts Down-Regulate Seed Fatty Acid Genes in Transgenic Soybean. *Plant J.* **2002,** *30* (2), 155–163.

Camacho, A.; Van Deynze, A.; Chi-Ham, C.; Bennett, A. B. Genetically Engineered Crops that Fly under the US Regulatory Radar. *Nat. Biotechnol.* **2014,** *32* (11), 1087–1091.

Canadian Biotechnology Action Network; Slater, A.; Holtslander, C. *Where in the World Are GM Crops and Foods?*, 2015.

Cao, D.; Hou, W.; Liu, W.; Yao, W.; Wu, C.; Liu, X.; Han, T. Overexpression of TaNHX2 Enhances Salt Tolerance of 'Composite' and Whole Transgenic Soybean Plants. *Plant Cell, Tissue Organ Cult.* **2011,** *107* (3), 541–552.

Carstens, M.; Vivier, M. A.; Pretorius, I. S. The *Saccharomyces cerevisiae* Chitinase, Encoded by the CTS1–2 Gene, Confers Antifungal Activity against *Botrytis cinerea* to Transgenic Tobacco. *Transgen. Res.* **2003,** *12* (4), 497–508.

Carter, III, J. E.; Langridge, W. H. R. Plant-Based Vaccines for Protection against Infectious and Autoimmune Diseases. *Crit. Rev. Plant Sci.* **2002,** *21* (2), 93–109.

Castle, L. A.; et al. Discovery and Directed Evolution of a Glyphosate Tolerance Gene. *Science* **2004,** *304* (5674), 1151–1154.

Chakraborty, S.; Chakraborty, N.; Datta, A. Increased Nutritive Value of Transgenic Potato by Expressing a Nonallergenic Seed Albumin Gene from *Amaranthus hypochondriacus*. *Proc. Nat. Acad. Sci.* **2000,** *97* (7), 3724–3729.

Champion, A.; Hébrard, E.; Parra, B.; Bournaud, C.; Marmey, P.; Tranchant, C.; Nicole, M. Molecular Diversity and Gene Expression of Cotton ERF Transcription Factors Reveal that Group IXa Members Are Responsive to Jasmonate, Ethylene and Xanthomonas. *Mol. Plant Pathol.* **2009,** *10* (4), 471–485.

Chan, Y.-L.; Yang, A.-H.; Chen, J.-T.; Yeh, K.-W.; Chan, M.-T. Heterologous Expression of Taro Cystatin Protects Transgenic Tomato against *Meloidogyne incognita* Infection by Means of Interfering Sex Determination and Suppressing Gall Formation. *Plant Cell Rep.* **2010,** *29* (3), 231–238.

Chang, M.-M.; Culley, D.; Choi, J. J.; Hadwiger, L. A. *Agrobacterium*-Mediated Co-transformation of a Pea β-1,3-Glucanase and Chitinase Genes in Potato (*Solanum tuberosum* L. cv. Russet Burbank) Using a Single Selectable Marker. *Plant Sci.* **2002,** *163* (1), 83–89.

Charity, J. A.; Hughes, P.; Anderson, M. A.; Bittisnich, D. J.; Whitecross, M.; Higgins, T. J. V. Pest and Disease Protection Conferred by Expression of Barley β-Hordothionin and *Nicotiana alata* Proteinase Inhibitor Genes in Transgenic Tobacco. *Funct. Plant Biol.* **2005,** *32* (1), 35–44.

Charlton, W. L.; Harel, H. Y. M.; Bakhetia, M.; Hibbard, J. K.; Atkinson, H. J.; McPherson, M. J. Additive Effects of Plant Expressed Double-Stranded RNAs on Root-Knot Nematode Development. *Int. J. Parasitol.* **2010,** *40* (7), 855–864.

Chatterjee, A.; Das, N. C.; Raha, S.; Babbit, R.; Huang, Q.; Zaitlin, D.; Maiti, I. B. Production of Xylanase in Transgenic Tobacco for Industrial Use in Bioenergy and Biofuel Applications. *In Vitro Cell. Dev. Biol.—Plant* **2010,** *46* (2), 198–209.

Chawla, R.; Shakya, R.; Rommens, C. M. Tuber-Specific Silencing of Asparagine Synthetase-1 Reduces the Acrylamide-Forming Potential of Potatoes Grown in the Field Without Affecting Tuber Shape and Yield. *Plant Biotechnol. J.* **2012,** *10* (8), 913–924.

Chen, D.-X.; Chen, X.-W.; Wang, Y.-P.; Zhu, L.-H.; Li, S.-G. Genetic Transformation of Rice with Pi-d2 Gene Enhances Resistance to Rice Blast Fungus *Magnaporthe oryzae*. *Rice Sci.* **2010,** *17* (1), 19–27.

Chen, F.; Dixon, R. A. Lignin Modification Improves Fermentable Sugar Yields for Biofuel Production. *Nat. Biotechnol.* **2007,** *25* (7), 759–761.

Chen, R.; Li, H.; Zhang, L.; Zhang, J.; Xiao, J.; Ye, Z. CaMi, a Root-Knot Nematode Resistance Gene from Hot Pepper (*Capsium annuum* L.) Confers Nematode Resistance in Tomato. *Plant Cell Rep.* **2007,** *26* (7), 895–905.

Chen, R.; Xue, G.; Chen, P.; Yao, B.; Yang, W.; Ma, Q.; Fan, Y.; Zhao, Z.; Tarczynski, M. C.; Shi, J. Transgenic Maize Plants Expressing a Fungal Phytase Gene. *Transgen. Res.* **2008,** *17* (4), 633–643.

Chen, S.; Yang, Y.; Shi, W.; Ji, Q.; He, F.; Zhang, Z.; Cheng, Z.; Liu, X.; Xu, M. Badh2, Encoding Betaine Aldehyde Dehydrogenase, Inhibits the Biosynthesis of 2-Acetyl-1-pyrroline, a Major Component in Rice Fragrance. *Plant Cell* **2008,** *20* (7), 1850–1861.

Chen, S-C.; Liu, A.-R.; Zou, Z.-R. Overexpression of Glucanase Gene and Defensin Gene in Transgenic Tomato Enhances Resistance to *Ralstonia solanacearum*. *Russ. J. Plant Physiol.* **2006,** *53* (5), 671–677.

Chen, Y.; Zhou, X.-R.; Zhang, Z.-J.; Dribnenki, P.; Singh, S.; Green, A. Development of High Oleic Oil Crop Platform in Flax through RNAi-Mediated Multiple FAD2 Gene Silencing. *Plant Cell Rep.* **2015,** *34* (4), 643–653.

Cheng, B.; Wu, G.; Vrinten, P.; Falk, K.; Bauer, J.; Qiu, X. Towards the Production of High Levels of Eicosapentaenoic Acid in Transgenic Plants: The Effects of Different Host Species, Genes and Promoters. *Transgen. Res.* **2010,** *19* (2), 221–229.

Cheng, Q.; Day, A.; Dowson-Day, M.; Shen, G.-F.; Dixon, R. The *Klebsiella pneumoniae* Nitrogenase Fe Protein Gene (nifH) Functionally Substitutes for the ChlL Gene in *Chlamydomonas reinhardtii*. *Biochem. Biophys. Res. Commun.* **2005,** *329* (3), 966–975.

Cheng, Z.; Targolli, J.; Huang, X.; Wu, R. Wheat LEA Genes, PMA80 and PMA1959, Enhance Dehydration Tolerance of Transgenic Rice (*Oryza sativa* L.). *Mol. Breed.* **2002,** *10* (1–2), 71–82.

Cheong, Y. H.; Kim, C. Y.; Chun, H. J.; Moon, B. C.; Park, H. C.; Kim, J. K.; Lee, S.-H.; Han, C.-D.; Lee, S. Y.; Cho, M. J. Molecular Cloning of a Soybean Class III β-1,3-Glucanase Gene That is Regulated both Developmentally and in Response to Pathogen Infection. *Plant Sci.* **2000,** *154* (1), 71–81.

Chikwamba, R.; McMurray, J.; Shou, H.; Frame, B.; Pegg, S. E.; Scott, P.; Mason, H.; Wang, K. Expression of a Synthetic *E. coli* Heat-Labile Enterotoxin B Sub-unit (LT-B) in Maize. *Mol. Breed.* **2005,** *10* (4), 253–265.

Cho, E. A.; Lee, C. A.; Kim, Y. S.; Baek, S. H.; de los Reyes, B. G.; Yun, S. J. Expression of Gamma-Tocopherol Methyltransferase Transgene Improves Tocopherol Composition in Lettuce (*Latuca sativa* L.). *Mol. Cells,* **2005,** *19* (1), 16.

Choi, J. Y.; Seo, Y. S.; Kim, S. J.; Kim, W. T.; Shin, J. S. Constitutive Expression of CaXTH3, a Hot Pepper Xyloglucan Endotransglucosylase/Hydrolase, Enhanced Tolerance to Salt and Drought Stresses Without Phenotypic Defects in Tomato Plants (*Solanum lycopersicum* cv. Dotaerang). *Plant Cell Rep.* **2011,** *30* (5), 867–877.

Chu, P.; Chen, H.; Zhou, Y.; Li, Y.; Ding, Y.; Jiang, L.; Tsang, E. W. T.; Wu, K.; Huang, S. Proteomic and Functional Analyses of *Nelumbo nucifera* Annexins Involved in Seed Thermotolerance and Germination Vigor. *Planta* **2012,** *235* (6), 1271–1288.

Chuck, G. S.; Tobias, C.; Sun, L.; Kraemer, F.; Li, C.; Dibble, D.; Arora, R.; et al. Overexpression of the Maize Corngrass1 MicroRNA Prevents Flowering, Improves Digestibility, and Increases Starch Content of Switchgrass. *Proc. Nat. Acad. Sci.* **2011,** *108* (42), 17550–17555.

Chye, M.-L.; Zhao, K.-J.; He, Z.-M.; Ramalingam, S.; Fung, K.-L. An Agglutinating Chitinase with Two Chitin-Binding Domains Confers Fungal Protection in Transgenic Potato. *Planta* **2005,** *220* (5), 717–730.

Clarke, J. L.; Daniell, H. Plastid Biotechnology for Crop Production: Present Status and Future Perspectives. *Plant Mol. Biol.* **2011,** *76* (3–5), 211–220.

Coates, C. J.; Kaminski, J. M.; Summers, J. B.; Segal, D. J.; Miller, A. D.; Kolb, A. F. Site-Directed Genome Modification: Derivatives of DNA-Modifying Enzymes as Targeting Tools. *Trends Biotechnol.* **2005,** *23* (8), 407–419.

Collinge, D. B.; Jørgensen, H. J. L.; Lund, O. S.; Lyngkjær, M. F. Engineering Pathogen Resistance in Crop Plants: Current Trends and Future Prospects. *Annu. Rev. Phytopathol.* **2010,** *48*, 269–291.

Connolly, E. L. Raising the Bar for Biofortification: Enhanced Levels of Bioavailable Calcium in Carrots. *Trends Biotechnol.* **2008,** *26* (8), 401–403.

Conrad, U. Polymers from Plants to Develop Biodegradable Plastics. *Trends Plant Sci.* **2005,** *10* (11), 511–512.

Cortina, C.; Culiáñez-Macià, F. A. Tomato Abiotic Stress Enhanced Tolerance by Trehalose Biosynthesis. *Plant Sci.* **2005,** *169* (1), 75–82.

Coutos-Thévenot, P.; Poinssot, B.; Bonomelli, A.; Yean, H.; Breda, C.; Buffard, D.; Esnault, R.; Hain, R.; Boulay, M. In Vitro Tolerance to *Botrytis cinerea* of Grapevine 41B Rootstock in Transgenic Plants Expressing the Stilbene Synthase Vst1 Gene under the Control of a Pathogen-Inducible PR 10 Promoter. *J. Exp. Bot.* **2001,** *52* (358), 901–910.

Cuellar, W.; Gaudin, A.; Solorzano, D.; Casas, A.; Nopo, L.; Chudalayandi, P.; Medrano, G.; Kreuze, J.; Ghislain, M. Self-Excision of the Antibiotic Resistance Gene nptII Using a Heat Inducible Cre-loxP System from Transgenic Potato. *Plant Mol. Biol.* **2006,** *62* (1–2), 71–82.

Cunha, W. G.; Tinoco, M. L. P.; Pancoti, H. L.; Ribeiro, R. E.; Aragão, F. J. L. High Resistance to *Sclerotinia sclerotiorum* in Transgenic Soybean Plants Transformed to Express an Oxalate Decarboxylase Gene. *Plant Pathol.* **2010,** *59* (4), 654–660.

D'Aoust, M.-A.; Couture, M. M.-J.; Charland, N.; Trépanier, S.; Landry, N.; Ors, F.; Vézina, L.-P. The Production of Hemagglutinin-Based Virus-Like Particles in Plants: A Rapid, Efficient and Safe Response to Pandemic Influenza. *Plant Biotechnol. J.* **2010,** *8* (5), 607–619.

Dafny-Yelin, M.; Tzfira, T. Delivery of Multiple Transgenes to Plant Cells. *Plant Physiol.* **2007,** *145* (4), 1118–1128.

Daniell, H.; Wiebe, P. O.; Millan, A. F.-S. Antibiotic-Free Chloroplast Genetic Engineering—An Environmentally Friendly Approach. *Trends Plant Sci.* **2001b,** *6* (6), 237–239.

Daniell, H.; Lee, S.-B.; Panchal, T.; Wiebe, P. O. Expression of the Native Cholera Toxin B Subunit Gene and Assembly as Functional Oligomers in Transgenic Tobacco Chloroplasts. *J. Mol. Biol.* **2001a,** *311* (5), 1001–1009.

Daniell, H.; Kumar, S.; Dufourmantel, N. Breakthrough in Chloroplast Genetic Engineering of Agronomically Important Crops. *Trends Biotechnol.* **2005,** *23* (5), 238–245.

Datta, K.; Tu, J.; Oliva, N.; Ona, I.; Velazhahan, R.; Mew, T. W.; Muthukrishnan, S.; Datta, S. K. Enhanced Resistance to Sheath Blight by Constitutive Expression of Infection-Related Rice Chitinase in Transgenic Elite Indica Rice Cultivars. *Plant Sci.* **2001,** *160* (3), 405–414.

De la Garza, R. I. D.; Gregory, 3rd, J. F.; Hanson, A. D. Folate Biofortification of Tomato Fruit. *Proc. Nat. Acad. Sci. USA* **2007,** *104* (10), 4218–4222.

De Marchis, F.; Wang, Y.; Stevanato, P.; Arcioni, S.; Bellucci, M. Genetic Transformation of the Sugar Beet Plastome. *Transgen. Res.* **2009,** *18* (1), 17–30.

De Neve, M.; Buck, S.; Jacobs, A.; Montagu, M.; Depicker, A. T-DNA Integration Patterns in Co-transformed Plant Cells Suggest that T-DNA Repeats Originate from Co-integration of Separate T-DNAs. *Plant J.* **1997,** *11* (1), 15–29.

De Ronde, J. A.; Cress, W. A.; Krüger, G. H. J.; Strasser, R. J.; Van Staden, J. Photosynthetic Response of Transgenic Soybean Plants, Containing an *Arabidopsis* P5CR Gene, During Heat and Drought Stress. *J. Plant Physiol.* **2004,** *161* (11), 1211–1224.

de Vetten, N.; Anne-Marie, W.; Raemakers, K.; van der Meer, I.; ter Stege, R.; Heeres, E.; Heeres, P.; Visser, R. A Transformation Method for Obtaining Marker-Free Plants of a Cross-Pollinating and Vegetatively Propagated Crop. *Nat. Biotechnol.* **2003,** *21* (4), 439–442.

De Wilde, C.; Peeters, K.; Jacobs, A.; Peck, I.; Depicker, A. Expression of Antibodies and Fab Fragments in Transgenic Potato Plants: A Case Study for Bulk Production in Crop Plants. *Mol. Breed.* **2002,** *9* (4), 271–282.

DeFrancesco, L. How Safe Does Transgenic Food Need To Be? *Nat. Biotechnol.* **2013,** *31* (9), 794–802.

DeGray, G.; Rajasekaran, K.; Smith, F.; Sanford, J.; Daniell, H. Expression of an Antimicrobial Peptide via the Chloroplast Genome to Control Phytopathogenic Bacteria and Fungi. *Plant Physiol.* **2001,** *127* (3), 852–862.

Dhekney, S. A.; Li, Z. T.; Gray, D. J. Grapevines Engineered to Express Cisgenic *Vitis vinifera* Thaumatin-Like Protein Exhibit Fungal Disease Resistance. *In Vitro Cell. Dev. Biol.—Plant* **2011,** *47* (4), 458–466.

Di, H.; Tian, Y.; Zu, H.; Meng, X.; Zeng, X.; Wang, Z. Enhanced Salinity Tolerance in Transgenic Maize Plants Expressing a BADH Gene from *Atriplex micrantha*. *Euphytica* **2015,** *206* (3), 775–783.

Dinh, P. T. Y.; Zhang, L.; Brown, C. R.; Elling, A. A. Plant-Mediated RNA Interference of Effector Gene Mc16D10L Confers Resistance against *Meloidogyne chitwoodi* in Diverse Genetic Backgrounds of Potato and Reduces Pathogenicity of Nematode Offspring. *Nematology* **2014,** *16* (6), 669–682.

Diretto, G.; Al-Babili, S.; Tavazza, R.; Papacchioli, V.; Beyer, P.; Giuliano, G. Metabolic Engineering of Potato Carotenoid Content through Tuber-Specific Overexpression of a Bacterial Mini-pathway. *PLoS ONE* **2007,** *2* (4), e350.

Dixit, P.; Mukherjee, P. K.; Ramachandran, V.; Eapen, S. Glutathione Transferase from *Trichoderma virens* Enhances Cadmium Tolerance without Enhancing its Accumulation in Transgenic *Nicotiana tabacum*. *PLoS ONE* **2011,** *6* (1), e16360.

Djoussé, L.; Biggs, M. L.; Lemaitre, R. N.; King, I. B.; Song, X.; Ix, J. H.; Mukamal, K. J.; Siscovick, D. S.; Mozaffarian, D. Plasma Omega-3 Fatty Acids and Incident Diabetes in Older Adults. *Am. J. Clin. Nutr.* **2011,** *94* (2), 527–533.

Duan, X.; Li, X.; Xue, Q.; Abo-El-Saad, M.; Xu, D.; Wu, R. Transgenic Rice Plants(2008) Plant triacylglycerols as feedstocks for the production of biofuels. Plant J 54: 593–607.

Dufourmantel, N.; Pelissier, B.; Garcon, F.; Peltier, G.; Ferullo, J.-M.; Tissot, G. Generation of Fertile Transplastomic Soybean. *Plant Mol. Biol.* **2004,** *55* (4), 479–489.

Dufourmantel, N.; Dubald, M.; Matringe, M.; Canard, H.; Garcon, F.; Job, C.; Kay, E.; et al. Generation and Characterization of Soybean and Marker-Free Tobacco Plastid Transformants Over-expressing a Bacterial 4-Hydroxyphenylpyruvate Dioxygenase Which Provides Strong Herbicide Tolerance. *Plant Biotechnol. J.* **2007,** *5* (1), 118–133.

Dunse, K. M.; Stevens, J. A.; Lay, F. T.; Gaspar, Y. M.; Heath, R. L.; Anderson, M. A. Coexpression of Potato Type I and II Proteinase Inhibitors Gives Cotton Plants Protection against Insect Damage in the Field. *Proc. Nat. Acad. Sci.* **2010,** *107* (34), 15011–15015.

Düring, K.; Hippe, S.; Kreuzaler, F.; Schell, J. Synthesis and Self-assembly of a Functional Monoclonal Antibody in Transgenic *Nicotiana tabacum*. *Plant Mol. Biol.* **1990,** *15* (2), 281–293.

Dyer, J. M.; Stymne, S.; Green, A. G.; Carlsson, A. S. High-Value Oils from Plants. *Plant J.* **2008,** *54* (4), 640–655.

Ebinuma, H.; Komamine, A. MAT (Multi-Auto-Transformation) Vector System. The Oncogenes of *Agrobacterium* as Positive Markers for Regeneration and Selection of Marker-Free Transgenic Plants. *In Vitro Cell. Dev. Biol.—Plant* **2001,** *37* (2), 103–113.

Ebinuma, H.; Sugita, K.; Matsunaga, E.; Yamakado, M. Selection of Marker-Free Transgenic Plants Using the Isopentenyl Transferase Gene. *Proc. Nat. Acad. Sci.* **1997,** *94* (6), 2117–2121.

Elías-López, A. L.; Marquina, B.; Gutiérrez-Ortega, A.; Aguilar, D.; Gomez-Lim, M.; Hernández-Pando, R. Transgenic Tomato Expressing Interleukin-12 Has a Therapeutic Effect in a Murine Model of Progressive Pulmonary Tuberculosis. *Clin. Exp. Immunol.* **2008,** *154* (1), 123–133.

Enfissi, E.; Fraser, P. D.; Lois, L.-M.; Boronat, A.; Schuch, W.; Bramley, P. M. Metabolic Engineering of the Mevalonate and Non-mevalonate Isopentenyl Diphosphate-Forming Pathways for the Production of Health-Promoting Isoprenoids in Tomato. *Plant Biotechnol. J.* **2005,** *3* (1), 17–27.

Ernst, K.; Kumar, A.; Kriseleit, D.; Kloos, D.-U.; Phillips, M. S.; Ganal, M. W. The Broad-Spectrum Potato Cyst Nematode Resistance Gene (Hero) from Tomato is the Only Member of a Large Gene Family of NBS-LRR Genes with an Unusual Amino Acid Repeat in the LRR Region. *Plant J.* **2002,** *31* (2), 127–136.

Fan, W.; Zhang, M.; Zhang, H.; Zhang, P. Improved Tolerance to Various Abiotic Stresses in Transgenic Sweet Potato (*Ipomoea batatas*) Expressing Spinach Betaine Aldehyde Dehydrogenase. *PLoS ONE* **2012,** *7* (5), e37344.

Fairbairn, D. J.; Cavallaro, A. S.; Bernard, M.; Mahalinga-Iyer, J.; Graham, M. W.; Botella, J. R. Host-delivered RNAi: an effective strategy to silence genes in plant parasitic nematodes. Planta, **2007**, 226, 1525–1533.

Farre, G.; Twyman, R. M.; Zhu, C.; Capell, T.; Christou, P. Nutritionally Enhanced Crops and Food Security: Scientific Achievements versus Political Expediency. *Curr. Opin. Biotechnol.* **2011,** *22* (2), 245–251.

Fillatti, J. J.; Kiser, J.; Rose, R.; Comai, L. Efficient Transfer of a Glyphosate Tolerance Gene into Tomato Using a Binary *Agrobacterium tumefaciens* Vector. *Nat. Biotechnol.* **1987,** *5* (7), 726–730.

Fischer, R.; Stoger, E.; Schillberg, S.; Christou, P.; Twyman, R. M. Plant-Based Production of Biopharmaceuticals. *Curr. Opin. Plant Biol.* **2004,** *7* (2), 152–158.

Fitches, E. C.; Bell, H. A.; Powell, M. E.; Back, E.; Sargiotti, C.; Weaver, R. J.; Gatehouse, J. A. Insecticidal Activity of Scorpion Toxin (ButaIT) and Snowdrop Lectin (GNA) Containing Fusion Proteins towards Pest Species of Different Orders. *Pest Manage. Sci.* **2010,** *66* (1), 74–83.

Flachowsky, H.; Szankowski, I.; Fischer, T. C.; Richter, K.; Peil, A.; Höfer, M.; Dörschel, C.; et al. Transgenic Apple Plants Overexpressing the Lc Gene of Maize Show an Altered Growth Habit and Increased Resistance to Apple Scab and Fire Blight. *Planta* **2010,** *231* (3), 623–635.

Food and Agriculture Organisation (FAO) of United Nations. *Declaration of the World Summit on Food Security*, Rome, 16–18 November 2009.

Frizzi, A.; Huang, S.; Gilbertson, L. A.; Armstrong, T. A.; Luethy, M. H.; Malvar, T. M. Modifying Lysine Biosynthesis and Catabolism in Corn with a Single Bifunctional Expression/Silencing Transgene Cassette. *Plant Biotechnol. J.* **2008,** *6* (1), 13–21.

Fujisawa, M.; Takita, E.; Harada, H.; Sakurai, N.; Suzuki, H.; Ohyama, K.; Shibata, D.; Misawa, N. Pathway Engineering of *Brassica napus* Seeds Using Multiple Key Enzyme Genes Involved in Ketocarotenoid Formation. *J. Exp. Bot.* **2009,** *60* (4), 1319–1332.

Fujiwara, Y.; Aiki, Y.; Yang, L.; Takaiwa, F.; Kosaka, A.; Tsuji, N. M.; Shiraki, K.; Sekikawa, K. Extraction and Purification of Human Interleukin-10 from Transgenic Rice Seeds. *Protein Express. Purif.* **2010,** *72* (1), 125–130.

Fuller, V. L.; Lilley, C. J.; Urwin, P. E. Nematode Resistance. *New Phytol.* **2008,** *180* (1), 27–44.

Gadaleta, A.; Giancaspro, A.; Blechl, A. E.; Blanco, A. A Transgenic Durum Wheat Line That Is Free of Marker Genes and Expresses 1Dy10. *J. Cereal Sci.* **2008,** *48* (2), 439–445.

Gahan, L. J.; Pauchet, Y.; Vogel, H.; Heckel, D. G. An ABC Transporter Mutation is Correlated with Insect Resistance to *Bacillus thuringiensis* Cry1Ac Toxin. **2010,** *6* (12), e1001248.

Gaines, T. A.; Zhang, W.; Wang, D.; Bukun, B.; Chisholm, S. T.; Shaner, D. L.; Nissen, S. J.; et al. Gene Amplification Confers Glyphosate Resistance in *Amaranthus palmeri*. *Proc. Nat. Acad. Sci.* **2010,** *107* (3), 1029–1034.

Gao, C. Genome Editing in Crops: From Bench to Field. *Natl. Sci. Rev.* **2015,** *2* (1), 13–15.

Gao, N.; Su, Y.; Min, J.; Shen, W.; Shi, W. Transgenic Tomato Overexpressing ath-miR399d Has Enhanced Phosphorus Accumulation through Increased Acid Phosphatase and Proton Secretion as well as Phosphate Transporters. *Plant Soil* **2010,** *334* (1–2), 123–136.

Gao, S.; Yu, B.; Yuan, L.; Zhai, H.; He, S.-Z.; Liu, Q.-C. Production of Transgenic Sweetpotato Plants Resistant to Stem Nematodes Using Oryzacystatin-I Gene. *Sci. Hortic.* **2011a,** *128* (4), 408–414.

Gao, S.; Yuan, L.; Zhai, H.; Liu, C.; He, S.; Liu, Q. Transgenic Sweetpotato Plants Expressing an LOS5 Gene Are Tolerant to Salt Stress. *Plant Cell, Tissue Organ Cult.* **2011b,** *107* (2), 205–213.

Gao, S.-Q.; Chen, M.; Xu, Z.-S.; Zhao, C.-P.; Li, L.; Xu, H.-J.; Tang, Y.-M.; Zhao, X.; Ma, Y.-Z. The Soybean GmbZIP1 Transcription Factor Enhances Multiple Abiotic Stress Tolerances in Transgenic Plants. *Plant Mol. Biol.* **2011c,** *75* (6), 537–553.

Gao, T.; Wu, Y.; Zhang, Y.; Liu, L.; Ning, Y.; Wang, D.; Tong, H.; Chen, S.; Chu, C.; Xie, Q. OsSDIR1 Overexpression Greatly Improves Drought Tolerance in Transgenic Rice. *Plant Mol. Biol.* **2011d,** *76* (1–2), 145–156.

Gao, X.; Zhou, J.; Li, J.; Zou, X.; Zhao, J.; Li, Q.; Xia, R.; et al. Efficient Generation of Marker-Free Transgenic Rice Plants Using an Improved Transposon-Mediated Transgene Reintegration Strategy. *Plant Physiol.* **2015,** *167* (1), 11–24.

García-Casal, M. N. Carotenoids Increase Iron Absorption from Cereal-Based Food in the Human. *Nutr. Res.* **2006,** *26* (7), 340–344.

Garg, A. K.; Kim, J.-K.; Owens, T. G.; Ranwala, A. P.; Choi, Y. D.; Kochian, L. V.; Wu, R. J. Trehalose Accumulation in Rice Plants Confers High Tolerance Levels to Different Abiotic Stresses. *Proc. Nat. Acad. Sci.* **2002,** *99* (25), 15898–15903.

Ge, L.-F.; Chao, D.-Y.; Shi, M.; Zhu, M.-Z.; Gao, J.-P.; Lin, H.-X. Overexpression of the Trehalose-6-Phosphate Phosphatase Gene OsTPP1 Confers Stress Tolerance in Rice and Results in the Activation of Stress Responsive Genes. *Planta* **2008,** *228* (1), 191–201.

Gibson, R. S. The Role of Diet- and Host-Related Factors in Nutrient Bioavailability and Thus in Nutrient-Based Dietary Requirement Estimates. *Food Nutr. Bull.* **2007,** *28* (Suppl. 1), 77S–100S.

Giorgi, C.; Franconi, R.; Rybicki, E. P. Human Papillomavirus Vaccines in Plants. **2010,** *9* (8), 913–924.

Girhepuje, P. V.; Shinde, G. B. Transgenic Tomato Plants Expressing a Wheat Endochitinase Gene Demonstrate Enhanced Resistance to *Fusarium oxysporum* f. sp. *lycopersici*. *Plant Cell, Tissue Organ Cult.* **2011,** *105* (2), 243–251.

Gleave, A. P.; Mitra, D. S.; Mudge, S. R.; Morris, B. A. M. Selectable Marker-Free Transgenic Plants Without Sexual Crossing: Transient Expression of Cre Recombinase and Use of a Conditional Lethal Dominant Gene. *Plant Mol. Biol.* **1999,** *40* (2), 223–235.

Goel, D.; Singh, A. K.; Yadav, V.; Babbar, S. B.; Bansal, K. C. Overexpression of Osmotin Gene Confers Tolerance to Salt and Drought Stresses in Transgenic Tomato (*Solanum lycopersicum* L.). *Protoplasma* **2010,** *245* (1–4), 133–141.

Goggin, F. L.; Jia, L.; Shah, G.; Hebert, S.; Williamson, V. M.; Ullman, D. E. Heterologous Expression of the Mi-1.2 Gene from Tomato Confers Resistance against Nematodes But Not Aphids in Eggplant. *Mol. Plant–Microbe Interact.* **2006,** *19* (4), 383–388.

Gómez-Galera, S.; Sudhakar, D.; Pelacho, A. M.; Capell, T.; Christou, P. Constitutive Expression of a Barley Fe Phytosiderophore Transporter Increases Alkaline Soil Tolerance and Results in Iron Partitioning between Vegetative and Storage Tissues under Stress. *Plant Physiol. Biochem.* **2012,** *53*, 46–53.

Gómez-Galera, S.; Rojas, E.; Sudhakar, D.; Zhu, C.; Pelacho, A. M.; Capell, T.; Christou, P. Critical Evaluation of Strategies for Mineral Fortification of Staple Food Crops. *Transgen. Res.* **2010,** *19* (2), 165–180.

Goodman, R. E.; Tetteh, A. O. Suggested Improvements for the Allergenicity Assessment of Genetically Modified Plants Used in Foods. *Curr. Allergy Asthma Rep.* **2011,** *11* (4), 317–324.

Gosal, S. S.; Wani, S. H.; Kang, M. S. Biotechnology and Drought Tolerance. *J. Crop Improv.* **2009,** *23* (1), 19–54.

Gouiaa, S.; Khoudi, H.; Leidi, E. O.; Pardo, J. M.; Masmoudi, K. Expression of Wheat Na^+/H^+ Antiporter TNHXS1 and H^+-Pyrophosphatase TVP1 Genes in Tobacco from a Bicistronic Transcriptional Unit Improve Salt Tolerance. *Plant Mol. Biol.* **2012,** *79* (1–2), 137–155.

Grabber, J. H.; Hatfield, R. D.; Lu, F.; Ralph, J. Coniferyl Ferulate Incorporation into Lignin Enhances the Alkaline Delignification and Enzymatic Degradation of Cell Walls. *Biomacromolecules* **2008,** *9* (9), 2510–2516.

Grennan, A. K. Ethylene Response Factors in Jasmonate Signaling and Defense Response. *Plant Physiol.* **2008,** *146* (4), 1457–1458.

Gressel, J.; Levy, A. A. Use of Multicopy Transposons Bearing Unfitness Genes in Weed Control: Four Example Scenarios. *Plant Physiol.* **2014,** *166* (3), 1221–1231.

Grover, A.; Mittal, D.; Negi, M.; Lavania, D. Generating High Temperature Tolerant Transgenic Plants: Achievements and Challenges. *Plant Sci.* **2013,** *205*, 38–47.

Guan, Q.; Lu, X.; Zeng, H.; Zhang, Y.; Zhu, J. Heat Stress Induction of miR398 Triggers a Regulatory Loop that Is Critical for Thermotolerance in *Arabidopsis*. *Plant J.* **2013,** *74* (5), 840–851.

Gudmundsson, G. H.; Agerberth, B.; Odeberg, J.; Bergman, T.; Olsson, B.; Salcedo, R. The Human Gene FALL39 and Processing of the Cathelin Precursor to the Antibacterial Peptide LL-37 in Granulocytes. *Eur. J. Biochem.* **1996,** *238* (2), 325–332.

Guo, W.-C.; Wang, Z.-A.; Luo, X.-L.; Jin, X.; Chang, J.; He, J.; Tu, E.-X.; Tian, Y.-C.; Si, H.-J.; Wu, J.-H. Development of Selectable Marker-Free Transgenic Potato Plants Expressing cry3A against the Colorado Potato Beetle (*Leptinotarsa decemlineata* Say). *Pest Manage. Sci.* **2015,** *72* (3), 497–504.

Guo, X.-H.; Jiang, J.; Lin, S.-J.; Wang, B.-C.; Wang, Y.-C.; Liu, G.-F.; Yang, C.-P. A ThCAP Gene from *Tamarix hispida* Confers Cold Tolerance in Transgenic *Populus* (*P. davidiana* × *P. bolleana*). *Biotechnol. Lett.* **2009,** *31* (7), 1079–1087.

Guo, Y,; Feng, Y.; Ge, Y.; Tetreau, G.; Chen, X.; Dong, X.; Shi, W. The Cultivation of Bt Corn Producing Cry1Ac Toxins Does Not Adversely Affect Non-Target Arthropods. **2014,** *PLoS ONE* 9(12), e114228.

Gururani, M. A.; Ganesan, M.; Song, I.-J.; Han, Y.; Kim, J.-I.; Lee, H.-Y.; Song, P.-S. Transgenic Turfgrasses Expressing Hyperactive Ser599Ala Phytochrome A Mutant Exhibit Abiotic Stress Tolerance. *J. Plant Growth Regul.* **2015,** *35*, 1–11.

Hallwass, M.; Oliveira, A. S.; Dianese, E. C.; Lohuis, D.; Boiteux, L. S.; Inoue-Nagata, A. K.; Resende, R. O.; Kormelink, R. The Tomato Spotted Wilt Virus Cell-to-Cell Movement Protein (NSM) Triggers a Hypersensitive Response in Sw-5-Containing Resistant Tomato Lines and in *Nicotiana benthamiana* Transformed with the Functional Sw-5b Resistance Gene Copy. *Mol. Plant Pathol.* **2014,** *15* (9), 871–880.

Halpin, C. Gene Stacking in Transgenic Plants—The Challenge for 21st Century Plant Biotechnology. *Plant Biotechnol. J.* **2005,** *3* (2), 141–155.

Hambidge, M. K.; Krebs, N. F. Zinc Deficiency: A Special Challenge. *J. Nutr.* **2007,** *137* (4), 1101–1105.

Han, Y.; Chen, Y.; Yin, S.; Zhang, M.; Wang, W. Over-expression of TaEXPB23, a Wheat Expansin Gene, Improves Oxidative Stress Tolerance in Transgenic Tobacco Plants. *J. Plant Physiol.* **2015,** *173*, 62–71.

Harrison, E. H. Mechanisms of Digestion and Absorption of Dietary Vitamin A. *Annu. Rev. Nutr.* **2005,** *25*, 87–103.

Hartl, M.; Giri, A. P.; Kaur, H.; Baldwin, I. T. Serine Protease Inhibitors Specifically Defend *Solanum nigrum* against Generalist Herbivores But Do Not Influence Plant Growth and Development. *Plant Cell* **2010,** *22* (12), 4158–4175.

Hartung, F.; Schiemann, J. Precise Plant Breeding Using New Genome Editing Techniques: Opportunities, Safety and Regulation in the EU. *Plant J.* **2014,** *78* (5), 742–752.

Harvey, A. J.; Speksnijder, G.; Baugh, L. R.; Morris, J. A.; Ivarie, R. Expression of Exogenous Protein in the Egg White of Transgenic Chickens. *Nat. Biotechnol.* **2002,** *20* (4), 396–399.

Hassan, F.; Meens, J.; Jacobsen, H.-J.; Kiesecker, H. A Family 19 Chitinase (Chit30) from *Streptomyces olivaceoviridis* ATCC 11238 Expressed in Transgenic Pea Affects the Development of *T. harzianum* In Vitro. *J. Biotechnol.* **2009,** *143* (4), 302–308.

Haverkort, A. J.; Struik, P. C.; Visser, R. G. F.; Jacobsen, E. Applied Biotechnology to Combat Late Blight in Potato Caused by *Phytophthora infestans*. *Potato Res.* **2009,** *52* (3), 249–264.

He, C.; He, Y.; Liu, Q.; Liu, T.; Liu, C.; Wang, L.; Zhang, J. Co-expression of Genes ApGSMT2 and ApDMT2 for Glycinebetaine Synthesis in Maize Enhances the Drought Tolerance of Plants. *Mol. Breed.* **2013,** *31* (3), 559–573.

He, X.; Miyasaka, S. C.; Fitch, M. M. M.; Moore, P. H.; Zhu, Y. J. *Agrobacterium tumefaciens*-Mediated Transformation of Taro (*Colocasia esculenta* (L.) Schott) with a Rice Chitinase Gene for Improved Tolerance to a Fungal Pathogen *Sclerotium rolfsii*. *Plant Cell Rep.* **2008a,** *27* (5), 903–909.

He, Z.-M.; Jiang, X.-L.; Qi, Y.; Luo, D.-Q. Assessment of the Utility of the Tomato Fruit-Specific E8 Promoter for Driving Vaccine Antigen Expression. *Genetica* **2008b,** *133* (2), 207–214.

Hiatt, A.; Caffferkey, R.; Bowdish, K. Production of Antibodies in Transgenic Plants. *Nature* **1989,** *342* (6245), 76–78.

Hiei, Y.; Ohta, S.; Komari, T.; Kumashiro, T. Efficient Transformation of Rice (*Oryza sativa* L.) Mediated by *Agrobacterium* and Sequence Analysis of the Boundaries of the T-DNA. *Plant J.* **1994,** *6* (2), 271–282.

Hoddinott, J.; Maluccio, J. A.; Behrman, J. R.; Flores, R.; Martorell, R. Effect of a Nutrition Intervention during Early Childhood on Economic Productivity in Guatemalan Adults. *Lancet* **2008,** *371* (9610), 411–416.

Hoffmann, M.; Wagner, M.; Abbadi, A.; Fulda, M.; Feussner, I. Metabolic Engineering of ω3-Very Long Chain Polyunsaturated Fatty Acid Production by an Exclusively acyl-CoA-Dependent Pathway. *J. Biol. Chem.* **2008,** *283* (33), 22352–22362.

Holme, I. B.; Dionisio, G.; Brinch-Pedersen, H.; Wendt, T.; Madsen, C. K.; Vincze, E.; Holm, P. B. Cisgenic Barley with Improved Phytase Activity. *Plant Biotechnol. J.* **2012,** *10* (2), 237–247.

Holme, I. B.; Wendt, T.; Holm, P.B. Intragenesis and cisgenesis as alternatives to transgenic crop development. *Plant Biotechnol J*, **2013,** *11*, 395-407.

Holmström, K.-O.; Somersalo, S.; Mandal, A.; Palva, T. E.; Welin, B. Improved Tolerance to Salinity and Low Temperature in Transgenic Tobacco Producing Glycine Betaine. *J. Exp. Bot.* **2000,** *51* (343), 177–185.

Hong, H.; Datla, N.; Reed, D. W.; Covello, P. S.; MacKenzie, S. L.; Qiu, X. High-Level Production of γ-Linolenic Acid in *Brassica juncea* Using a Δ6 Desaturase from *Pythium irregulare*. *Plant Physiol.* **2002,** *129* (1), 354–362.

Hood, E. E.; Witcher, D. R.; Maddock, S.; Meyer, T.; Baszczynski, C.; Bailey, M.; Flynn, P.; et al. Commercial Production of Avidin from Transgenic Maize: Characterization of Transformant, Production, Processing, Extraction and Purification. *Mol. Breed.* **1997,** *3* (4), 291–306.

Horsch, R. B.; Fry, J. E.; Hoffmann, N. L.; Eichholtz, D.; Rogers, S. G.; Fraley, R. T. A Simple and General Method for Transferring Genes into Plants. *Science* **1985,** *227*, 1229–1231.

Horvath, D. M.; Stall, R. E.; Jones, J. B.; Pauly, M. H.; Vallad, G. E.; Dahlbeck, D.; Staskawicz, B. J.; Scott, J. W. Transgenic Resistance Confers Effective Field Level Control of Bacterial Spot Disease in Tomato. *PLoS ONE* **2012,** *7* (8), e42036.

Hossain, M. A.; Maiti, M. K.; Basu, A.; Sen, S.; Ghosh, A. K.; Sen, S. K. Transgenic Expression of Onion Leaf Lectin Gene in Indian Mustard Offers Protection against Aphid Colonization. *Crop Sci.* **2006,** *46* (5), 2022–2032.

Hou, B.-K.; Zhou, Y.-H.; Wan, L.-H.; Zhang, Z.-L.; Shen, G.-F.; Chen, Z.-H.; Hu, Z.-M. Chloroplast Transformation in Oilseed Rape. *Transgen. Res.* **2003,** *12* (1), 111–114.

Huang, F-C.; Klaus, S.; Herz, S.; Zou, Z.; Koop, H.-U.; Golds, T. Efficient Plastid Transformation in Tobacco Using the aphA-6 Gene and Kanamycin Selection. *Mol. Genet. Genomics* **2002,** *268* (1), 19–27.

Huang, N.; Bethell, D.; Card, C.; Cornish, J.; Marchbank, T.; Wyatt, D.; Mabery, K.; Playford, R. Bioactive Recombinant Human Lactoferrin, Derived from Rice, Stimulates Mammalian Cell Growth. *In Vitro Cell. Dev. Biol.—Animal* **2008,** *44* (10), 464–471.

Ibrahim, H. M. M.; Alkharouf, N. W.; Meyer, S. L. F.; Aly, M. A. M.; El Kader, Y. A.; Hussein, E. H. A.; Matthews, B. F. Post-transcriptional Gene Silencing of Root-Knot Nematode in Transformed Soybean Roots. *Exp. Parasitol.* **2011,** *127* (1), 90–99.

Ignacimuthu, S.; Ceasar, A. S. Development of Transgenic Finger Millet (*Eleusine coracana* (L.) Gaertn.) Resistant to Leaf Blast Disease. *J. Biosci.* **2012,** *37* (1), 135–147.

Ignacimuthu, S.; Prakash, S. *Agrobacterium*-Mediated Transformation of Chickpea with α-Amylase Inhibitor Gene for Insect Resistance. *J. Biosci.* **2006,** *31* (3), 339–345.

Ishimoto, M.; Rahman, S. M.; Hanafy, M. S.; Khalafalla, M. M.; El-Shemy, H. A.; Nakamoto, Y.; Kita, Y.; et al. Evaluation of Amino Acid Content and Nutritional Quality of Transgenic Soybean Seeds with High-Level Tryptophan Accumulation. *Mol. Breed.* **2010,** *25* (2), 313–326.

Islam, M. M.; Khalekuzzaman, M. Development of Transgenic Rice (*Oryza sativa* L.) Plant Using Cadmium Tolerance Gene (YCFI) through *Agrobacterium*-Mediated Transformation for Phytoremediation. *Asian J. Agric. Res.* **2015,** *9* (4), 139–154.

Ismaili, A.; Jalali-Javaran, M.; Rasaee, M. J.; Rahbarizadeh, F.; Forouzandeh-Moghadam, M.; Rajabi Memari, H. Production and Characterization of Anti- (Mucin MUC1) Single-Domain Antibody in Tobacco (*Nicotiana tabacum* cultivar Xanthi). *Biotechnol. Appl. Biochem.* **2007,** *47* (1), 11–19.

Iwai, T.; Kaku, H.; Honkura, R.; Nakamura, S.; Ochiai, H.; Sasaki, T.; Ohashi, Y. Enhanced Resistance to Seed-Transmitted Bacterial Diseases in Transgenic Rice Plants Overproducing an Oat Cell-Wall-Bound Thionin. *Mol. Plant–Microbe Interact.* **2002,** *15* (6), 515–521.

Jabeen, R.; Khan, M. S.; Zafar, Y.; Anjum, T. Codon Optimization of Cry1Ab Gene for Hyper Expression in Plant Organelles. *Mol. Biol. Rep.* **2010,** *37* (2), 1011–1017.

Jacobsen, S.-E.; Sørensen, M.; Marcus Pedersen, S.; Weiner, J. Feeding the World: Genetically Modified Crops versus Agricultural Biodiversity. *Agron. Sustain. Dev.* **2013,** *33* (4), 651–662.

Jain, A. K.; Nessler, C. L. Metabolic Engineering of an Alternative Pathway for Ascorbic Acid Biosynthesis in Plants. *Mol. Breed.* **2000,** *6* (1), 73–78.

James, R. A.; Blake, C.; Zwart, A. B.; Hare, R. A.; Rathjen, A. J.; Munns, R. Impact of Ancestral Wheat Sodium Exclusion Genes Nax1 and Nax2 on Grain Yield of Durum Wheat on Saline Soils. *Funct. Plant Biol.* **2012,** *39* (7), 609–618.

Jan, P.-S.; Huang, H.-Y.; Chen, H.-M. Expression of a Synthesized Gene Encoding Cationic Peptide Cecropin B in Transgenic Tomato Plants Protects against Bacterial Diseases. *Appl. Environ. Microbiol.* **2010,** *76* (3), 769–775.

Jang, I.-C.; Oh, S.-J.; Seo, J.-S.; Choi, W.-B.; Song, S. I.; Kim, C. H.; Kim, Y. S.; et al. Expression of a Bifunctional Fusion of the *Escherichia coli* Genes for Trehalose-6-Phosphate Synthase and Trehalose-6-Phosphate Phosphatase in Transgenic Rice Plants Increases Trehalose Accumulation and Abiotic Stress Tolerance without Stunting Growth. *Plant Physiol.* **2003,** *131* (2), 516–524.

Jeong, J.; Guerinot, M. L. Biofortified and Bioavailable: The Gold Standard for Plant-Based Diets. *Proc. Nat. Acad. Sci.* **2008,** *105* (6), 1777–1778.

Jeong, J. S.; Kim, Y. S.; Baek, K. H.; Jung, H.; Ha, S.-H.; Choi, Y. D.; Kim, M.; Reuzeau, C.; Kim, J.-K. Root-Specific Expression of OsNAC10 Improves Drought Tolerance and Grain Yield in Rice under Field Drought Conditions. *Plant Physiol.* **2010,** *153* (1), 185–197.

Jha, S.; Chattoo, B. B. Expression of a Plant Defensin in Rice Confers Resistance to Fungal Phytopathogens. *Transgen. Res.* **2010,** *19* (3), 373–384.

Jha, S.; Chattoo, B. B. Transgene Stacking and Coordinated Expression of Plant Defensins Confer Fungal Resistance in Rice. *Rice*, **2009,** *2* (4), 143–154.

Jha, S.; Tank, H. G.; Prasad, B. D.; Chattoo, B. B. Expression of Dm-AMP1 in Rice Confers Resistance to *Magnaporthe oryzae* and *Rhizoctonia solani*. *Transgen. Res.* **2009,** *18* (1), 59–69.

Ji, W.; Zhu, Y.; Li, Y.; Yang, L.; Zhao, X.; Cai, H.; Bai, X. Over-expression of a Glutathione S-Transferase Gene, GsGST, from Wild Soybean (*Glycine soja*) Enhances Drought and Salt Tolerance in Transgenic Tobacco. *Biotechnol. Lett.* **2010,** *32* (8), 1173–1179.

Jia, H.; Ren, H.; Gu, M.; Zhao, J.; Sun, S.; Zhang, X.; Chen, J.; Wu, P.; Xu, G. The Phosphate Transporter Gene OsPht1;8 Is Involved in Phosphate Homeostasis in Rice. *Plant Physiol.* **2011,** *156* (3), 1164–1175.

Jiang, W.; Zhou, H.; Bi, H.; Fromm, M.; Yang, B.; Weeks, D. P. Demonstration of CRISPR/Cas9/sgRNA-Mediated Targeted Gene Modification in *Arabidopsis*, Tobacco, Sorghum and Rice. *Nucleic Acids Res.* **2013,** *41* (20), e188.

Jiao, Y.; Wang, Y.; Xue, D.; Wang, J.; Yan, M.; Liu, G.;, Dong, G.; Zeng, D.; Lu, Z.; Zhu, X.; Qian, Q.; Li, J. Regulation of OsSPL14 by OsmiR156 defines ideal plant architecture in rice. *Nat Genet.* **2010,** *42* (6), 541–544.

Jin, T.; Chang, Q.; Li, W.; Yin, D.; Li, Z.; Wang, D.; Liu, B.; Liu, L. Stress-Inducible Expression of GmDREB1 Conferred Salt Tolerance in Transgenic Alfalfa. *Plant Cell, Tissue Organ Cult.* **2010,** *100* (2), 219–227.

Johnson, A. A. T.; Kyriacou, B.; Callahan, D. L.; Carruthers, L.; Stangoulis, J.; Lombi, E.; Tester, M. Constitutive Overexpression of the OsNAS Gene Family Reveals Single-Gene Strategies for Effective Iron- and Zinc-Biofortification of Rice Endosperm. *PLoS ONE* **2011,** *6* (9), e24476.

Jones, D.; Kroos, N.; Anema, R.; Van Montfort, B.; Vooys, A.; van der Kraats, S.; van der Helm, E.; et al. High-Level Expression of Recombinant IgG in the Human Cell Line PER. C6. *Biotechnol. Progr.* **2003,** *19* (1), 163–168.

Joshi, A.; Dang, H. Q.; Vaid, N.; Tuteja, N. Pea Lectin Receptor-Like Kinase Promotes High Salinity Stress Tolerance in Bacteria and Expresses in Response to Stress in Planta. *Glycoconj. J.* **2010,** *27* (1), 133–150.

Joshi, S. G.; Schaart, J. G.; Groenwold, R.; Jacobsen, E.; Schouten, H. J.; Krens, F. A. Functional Analysis and Expression Profiling of HcrVf1 and HcrVf2 for Development of Scab Resistant Cisgenic and Intragenic Apples. *Plant Mol. Biol.* **2011,** *75* (6), 579–591.

Jung, R.; Carl, F. Transgenic Corn with an Improved Amino Acid Composition. In: *8th International Symposium on Plant Seeds*, Institute of Plant Genetics and Crop Plant Research, Gatersleben, Germany, 2000.

Jung, Y.-J.; Lee, S.-Y.; Moon, Y.-S.; Kang, K.-K. Enhanced Resistance to Bacterial and Fungal Pathogens by Overexpression of a Human Cathelicidin Antimicrobial Peptide (hCAP18/LL-37) in Chinese Cabbage. *Plant Biotechnol. Rep.* **2012,** *6* (1), 39–46.

Kachroo, A.; He, Z.; Patkar, R.; Zhu, Q.; Zhong, J.; Li, D.; Ronald, P.; Lamb, C.; Chattoo, B. B. Induction of H_2O_2 in Transgenic Rice Leads to Cell Death and Enhanced Resistance to both Bacterial and Fungal Pathogens. *Transgen. Res.* **2003,** *12* (5), 577–586.

Kanchiswamy, C. N.; Sargent, D. J.; Velasco, R.; Maffei, M. E.; Malnoy, M. Looking Forward to Genetically Edited Fruit Crops. *Trends Biotechnol.* **2015,** *33* (2), 62–64.

Kanzaki, H.; Nirasawa, S.; Saitoh, H.; Ito, M.; Nishihara, M.; Terauchi, R.; Nakamura, I. Overexpression of the Wasabi Defensin Gene Confers Enhanced Resistance to Blast Fungus (*Magnaporthe grisea*) in Transgenic Rice. *Theor. Appl. Genet.* **2002,** *105* (6–7), 809–814.

Karim, S.; Aronsson, H.; Ericson, H.; Pirhonen, M.; Leyman, B.; Welin, B.; Mäntylä, E.; Tapio Palva, E.; Van Dijck, P.; Holmström, K.-O. Improved Drought Tolerance without Undesired Side Effects in Transgenic Plants Producing Trehalose. *Plant Mol. Biol.* **2007,** *64* (4), 371–386.

Karunanandaa, B.; Qi, Q.; Hao, M.; Baszis, S. R.; Jensen, P. K.; Wong, Y.-H. H.; Jiang, J.; et al. Metabolically Engineered Oilseed Crops with Enhanced Seed Tocopherol. *Metab. Eng.* **2005,** *7* (5), 384–400.

Kavitha, K.; George, S.; Venkataraman, G.; Parida, A. A Salt-Inducible Chloroplastic Monodehydroascorbate Reductase from Halophyte *Avicennia marina* Confers Salt Stress Tolerance on Transgenic Plants. *BioChimie* **2010,** *92* (10), 1321–1329.

Khalili, H.; Soudbakhsh, A.; Hajiabdolbaghi, M.; Dashti-Khavidaki, S.; Poorzare, A.; Saeedi, A. A.; Sharififar, R. Nutritional Status and Serum Zinc and Selenium Levels in Iranian HIV Infected Individuals. *BMC Infect. Dis.* **2008,** *8* (1), 165.

Khan, R. S.; Sjahril, R.; Nakamura, I.; Mii, M. Production of Transgenic Potato Exhibiting Enhanced Resistance to Fungal Infections and Herbicide Applications. *Plant Biotechnol. Rep.* **2008,** *2* (1), 13–20.

Khare, N.; Goyary, D.; Singh, N. K.; Shah, P.; Rathore, M.; Anandhan, S.; Sharma, D.; Arif, M.; Ahmed, Z. Transgenic Tomato cv. Pusa Uphar Expressing a Bacterial Mannitol-1-Phosphate Dehydrogenase Gene Confers Abiotic Stress Tolerance. *Plant Cell, Tissue Organ Cult.* **2010,** *103* (2), 267–277.

Kim, J.-K.; Jang, I.-C.; Wu, R.; Zuo, W.-N.; Boston, R. S.; Lee, Y.-H.; Ahn, I.-P.; Nahm, B. H. Co-expression of a Modified Maize Ribosome-Inactivating Protein and a Rice Basic Chitinase Gene in Transgenic Rice Plants Confers Enhanced Resistance to Sheath Blight. *Transgen. Res.* **2003,** *12* (4), 475–484.

Kim, W.-S.; Chronis, D.; Juergens, M.; Schroeder, A. C.; Hyun, S. W.; Jez, J. M.; Krishnan, H. B. Transgenic Soybean Plants Overexpressing *O*-Acetylserine Sulfhydrylase Accumulate Enhanced Levels of Cysteine and Bowman–Birk Protease Inhibitor in Seeds. *Planta* **2012,** *235* (1), 13–23.

Kim, Y. S.; Kim, J.-K. Rice Transcription Factor AP37 Involved in Grain Yield Increase under Drought Stress. *Plant Signal. Behav.* **2009,** *4* (8), 735–736.

Kim, Y.-H.; Lim, S.; Han, S.-H.; Lee, J. J.; Nam, K. J.; Jeong, J. C.; Lee, H.-S.; Kwak, S.-S. Expression of both CuZnSOD and APX in Chloroplasts Enhances Tolerance to Sulfur Dioxide in Transgenic Sweet Potato Plants. *Compt. Rendus Biol.* **2015,** *338* (5), 307–313.

Kinney, A.; Cahoon, E.; Damude, H.; Hitz, W.; Liu, Z.-B.; Kolar, C. Production of Very Long Chain Polyunsaturated Fatty Acids in Oilseed Plants. U.S. Patent Application 10/776,311, filed February 11, 2004.

Kishimoto, K.; Nishizawa, Y.; Tabei, Y.; Hibi, T.; Nakajima, M.; Akutsu, K. Detailed Analysis of Rice Chitinase Gene Expression in Transgenic Cucumber Plants Showing Different Levels of Disease Resistance to Gray Mold (*Botrytis cinerea*). *Plant Sci.* **2002,** *162* (5), 655–662.

Kissoudis, C.; Kalloniati, C.; Flemetakis, E.; Madesis, P.; Labrou, N. E.; Tsaftaris, A.; Nianiou-Obeidat, I. Stress-Inducible GmGSTU4 Shapes Transgenic Tobacco Plants Metabolome towards Increased Salinity Tolerance. *Acta Physiol. Plant.* **2015,** *37* (5), 1–11.

Klaus, S. M. J.; Huang, F.-C.; Eibl, C.; Koop, H.-U.; Golds, T. J. Rapid and Proven Production of Transplastomic Tobacco Plants by Restoration of Pigmentation and Photosynthesis. *Plant J.* **2003,** *35* (6), 811–821.

Klink, V. P.; Kim, K.-H.; Martins, V.; MacDonald, M. H.; Beard, H. S.; Alkharouf, N. W.; Lee, S.-K.; Park, S.-C.; Matthews, B. F. A Correlation between Host-Mediated Expression of Parasite Genes as Tandem Inverted Repeats and Abrogation of Development of female *Heterodera glycines* Cyst Formation during Infection of *Glycine max*. *Planta* **2009,** *230* (1), 53–71.

Knäblein, J. Plant-Based Expression of Biopharmaceuticals. In: *Encyclopedia of Molecular Cell Biology and Molecular Medicine*, 2nd ed.; Meyers, R. A.; Ed.; Wiley VCH Verlag GmbH: Weinheim, 2005; pp 386–407.

Kode, V.; Mudd, E. A.; Iamtham, S.; Day, A. Isolation of Precise Plastid Deletion Mutants by Homology-Based Excision: A Resource for Site-Directed Mutagenesis, Multi-Gene Changes and High-Throughput Plastid Transformation. *Plant J.* **2006,** *46* (5), 901–909.

Kong, X.; Pan, J.; Zhang, M.; Xing, X.; Zhou, Y.; Liu, Y.; Li, D.; Li, D. ZmMKK4, a Novel Group C Mitogen-Activated Protein Kinase Kinase in Maize (*Zea mays*), Confers Salt and Cold Tolerance in Transgenic *Arabidopsis*. *Plant, Cell, Environ.* **2011,** *34* (8), 1291–1303.

Koprowski, H. Vaccines and Sera through Plant Biotechnology. *Vaccine* **2005,** *23* (15), 1757–1763.

Koussevitzky, S.; Suzuki, N.; Huntington, S.; Armijo, L.; Sha, W.; Cortes, D.; Shulaev, V.; Mittler, R. Ascorbate Peroxidase 1 Plays a Key Role in the Response of *Arabidopsis thaliana* to Stress Combination. *J. Biol. Chem.* **2008,** *283* (49), 34197–34203.

Krämer, U. Phytoremediation: Novel Approaches to Cleaning Up Polluted Soils. *Curr. Opin. Biotechnol.* **2005,** *16* (2), 133–141.

Kumar, S. G. B.; Ganapathi, T. R.; Revathi, C. J.; Srinivas, L.; Bapat, V. A. Expression of Hepatitis B Surface Antigen in Transgenic Banana Plants. *Planta* **2005,** *222* (3), 484–493.

Kumar, K. K.; Poovannan, K.; Nandakumar, R.; Thamilarasi, K.; Geetha, C.; Jayashree, N.; Kokiladevi, E.; et al. A High Throughput Functional Expression Assay System for a Defence Gene Conferring Transgenic Resistance on Rice against the Sheath Blight Pathogen, *Rhizoctonia solani*. *Plant Sci.* **2003,** *165* (5), 969–976.

Kumar, S.; Dhingra, A.; Daniell, H. Stable Transformation of the Cotton Plastid Genome and Maternal Inheritance of Transgenes. *Plant Mol. Biol.* **2004,** *56* (2), 203–216.

Kumar, V.; Shriram, V.; Kavi Kishor, P. B.; Jawali, N.; Shitole, M. G. Enhanced Proline Accumulation and Salt Stress Tolerance of Transgenic Indica Rice by Over-Expressing P5CSF129A Gene. *Plant Biotechnol. Rep.* **2010,** *4* (1), 37–48.

Kumar, V.; Joshi, S. G.; Bell, A. A.; Rathore, K. S. Enhanced Resistance against *Thielaviopsis basicola* in Transgenic Cotton Plants Expressing *Arabidopsis* NPR1 Gene. *Transgen. Res.* **2013,** *22* (2), 359–368.

Kunze, I.; Ebneth, M.; Heim, U.; Geiger, M.; Sonnewald, U.; Herbers, K. 2-Deoxyglucose Resistance: A Novel Selection Marker for Plant Transformation. *Mol. Breed.* **2001,** *7* (3), 221–227.

Kuwano, M.; Mimura, T.; Takaiwa, F.; Yoshida, K. T. Generation of Stable 'Low Phytic Acid' Transgenic Rice through Antisense Repression of the 1d-Myo-inositol 3-Phosphate Synthase Gene (RINO1) Using the 18-kDa Oleosin Promoter. *Plant Biotechnol. J.* **2009,** *7* (1), 96–105.

Kwit, C.; Moon, H. S.; Warwick, S. I.; Neal Stewart, C. Transgene Introgression in Crop Relatives: Molecular Evidence and Mitigation Strategies. *Trends Biotechnol.* **2011,** *29* (6), 284–293.

Lacombe, S.; Rougon-Cardoso, A.; Sherwood, E.; Peeters, N.; Dahlbeck, D.; Peter Van Esse, H.; Smoker, M.; et al. Interfamily Transfer of a Plant Pattern-Recognition Receptor Confers Broad-Spectrum Bacterial Resistance. *Nat. Biotechnol.* **2010,** *28* (4), 365–369.

Lardizabal, K.; Effertz, R.; Levering, C.; Mai, J.; Pedroso, M. C.; Jury, T.; Aasen, E.; Gruys, K.; Bennett, K. Expression of *Umbelopsis ramanniana* DGAT2A in Seed Increases Oil in Soybean. *Plant Physiol.* **2008,** *148* (1), 89–96.

Lauterslager, T. G. M.; Florack, D. E. A.; Van der Wal, T. J.; Molthoff, J. W.; Langeveld, J. P. M.; Bosch, D.; Boersma, W. J. A.; Th Hilgers, L. A. Oral Immunisation of Naive and Primed Animals with Transgenic Potato Tubers Expressing LT-B. *Vaccine* **2001,** *19* (17), 2749–2755.

Lee, M. K.; Miles, P.; Chen, J.-S. Brush Border Membrane Binding Properties of Bacillus *thuringiensis* Vip3A Toxin to *Heliothis virescens* and *Helicoverpa zea* Midguts. *Biochem. Biophys. Res. Commun.* **2006a,** *339* (4), 1043–1047.

Lee, S. M.; Kang, K.; Chung, H.; Yoo, S. H.; Xu, X. M.; Lee, S.-B.; Cheong, J.-J.; Daniell, H.; Kim, M. Plastid Transformation in the Monocotyledonous Cereal Crop, Rice (*Oryza sativa*) and Transmission of Transgenes to their Progeny. *Mol. Cells* **2006b,** *21* (3), 401.

Lee, S.; Jeon, J.-S.; An, G. Iron Homeostasis and Fortification in Rice. *J. Plant Biol.* **2012,** *55* (4), 261–267.

Lelivelt, C. L. C.; McCabe, M. S.; Newell, C. A.; Bastiaan deSnoo, C.; Van Dun, K. M. P.; Birch-Machin, I.; Gray, J. C.; Mills, K. H. G.; Nugent, J. M. Stable Plastid Transformation in Lettuce (*Lactuca sativa* L.). *Plant Mol. Biol.* **2005,** *58* (6), 763–774.

Lentz, E. M.; Segretin, M. E.; Morgenfeld, M. M.; Wirth, S. A.; Dus Santos, M. J.; Valeria Mozgovoj, M.; Wigdorovitz, A.; Félix Bravo-Almonacid, F. High Expression Level of a Foot and Mouth Disease Virus Epitope in Tobacco Transplastomic Plants. *Planta* **2010,** *231* (2), 387–395.

Lermontova, I.; Grimm, B. Overexpression of Plastidic Protoporphyrinogen IX Oxidase Leads to Resistance to the Diphenyl-Ether Herbicide Acifluorfen. *Plant Physiol.* **2000,** *122* (1), 75–84.

Li, B.; Li, N.; Duan, X.; Wei, A.; Yang, A.; Zhang, J. Generation of Marker-Free Transgenic Maize with Improved Salt Tolerance Using the FLP/FRT Recombination System. *J. Biotechnol.* **2010a,** *145* (2), 206–213.

Li, G.; Xu, X.; Xing, H.; Zhu, H.; Fan, Q. Insect Resistance to *Nilaparvata lugens* and *Cnaphalocrocis medinalis* in Transgenic Indica Rice and the Inheritance of Gna+ Sbti Transgenes. *Pest Manage. Sci.* **2005,** *61* (4), 390–396.

Li, H. Y.; Zhu, Y. M.; Chen, Q.; Conner, R. L.; Ding, X. D.; Li, J.; Zhang, B. B. Production of Transgenic Soybean Plants with Two Anti-Fungal Protein genes via *Agrobacterium* and Particle Bombardment. *Biol. Plant.* **2004,** *48* (3), 367–374.

Li, H.; Flachowsky, H.; Fischer, T. C.; Hanke, M.-V.; Forkmann, G.; Treutter, D.; Schwab, W.; Hoffmann, T.; Szankowski, I. Maize Lc Transcription Factor Enhances Biosynthesis of Anthocyanins, Distinct Proanthocyanidins and Phenylpropanoids in Apple (*Malus domestica* Borkh.). *Planta* **2007a,** *226* (5), 1243–1254.

Li, J.; Todd, T. C.; Trick, H. N. Rapid in Planta Evaluation of Root Expressed Transgenes in Chimeric Soybean Plants. *Plant Cell Rep.* **2010b,** *29* (2), 113–123.

Li, J.; Todd, T. C.; Oakley, T. R.; Lee, J.; Trick, H. N. Host-Derived Suppression of Nematode Reproductive and Fitness Genes Decreases Fecundity of *Heterodera glycines* Ichinohe. *Planta* **2010c,** *232* (3), 775–785.

Li, J.; Yu, G.; Sun, X.; Liu, Y.; Liu, J.; Zhang, X.; Jia, C.; Pan, H. AcPIP2, a Plasma Membrane Intrinsic Protein from Halophyte *Atriplex canescens*, Enhances Plant Growth Rate and Abiotic Stress Tolerance when Overexpressed in *Arabidopsis thaliana*. *Plant Cell Rep.* **2015,** 34 (8), 1401–1415.

Li, L.; Steffens, J. C. Overexpression of Polyphenol Oxidase in Transgenic Tomato Plants Results in Enhanced Bacterial Disease Resistance. *Planta* **2002,** *215* (2), 239–247.

Li, Q.; Lawrence, C. B.; Xing, H.-Y.; Babbitt, R. A.; Troy Bass, W.; Maiti, I. B.; Everett, N. P. Enhanced Disease Resistance Conferred by Expression of an Antimicrobial Magainin Analog in Transgenic Tobacco. *Planta* **2001,** *212* (4), 635–639.

Li, X.-Q.; Tan, A.; Voegtline, M.; Bekele, S.; Chen, C.-S.; Aroian, R. V. Expression of Cry5B Protein from *Bacillus thuringiensis* in Plant Roots Confers Resistance to Root-Knot Nematode. *Biol. Control* **2008,** *47* (1), 97–102.

Li, X.-Q.; Wei, J.-Z.; Tan, A.; Aroian, R. V. Resistance to Root-Knot Nematode in Tomato Roots Expressing a Nematicidal *Bacillus thuringiensis* Crystal Protein. *Plant Biotechnol. J.* **2007b,** *5* (4), 455–464.

Li, Z.; Zhou, M.; Zhang, Z.; Ren, L.; Du, L.; Zhang, B.; Xu, H.; Xin, Z. Expression of a Radish Defensin in Transgenic Wheat Confers Increased Resistance to *Fusarium graminearum* and *Rhizoctonia cerealis*. *Funct. Integr. Genomics* **2011a,** *11* (1), 63–70.

Li, Z.; Gao, Q.; Liu, Y.; He, C.; Zhang, X.; Zhang, J. Overexpression of Transcription Factor ZmPTF1 Improves Low Phosphate Tolerance of Maize by Regulating Carbon Metabolism and Root Growth. *Planta* **2011b,** *233* (6), 1129–1143.

Li, Z. T.; Dhekney, S. A.; Gray, D. J. PR-1 Gene Family of Grapevine: A Uniquely Duplicated PR-1 Gene from a *Vitis* Interspecific Hybrid Confers High Level Resistance to Bacterial Disease in Transgenic Tobacco. *Plant Cell Rep.* **2011c,** *30* (1), 1–11.

Liang, H.; Zheng, J.; Duan, X.; Sheng, B.; Jia, S.; Wang, D.; Ouyang, J.; et al. A Transgenic Wheat with a Stilbene Synthase Gene Resistant to Powdery Mildew Obtained by Biolistic Method. *Chin. Sci. Bull.* **2000,** *45* (7), 634–638.

Liang, Z.; Zhang, K.; Chen, K.; Gao, C. Targeted Mutagenesis in *Zea mays* Using TALENs and the CRISPR/Cas System. *J. Genet. Genomics* **2014,** *41* (2), 63–68.

Lilley, C. J.; Wang, D.; Atkinson, H. J.; Urwin, P. E. Effective Delivery of a Nematode-Repellent Peptide Using a Root-Cap-Specific Promoter. *Plant Biotechnol. J.* **2011,** *9* (2), 151–161.

Lin, B.; Zhuo, K.; Wu, P.; Cui, R.; Zhang, L.-H.; Liao, J. A Novel Effector Protein, MJ-NULG1a, Targeted to Giant Cell Nuclei Plays a Role in *Meloidogyne javanica* Parasitism. *Mol. Plant–Microbe Interact.* **2013,** *26* (1), 55–66.

Lin, W.-C.; Lu, C.-F.; Wu, J.-W.; Cheng, M.-L.; Lin, Y.-M.; Yang, N.-S.; Black, L.; Green, S. K.; Wang, J.-F.; Cheng, C.-P. Transgenic Tomato Plants Expressing the *Arabidopsis* NPR1 Gene Display Enhanced Resistance to a Spectrum of Fungal and Bacterial Diseases. *Transgen. Res.* **2004,** *13* (6), 567–581.

Lin, Y.-H.; Huang, H.-E.; Wu, F.-S.; Ger, M.-J.; Liao, P.-L.; Chen, Y.-R.; Tzeng, K.-C.; Feng, T.-Y. Plant Ferredoxin-Like Protein (PFLP) Outside Chloroplast in *Arabidopsis* Enhances Disease Resistance against Bacterial Pathogens. *Plant Sci.* **2010,** *179* (5), 450–458.

Ling, H.-Y.; Pelosi, A.; Walmsley, A. M. Current Status of Plant-Made Vaccines for Veterinary Purposes. *Expert Rev. Vaccines* **2010,** 9 (8), 971–982.

Liu, C.-W.; Lin, C.-C.; Chen, J. J. W.; Tseng, M.-J. Stable Chloroplast Transformation in Cabbage (*Brassica oleracea* L. var. *capitata* L.) by Particle Bombardment. *Plant Cell Rep.* **2007,** *26* (10), 1733–1744.

Liu, D.; An, Z.; Mao, Z.; Ma, L.; Lu, Z. Enhanced Heavy Metal Tolerance and Accumulation by Transgenic Sugar Beets Expressing *Streptococcus thermophilus* STGCS-GS in the Presence of Cd, Zn and Cu Alone or in Combination. *PLoS ONE,* **2015a,** *10* (6), e0128824.

Liu, F.; Wang, X.-D.; Zhao, Y.-Y.; Li, Y.-J.; Liu, Y.-C.; Sun, J. Silencing the HaAK Gene by Transgenic Plant-Mediated RNAi Impairs Larval Growth of *Helicoverpa armigera*. *Int. J. Biol. Sci.* **2015b,** *11* (1), 67.

Liu, J.; Tjellström, H.; McGlew, K.; Shaw, V.; Rice, A.; Simpson, J.; Kosma, D.; et al. Field Production, Purification and Analysis of High-Oleic Acetyl-Triacylglycerols from Transgenic *Camelina sativa*. *Ind. Crops Prod.* **2015c,** *65*, 259–268.

Liu, M.; Sun, Z.-X.; Zhu, J.; Xu, T.; Harman, G. E.; Lorito, M. Enhancing Rice Resistance to

Liu, X.; Hong, L.; Li, X.-Y.; Yao, Y. A. O.; Hu, B.; Li, L. Improved Drought and Salt Tolerance in Transgenic *Arabidopsis* Overexpressing a NAC Transcriptional Factor from *Arachis hypogaea*. *Biosci., Biotechnol., Biochem.* **2011,** *75* (3), 443–450.

Lobell, D. B.; Gourdji, S. M. The Influence of Climate Change on Global Crop Productivity. *Plant Physiol.* **2012,** *160* (4), 1686–1697.

López, A. B.; Van Eck, J.; Conlin, B. J.; Paolillo, D. J.; O'Neill, J.; Li, L. Effect of the Cauliflower or Transgene on Carotenoid Accumulation and Chromoplast Formation in Transgenic Potato Tubers. *J. Exp. Bot.* **2008,** *59* (2), 213–223.

López, W. H.; Leenhardt, F.; Coudray, C.; Remesy, C. Minerals and Phytic Acid Interactions: Is It a Real Problem for Human Nutrition? *Int. J. Food Sci. Technol.* **2002,** *37* (7), 727–739.

Lu, B.-R.; Snow, A. A. Gene Flow from Genetically Modified Rice and Its Environmental Consequences. *BioScience* **2005,** *55* (8), 669–678.

Ma, J. K.-C.; Hikmat, B. Y.; Wycoff, K.; Vine, N. D.; Chargelegue, D.; Yu, L.; Hein, M. B.; Lehner, T. Characterization of a Recombinant Plant Monoclonal Secretory Antibody and Preventive Immunotherapy in Humans. *Nat. Med.* **1998,** *4* (5), 601–606.

Macková, H.; Hronková, M.; Dobrá, J.; Turečková, V.; Novák, O.; Lubovská, Z.; Motyka, V.; et al. Enhanced Drought and Heat Stress Tolerance of Tobacco Plants with Ectopically Enhanced Cytokinin Oxidase/Dehydrogenase Gene Expression. *J. Exp. Bot.* **2013,** *64* (10), 2805–2815.

Maclean, J.; Koekemoer, M.; Olivier, A. J.; Stewart, D.; Hitzeroth, I. I.; Rademacher, T.; Fischer, R.; Williamson A-L.; Rybicki, E. P. Optimization of Human Papillomavirus Type 16 (HPV-16) L1 Expression in Plants: Comparison of the Suitability of Different HPV-16 L1 Gene Variants and Different Cell-Compartment Localization. *J. Gen. Virol.* **2007,** *88* (5), 1460–1469.

Macrae, T. C.; Baur, M. E.; Boethel, D. J.; Fitzpatrick, B. J.; Gao, A.-G.; Gamundi, J. C.; Harrison, L. A.; et al. Laboratory and Field Evaluations of Transgenic Soybean Exhibiting High-Dose Expression of a Synthetic *Bacillus thuringiensis* cry1A Gene for Control of Lepidoptera. *J. Econ. Entomol.* **2005,** *98* (2), 577–587.

Mäe, A.; Montesano, M.; Koiv, V.; Tapio Palva, E. Transgenic Plants Producing the Bacterial Pheromone *N*-acyl-homoserine Lactone Exhibit Enhanced Resistance to the Bacterial Phytopathogen *Erwinia carotovora*. *Mol. Plant–Microbe Interact.* **2001,** *14* (9), 1035–1042.

Malnoy, M.; Venisse, J. S.; Chevreau, E. Expression of a Bacterial Effector, Harpin N, Causes Increased Resistance to Fire Blight in *Pyrus communis*. *Tree Genet. Genomes* **2005,** *1* (2), 41–49.

Mao, Y.-B.; Tao, X.-Y.; Xue, X.-Y.; Wang, L.-J.; Chen, X.-Y. Cotton Plants Expressing CYP6AE14 Double-Stranded RNA Show Enhanced Resistance to Bollworms. *Transgen. Res.* **2011,** *20* (3), 665–673.

Mason, H. S.; Ball, J. M.; Shi, J.-J.; Jiang, X.; Estes, M. K.; Arntzen, C. J. Expression of Norwalk Virus Capsid Protein in Transgenic Tobacco and Potato and Its Oral Immunogenicity in Mice. *Proc. Nat. Acad. Sci.* **1996,** *93* (11), 5335–5340.

Matros, A.; Mock, H.-P. Ectopic Expression of a UDP-Glucose: Phenylpropanoid Glucosyltransferase Leads to Increased Resistance of Transgenic Tobacco Plants against Infection with Potato Virus Y. *Plant Cell Physiol.* **2004,** *45* (9), 1185–1193.

Matsumoto, K.; Murata, T.; Nagao, R.; Nomura, C. T.; Arai, S.; Arai, Y.; Takase, K.; Nakashita, H.; Taguchi, S.; Shimada, H. Production of Short-Chain-Length/Medium-Chain-Length Polyhydroxyalkanoate (PHA) Copolymer in the Plastid of *Arabidopsis thaliana* Using an Engineered 3-Ketoacyl-Acyl Carrier Protein Synthase III. *Biomacromolecules* **2009,** *10* (4), 686–690.

McCarter, J. P. Molecular Approaches toward Resistance to Plant-Parasitic Nematodes. *Cell Biology of Plant Nematode Parasitism*; Springer: Berlin-Heidelberg, 2009; pp 239–267.

McLean, M. D.; Yevtushenko, D. P.; Deschene, A.; Van Cauwenberghe, O. R.; Makhmoudova, A.; Potter, J. W.; Bown, A. W.; Shelp, B. J. Overexpression of Glutamate Decarboxylase in Transgenic Tobacco Plants Confers Resistance to the Northern Root-Knot Nematode. *Mol. Breed.* **2003,** *11* (4), 277–285.

Mehrotra, M.; Singh, A. K.; Sanyal, I.; Altosaar, I.; Amla, D. V. Pyramiding of Modified cry1Ab and cry1Ac Genes of *Bacillus thuringiensis* in Transgenic Chickpea (*Cicer arietinum* L.) for Improved Resistance to Pod Borer Insect *Helicoverpa armigera. Euphytica* **2011,** *182* (1), 87–102.

Mei, C.; Wassom, J. J.; Widholm, J. M. Expression Specificity of the Globulin-1 Promoter Driven Transgene (Chitinase) in Maize Seed Tissues. *Maydica* **2004,** *49* (4), 255–265.

Mei, C.; Park, S.-H.; Sabzikar, R.; Ransom, C.; Qi, C.; Sticklen, M. Green Tissue-Specific Production of a Microbial Endo-Cellulase in Maize (*Zea mays* L.) Endoplasmic-Reticulum and Mitochondria Converts Cellulose into Fermentable Sugars. *J. Chem. Technol. Biotechnol.* **2009,** *84* (5), 689–695.

Melander, M.; Kamnert, I.; Happstadius, I.; Liljeroth, E.; Bryngelsson, T. Stability of Transgene Integration and Expression in Subsequent Generations of Doubled Haploid Oilseed Rape Transformed with Chitinase and β-1,3-Glucanase Genes in a Double-Gene Construct. *Plant Cell Rep.* **2006,** *25* (9), 942–952.

Mendes, B. M. J.; Cardoso, S. C.; Boscariol-Camargo, R. L.; Cruz, R. B.; Mourão Filho, F. A. A.; Bergamin Filho, A. Reduction in Susceptibility to *Xanthomonas axonopodis* pv. *citri* in Transgenic *Citrus sinensis* Expressing the Rice Xa21 Gene. *Plant Pathol.* **2010,** *59* (1), 68–75.

Menkir, A.; Chikoye, D.; Lum, F. Incorporating an Herbicide Resistance Gene into Tropical Maize with Inherent Polygenic Resistance to Control *Striga hermonthica* (Del.) Benth. *Plant Breed.* **2010,** *129* (4), 385–392.

Mentag, R.; Luckevich, M.; Morency, M.-J.; Seguin, A. Bacterial Disease Resistance of Transgenic Hybrid Poplar Expressing the Synthetic Antimicrobial Peptide D4E1. *Tree Physiol.* **2003,** *23* (6), 405–411.

Mercer, K. L.; Shaw, R. G.; Wyse, D. L. Increased Germination of Diverse Crop-Wild Hybrid Sunflower Seeds. *Ecol. Appl.* **2006,** *16* (3), 845–854.

Mercedes Dana, M.; Pintor-Toro, J. A.; Cubero, B. Transgenic tobacco plants overexpressing chitinases of fungal origin show enhanced resistance to biotic and abiotic stress agents. *Plant Physiol.* **2006**, 142, 722–730.

Meyers, A.; Chakauya, E.; Shephard, E.; Tanzer, F. L.; Maclean, J.; Lynch, A.; Williamson, A.-L.; Rybicki, E. P. Expression of HIV-1 Antigens in Plants as Potential Subunit Vaccines. *BMC Biotechnol.* **2008,** *8* (1), 53.

Mian, A.; Oomen, R. J. F. J.; Isayenkov, S.; Sentenac, H.; Maathuis, F. J. M.; Véry, A.-A. Over-expression of an Na+-and K+-Permeable HKT Transporter in Barley Improves Salt Tolerance. *Plant J.* **2011,** *68* (3), 468–479.

Miller, M.; Tagliani, L.; Wang, N.; Berka, B.; Bidney, D.; Zhao, Z.-Y. High Efficiency Transgene Segregation in Co-transformed Maize Plants Using an *Agrobacterium tumefaciens* 2 T-DNA Binary System. *Transgen. Res.* **2002,** *11* (4), 381–396.

Milligan, S. B.; Bodeau, J.; Yaghoobi, J.; Kaloshian, I.; Zabel, P.; Williamson, V. M. The Root Knot Nematode Resistance Gene Mi from Tomato is a Member of the Leucine Zipper, Nucleotide Binding, Leucine-Rich Repeat Family of Plant Genes. *Plant Cell* **1998,** *10* (8), 1307–1319.

Miranda, J. A.; Avonce, N.; Suárez, R.; Thevelein, J. M.; Van Dijck, P.; Iturriaga, G. A Bifunctional TPS–TPP Enzyme from Yeast Confers Tolerance to Multiple and Extreme Abiotic-Stress Conditions in Transgenic Arabidopsis. *Planta* **2007,** *226* (6), 1411–1421.

Mishra, S.; Yadav, D. K.; Tuli, R. Ubiquitin Fusion Enhances Cholera Toxin B Subunit Expression in Transgenic Plants and the Plant-Expressed Protein Binds GM1 Receptors More Efficiently. *J. Biotechnol.* **2006,** *127* (1), 95–108.

Mishra, K. B.; Iannacone, R.; Petrozza, A.; Mishra, A.; Armentano, N.; La Vecchia, G.; Trtilek, M.; Cellini, F.; Nedbal, L. Engneering drought tolerance in tomato plants is reflected in chlorophyll fluorescence emission. *Plant Sci.* **2012**, 182, 79–86.

Mitani, N.; Kobayashi, S.; Nishizawa, Y.; Kuniga, T.; Matsumoto, R. Transformation of Trifoliate Orange with Rice Chitinase Gene and the Use of the Transformed Plant as a Rootstock. *Sci. Hortic.* **2006,** *108* (4), 439–441.

Mitsuhara, I.; Matsufuru, H.; Ohshima, M.; Kaku, H.; Nakajima, Y.; Murai, N.; Natori, S.; Ohashi, Y. Induced Expression of Sarcotoxin IA Enhanced Host Resistance against Both Bacterial and Fungal Pathogens in Transgenic Tobacco. *Mol. Plant–Microbe Interact.* **2000,** *13* (8), 860–868.

Mittler, R.; Blumwald, E. Genetic Engineering for Modern Agriculture: Challenges and Perspectives. *Annu. Rev. Plant Biol.* **2010,** *61*, 443–462.

Mohanty, A.; Kathuria, H.; Ferjani, A.; Sakamoto, A.; Mohanty, P.; Murata, N.; Tyagi, A. Transgenics of an Elite Indica Rice Variety Pusa Basmati 1 Harbouring the codA Gene are Highly Tolerant to Salt Stress. *Theor. Appl. Genet.* **2002,** *106* (1), 51–57.

Mondal, K. K.; Bhattacharya, R. C.; Koundal, K. R.; Chatterjee, S. C. Transgenic Indian Mustard (*Brassica juncea*) Expressing Tomato Glucanase Leads to Arrested Growth of *Alternaria brassicae. Plant Cell Rep.* **2007,** *26* (2), 247–252.

Moravčíková, J.; Matušíková, I.; Libantova, J.; Bauer, M.; Mlynárová, L. Expression of a Cucumber Class III Chitinase and *Nicotiana plumbaginifolia* Class I Glucanase Genes in Transgenic Potato Plants. *Plant Cell, Tissue Organ Cult.* **2004,** *79* (2), 161–168.

Moravčíková, J.; Libantová, J.; Heldák, J.; Salaj, J.; Bauer, M.; Matušíková, I.; Gálová, Z.; Mlynárová, Ľ. Stress-Induced Expression of Cucumber Chitinase and *Nicotiana plumbaginifolia* β-1,3-Glucanase Genes in Transgenic Potato Plants. *Acta Physiol. Plant.* **2007,** *29* (2), 133–141.

Moravec, T.; Schmidt, M. A.; Herman, E. M.; Woodford-Thomas, T. Production of *Escherichia coli* Heat Labile Toxin (LT) B Subunit in Soybean Seed and Analysis of its Immunogenicity as an Oral Vaccine. *Vaccine* 25 (9), 1647–1657.

Morton, R. L.; Schroeder, H. E.; Bateman, K. S.; Chrispeels, M. J.; Armstrong, E.; Higgins, T. J. V. Bean α-Amylase Inhibitor 1 in Transgenic Peas (*Pisum sativum*) Provides Complete Protection from Pea Weevil (*Bruchus pisorum*) under Field Conditions. *Proc. Nat. Acad. Sci.* **2000,** *97* (8), 3820–3825.

Moshelion, M.; Altman, A. Current Challenges and Future Perspectives of Plant and Agricultural Biotechnology. *Trends Biotechnol.* **2015,** *33* (6), 337–342.

Movahedi, S.; Sayed Tabatabaei, B. E.; Alizade, H.; Ghobadi, C.; Yamchi, A.; Khaksar, G. Constitutive Expression of *Arabidopsis* DREB1B in Transgenic Potato Enhances Drought and Freezing Tolerance. *Biol. Plant.* **2012,** *56* (1), 37–42.

Mukhopadhyay, A.; Vij, S.; Tyagi, A. K. Overexpression of a Zinc-Finger Protein Gene from Rice Confers Tolerance to Cold, Dehydration, and Salt Stress in Transgenic Tobacco. *Proc. Nat. Acad. Sci. USA* **2004,** *101* (16), 6309–6314.

Munis, F. H. M.; Tu, L.; Deng, F.; Tan, J.; Xu, L.; Xu, S.; Long, L.; Zhang, X. A Thaumatin-Like Protein Gene Involved in Cotton Fiber Secondary Cell Wall Development Enhances Resistance against *Verticillium dahliae* and Other Stresses in Transgenic Tobacco. *Biochem. Biophys. Res. Commun.* **2010,** *393* (1), 38–44.

Munns, R.; Tester, M. Mechanisms of Salinity Tolerance. *Annu. Rev. Plant Biol.* **2008,** *59*, 651–681.

Munns, R.; James, R. A.; Xu, B.; Athman, A.; Conn, S. J.; Jordans, C.; Byrt, C. S.; et al. Wheat Grain Yield on Saline Soils Is Improved by an Ancestral Na+ Transporter Gene. *Nat. Biotechnol.* **2012,** *30* (4), 360–364.

Musa, T. A.; Hung, C.-Y.; Darlington, D. E.; Sane, D. C.; Xie, J. Overexpression of Human Erythropoietin in Tobacco Does Not Affect Plant Fertility or Morphology. *Plant Biotechnol. Rep.* **2009,** *3* (2), 157–165.

Muthurajan, R.; Balasubramanian, P. Pyramiding Genes for Enhancing Tolerance to Abiotic and Biotic Stresses. *Molecular Techniques in Crop Improvement.* Springer: Netherlands, 2009; pp 163–184.

Nagadhara, D.; Ramesh, S.; Pasalu, I. C.; Kondala Rao, Y.; Sarma, N. P.; Reddy, V. D.; Rao, K. V. Transgenic Rice Plants Expressing the Snowdrop Lectin Gene (gna) Exhibit High-Level Resistance to the Whitebacked Planthopper (*Sogatella furcifera*). *Theor. Appl. Genet.* **2004,** *109* (7), 1399–1405.

Nagadhara, D.; Ramesh, S.; Pasalu, I. C.; Kondala Rao, Y.; Krishnaiah, N. V.; Sarma, N. P.; Bown, D. P.; Gatehouse, J. A.; Reddy, V. D.; Rao, K. V. Transgenic Indica Rice Resistant to Sap-Sucking Insects. *Plant Biotechnol. J.* **2003,** *1* (3), 231–240.

Namukwaya, B.; Tripathi, L.; Tripathi, J. N.; Arinaitwe, G.; Mukasa, S. B.; Tushemereirwe, W. K. Transgenic Banana Expressing Pflp Gene Confers Enhanced Resistance to Xanthomonas Wilt Disease. *Transgen. Res.* **2012,** *21* (4), 855–865.

Nanto, K.; Ebinuma, H. Marker-Free Site-Specific Integration Plants. *Transgen. Res.* **2008,** *17* (3), 337–344.

Napier, J. A. The Production of Unusual Fatty Acids in Transgenic Plants. *Annu. Rev. Plant Biol.* **2007,** *58*, 295–319.

Napier, J. A.; Usher, S.; Haslam, R.; Ruiz-López, N.; Sayanova, O. Transgenic Plants as a Sustainable, Terrestrial Source of Fish Oils. *Eur. J. Lipid Sci. Technol.* **2015,** *117* (9), 1317–1324.

Naqvi, S.; Zhu, C.; Farre, G.; Ramessar, K.; Bassie, L.; Breitenbach, J.; Conesa, D. P.; et al. Transgenic Multivitamin Corn through Biofortification of Endosperm with Three Vitamins Representing Three Distinct Metabolic Pathways. *Proc. Nat. Acad. Sci.* **2009,** *106* (19), 7762–7767.

Naqvi, S.; Farré, G.; Zhu, C.; Sandmann, G.; Capell, T.; Christou, P. Simultaneous Expression of Arabidopsis ρ-Hydroxyphenylpyruvate Dioxygenase and MPBQ Methyltransferase in Transgenic Corn Kernels Triples the Tocopherol Content. *Transgen. Res.* **2011,** *20* (1), 177–181.

Namukwaya, N.; Tripathi, L.; Tripathi, J. N.; Arinaitwe, G.; Mukasa, S. B.; Tushemereirwe, W. K. Transgenic banana expressing Pflp gene confers enhanced resistance to Xanthomonas wilt disease. *Transgenic Res.* **2012**, 21(4), 855–865.

Naqvi, S.; Farré, G.; Sanahuja, G.; Capell, T.; Zhu, C.; Christou, P. When More Is Better: Multigene Engineering in Plants. *Trends Plant Sci.* **2010,** *15* (1), 48–56.

National Academies. The Impact of Genetically Engineered Crops on Farm Sustainability in the United States, 2010. http://www.nationalacademies.org/.

Nguyen, T. X.; Nguyen, T.; Alameldin, H.; Goheen, B.; Loescher, W.; Sticklen, M. Transgene Pyramiding of the HVA1 and mtlD in T3 Maize (*Zea mays* L.) Plants Confers Drought and Salt Tolerance, along with an Increase in Crop Biomass. *Int. J. Agron.* **2013,** *2013*, 10 pp.

Nicholson, L.; Gonzalez-Melendi, P.; Van Dolleweerd, C.; Tuck, H.; Perrin, Y.; Ma, J. K.-C.; Fischer, R.; Christou, P.; Stoger, E. A Recombinant Multimeric Immunoglobulin Expressed in Rice Shows Assembly-Dependent Subcellular Localization in Endosperm Cells. *Plant Biotechnol. J.* **2005,** *3* (1), 115–127.

Niu, C.-F.; Wei, W.; Zhog, Q.-Y.; Tian, A.-G.; Hao, Y.-J.; Zhang, W.-K.; Ma, B.; et al. Wheat WRKY Genes TaWRKY2 and TaWRKY19 Regulate Abiotic Stress Tolerance in Transgenic Arabidopsis Plants. *Plant, Cell Environ.* **2012,** *35* (6), 1156–1170.

Nochi, T.; Takagi, H.; Yuki, Y.; Yang, L.; Masumura, T.; Mejima, M.; Nakanishi, U.; et al. Rice-Based Mucosal Vaccine as a Global Strategy for Cold-Chain- and Needle-Free Vaccination. *Proc. Nat. Acad. Sci.* **2007,** *104* (26), 10986–10991.

Ntui, V. O.; Kynet, K.; Khan, R. S.; Ohara, M.; Goto, Y.; Watanabe, M.; Fukami, M.; Nakamura, I.; Mii, M. Transgenic Tobacco Lines Expressing Defective CMV Replicase-Derived dsRNA Are Resistant to CMV-O and CMV-Y. *Mol. Biotechnol.* **2014,** *56* (1), 50–63.

Nugent, G. D.; Ten Have, M.; van der Gulik, A.; Dix, P. J.; Uijtewaal, B. A.; Mordhorst, A. P. Plastid Transformants of Tomato Selected Using Mutations Affecting Ribosome Structure. *Plant Cell Rep.* **2005,** *24* (6), 341–349.

Nugent, G. D.; Coyne, S.; Nguyen, T. T.; Kavanagh, T. A.; Dix, P. J. Nuclear and Plastid Transformation of *Brassica oleracea* var. *botrytis* (cauliflower) Using PEG-Mediated Uptake of DNA into Protoplasts. *Plant Sci.* **2006,** *170* (1), 135–142.

Nunes, A. C. S.; Vianna, G. R.; Cuneo, F.; Amaya-Farfán, J.; de Capdeville, G.; Rech, E. L.; Aragão, F. J. L. RNAi-Mediated Silencing of the Myo-inositol-1-phosphate Synthase Gene (GmMIPS1) in Transgenic Soybean Inhibited Seed Development and Reduced Phytate Content. *Planta* **2006,** *224* (1), 125–132.

Nykiforuk, C. L.; Shewmaker, C.; Harry, I.; Yurchenko, O. P.; Zhang, M.; Reed, C.; Oinam, G. S.; et al. High Level Accumulation of Gamma Linolenic Acid (C18: 3Δ6. 9, 12 cis) in Transgenic Safflower (*Carthamus tinctorius*) Seeds. *Transgen. Res.* **2012,** *21* (2), 367–381.

Nykiforuk, C. L.; Boothe, J. G.; Murray, E. W.; Keon, R. G.; Joseph Goren, H.; Markley, N. A.; Moloney, M. M. Transgenic Expression and Recovery of Biologically Active Recombinant Human Insulin from *Arabidopsis thaliana* Seeds. *Plant Biotechnol. J.* **2006,** *4* (1), 77–85.

O'Callaghan, M.; Glare, T. R.; Burgess, E. P. J.; Malone, L. A. Effects of Plants Genetically Modified for Insect Resistance on Nontarget Organisms. *Annu. Rev. Entomol.* **2005,** *50,* 271–292.

Oh, S.-J.; Song, S. I.; Kim, Y. S.; Jang, H.-J.; Kim, S. Y.; Kim, M.; Kim, Y.-K.; Nahm, B. H.; Kim, J.-K.. Arabidopsis CBF3/DREB1A and ABF3 in Transgenic Rice Increased Tolerance to Abiotic Stress without Stunting Growth. *Plant Physiol.* **2005,** *138* (1), 341–351.

Okkels, F. T.; Ward, J. L.; Joersbo, M. Synthesis of Cytokinin Glucuronides for the Selection of Transgenic Plant Cells. *Phytochemistry* **1997,** *46* (5), 801–804.

Oliveira, A. R.; Castro, T. R.; Capalbo, D. M. F.; Delalibera, Jr., I. Toxicological Evaluation of Genetically Modified Cotton (Bollgard®) and Dipel® WP on the Non-target Soil Mite *Scheloribates praeincisus* (Acari: Oribatida). *Exp. Appl. Acarol.* **2007,** *41* (3), 191–201.

Paine, J. A.; Shipton, C. A.; Chaggar, S.; Howells, R. M.; Kennedy, M. J.; Vernon, G.; Wright, S. Y.; et al. Improving the Nutritional Value of Golden Rice through Increased Pro-vitamin A Content. *Nat. Biotechnol.* **2005,** *23* (4), 482–487.

Palmgren, M. G.; Clemens, S.; Williams, L. E.; Krämer, U.; Borg, S.; Schjørring, J. K.; Sanders, D. Zinc Biofortification of Cereals: Problems and Solutions. *Trends Plant Sci.* **2008,** *13* (9), 464–473.

Park, B.-J.; Liu, Z.; Kanno, A.; Kameya, T. Increased Tolerance to Salt- and Water-Deficit Stress in Transgenic Lettuce (*Lactuca sativa* L.) by Constitutive Expression of LEA. *Plant Growth Regul.* **2005a,** *45* (2), 165–171.

Park, B.-J.; Liu, Z.; Kanno, A.; Kameya, T. Genetic Improvement of Chinese Cabbage for Salt and Drought Tolerance by Constitutive Expression of a *B. napus* LEA gene. *Plant Sci.* **2005b,** *169* (3), 553–558.

Park, E.-J.; Jeknic, Z.; Pino, M.-T.; Murata, N.; Chen, T. H.-H. Glycinebetaine Accumulation is More Effective in Chloroplasts than in the Cytosol for Protecting Transgenic Tomato Plants against Abiotic Stress. *Plant, Cell Environ.* **2007,** *30* (8), 994–1005.

Park, J. M.; Park, C.-J.; Lee, S.-B.; Ham, B.-K.; Shin, R.; Paek, K.-H. Overexpression of the Tobacco Tsi1 Gene Encoding an EREBP/AP2-type Transcription Factor Enhances Resistance against Pathogen Attack and Osmotic Stress in Tobacco. *Plant Cell* **2001,** *13* (5), 1035–1046.

Park, T.-H.; Gros, J.; Sikkema, A.; Vleeshouwers, V. G. A. A.; Muskens, M.; Allefs, S.; Jacobsen, E.; Visser, R. G. F.; van der Vossen, E. A. G. The Late Blight Resistance Locus Rpi-blb3 from *Solanum bulbocastanum* Belongs to a Major Late Blight R Gene Cluster on Chromosome 4 of Potato. *Mol. Plant–Microbe Interact.* **2005c,** *18* (7), 722–729.

Parkhi, V.; Kumar, V.; Campbell, L. M.; Bell, A. A.; Shah, J.; Rathore, K. S. Resistance against Various Fungal Pathogens and Reniform Nematode in Transgenic Cotton Plants Expressing Arabidopsis NPR1. *Transgen. Res.* **2010,** *19* (6), 959–975.

Pasquali, G.; Biricolti, S.; Locatelli, F.; Baldoni, E.; Mattana, M. Osmyb4 Expression Improves Adaptive Responses to Drought and Cold Stress in Transgenic Apples. *Plant Cell Rep.* **2008,** *27* (10), 1677–1686.

Pasapula, V.; Shen, G.; Kuppu, S.; Paez-Valencia, J.; Mendoza, M.; Hou, P.; Chen, J.; Qiu, X.; Zhu, L.; Zhang, X.; Auld, D.; Blumwald, E.; Zhang, H.; Gaxiola, R.; Payton, P.; Expression of an Arabidopsis vacuolar H+-pyrophosphatase gene (AVP1) in cotton improves drought- and salt tolerance and increases fibre yield in the field conditions. *Plant Biotechnol J.* **2011,** *9* (1), 88–99.

Patel, N.; Hamamouch, N.; Li, C.; Hussey, R.; Mitchum, M.; Baum, T.; Wang, X.; Davis, E. L. Similarity and Functional Analyses of Expressed Parasitism Genes in *Heterodera schachtii* and *Heterodera glycines*. *J. Nematol.* **2008,** *40* (4), 299.

Patel, N.; Hamamouch, N.; Li, C.; Hewezi, T.; Hussey, R. S.; Baum, T. J.; Mitchum, M. G.; Davis, E. L. A Nematode Effector Protein Similar to Annexins in Host Plants. *J. Exp. Bot.* **2010,** *61* (1), 235–248.

Patkar, R. N.; Chattoo, B. B. Transgenic Indica Rice Expressing ns-LTP-like Protein Shows Enhanced Resistance to both Fungal and Bacterial Pathogens. *Mol. Breed.* **2006,** *17* (2), 159–171.

Pauly, M.; Keegstra, K. Cell-Wall Carbohydrates and their Modification as a Resource for Biofuels. *Plant J.* **2008,** *54* (4), 559–568.

Peiró, A.; Carmen Cañizares, M.; Rubio, L.; López, C.; Moriones, E.; Aramburu, J.; Sánchez-Navarro, J. The Movement Protein (NSm) of Tomato Spotted Wilt Virus Is the Avirulence Determinant in the Tomato Sw-5 Gene-Based Resistance. *Mol. Plant Pathol.* **2014,** *15* (8), 802–813.

Pel, M. A.; Foster, S. J.; Park, T.-H.; Rietman, H.; van Arkel, G.; Jones, J. D. G.; Van Eck, H. J.; Jacobsen, E.; Visser, R. G. F.; Van der Vossen, E. A. G. Mapping and Cloning of Late Blight Resistance Genes from *Solanum venturii* Using an Interspecific Candidate Gene Approach. *Mol. Plant–Microbe Interact.* **2009,** *22* (5), 601–615.

Pellegrineschi, A.; Reynolds, M.; Pacheco, M.; Maria Brito, R.; Almeraya, R.; Yamaguchi-Shinozaki, K.; Hoisington, D. Stress-Induced Expression in Wheat of the *Arabidopsis thaliana* DREB1A Gene Delays Water Stress Symptoms under Greenhouse Conditions. *Genome* **2004,** *47* (3), 493–500.

Pereira, J. F.; Zhou, G.; Delhaize, E.; Richardson, T.; Zhou, M.; Ryan, P. R. Engineering Greater Aluminium Resistance in Wheat by Over-expressing TaALMT1. *Ann. Bot.* **2010,** *106* (1), 205–214.

Pérez-Massot, E.; Banakar, R.; Gómez-Galera, S.; Zorrilla-López, U.; Sanahuja, G.; Arjó, G.; Miralpeix, B.; et al. The Contribution of Transgenic Plants to Better Health through Improved Nutrition: Opportunities and Constraints. *Genes Nutr.* **2013,** *8* (1), 29–41.

Petrie, J. R.; Shrestha, P.; Mansour, M. P.; Nichols, P. D.; Liu, Q.; Singh, S. P. Metabolic Engineering of Omega-3 Long-Chain Polyunsaturated Fatty Acids in Plants Using an Acyl-CoA Δ6-Desaturase with ω3-Preference from the Marine Microalga *Micromonas pusilla. Metab. Eng.* **2010a,** *12* (3), 233–240.

Petrie, J. R.; Shrestha, P.; Liu, Q.; Mansour, M. P.; Wood, C. C.; Zhou, X.-R.; Nichols, P. D.; Green, A. G.; Singh, S. P. Rapid Expression of Transgenes Driven by Seed-Specific Constructs in Leaf Tissue: DHA Production. *Plant Methods* **2010b,** *6* (1), 8.

Polkinghorne, I.; Hamerli, D.; Cowan, P.; Duckworth, J. Plant-Based Immunocontraceptive Control of Wildlife—Potentials, Limitations, and Possums. *Vaccine* **2005,** *23* (15), 1847–1850.

Powell, A. L. T.; van Kan, J.; ten Have, A.; Visser, J.; Carl Greve, L.; Bennett, A. B.; Labavitch, J. M. Transgenic Expression of Pear PGIP in Tomato Limits Fungal Colonization. *Mol. Plant–Microbe Interact.* **2000,** *13* (9), 942–950.

Pradeep, K.; Satya, V. K.; Selvapriya, M.; Vijayasamundeeswari, A.; Ladhalakshmi, D.; Paranidharan, V.; Rabindran, R.; Samiyappan, R.; Balasubramanian, P.; Velazhahan, R. Engineering Resistance against Tobacco Streak Virus (TSV) in Sunflower and Tobacco Using RNA Interference. *Biol. Plant.* **2012,** *56* (4), 735–741.

Prasad, B. D.; Creissen, G.; Lamb, C.; Chattoo, B. B. Overexpression of Rice (*Oryza sativa* L.) OsCDR1 Leads to Constitutive Activation of Defense Responses in Rice and Arabidopsis. *Mol. Plant–Microbe Interact.* **2009,** *22* (12), 1635–1644.

Prasad, B. D.; Jha, S.; Chattoo, B. B. Transgenic Indica Rice Expressing Mirabilis *jalapa* Antimicrobial Protein (Mj-AMP2) Shows Enhanced Resistance to the Rice Blast Fungus *Magnaporthe oryzae. Plant Sci.* **2008,** *175* (3), 364–371.

Prins, M.; Laimer, M.; Noris, E.; Schubert, J.; Wassenegger, M.; Tepfer, M. Strategies for Antiviral Resistance in Transgenic Plants. *Mol. Plant Pathol.* **2008,** *9* (1), 73–83.

Priya, B. D.; Somasekhar, N.; Prasad, J. S.; Kirti, P. B. Transgenic Tobacco Plants Constitutively Expressing Arabidopsis NPR1 Show Enhanced Resistance to Root-Knot Nematode, *Meloidogyne incognita. BMC Res. Notes,* **2011,** *4* (1), 231.

Qi, B.; Fraser, T.; Mugford, S.; Dobson, G.; Sayanova, O.; Butler, J.; Napier, J. A.; Keith Stobart, A.; Lazarus, C. M. Production of Very Long Chain Polyunsaturated Omega-3 and Omega-6 Fatty Acids in Plants. *Nat. Biotechnol.* **2004,** *22* (6), 739–745.

Qi, Y. C.; Liu, W. Q.; Qiu, L.Y.; Zhang, S. M.; Ma, L.; Zhang, H.; Overexpression of glutathione S-transferase gene increases salt tolerance of Arabidopsis. *Russ. J. Plant Physiol.* **2010,** *57*, 233–240.

Qi, Y.; Chen, L.; He, X.; Jin, Q.; Zhang, X.; He, Z. Marker-free, Tissue-specific Expression of Cry1Ab as a Safe Transgenic Strategy for Insect Resistance in Rice Plants. *Pest Manag. Sci.* **2012,** *69* (1), 135–141.

Qin, F.; Kakimoto, M.; Sakuma, Y.; Maruyama, K.; Osakabe, Y.; Phan Tran, L.-S.; Shinozaki, K.; Yamaguchi-Shinozaki, K. Regulation and Functional Analysis of ZmDREB2A in Response to Drought and Heat Stresses in *Zea mays* L. *Plant J.* **2007,** *50* (1), 54–69.

Qiu, X.; Hong, H.; Datla, N.; MacKenzie, S. L.; Taylor, D. C.; Thomas, T. L. Expression of Borage Δ6 Desaturase in *Saccharomyces cerevisiae* and Oilseed Crops. *Can. J. Bot.* **2002,** *80* (1), 42–49.

Qu, J.; Mao, H.-Z.; Chen, W.; Gao, S.-Q.; Bai, Y.-N.; Sun, Y.-W.; Geng, Y.-F.; Ye, J. Development of Marker-Free Transgenic Jatropha Plants with Increased Levels of Seed Oleic Acid. *Biotechnol. Biofuels* **2012,** 29 (5), 10.

Quan, R.; Shang, M.; Zhang, H.; Zhao, Y.; Zhang, J. Engineering of Enhanced Glycine Betaine Synthesis Improves Drought Tolerance in Maize. *Plant Biotechnol. J.* **2004,** *2* (6), 477–486.

Raclaru, M.; Gruber, J.; Kumar, R.; Sadre, R.; Lühs, W.; Karim Zarhloul, M.; Friedt, W.; Frentzen, M.; Weier, D. Increase of the Tocochromanol Content in Transgenic *Brassica napus* Seeds by Overexpression of Key Enzymes Involved in Prenylquinone Biosynthesis. *Mol. Breed.* **2006,** *18* (2), 93–107.

Rahbé, Y.; Deraison, C.; Bonadé-Bottino, M.; Girard, C.; Nardon, C.; Jouanin, L. Effects of the Cysteine Protease Inhibitor Oryzacystatin (OC-I) on Different Aphids and Reduced Performance of *Myzus persicae* on OC-I Expressing Transgenic Oilseed Rape. *Plant Sci.* **2003,** *164* (4), 441–450.

Ramessar, K.; Sabalza, M.; Capell, T.; Christou, P. Maize Plants: An Ideal Production Platform for Effective and Safe Molecular Pharming. *Plant Sci.* **2008a,** *174* (4), 409–419.

Ramessar, K.; Rademacher, T.; Sack, M.; Stadlmann, J.; Platis, D.; Stiegler, G.; Labrou, N.; et al. Cost-effective Production of a Vaginal Protein Microbicide to Prevent HIV Transmission. *Proc. Nat. Acad. Sci.* **2008b,** *105* (10), 3727–3732.

Rascón-Cruz, Q.; Sinagawa-García, S.; Osuna-Castro, J. A.; Bohorova, N.; Paredes-López, O. Accumulation, Assembly, and Digestibility of Amarantin Expressed in Transgenic Tropical Maize. *Theor. Appl. Genet.* **2004,** *108* (2), 335–342.

Ravanello, M. P.; Ke, D.; Alvarez, J.; Huang, B.; Shewmaker, C. K. Coordinate Expression of Multiple Bacterial Carotenoid Genes in Canola Leading to Altered Carotenoid Production. *Metab. Eng.* **2003,** *5* (4), 255–263.

Reddy, A. S.; Thomas, T. L. Expression of a Cyanobacterial Δ6-Desaturase Gene Results in γ-Linolenic Acid Production in Transgenic Plants. *Nat. Biotechnol.* **1996,** *14* (5), 639–642.

Reeck, G. R.; Krämer, K. J.; Baker, J. E.; Kanost, M. R.; Fabrick, J. A.; Behnke, C. A. Proteinase Inhibitors and Resistance of Transgenic Plants to Insects. *Advances in Insect Control: The Role of Transgenic Plants*; Taylor and Francis: London, 1997; pp 157–183.

Ribeiro, A. P. O.; Pereira, E. J. G.; Galvan, T. L.; Picanco, M. C.; Picoli, E. A. T.; da Silva, D. J. H.; Fari, M. G.; Otoni, W. C. Effect of Eggplant Transformed with Oryzacystatin Gene on *Myzus persicae* and *Macrosiphum euphorbiae*. *J. Appl. Entomol.* **2006,** *130* (2), 84–90.

Richter, L. J.; Thanavala, Y.; Arntzen, C. J.; Mason, H. S. Production of Hepatitis B Surface Antigen in Transgenic Plants for Oral Immunization. *Nat. Biotechnol.* **2000,** *18* (11), 1167–1171.

Roderick, H.; Tripathi, L.; Babirye, A.; Wang, D.; Tripathi, J.; Urwin, P. E.; Atkinson, H. J. Generation of Transgenic Plantain (*Musa* spp.) with Resistance to Plant Pathogenic Nematodes. *Mol. Plant Pathol.* **2012,** *13* (8), 842–851.

Rohini, V. K.; Sankara Rao, K. Transformation of Peanut (*Arachis hypogaea* L.) with Tobacco Chitinase Gene: Variable Response of Transformants to Leaf Spot Disease. *Plant Sci.* **2001,** *160* (5), 889–898.

Rommens, C. M.; Ye, J.; Richael, C.; Swords, K. Improving Potato Storage and Processing Characteristics through All-Native DNA Transformation. *J. Agric. Food Chem.* **2006,** *54* (26), 9882–9887.

Rosales-Mendoza, S.; Alpuche-Solís, Á. G.; Soria-Guerra, R. E.; Moreno-Fierros, L.; Martínez-González, L.; Herrera-Díaz, A.; Korban, S. S. Expression of an *Escherichia coli* Antigenic Fusion Protein Comprising the Heat Labile Toxin B Subunit and the Heat Stable Toxin, and Its Assembly as A Functional Oligomer in Transplastomic Tobacco Plants. *Plant J.* **2009,** *57*, 45–54.

Rosales-Mendoza, S.; Soria-Guerra, R. E.; Moreno-Fierros, L.; Alpuche-Solís, Á. G.; Martínez-González, L.; Korban, S. S. Expression of an Immunogenic F1-V Fusion Protein in Lettuce as a Plant-Based Vaccine against Plague. *Planta* **2010,** *232* (2), 409–416.

Roosens, N.H., Bitar, F.A., Loenders, K.; Angenon, G.; Jacobs, M. Overexpression of Ornithine-δ-aminotransferase Increases Proline Biosynthesis and Confers Osmotolerance in Transgenic Plants. *Mole. Breed* **2002,** *9* (2), 73–80.

Roxas, V. P.; Lodhi, S. A.; Garrett, D. K.; Mahan, J. R.; Allen, R. D. Stress Tolerance in Transgenic Tobacco Seedlings that Overexpress Glutathione S-Transferase/Glutathione Peroxidase. *Plant Cell Physiol.* **2000,** *41* (11), 1229–1234.

Roy, S. J.; Negrão, S.; Tester, M. Salt Resistant Crop Plants. *Curr. Opin. Biotechnol.* **2014,** *26*, 115–124.

Roy-Barman, S.; Sautter, C.; Chattoo, B. B. Expression of the Lipid Transfer Protein Ace-AMP1 in Transgenic Wheat Enhances Antifungal Activity and Defense Responses. *Transgen. Res.* **2006,** *15* (4), 435–446.

Ruf, S.; Hermann, M.; Berger, I. J.; Carrer, H.; Bock, R. Stable Genetic Transformation of Tomato Plastids and Expression of a Foreign Protein in Fruit. *Nat. Biotechnol.* **2001,** *19* (9), 870–875.

Rugh, C. L.; Senecoff, J. F.; Meagher, R. B.; Merkle, S. A. Development of transgenic yellow poplar for mercury phytoremediation. *Nature Biotechnol.* **1998**, 16, 925–928.

Ruhlman, T.; Verma, D.; Samson, N.; Daniell, H. The Role of Heterologous Chloroplast Sequence Elements in Transgene Integration and Expression. *Plant Physiol.* **2010,** *152* (4), 2088–2104.

Ruiz-López, N.; Haslam, R. P.; Napier, J. A.; Sayanova, O. Successful High-Level Accumulation of Fish Oil Omega-3 Long-Chain Polyunsaturated Fatty Acids in a Transgenic Oilseed Crop. *Plant J.* **2014,** *77* (2), 198–208.

Ruiz-López, N.; Haslam, R. P.; Venegas-Calerón, M.; Larson, T. R.; Graham, I. A.; Napier, J. A.; Sayanova, O. The Synthesis and Accumulation of Stearidonic Acid in Transgenic Plants: A Novel Source of 'Heart-Healthy' Omega-3 Fatty Acids. *Plant Biotechnol. J.* **2009,** *7* (7), 704–716.

Ruiz-López, N.; Usher, S.; Sayanova, O. V.; Napier, J. A.; Haslam, R. P. Modifying the Lipid Content and Composition of Plant Seeds: Engineering the Production of LC-PUFA. *Appl. Microbiol. Biotechnol.* **2015,** *99* (1), 143–154.

Rybicki, E. P. Plant-Made Vaccines for Humans and Animals. *Plant Biotechnol. J.* **2010,** *8* (5), 620–637.

Rybicki, E. P. Plant-Produced Vaccines: Promise and Reality. *Drug Discov. Today* **2009,** *14* (1), 16–24.

Saha, P.; Majumder, P.; Dutta, I.; Ray, T.; Roy, S. C.; Das, S. Transgenic Rice Expressing *Allium sativum* Leaf Lectin with Enhanced Resistance against Sap-Sucking Insect Pests. *Planta* **2006,** *223* (6), 1329–1343.

Sahoo, R. K.; Ansari, M. W.; Tuteja, R.; Tuteja, N. Salt Tolerant SUV3 Overexpressing Transgenic Rice Plants Conserve Physicochemical Properties and Microbial Communities of Rhizosphere. *Chemosphere* **2015,** *119*, 1040–1047.

Samac, D. A.; Smigocki, A. C. Expression of Oryzacystatin I and II in Alfalfa Increases Resistance to the Root-Lesion Nematode. *Phytopathology* **2003,** *93* (7), 799–804.

Sandal, I.; Koul, R.; Saini, U.; Mehta, M.; Dhiman, N.; Kumar, N.; Singh Ahuja, P.; Bhattacharya, A. Development of Transgenic Tea Plants from Leaf Explants by the Biolistic Gun Method and their Evaluation. *Plant Cell, Tissue Organ Cult.* **2015,** *123* (2), 245–255.

Sandhu, J. S.; Krasnyanski, S. F.; Domier, L. L.; Korban, S. S.; Osadjan, M. D.; Buetow, D. E. Oral Immunization of Mice with Transgenic Tomato Fruit Expressing Respiratory Syncytial Virus-F Protein Induces a Systemic Immune Response. *Transgen. Res.* **2000,** *9* (2), 127–135.

Santi, L.; Huang, Z.; Mason, H. Virus-Like Particles Production in Green Plants. *Methods* **2006,** *40* (1), 66–76.

Sanyal, I.; Singh, A. K.; Kaushik, M.; Amla, D. V. Agrobacterium-Mediated Transformation of Chickpea (*Cicer arietinum* L.) with *Bacillus thuringiensis* cry1Ac Gene for Resistance against Pod Borer Insect *Helicoverpa armigera*. *Plant Sci.* **2005,** *168* (4), 1135–1146.

Sato, S.; Xing, A.; Ye, X.; Schweiger, B.; Kinney, A.; Graef, G.; Clemente, T. Production of γ-Linolenic Acid and Stearidonic Acid in Seeds of Marker-Free Transgenic Soybean. *Crop Sci.* **2004,** *44* (2), 646–652.

Sato, Y.; Masuta, Y.; Saito, K.; Murayama, S.; Ozawa, K. Enhanced Chilling Tolerance at the Booting Stage in Rice by Transgenic Overexpression of the Ascorbate Peroxidase Gene, OsAPXa. *Plant Cell Rep.* **2011,** *30* (3), 399–406.

Satyavathi, V.; Prasad, V.; Shaila, M.; Sita, L. G. Expression of Hemagglutinin Protein of Rinderpest Virus in Transgenic Pigeon Pea [*Cajanus cajan* (L.) Millsp.] Plants. *Plant Cell Rep.* **2003,** *21* (7), 651–658.

Sayanova, O.; Napier, J. A. Transgenic Oilseed Crops as an Alternative to Fish Oils. *Prostagland., Leukotrienes Essen. Fatty Acids (PLEFA)* **2011,** *85* (5), 253–260.

Scholl, T. O.; Johnson, W. G. Folic Acid: Influence on the Outcome of Pregnancy. *Am. J. Clin. Nutr.* **2000,** *71* (5), 1295s–1303s.

Schubert, C. Can Biofuels Finally Take Center Stage? *Nat. Biotechnol.* **2006,** *24* (7), 777–784.

Sciutto, E.; Fragoso, G.; Manoutcharian, K.; Gevorkian, G.; Rosas-Salgado, G.; Hernández-Gonzalez, M.; Herrera-Estrella, L.; et al. New Approaches to Improve a Peptide Vaccine against Porcine *Taenia solium* Cysticercosis. *Arch. Med. Res.* **2002,** *33* (4), 371–378.

Segal, G.; Song, R.; Messing, J. A New Opaque Variant of Maize by a Single Dominant RNA-Interference-Inducing Transgene. *Genetics* **2003,** *165* (1), 387–397.

Sekhar, K.; Priyanka, B.; Reddy, V. D.; Rao, K. V. Metallothionein 1 (CcMT1) of Pigeonpea (*Cajanus cajan*, L.) Confers Enhanced Tolerance to Copper and Cadmium in *Escherichia coli* and *Arabidopsis thaliana*. *Environ. Exp. Bot.* **2011,** *72* (2), 131–139.

Sengupta, S.; Chakraborti, D.; Mondal, H. A.; Das, S. Selectable Antibiotic Resistance Marker Gene-Free Transgenic Rice Harbouring the Garlic Leaf Lectin Gene Exhibits Resistance to Sap-Sucking Planthoppers. *Plant Cell Rep.* **2010,** *29* (3), 261–271.

Senthil-Kumar, M.; Wang, K.; Mysore, K. S. AtCYP710A1 Gene-Mediated Stigmasterol Production Plays a Role in Imparting Temperature Stress Tolerance in *Arabidopsis thaliana*. *Plant Signal. Behav.* **2013,** *8* (2), e23142.

Senthilkumar, R.; Cheng, C.-P.; Yeh, K.-W. Genetically Pyramiding Protease-Inhibitor Genes for Dual Broad-Spectrum Resistance against Insect and Phytopathogens in Transgenic Tobacco. *Plant Biotechnol. J.* **2010,** *8* (1), 65–75.

Seo, Y. J.; Park, J.-B.; Cho, Y.-J.; Jung, C.; Seo, H. S.; Park, S.-K.; Nahm, B. H.; Song, J. T. Overexpression of the Ethylene-Responsive Factor Gene BrERF4 from *Brassica rapa* Increases Tolerance to Salt and Drought in Arabidopsis Plants. *Mol. Cells* **2010,** *30* (3), 271–277.

Shaaltiel, Y.; Bartfeld, D.; Hashmueli, S.; Baum, G.; Brill-Almon, E.; Galili, G.; Dym, O.; et al. Production of Glucocerebrosidase with Terminal Mannose Glycans for Enzyme Replacement Therapy of Gaucher's Disease Using a Plant Cell System. *Plant Biotechnol. J.* **2007,** *5* (5), 579–590.

Shan, D.-P.; Huang, J.-G.; Yang, Y.-T.; Guo, Y.-H.; Wu, C.-A.; Yang, G.-D.; Gao, Z.; Zheng, C.-C. Cotton GhDREB1 Increases Plant Tolerance to Low Temperature and Is Negatively Regulated by Gibberellic Acid. *New Phytol.* **2007,** *176* (1), 70–81.

Shan, Q.; Wang, Y.; Li, J.; Zhang, Y.; Chen, K.; Liang, Z.; Zhang, K.; et al. Targeted Genome Modification of Crop Plants Using a CRISPR–Cas System. *Nat. Biotechnol.* **2013a,** *31* (8), 686–688.

Shan, Q.; Wang, Y.; Chen, K.; Liang, Z.; Li, J.; Zhang, Y.; Zhang, K.; et al. Rapid and Efficient Gene Modification in Rice and *Brachypodium* Using TALENs. *Mol. Plant* **2013b,** *6* (4), 1365–1368.

Sharma, H. C.; Sharma, K. K.; Crouch, J. H. Genetic Transformation of Crops for Insect Resistance: Potential and Limitations. *Crit. Rev. Plant Sci.* **2004,** *23* (1), 47–72.

Sheehy, J. E.; Mitchell, P. L. Rice and Global Food Security: The Race between Scientific Discovery and Catastrophe. *Access Not Excess—The Search for Better Nutrition*; Smith-Gordon: Cambridgeshire, 2011; pp 81–90.

Shekhawat, U. K. S.; Ganapathi, T. R.; Hadapad, A. B. Transgenic Banana Plants Expressing Small Interfering RNAs Targeted against Viral Replication Initiation Gene Display High-Level Resistance to Banana Bunchy Top Virus Infection. *J. Gen. Virol.* **2012,** *93* (Pt 8), 1804–1813.

Shen, B.; Allen, W. B.; Zheng, P.; Li, C.; Glassman, K.; Ranch, J.; Nubel, D.; Tarczynski, M. C. Expression of ZmLEC1 and ZmWRI1 Increases Seed Oil Production in Maize. *Plant Physiol.* **2010,** *153* (3), 980–987.

Shen, Y.-G.; Du, B.-X.; Zhang, W.-K.; Zhang, J.-S.; Chen, S.-Y. AhCMO, Regulated by Stresses in *Atriplex hortensis*, Can Improve Drought Tolerance in Transgenic Tobacco. *Theor. Appl. Genet.* **2002,** *105* (6–7), 815–821.

Shimamura, C.; Ohno, R.; Nakamura, C.; Takumi, S. Improvement of Freezing Tolerance in Tobacco Plants Expressing a Cold-Responsive and Chloroplast-Targeting Protein WCOR15 of Wheat. *J. Plant Physiol.* **2006,** *163* (2), 213–219.

Shimizu, T.; Ogamino, T.; Hiraguri, A.; Nakazono-Nagaoka, E.; Uehara-Ichiki, T.; Nakajima, M.; Akutsu, K.; Omura, T.; Sasaya, T. Strong Resistance against Rice Grassy Stunt Virus is Induced in Transgenic Rice Plants Expressing Double-Stranded RNA of the Viral Genes for Nucleocapsid or Movement Proteins as Targets for RNA Interference. *Phytopathology* **2013,** *103* (5), 513–519.

Shin, R.; Park, J. M.; An, J.-M.; Paek, K.-H. Ectopic Expression of Tsi1 in Transgenic Hot Pepper Plants Enhances Host Resistance to Viral, Bacterial, and Oomycete Pathogens. *Mol. Plant–Microbe Interact.* **2002,** *15* (10), 983–989.

Shirasawa, K.; Takabe, T.; Takabe, T.; Kishitani, S. Accumulation of Glycinebetaine in Rice Plants that Overexpress Choline Monooxygenase from Spinach and Evaluation of their Tolerance to Abiotic Stress. *Ann. Bot.* **2006,** *98* (3), 565–571.

Shukla, V. K.; Doyon, Y.; Miller, J. C.; DeKelver, R. C.; Moehle, E. A.; Worden, S. E.; Mitchell, J. C.; et al. Precise Genome Modification in the Crop Species *Zea mays* Using Zinc-Finger Nucleases. *Nature* **2009,** *459* (7245), 437–441.

Siehl, D. L.; Castle, L. A.; Gorton, R.; Keenan, R. J. The Molecular Basis of Glyphosate Resistance by an Optimized Microbial Acetyltransferase. *J. Biol. Chem.* **2007,** *282* (15), 11446–11455.

Sijmons, P. C.; Dekker, B. M. M.; Schrammeijer, B.; Verwoerd, T. C.; Van Den Elzen, P. J. M.; Hoekema, A. Production of Correctly Processed Human Serum Albumin in Transgenic Plants. *Nat. Biotechnol.* **1990,** *8* (3), 217–221.

Simón-Mateo, C.; García, J. A. Antiviral Strategies in Plants Based on RNA Silencing. *Biochim. Biophys. Acta (BBA)—Gene Regul. Mech.* **2011,** *1809* (11), 722–731.

Sindhu, A. S.; Maier, T. R.; Mitchum, M. G.; Hussey, R. S.; Davis, E. L.; Baum, T. J. Effective and Specific in Planta RNAi in Cyst Nematodes: Expression Interference of Four Parasitism Genes Reduces Parasitic Success. *J. Exp. Bot.* **2009,** *60* (1), 315–324.

Sindhu, A. S.; Zheng, Z.; Murai, N. The Pea Seed Storage Protein Legumin Was Synthesized, Processed, and Accumulated Stably in Transgenic Rice Endosperm. *Plant Sci.* **1997,** *130* (2), 189–196.

Singh, A. K.; Verma, S. S.; Bansal, K. C. Plastid Transformation in Eggplant (*Solanum melongena* L.). *Transgen. Res.* **2010,** *19* (1), 113–119.

Singh, A.; Taneja, J.; Dasgupta, I.; Mukherjee, S. K. Development of Plants Resistant to Tomato Geminiviruses Using Artificial Trans-acting Small Interfering RNA. *Mol. Plant Pathol.* **2015,** *16* (7), 724–734.

Singh, S.; Rathore, M.; Goyary, D.; Singh, R. K.; Anandhan, S.; Sharma, D. K.; Ahmed, Z. Induced Ectopic Expression of At-CBF1 in Marker-Free Transgenic Tomatoes Confers Enhanced Chilling Tolerance. *Plant Cell Rep.* **2011,** *30* (6), 1019–1028.

Sirko, A.; Błaszczyk, A.; Liszewska, F. Overproduction of SAT and/or OASTL in Transgenic Plants: A Survey of Effects. *J. Exp. Bot.* **2004,** *55* (404), 1881–1888.

Sivamani, E.; Bahieldin, A.; Wraith, J. M.; Al-Niemi, T.; Dyer, W. E.; Ho, T.-H. D.; Qu, R. Improved Biomass Productivity and Water Use Efficiency under Water Deficit Conditions in Transgenic Wheat Constitutively Expressing the Barley HVA1 Gene. *Plant Sci.* **2000,** *155* (1), 1–9.

Sobczak, M.; Avrova, A.; Jupowicz, J.; Phillips, M. S.; Ernst, K.; Kumar, A. Characterization of Susceptibility and Resistance Responses to Potato Cyst Nematode (*Globodera* spp.) Infection of Tomato Lines in the Absence and Presence of the Broad-Spectrum Nematode Resistance Hero Gene. *Mol. Plant–Microbe Interact.* **2005,** *18* (2), 158–168.

Sojikul, P.; Buehner, N.; Mason, H. S. A Plant Signal Peptide–Hepatitis B Surface Antigen Fusion Protein with Enhanced Stability and Immunogenicity Expressed in Plant Cells. *Proc. Nat. Acad. Sci.* **2003,** *100* (5), 2209–2214.

Somleva, M. N.; Snell, K. D.; Beaulieu, J. J.; Peoples, O. P.; Garrison, B. R.; Patterson, N. A. Production of Polyhydroxybutyrate in Switchgrass, a Value-Added Co-product in an Important Lignocellulosic Biomass Crop. *Plant Biotechnol. J.* **2008,** *6* (7), 663–678.

Sreedharan, S.; Singh Shekhawat, U. K.; Ganapathi, T. R. Constitutive and Stress-Inducible Overexpression of a Native Aquaporin Gene (MusaPIP2; 6) in Transgenic Banana Plants Signals its Pivotal Role in Salt Tolerance. *Plant Mol. Biol.* **2015,** *88* (1–2), 41–52.

Sreekala, C.; Wu, L.; Gu, K.; Wang, D.; Tian, D.; Yin, Z. Excision of a Selectable Marker in Transgenic Rice (*Oryza sativa* L.) Using a Chemically Regulated Cre/loxP System. *Plant Cell Rep.* **2005,** *24* (2), 86–94.

Sripriya, R.; Raghupathy, V.; Veluthambi, K. Generation of Selectable Marker-Free Sheath Blight Resistant Transgenic Rice Plants by Efficient Co-transformation of a Cointegrate Vector T-DNA and a Binary Vector T-DNA in One *Agrobacterium tumefaciens* Strain. *Plant Cell Rep.* **2008,** *27* (10), 1635–1644.

Stefanova, G.; Vlahova, M.; Atanassov, A. Production of Recombinant Human Lactoferrin from Transgenic Plants. *Biol. Plant.* **2008,** *52* (3), 423–428.

Stein, J. C.; Hansen, G. Mannose Induces an Endonuclease Responsible for DNA Laddering in Plant Cells. *Plant Physiol.* **1999,** *121* (1), 71–80.

Stewart, C. N.; Richards, H. A.; Halfhill, M. D. Transgenic Plants and Biosafety: Science, Misconceptions and Public Perceptions. *Biotechniques* **2000,** *29* (4), 832–843.

Sticklen, M. B. Plant Genetic Engineering for Biofuel Production: Towards Affordable Cellulosic Ethanol. *Nat. Rev. Genet.* **2008,** *9* (6), 433–443.

Stöger, E.; Parker, M.; Christou, P.; Casey, R. Pea Legumin Overexpressed in Wheat Endosperm Assembles into an Ordered Paracrystalline Matrix. *Plant Physiol.* **2001,** *125* (4), 1732–1742.

Storozhenko, S.; De Brouwer, V.; Volckaert, M.; Navarrete, O.; Blancquaert, D.; Zhang, G.-F.; Lambert, W.; Van Der Straeten, D. Folate Fortification of Rice by Metabolic Engineering. *Nat. Biotechnol.* **2007,** *25* (11), 1277–1279.

Su, J.; Hirji, R.; Zhang, L.; He, C.; Selvaraj, G.; Wu, R. Evaluation of the Stress-Inducible Production of Choline Oxidase in Transgenic Rice as a Straegy for Producing the Stress-Protectant Glycine Betaine. *J. Exp. Bot.* **2006,** *57* (5), 1129–1135.

Suárez, R.; Calderón, C.; Iturriaga, G. Enhanced Tolerance to Multiple Abiotic Stresses in Transgenic Alfalfa Accumulating Trehalose. *Crop Sci.* **2009,** *49* (5), 1791–1799.

Subramanyam, K.; Sailaja, K. V.; Subramanyam, K.; Muralidhara Rao, D.; Lakshmidevi, K. Ectopic Expression of an Osmotin Gene Leads to Enhanced Salt Tolerance in Transgenic Chilli Pepper (*Capsicum annum* L.). *Plant Cell, Tissue Organ Cult.* **2011,** *105* (2), 181–192.

Sugita, K.; Kasahara, T.; Matsunaga, E.; Ebinuma, H. A Transformation Vector for the Production of Marker-Free Transgenic Plants Containing a Single Copy Transgene at High Frequency. *Plant J.* **2000,** 22 (5), 461–469.

Sujatha, M.; Papi Reddy, T.; Mahasi, M. J. Role of Biotechnological Interventions in the Improvement of Castor (*Ricinus communis* L.) and *Jatropha curcas* L. *Biotechnol. Adv.* **2008,** *26* (5), 424–435.

Sun, J.; Hu, W.; Zhou, R.; Wang, L.; Wang, X.; Wang, Q.; et al. The Brachypodium distachyon BdWRKY36 gene confers tolerance to drought stress in transgenic tobacco plants. *Plant Cell Rep.* **2015**, 34, 23- 35. 35.

Sun, S. S. M. Application of Agricultural Biotechnology to Improve Food Nutrition and Healthcare Products. *Asia Pac. J. Clin. Nutr.* **2008,** *17* (S1), 87–90.

Sunilkumar, G.; Rathore, K. S. Transgenic Cotton: Factors Influencing *Agrobacterium*-Mediated Transformation and Regeneration. *Mol. Breed.* **2001,** *8* (1), 37–52.

Tai, T. H.; Dahlbeck, D.; Clark, E. T.; Gajiwala, P.; Pasion, R.; Whalen, M. C.; Stall, R. E.; Staskawicz, B. J. Expression of the Bs2 Pepper Gene Confers Resistance to Bacterial Spot Disease in Tomato. *Proc. Nat. Acad. Sci.* **1999,** *96* (24), 14153–14158.

Takakura, Y.; Ito, T.; Saito, H.; Inoue, T.; Komari, T.; Kuwata, S. Flower-Predominant Expression of a Gene Encoding a Novel Class I Chitinase in Rice (*Oryza sativa* L.). *Plant Mol. Biol.* **2000,** *42* (6), 883–897.

Tamás, C.; Kisgyörgy, B. N.; Rakszegi, M.; Wilkinson, M. D.; Yang, M.-S. Láng, L.; Tamás, L.; Bedő, Z. Transgenic Approach to Improve Wheat (*Triticum aestivum* L.) Nutritional Quality. *Plant Cell Rep.* **2009,** *28* (7), 1085–1094.

Tang, G.; Qin, J.; Dolnikowski, G. G.; Russell, R. M.; Grusak, M. A. Golden Rice is an Effective Source of Vitamin A. *Am. J. Clin. Nutr.* **2009,** *89* (6), 1776–1783.

Tang, K.; Zhao, E.; Sun, X.; Wan, B.; Qi, H.; Lu, X. Production of Transgenic Rice Homozygous Lines with Enhanced Resistance to the Rice Brown Planthopper. *Acta Biotechnol.* **2001,** *21* (2), 117–128.

Tang, M.; Liu, X.; Deng, H.; Shen, S. Over-expression of JcDREB, a Putative AP2/EREBP Domain-Containing Transcription Factor Gene in Woody Biodiesel Plant *Jatropha curcas*, Enhances Salt and Freezing Tolerance in Transgenic *Arabidopsis thaliana*. *Plant Sci.* **2011,** *181* (6), 623–631.

Tavva, V. S.; Kim, Y.-H.; Kagan, I. A.; Dinkins, R. D.; Kim, K.-H.; Collins, G. B. Increased α-Tocopherol Content in Soybean Seed Overexpressing the *Perilla frutescens* γ-Tocopherol Methyltransferase Gene. *Plant Cell Rep.* **2007,** *26* (1), 61–70.

Teixeira, E. I.; Fischer, G.; van Velthuizen, H.; Walter, C.; Ewert, F. Global Hot-Spots of Heat Stress on Agricultural Crops Due to Climate Change. *Agric. For. Meteorol.* **2013,** *170*, 206–215.

Tester, M.; Langridge, P. Breeding Technologies to Increase Crop Production in a Changing World. *Science* **2010,** *327* (5967), 818–822.

Tian, J.; Wang, X.; Tong, Y.; Chen, X.; Liao, H. Bioengineering and Management for Efficient Phosphorus Utilization in Crops and Pastures. *Curr. Opin. Biotechnol.* **2012,** *23* (6), 866–871.

Tian, Y.-S.; Xu, J.; Xing, X.-J.; Zhao, W.; Fu, X.-Y.; Peng, R.-H.; Yao, Q.-H. Improved Glyphosate Resistance of 5-Enolpyruvylshikimate-3-Phosphate Synthase from *Vitis vinifera* in Transgenic Arabidopsis and Rice by DNA Shuffling. *Mol. Breed.* **2015,** *35* (7), 1–11.

Tilman, D.; Socolow, R.; Foley, J. A.; Hill, J.; Larson, E.; Lynd, L.; Pacala, S.; et al. Beneficial Biofuels—The Food, Energy, and Environment Trilemma. *Science* **2009,** *325* (5938), 270.

Tiwari, L. D.; Mittal, D.; Mishra, C. R.; Grover, A. Constitutive Overexpression of Rice Chymotrypsin Protease Inhibitor Gene ocpi2 Results in Enhanced Growth, Salinity and Osmotic Stress Tolerance of The Transgenic Arabidopsis Plants. *Plant Physiol. Biochem.* **2015,** *92*, 48–55.

Tobias, D. J.; Manoharan, M.; Pritsch, C.; Dahleen, L. S. Co-bombardment, Integration and Expression of Rice Chitinase and Thaumatin-Like Protein Genes in Barley (*Hordeum vulgare* cv. Conlon). *Plant Cell Rep.* **2007,** *26* (5), 631–639.

Tohidfar, M.; Khosravi, S. Transgenic Crops with an Improved Resistance to Biotic Stresses. A Review. *Base* **2015,** *19* (1), 62–70.

Tohidfar, M.; Ghareyazie, B.; Mosavi, M.; Yazdani, S.; Golabchian, R. Agrobacterium-Mediated Transformation of Cotton (*Gossypium hirsutum*) Using a Synthetic cry1Ab Gene for Enhanced Resistance against *Heliothis armigera*. *Iran. J. Biotechnol.* **2008,** *6* (3), 164–173.

Tohidfar, M.; Mohammadi, M.; Ghareyazie, B. *Agrobacterium*-Mediated Transformation of Cotton (*Gossypium hirsutum*) Using a Heterologous Bean Chitinase Gene. *Plant Cell, Tissue Organ Cult.* **2005,** *83* (1), 83–96.

Tohidfar, M.; Zare, N.; Jouzani, G. S.; Eftekhari, S. M. *Agrobacterium*-Mediated Transformation of Alfalfa (*Medicago sativa*) Using a Synthetic cry3a Gene to Enhance Resistance against Alfalfa Weevil. *Plant Cell Tissue Organ Cult.* **2013,** *113* (2), 227–235.

Tougou, M.; Yamagishi, Noriyuki Furutani, N., Shizukawa, Y.; Takahata, Y.; Soh Hidaka. Soybean Dwarf Virus-Resistant Transgenic Soybeans with the Sense Coat Protein Gene. *Plant Cell Rep.* **2007,** *26* (11), 1967–1975.

Townsend, J. A.; Wright, D. A.; Winfrey, R. J.; Fu, F.; Maeder, M. L.; Keith Joung, J.; Voytas, D. F. High-Frequency Modification of Plant Genes Using Engineered Zinc-Finger Nucleases. *Nature* **2009,** *459* (7245), 442–445.

Trivedi, P. K.; Nath, P. MaExp1, an Ethylene-Induced Expansin from Ripening Banana Fruit. *Plant Sci.* **2004,** *167* (6), 1351–1358.

Tuboly, T.; Yu, W.; Bailey, A.; Degrandis, S.; Du, S.; Erickson, L.; Nagy, E. Immunogenicity of Porcine Transmissible Gastroenteritis Virus Spike Protein Expressed in Plants. *Vaccine* **2000,** *18* (19), 2023–2028.

Tuteja, N.; Verma, S.; Sahoo, R.K.; Raveendar, S. and Reddy, I.N. Recent advances in development of marker-free transgenic plants: regulation and biosafety concern. *J. Biosci.* **2012,** *37*, 167–197.

Tuteja, N.; Sahoo, R. K.; Garg, B.; Tuteja, R. OsSUV3 dual helicase functions in salinity stress tolerance by maintaining photosynthesis and antioxidant machinery in rice (Oryza sativa L. cv. IR64). *Plant J.* **2013,** *76*, 115–127.

Twyman, R. M.; Christou, P.; Stoger, E. Genetic Transformation of Plants and their Cells. *Plant Biotechnology and Transgenic Plants*; Marcel Dekker: New York, 2002; pp 111–141.

Upadhyaya, C. P.; Akula, N.; Young, K. E.; Chun, S. C.; Kim, D. H.; Park, S. W. Enhanced Ascorbic Acid Accumulation in Transgenic Potato Confers Tolerance to Various Abiotic Stresses. *Biotechnol. Lett.* **2010,** *32* (2), 321–330.

Usher, S.; Haslam, R. P.; Ruiz-López, N.; Sayanova, O.; Napier, J. A. Field Trial Evaluation of the Accumulation of Omega-3 Long Chain Polyunsaturated Fatty Acids in Transgenic *Camelina sativa*: Making Fish Oil Substitutes in Plants. *Metab. Eng. Commun.* **2015,** *2*, 93–98.

Van Aken, B.; Correa, P. A.; Schnoor, J. L. Phytoremediation of Polychlorinated Biphenyls: New Trends and Promises. *Environ. Sci. Technol.* **2009,** *44* (8), 2767–2776.

Vanblaere, T.; Szankowski, I.; Schaart, J.; Schouten, H.; Flachowsky, H.; Broggini, G. A. L.; Gessler, C. The Development of a Cisgenic Apple Plant. *J. Biotechnol.* **2011,** *154* (4), 304–311.

Vega-Sánchez, M. E.; Ronald, P. C. Genetic and Biotechnological Approaches for Biofuel Crop Improvement. *Curr. Opin. Biotechnol.* **2010,** *21* (2), 218–224.

Vellicce, G. R.; Díaz Ricci, J. C.; Hernández, L.; Castagnaro, A. P. Enhanced Resistance to *Botrytis cinerea* Mediated by the Transgenic Expression of the Chitinase Gene ch5B in Strawberry. *Transgen. Res.* **2006,** *15* (1), 57–68.

Vigeolas, H.; Waldeck, P.; Zank, T.; Geigenberger, P. Increasing Seed Oil Content in Oil-Seed Rape (*Brassica napus* L.) by Over-expression of a Yeast Glycerol-3-Phosphate

Dehydrogenase under the Control of a Seed-Specific Promoter. *Plant Biotechnol. J.* **2007,** *5* (3), 431–441.

Villani, M. E.; Morgun, B.; Brunetti, P.; Marusic, C.; Lombardi, R.; Pisoni, I.; Bacci, C.; Desiderio, A.; Benvenuto, E.; Donini, M. Plant Pharming of a Full-Sized, Tumour-Targeting Antibody Using Different Expression Strategies. *Plant Biotechnol. J.* **2009,** *7* (1), 59–72.

Vishnevetsky, J.; White, Jr., T. L.; Palmateer, A. J.; Flaishman, M.; Cohen, Y.; Elad, Y.; Velcheva, M.; et al. Improved Tolerance Toward Fungal Diseases in Transgenic *Cavendish banana* (*Musa* spp. AAA Group) cv. Grand Nain. *Transgen. Res.* **2011,** *20* (1), 61–72.

Vishnudasan, D.; Tripathi, M. N.; Rao, U.; Khurana, P. Assessment of Nematode Resistance in Wheat Transgenic Plants Expressing Potato Proteinase Inhibitor (PIN2) Gene. *Transgen. Res.* **2005,** *14* (5), 665–675.

Vleeshouwers, V. G. A. A.; Rietman, H.; Krenek, P.; Champouret, N.; Young, C.; Oh, S.-K.; Wang, M.; et al. Effector Genomics Accelerates Discovery and Functional Profiling of Potato Disease Resistance and *Phytophthora infestans* Avirulence Genes. *PLoS ONE* **2008,** *3* (8), e2875.

Voytas, D. F.; Gao, C. Precision Genome Engineering and Agriculture: Opportunities and Regulatory Challenges. *PLoS Biol.* **2014,** *12* (6), e1001877.

Waie, B.; Rajam, M. V. Effect of Increased Polyamine Biosynthesis on Stress Responses in Transgenic Tobacco by Introduction of Human S-Adenosylmethionine Gene. *Plant Sci.* **2003,** *164* (5), 727–734.

Wakasa, K.; Hasegawa, H.; Nemoto, H.; Matsuda, F.; Miyazawa, H.; Tozawa, Y.; Morino, K.; et al. High-level Tryptophan Accumulation in Seeds of Transgenic Rice and its Limited Effects on Agronomic Traits and Seed Metabolite Profile. *J. Exp. Bot.* **2006,** *57* (12), 3069–3078.

Wally, O.; Punja, Z. K. Genetic Engineering for Increasing Fungal and Bacterial Disease Resistance in Crop Plants. *GM Crops* **2010,** *1* (4), 199–206.

Wally, O.; Jayaraj, J.; Punja, Z. Comparative Resistance to Foliar Fungal Pathogens in Transgenic Carrot Plants Expressing Genes Encoding for Chitinase, β-1,3-Glucanase and Peroxidise. *Eur. J. Plant Pathol.* **2009,** *123* (3), 331–342.

Walmsley, A. M.; Arntzen, C. J. Plants for Delivery of Edible Vaccines. *Curr. Opin. Biotechnol.* **2000,** *11* (2), 126–129.

Wang, C.; Ying, S.; Huang, H.; Li, K.; Wu, P.; Shou, H. Involvement of OsSPX1 in Phosphate Homeostasis in Rice. *Plant J.* **2009,** *57* (5), 895–904.

Wang, H.-W.; Zhang, B.; Hao, Y.-J.; Huang, J.; Tian, A.-G.; Liao, Y.; Zhang, J.-S.; Chen, S.-Y. The Soybean Dof-Type Transcription Factor Genes, GmDof4 and GmDof11, Enhance Lipid Content in the Seeds of Transgenic Arabidopsis Plants. *Plant J.* **2007,** *52* (4), 716–729.

Wang, Y.; Ren, H.; Pan, H.; Liu, J.; Zhang, L. Enhanced Tolerance and Remediation to Mixed Contaminates of PCBs and 2,4-DCP by Transgenic Alfalfa Plants Expressing the 2,3-Dihydroxybiphenyl-1,2-Dioxygenase. *J. Hazard. Mater.* **2015,** *286*, 269–275.

Wang, Y.; Cheng, X.; Shan, Q.; Zhang, Y.; Liu, J.; Gao, C.; Qiu, J.-L. Simultaneous Editing of Three Homoeoalleles in Hexaploid Bread Wheat Confers Heritable Resistance to Powdery Mildew. *Nat. Biotechnol.* **2014,** *32* (9), 947–951.

Wankhede, D. P.; Misra, M.; Singh, P.; Sinha, A. K. Rice Mitogen Activated Protein Kinase Kinase and Mitogen Activated Protein Kinase Interaction Network Revealed by In-Silico Docking and Yeast Two-Hybrid Approaches. *PLoS ONE* **2013,** *8* (5), e65011.

Weeks, T. J.; Ye, J.; Rommens, C. M. Development of an In Planta Method for Transformation of Alfalfa (*Medicago sativa*). *Transgen. Res.* **2008,** *17* (4), 587–597.

Wei, A. Y.; He, C. M.; Li, B.; Li, N.; Zhang, J. R. The Pyramid of Transgenes TsVP and BetA Effectively Enhances the Drought Tolerance of Maize Plants. *Plant Biotechnol. J.* **2011,** *9* (2), 216–229.

Weise, A.; Altmann, F.; Rodriguez-Franco, M.; Sjoberg, E. R.; Bäumer, W.; Launhardt, H.; Kietzmann, M.; Gorr, G. High-Level Expression of Secreted Complex Glycosylated Recombinant Human Erythropoietin in the Physcomitrella Δ-fuc-t Δ-xyl-t Mutant. *Plant Biotechnol. J.* **2007,** *5* (3), 389–401.

Welsch, R.; Arango, J.; Bär, C.; Salazar, B.; Al-Babili, S.; Beltrán, J.; Chavarriaga, P.; Ceballos, H.; Tohme, J.; Beyer, P. Provitamin A Accumulation in Cassava (*Manihot esculenta*) Roots Driven by a Single Nucleotide Polymorphism in a Phytoene Synthase Gene. *Plant Cell* **2010,** *22* (10), 3348–3356.

Wen, X. P.; Pang, X. M.; Matsuda, N.; Kita, M.; Inoue, H.; Hao, Y. J.; Honda, C.; Moriguchi, T. Over-expression of the apple spermidine synthase gene in pear confers multiple abiotic stress tolerance by altering polyamine titers. *Transgenic Res.* **2008**, 17, 251–263.

Weng, J.-K.; Li, X.; Bonawitz, N. D.; Chapple, C. Emerging Strategies of Lignin Engineering and Degradation for Cellulosic Biofuel Production. *Curr. Opin. Biotechnol.* **2008,** *19* (2), 166–172.

Weng, L.-X.; Deng, H.-H.; Xu, J.-L.; Li, Q.; Zhang, Y.-Q.; Jiang, Z.-D.; Li, Q.-W.; Chen, J.-W.; Zhang, L.-H. Transgenic Sugarcane Plants Expressing High Levels of Modified cry1Ac Provide Effective Control against Stem Borers in Field Trials. *Transgen. Res.* **2011,** *20* (4), 759–772.

Wirth, S.; Calamante, G.; Mentaberry, A.; Bussmann, L.; Lattanzi, M.; Barañao, L.; Bravo-Almonacid, F. Expression of Active Human Epidermal Growth Factor (hEGF) in Tobacco Plants by Integrative and Non-integrative Systems. *Mol. Breed.* **2004,** *13* (1), 23–35.

Woo, H.-J.; Qin, Y.; Park, S.-Y.; Park, S. K.; Cho, Y.-G.; Shin, K.-S.; Lim, M.-H.; Cho, H.-S. Development of Selectable Marker-Free Transgenic Rice Plants with Enhanced Seed Tocopherol Content through FLP/FRT-Mediated Spontaneous Auto-excision. *PLoS ONE* **2015,** *10* (7), e0132667.

World Health Organization. The World Health Report 2007—A Safer Future: Global Public Health Security in the 21st Century, 2007.

Wright, T. R.; Shan, G.; Walsh, T. A.; Lira, J. M.; Cui, C.; Song, P.; Zhuang, M.; et al. Robust Crop Resistance to Broadleaf and Grass Herbicides Provided by Aryloxyalkanoate Dioxygenase Transgenes. *Proc. Nat. Acad. Sci.* **2010,** *107* (47), 20240–20245.

Wróbel-Kwiatkowska, M.; Lorenc-Kukula, K.; Starzycki, M.; Oszmiański, J.; Kepczyńska, E.; Szopa, J. Expression of β-1,3-Glucanase in Flax Causes Increased Resistance to Fungi. *Physiol. Mol. Plant Pathol.* **2004,** *65* (5), 245–256.

Wu, F. Mycotoxin Reduction in Bt Corn: Potential Economic, Health, and Regulatory Impacts. *Transgen. Res.* **2006,** *15* (3), 277–289.

Wu, G.; Truksa, M.; Datla, N.; Vrinten, P.; Bauer, J.; Zank, T.; Cirpus, P.; Heinz, E.; Qiu, X. Stepwise Engineering to Produce High Yields of Very Long-Chain Polyunsaturated Fatty Acids in Plants. *Nat. Biotechnol.* **2005,** *23* (8), 1013–1017.

Wu, X. R.; Kenzior, A.; Willmot, D.; Scanlon, S.; Chen, Z.; Topin, A.; He, S. H.; Acevedo, A.; Folk, W. R. Altered Expression of Plant Lysyl tRNA Synthetase Promotes tRNA Misacylation and Translational Recoding of Lysine. *Plant J.* **2007,** *50* (4), 627–636.

Würdig, J.; Flachowsky, H.; Saß, A.; Peil, A.; Hanke, M.-V. Improving Resistance of Different Apple Cultivars Using the Rvi6 Scab Resistance Gene in a Cisgenic Approach Based on the Flp/FRT Recombinase System. *Mol. Breed.* **2015,** *35* (3), 1–18.

Xia, H.; Chi, X.; Yan, Z.; Cheng, W. Enhancing Plant Uptake of Polychlorinated Biphenyls and Cadmium Using Tea Saponin. *Bioresour. Technol.* **2009,** *100* (20), 4649–4653.

Xiao, B.; Huang, Y.; Tang, N.; Xiong, L. Over-expression of a LEA Gene in Rice Improves Drought Resistance under the Field Conditions. *Theor. Appl. Genet.* **2007,** *115* (1), 35–46.

Xiong, Y.; Zeng, H.; Zhang, Y.; Xu, D.; Qiu, D. Silencing the HaHR3 Gene by Transgenic Plant-Mediated RNAi to Disrupt *Helicoverpa armigera* Development. *Int. J. Biol. Sci.* **2013,** *9* (4), 370.

Xu, J.; Dolan, M. C.; Medrano, G.; Cramer, C. L.; Weathers, P. J. Green Factory: Plants as Bioproduction Platforms for Recombinant Proteins. *Biotechnol. Adv.* **2012,** *30* (5), 1171–1184.

Yadav, J. S.; Ogwok, E.; Wagaba, H.; Patil, B. L.; Bagewadi, B.; Alicai, T.; Gaitan-Solis, E.; Taylor, N. J.; Fauquet, C. M. RNAi-Mediated Resistance to *Cassava brown* Streak Uganda Virus in Transgenic Cassava. *Mol. Plant Pathol.* **2011a,** *12* (7), 677–687.

Yadav, B. C.; Veluthambi, K.; Subramaniam, K. Host-Generated Double Stranded RNA Induces RNAi in Plant-Parasitic Nematodes and Protects the Host from Infection. *Mol. Biochem. Parasitol.* **2006,** *148* (2), 219–222.

Yadav, S. K.; Singla-Pareek, S. L.; Reddy, M. K.; Sopory, S. K. Transgenic Tobacco Plants Overexpressing Glyoxalase Enzymes Resist an Increase in Methylglyoxal and Maintain Higher Reduced Glutathione Levels under Salinity Stress. *FEBS Lett.* **2005,** *579* (27), 6265–6271.

Yamada, M.; Morishita, H.; Urano, K.; Shiozaki, N.; Yamaguchi-Shinozaki, K.; Shinozaki, K.; Yoshiba, Y. Effects of Free Proline Accumulation in Petunias under Drought Stress. *J. Exp. Bot.* **2005a,** *56* (417), 1975–1981.

Yamada, T.; Tozawa, Y.; Hasegawa, H.; Terakawa, T.; Ohkawa, Y.; Wakasa, K. Use of a Feed-back-Insensitive α Subunit of Anthranilate Synthase as a Selectable Marker for Transformation of Rice and Potato. *Mol. Breed.* **2005b,** *14* (4), 363–373.

Yamaguchi, T.; Blumwald, E. Developing Salt-Tolerant Crop Plants: Challenges and Opportunities. *Trends Plant Sci.* **2005,** *10* (12), 615–620.

Yamamoto, T.; Iketani, H.; Ieki, H.; Nishizawa, Y.; Notsuka, K.; Hibi, T.; Hayashi, T.; Matsuta, N. Transgenic Grapevine Plants Expressing a Rice Chitinase with Enhanced Resistance to Fungal Pathogens. *Plant Cell Rep.* **2000,** *19* (7), 639–646.

Yang, J.; Chen, Z.; Wu, S.; Cui, Y.; Zhang, L.; Dong, H.; Yang, C.; Li, C. Overexpression of the *Tamarix hispida* ThMT3 Gene Increases Copper Tolerance and Adventitious Root Induction in *Salix matsudana* Koidz. *Plant Cell, Tissue Organ Cult.* **2015,** *121* (2), 469–479.

Yang, H.; Knapp, J.; Koirala, P.; Rajagopal, D.; Peer, W. A.; Silbart, L. K.; Murphy, A.; Gaxiola, R. A.. Enhanced phosphorus nutrition in monocots and dicots over-expressing a phosphorus-responsive type I H+-pyrophosphatase. *Plant Biotechnol. J.* **2007**, 5, 735–745.

Yang, X.; Xia, H.; Wang, W.; Wang, F.; Su, J.; Snow, A. A.; Lu, B.-R. Transgenes for Insect Resistance Reduce Herbivory and Enhance Fecundity in Advanced Generations of Crop–Weed Hybrids of Rice. *Evol. Appl.* **2011a,** *4* (5), 672–684.

Yang, Y.; Jittayasothorn, Y.; Chronis, D.; Wang, X.; Cousins, P.; Zhong, G.-Y. Molecular Characteristics and Efficacy of 16D10 siRNAs in Inhibiting Root-Knot Nematode Infection in Transgenic Grape Hairy Roots. *PLoS ONE* **2013,** *8* (7), e69463.

Yang, Z.; Wu, Y.; Li, Y.; Ling, H.-Q.; Chu, C. OsMT1a, a Type 1 Metallothionein, Plays the pivotal role in Zinc Homeostasis and Drought Tolerance in Rice. *Plant Mol. Biol.* **2009,** *70* (1–2), 219–229.

Yang, Z.; Chen, H.; Tang, W.; Hua, H.; Lin, Y. Development and Characterisation of Transgenic Rice Expressing two *Bacillus thuringiensis* Genes. *Pest Manage. Sci.* **2011b,** *67* (4), 414–422.

Yarra, R.; He, S.-J.; Abbagani, S.; Ma, B.; Bulle, M.; Zhang, W.-K. Overexpression of a Wheat Na^+/H^+ Antiporter Gene (TaNHX2) Enhances Tolerance to Salt Stress in Transgenic Tomato Plants (*Solanum lycopersicum* L.). *Plant Cell, Tissue Organ Cult.* **2012,** *111* (1), 49–57.

Ye, G.-N.; Colburn, S. M.; Xu, C. W.; Hajdukiewicz, P. T. J.; Staub, J. M. Persistence of Unselected Transgenic DNA During a Plastid Transformation and Segregation Approach to Herbicide Resistance. *Plant Physiol.* **2003,** *133* (1), 402–410.

Ye, X.; Al-Babili, S.; Klöti, A.; Zhang, J.; Lucca, P.; Beyer, P.; Potrykus, I. Engineering the Provitamin A (β-Carotene) Biosynthetic Pathway into (Carotenoid-Free) Rice Endosperm. *Science* **2000,** *287* (5451), 303–305.

Yi, K.; Wu, Z.; Zhou, J.; Du, L.; Guo, L.; Wu, Y.; Wu, P. OsPTF1, a Novel Transcription Factor Involved in Tolerance to Phosphate Starvation in Rice. *Plant Physiol.* **2005,** *138* (4), 2087–2096.

Yi, D.; Cui, L.; Wang, L.; Liu, Y.; Zhuang, M.; Zhang, Y.; Yang, L. Pyramiding of Bt cry1Ia8 and cry1Ba3 genes into cabbage (*Brassica oleracea* L. var. capitata) confers effective control against diamondback moth. *Plant Cell Tissue and Organ Culture,* **2013**, 115, 419–428.

Yu, J.; Peng, P.; Zhang, X.; Zhao, Q.; Zhy, D.; Sun, X.; Liu, J.; Ao, G. Seed-Specific Expression of a Lysine Rich Protein sb401 Gene Significantly Increases both Lysine and Total Protein Content in Maize Seeds. *Mol. Breed.* **2004,** *14* (1), 1–7.

Yu, L.; Chen, X.; Wang, Z.; Wang, S.; Wang, Y.; Zhu, Q.; Li, S.; Xiang, C. Arabidopsis Enhanced Drought Tolerance 1/HOMEODOMAIN GLABROUS11 Confers Drought Tolerance in Transgenic Rice without Yield Penalty. *Plant Physiol.* **2013,** *162* (3), 1378–1391.

Yue, Y.; Zhang, M.; Zhang, J.; Duan, L.; Li, Z. Arabidopsis LOS5/ABA3 Overexpression in Transgenic Tobacco (*Nicotiana tabacum* cv. *Xanthi-nc*) Results Inenhanced Drought Tolerance. *Plant Sci.* **2011,** *181* (4), 405–411.

Zang, N.; Zhai, H.; Gao, S.; Chen, W.; He, S.; Liu, Q. Efficient Production of Transgenic Plants Using the Bar Gene for Herbicide Resistance in Sweetpotato. *Sci. Hortic.* **2009,** *122* (4), 649–653.

Zasloff, M. Magainins, a Class of Antimicrobial Peptides from Xenopus Skin: Isolation, Characterization of Two Active Forms, and Partial cDNA Sequence of a Precursor. *Proc. Nat. Acad. Sci.* **1987,** *84* (15), 5449–5453.

Zeeb, B. A.; Amphlett, J. S.; Rutter, A.; Reimer, K. J. Potential for Phytoremediation of Polychlorinated Biphenyl-(PCB)-Contaminated Soil. *Int. J. Phytoremed.* **2006,** *8* (3), 199–221.

Zha, W.; Peng, X.; Chen, R.; Du, B.; Zhu, L.; He, G. Knockdown of Midgut Genes by dsRNA-Transgenic Plant-Mediated RNA Interference in the Hemipteran Insect *Nilaparvata lugens*. *PLoS ONE* **2011,** *6* (5), e20504.

Zhang, F.; Wen, Y.; Guo, X. CRISPR/Cas9 for Genome Editing: Progress, Implications and Challenges. *Hum. Mol. Genet.* **2014,** *23* (R1), R40–R46.

Zhang, H.; Dong, H.; Li, W.; Sun, Y.; Chen, S.; Kong, X. Increased Glycine Betaine Synthesis and Salinity Tolerance in AhCMO Transgenic Cotton Lines. *Mol. Breed.* **2009,** *23* (2), 289–298.

Zhang, K.; Wang, J.; Lian, L.; Fan, W.; Guo, N.; Lv, S. Increased Chilling Tolerance Following Transfer of a betA Gene Enhancing Glycinebetaine Synthesis in Cotton (*Gossypium hirsutum* L.). *Plant Mol. Biol. Rep.* **2012,** *30* (5), 1158–1171.

Zhang, N.; Si, H.-J.; Wen, G.; Du, H.-H.; Liu, B.-L.; Wang, D. Enhanced Drought and Salinity Tolerance in Transgenic Potato Plants with a BADH Gene from Spinach. *Plant Biotechnol. Rep.* **2011,** *5* (1), 71–77.

Zhang, P.; Vanderschuren, H.; Fütterer, J.; Gruissem, W. Resistance to Cassava Mosaic Disease in Transgenic Cassava Expressing Antisense RNAs Targeting Virus Replication Genes. *Plant Biotechnol. J.* **2005,** *3* (4), 385–397.

Zhang, S.; Li, N.; Gao, F.; Yang, A.; Zhang, J. Over-expression of TsCBF1 Gene Confers Improved Drought Tolerance in Transgenic Maize. *Mol. Breed.* **2010a,** *26* (3), 455–465.

Zhang, X.; Guo, X.; Lei, C.; Cheng, Z.; Lin, Q.; Wang, J.; Wu, F.; Wang, J.; Wan, J. Over-expression of SlCZFP1, a Novel TFIIIA-Type Zinc Finger Protein from Tomato, Confers Enhanced Cold Tolerance in Transgenic Arabidopsis and Rice. *Plant Mol. Biol. Rep.* **2011a,** *29* (1), 185–196.

Zhang, X.; Sato, S.; Ye, X.; Dorrance, A. E.; Jack Morris, T.; Clemente, T. E.; Qu, F. Robust RNAi-Based Resistance to Mixed Infection of Three Viruses in Soybean Plants Expressing Separate Short Hairpins from a Single Transgene. *Phytopathology* **2011b,** *101* (11), 1264–1269.

Zhang, X.; Francis, M. I.; Dawson, W. O.; Graham, J. H.; Orbović, V.; Triplett, E. W.; Mou, Z. Over-expression of the Arabidopsis NPR1 Gene in Citrus Increases Resistance to Citrus Canker. *Eur. J. Plant Pathol.* **2010b,** *128* (1), 91–100.

Zhang, Z.; Huang, R. Enhanced Tolerance to Freezing in Tobacco and Tomato Overexpressing Transcription Factor TERF2/LeERF2 Is Modulated by Ethylene Biosynthesis. *Plant Mol. Biol.* **2010c,** *73* (3), 241–249.

Zhao, B.; Lin, X.; Poland, J.; Trick, H.; Leach, J.; Hulbert, S. A Maize Resistance Gene Functions against Bacterial Streak Disease in Rice. *Proc. Nat. Acad. Sci.* U. S. A. **2005,** *102* (43), 15383–15388.

Zhu, C.; Naqvi, S.; Breitenbach, J.; Sandmann, G.; Christou, P.; Capella, T. Combinatorial genetic transformation generates a library of metabolic phenotypes for the carotenoid pathway in maize. *Proc. Natl. Acad. Sci. U. S. A.* **2008**, 105(47), 18232–18237.

Zheng, L.; Cheng, Z.; Ai, C.; Jiang, X.; Bei, X.; Zheng, Y.; Glahn, R. P.; et al. Nicotianamine, a Novel Enhancer of Rice Iron Bioavailability to Humans. *PLoS ONE* **2010,** *5* (4), e10190.

Zhou, J.; Jiao, F. C.; Wu, Z.; Li, Y.; Wang, X.; He, X.; Zhong, W.; Wu, P. OsPHR2 is Involved in Phosphate-Starvation Signaling and Excessive Phosphate Accumulation in Shoots of Plants. *Plant Physiol.* **2008,** *146* (4), 1673–1686.

Zhou, M.; Luo, H. MicroRNA-mediated gene regulation: potential applications for plant genetic engineering. *Plant Mol Biol.* **2013,** *83*(1-2):59–75.

Zhu, C.; Naqvi, S.; Breitenbach, J.; Sandmann, G.; Christou, P.; Capell, T. Combinatorial Genetic Transformation Generates a Library of Metabolic Phenotypes for the Carotenoid Pathway in Maize. *Proc. Nat. Acad. Sci.* **2008,** *105* (47), 18232–18237.

Zhu, C.; Naqvi, S.; Gomez-Galera, S.; Pelacho, A. M.; Capell, T.; Christou, P. Transgenic Strategies for the Nutritional Enhancement of Plants. *Trends Plant Sci.* **2007a,** *12* (12), 548–555.

Zhu, H. C.; Xu, X. P.; Xiao, G. Y.; Yuan, L. P.; Li, B. J. Enhancing Disease Resistances of Super Hybrid Rice with Four Antifungal Genes. *Sci. China Ser. C: Life Sci.* **2007b,** *50* (1), 31–39.

Zhu, J.-Q.; Liu, S.; Ma, Y.; Zhang, J.-Q.; Qi, H.-S.; Wei, Z.-J.; Yao, Q.; Zhang, W.-Q.; Li, S. Improvement of Pest Resistance in Transgenic Tobacco Plants Expressing dsRNA of an Insect-Associated Gene EcR. *PLoS ONE* **2012,** *7* (6), e38572.

Zhu, Z.; Hughes, K. W.; Huang, L.; Sun, B.; Liu, C.; Li, Y. Expression of Human α-Interferon cDNA in Transgenic Rice Plants. *Plant Cell, Tissue Organ Cult.* **1994,** *36* (2), 197–204.

Zimmermann, M. B.; Hurrell, R. F. Improving Iron, Zinc and Vitamin A Nutrition through Plant Biotechnology. *Curr. Opin. Biotechnol.* **2002,** *13* (2), 142–145.

Zuo, J.; Niu, Q.-W.; Møller, S. G.; Chua, N.-H. Chemical-Regulated, Site-Specific DNA Excision in Transgenic Plants. *Nat. Biotechnol.* **2001,** *19* (2), 157–161.

PART IV
Microbial Biotechnology

CHAPTER 11

MICROBIAL BIOTECHNOLOGY: ROLE OF MICROBES IN SUSTAINABLE AGRICULTURE

DEEPAK KUMAR[1], MAHESH KUMAR[2], P. VERMA[3], and MD. SHAMIM[2*]

[1]*Research and Development Division, Shri Ram Solvent Extractions Pvt. Ltd., Jaspur 244712, Uttarakhand, India*

[2]*Department of Molecular Biology and Genetic Engineering, Bihar Agricultural University, Sabour 813219, Bihar, India*

[3]*Division of Crop Protection, ICAR—Central Institute of Cotton Research, Nagpur 440010, Maharashtra, India*

[*]*Corresponding author. E-mail: shamimnduat@gmail.com*

CONTENTS

ABSTRACT

Biotechnology is a very fast growing division in biological sciences in present days and it also expanded in sustainable agriculture production. Beneficial microbes in agricultural systems are used as biofertilizers, bio-pesticides, bio-herbicides, bioinsecticides, fungal based bioinsecticides, bacterial based bioisecticides and several viral based bioinsecticides for the enhance production and protection of cereals and other plants. With the help of new biotechnological tools, exploitation of microbial genomes as well as invention of many new, creative ways are utilized for the production of new important beneficial microbes. The research challenge is to meet sustainable environmental and economical issues without compromising yields. In this context, exploiting the agro-ecosystem services of soil microbial communities appears as a promising effective approach. In this chapter, role of microbial biotechnology in sustainable agriculture are briefly discussed. The purpose of this chapter is to convey the impact, the extraordinary breadth of applications, and the multidisciplinary nature of microbial biotechnology.

11.1 INTRODUCTION

In current scenario, a lot of attention has been paid to the promotion of sustainable agriculture in which the elevated productivities of agricultural crops are possible by the use of their natural adaptive potentials, with a negligible disturbance of the environment without compromising yields. Under this circumstance, exploiting the agro-ecosystem services of soil microbial communities appears as a promising approach. By 2050, agricultural production is expected to increase by at least 70%. At the same time, people are becoming conscious that sustainable agricultural exercises are fundamental to gathering the future world's agricultural stipulates (Altieri, 2004). This is why present agriculture is being realized on a global magnitude, and various research approaches are being undertaken to tackle environmental and economical sustainability issues. Thus, a recommended approach based on exploiting the role of soil microbial communities for a sustainable and healthy crop production is needed for preserving the biosphere. It is well known that soil microorganisms play an important role (microbial services) in agriculture by improving plant nutrition and health, as well as soil quality and other parameters (Barea et al., 2013; Lugtenberg, 2015). Consequently, a number of approaches for a more successful utilization of beneficial microbial services, as a low input in the field of biotechnology which helps sustainable and environment-friendly agrotechnological practices have been proposed. The final goal of biotechnology is to boost the role of the root-associated microbiome in nutrient supply and plant protection in sustainable agriculture practices (Raaijmakers and Lugtenberg, 2013). Interactions between the important microbial communities and crops are influenced by varied ecological features and agronomic supervisions; the collision of environmental stress features must be deemed meticulously in the existing situation of universal amend, as they involve an appropriate supervision of the crop–microbiome interactions (Zolla et al., 2013).

In the present scenario, use of advantageous microorganisms in as the replacement or the reduction of chemicals has been so far demonstrated by the agricultural researchers (Burdman et al., 2000; Dobbelaere et al., 2003). Several beneficial microorganisms such as diazotrophs bacteria, biological control agents (BCAs), plant-growth-promoting rhizobacteria (PGPRs), and fungi (PGPFs) can play an important role in this major confront, as they accomplish important ecosystem gatherings for plants and soil (Hermosa et al., 2011; Raaijmakers et al., 2009). Moreover, current agriculture, supported by the cultivation of a very large number of crop species and cultivars, is at risk to epidemic diseases, conventionally featured through the use of chemicals fertilizers. Presently, for the most crops, there are no effective fungicides available against a large number of fungal diseases. Plant growth motivation and crop safety may get better by the direct application of a number of microorganisms known to act as biofertilizers and/or bioprotectors. How these beneficial microorganism benefits to the plants as well as soil is under progress, however full mechanism is only partially known yet. There are several molecular interaction involve in between the PGPR and plants such as (1) production of metabolites connected to root improvement and growth, pests, and pathogen control (phytohormones, antimicrobials, antibiotics, growth repellent, insecticidal) by the microorganism and (2) the difficulty to discriminate the straight effects on the specific/total actions and the oblique effects due to the improved availability of nutrients and growth regulators.

This chapter is a summary of the strategies addressed to an effective exploitation of beneficial microbial packages in sustainable agriculture.

11.2 WHAT IS BIOFERTILIZERS?

Biofertilizers are basically live formulates which include living microorganisms which, when pertained to seed, plant surfaces, root, or soil, inhabit around the rhizosphere and boost the bioavailability of nutrients and escalating the microflora through their biological activities, and thereby promoting plant's growth. Biofertilizers are formulations that readily progress the fertility of land using biological agents (Babalola, 2010; Schoebitz et al., 2014). Biofertilizers are collected and prepared from biological wastes and are not hazardous to soil. Biofertilizers are not only valuable for the enriching soil quality but also facilitate to fight with the pathogens.

Besides accessing nutrients to the plants, for current intake as well as residual, different biofertilizers also supply growth-promoting aspects to plants and a little have been successfully facilitating composting and effective recycling of solid wastes. Biofertilizers, depending on accessible or present microorganisms, have come up as a replacement for chemical fertilizers to augment soil fertility and crop yield in sustainable agriculture. Symbiotic, free-living soil bacteria are named PGPR. They are engaged in significant ecosystem developments, and their performance includes biological control of plant pathogens, nitrogen fixation, mineralization of nutrients, and phytohormones production. The capability of microorganism of the above-mentioned qualities, they occupy a unique place in the sustainability of agroecosystems.

11.3 MICROORGANISM USED IN BIOFERTILIZERS

Microorganisms that are usually employed as biofertilizers constituent are nitrogen fixers (N-fixer), potassium solubilizer (K-solubilizer), and phosphorus solubilizer (P-solubilizer), or with the mixture of molds or fungi. Most of the bacteria included in biofertilizer have close association with plant roots. Rhizobium has symbiotic interaction with legume roots, and rhizobacteria reside on the root surface or in rhizosphere soil. The chief resources of biofertilizers are bacteria, fungi, and cyanobacteria (blue-green algae). The association of these organisms have with plants is referred to as symbiosis. In this case, both collaborators derive benefits from each other (Babalola and Glick, 2012a; Simmons et al., 2014). It is also very important to state that there are few instances when they have least or no effects. Table 11.1 demonstrates the names, description, and probable locations of some microbes that are useful as biofertilizers.

11.3.1 BACTERIAL BIOFERTILIZERS USED FOR NITROGEN FIXATION

Nitrogen is an individual key nutrient which is very essential for development and growth of crops. Atmosphere holds about 80% of nitrogen volume

TABLE 11.1 The Names, Descriptions, and Habitats of Important Biofertilizers (Adapted from Lawal and Babalola 2014).

Microbes names	Properties of the microbes	Habitats	References
Azotobacter	*Azotobacter* are motile and it can be oval or spherical in shape. They are aerobic; good in nitrogen fixation. It is used ss food additives and biopolymers, and also used as a sources of antibodies production.	Soil	Oldroyd and Dixon 2014
Azospirillum	These are gram-negative aerobic bacteria helps in nitrogen fixations and PGPRs.	Soil	Serelis et al., 2013
Bacillus	Gram-positive rod shape bacteria and a member of the phylum *Firmicutes* basically used as biopesticides.	Nonsterile soil shaped	Kumari and Sarkar 2014
Burkholderia	It is a genus of proteobacteria genus, virtually ubiquitous gram-negative, motile, obligately aerobic rod-shaped bacteria. These groups of bacteria are known to enhance disease resistance and nitrogen fixation.	In wet soil	Vandamme et al., 2002
Cyanobacteria	Also known as blue-green bacteria, blue-green algae or cyanophyta. They are photosynthetic nitrogen fixers ensure soil fertility and plant growth.	Soil, ocean, and fresh	Taylor et al., 2014
Enterobacter	Gram-negative, anaerobic, shaped bacterium which promotes the germination and growth of cereals	In water and soil	Jin et al., 2014
Gluconobacter	The acetic acid bacteria are usually airborne and are ubiquitous in nature, plays crucial role in growth promotion of plants	Air	Guo et al., 2013
Herbaspirillum	Gram-negative soil and water-based bacteria that rarely cause human infections and produce phytohormone.	Soil and water	Lubambo et al., 2013
Klebsiella	Gram-negative, nonmotile, encapsulated, lactose fermenting, facultative anaerobic, rod-shaped bacterium involves in N_2-fixation in nonlegumes	Soil	Ku et al., 2014
Mycorrhiza	They are a group of fungi that include a number of types based on different structures formed inside or outside the root. Inoculation with mycorrhizal fungi with plants increases plant growth and yield, and resistance against climatic and edaphic stresses, pathogens and pests.	Soil	Vos et al., 2013

TABLE 11.1 *(Continued)*

Microbes names	Properties of the microbes	Habitats	References
Pantoea agglomerans	Gram-negative bacteria formerly called *Enterobacter agglomerans*. It is a phosphate-solubilizing bacteria used as a plant growth promoting agent and bioremediation purpose.	Plant surfaces and animal feces	Kouvoutsakis et al., 2014
Pseudomonas putida:	Gram-negative rod-shaped saprophytic soil bacterium used as soil inoculants as a bioremediation agents.	Soil	Annesini et al., 2014
Rhizobium:	Genus of Gram-negative fast growing soil bacteria. It fixes nitrogen (diazotroph) after becoming established inside root nodules of legumes (Fabaceae).	Root nodules and stem nodules	Qin et al., 2014
Trichoderma	A genus of microscopic fungi . They are opportunistic avirulent plant symbionts andalso improves plant growth, crop yield, and nutritional quality	Soil	Nawrocka and Maolepsza 2013

in free state on the earth. The main part of the elemental nitrogen that locates its means into the soil is completely due to its fixation by definite specialized set of microorganisms. Biological nitrogen fixation (BNF) is regarded to be an important process which determines nitrogen stability in soil ecosystem. Nitrogen inputs through BNF maintain sustainable environmentally sound agricultural construction. The rate of nitrogen-fixing legumes in improving and higher yield of legumes and other crops can be accomplished by the submission of biofertilizers (Kannaiyan, 2002). Biological nitrogen fixation is one of the methods of switching elemental nitrogen into plant exploitable form (Gothwal et al., 2007). Nitrogen-fixing bacteria (NFB) that function transform inert atmospheric N_2 to organic composites (Bakulin et al., 2007). Nitrogen fixer or N-fixer organisms are employed in biofertilizer as a living fertilizer invented of microbial inoculants or groups of microorganisms which are able to fix atmospheric nitrogen. These microbes are grouped into free-living bacteria (*Azotobacter* and *Azospirillium*) and the blue-green algae and symbionts for instance *Rhizobium*, *Frankia*, and *Azolla* (Gupta, 2004). The list of NFB associated with nonlegumes contained species of *Achromobacter*, *Alcaligenes*, *Arthrobacter*, *Acetobacter*, *Azomonas*, *Beijerinckia*, *Bacillus*, *Clostridium*, *Enterobacter*, *Erwinia*, *Derxia*, *Desulfovibrio*, *Corynebacterium*, *Campylobacter*, *Herbaspirillum*, *Klebsiella*, *Lignobacter*, *Mycobacterium*, *Rhodospirillum*, *Rhodo-pseudomonas*, *Xanthobacter*, *Mycobacterium*, and *Methylosinus* (Wani, 1990). Although numerous genera and species of NFB are isolated from the rhizosphere of diverse cereals, mainly part of *Azotobacter* and *Azospirillum* genera have been broadly tested to boost yield of cereals and legumes under field conditions. *Rhizobium* inoculation is well recognized as agronomic exercise to certify adequate nitrogen of legumes instead of N-fertilizer (Gupta, 2004). In root nodules, the O_2 level is legalized by particular hemoglobin called leg-hemoglobin. The globin protein is encoded by plant genes but the heme cofactor is built up by the symbiotic bacteria. However, this is only manufactured, when the plant is infected with *Rhizobium*. The plant root cells renovate sugar to organic acids which they deliver to the bacteroids. In exchange, the plant will accept amino acids rather than free ammonia. *Azolla* biofertilizer is used for rice cultivation in various countries such as Vietnam, China, Thailand, and Philippines. Field trial specified that rice yields are raised by 0.5–2 t/ha due to *Azolla* application (Gupta, 2004). In several crops, *Azobacter* and *Azospirillum* can fix atmospheric nitrogen without any symbiosis; however, in rice and banana plantation, blue-green algae have been found to be extremely successful (Gupta, 2004). El-Komy (2005) confirmed the beneficial influence of co-inoculation of *Azospirillum lipoferum* and *Bacillus megaterium* for supplying balanced nitrogen and phosphorus nutrition of wheat plants. The inoculation with bacterial combinations provided an extra balanced nutrition for the plants and the improvement in root uptake of nitrogen and phosphorus was the main mechanism of interaction between plants and bacteria. Co-inoculation of some *Pseudomonas* and *Bacillus* strains together with effective *Rhizobium* spp. is shown to stimulate chickpea growth, nodulation, and nitrogen fixation. Mohammadi et al. (2010) reported that the highest sugar, protein, starch contents, seed weight, nitrogen, potassium, phosphorus of chickpea were obtained from combined use of phosphate-solubilizing bacteria (PSB), *Trichoderma* and *Rhizobium* fungus. Shanmugam and Veeraputhran (2000) stated that use of green manure and biofertilizer encouraged the growth of plants with additional number of tillers and broader leaves in rice that could be the probable motive for the enlarged leaf area. Submission of biofertilizer increased the number of leaves in betel vine, and this could be due to correctly colonized roots, increased mineral and water uptake from the soil, and biological nitrogen fixation (Okon, 1984). It could also be ascribed to the production of indole-3-acetic acid (IAA), gibberellins, and cytokinins like substances produced by the bacterium as apparent from the findings in banana by Jeeva (1987).

11.3.2 BACTERIAL BIOFERTILIZERS USED FOR PHOSPHORUS SOLUBILIZATION

PSB solubilized phosphorus in the soil have the ability to convert inorganic unavailable phosphorus form to soluble forms HPO_4^{2-} and H_2PO_4 through the process of organic acid production, chelation, and ion exchange reactions and make them accessible to plants. Use of PSB in agricultural practice would not only compensate the elevated price of manufacturing phosphate fertilizers but would also activate insoluble form of phosphorus present in the fertilizers and soils to which they are pertained in the soluble form of phosphorus (Banerjee et al., 2010; Chang and Yang, 2009). Confirmation of naturally occurring rhizospheric phosphorus-solubilizing microorganism (PSM) dates back to 1903 (Khan et al., 2007). Bacteria are added successfully in phosphorus solubilization than fungi (Alam et al., 2002). Among the entire microbial population in soil, PSB make up to 1–50%, whereas phosphorus-solubilizing fungi are only 0.1–0.5% (Chen et al., 2006). Among the soil bacterial communities, ectorhizospheric strains from *Bacillus*, *Pseudomonas*, and endosymbiotic *rhizobia* have been explained as effective phosphate solubilizers (Igual et al., 2001).

Several bacterial strains from their effective genera, that is, *Pseudomonas*, *Bacillus*, *Rhizobium*, and *Enterobacter* along with *Penicillium* and *Aspergillus* fungi are the largely significant P solubilizers (Whitelaw, 2000). *B. megaterium*, *Bacillus circulans*, *Bacillus subtilis*, *Bacillus polymyxa*, *Bacillus sircalmous*, *Pseudomonas striata*, and *Enterobacter* could be referred as the most significant strains (Subbarao, 1988). A nematofungus *Arthrobotrys oligospora* also has the ability to solubilize the phosphate rocks (Duponnois et al., 2006). Elevated quantity of PSM is concentrated in the rhizosphere, and they are metabolically added actively than from other basis (Vazquez et al., 2000). Usually, 1 g of fertile soil has 101–1010 bacteria, and their live weight may surpass 2000 kg/ha. Soil bacteria are in cocci (sphere, 0.5 μm), bacilli (rod, 0.5–0.3 μm), or spiral (1–100 μm) shapes. *Bacilli* are widespread in soil, whereas spirilli are very uncommon in natural environments (Baudoin et al., 2002). The PSB are ubiquitous with disparity in forms and population in diverse soils. Population of PSB depends on different soil belongings (physical and chemical properties, organic matter, and P content) and cultural activities (Kim et al., 1998). A number of bacterial species have the capacity of mineralization and solubilization potential for organic and inorganic phosphorus, correspondingly (Hilda and Fraga, 2000; Khiari and Parent, 2005). Phosphorus-solubilizing motion is determined by the aptitude of microbes to discharge metabolites such as organic acids, which through their hydroxyl and carboxyl groups chelate the cation leap to phosphate, the latter being rehabilitated to soluble forms (Sagoe et al., 1998). Phosphate solubilization is acquired through the diverse microbial processes/mechanisms including organic acid production and proton extrusion (Dutton and Evans, 1996; Nahas, 1996).

A wide range of microbial P solubilization mechanisms subsist in nature and much of the global cycling of insoluble organic and inorganic soil phosphates is attributed to several bacteria and fungi (Banik and Dey, 1982). Phosphorus solubilization is carried out by a large number of saprophytic bacteria and fungi performing on sparingly soluble soil phosphates, mainly by chelation-mediated mechanisms (Whitelaw, 2000). Some organic acids and enzymes are secreted by PSM that act on insoluble phosphates and renovate it into soluble appearance, thus, proving P to plants (Ponmurugan and Gopi, 2006). Inorganic P is solubilized by the achievement of inorganic and organic acids secreted by PSB in which hydroxyl and carboxyl groups of acids chelate cations (Al, Fe, Ca) and reduce the pH in basic soils. The PSB liquefy the soil P through fabrication of low molecular weight organic acids mostly gluconic and ketogluconic acids (Deubel et al., 2000), in addition, to lowering the pH of rhizosphere. The pH of rhizosphere is decreased through the biotical production of proton/bicarbonate discharge (anion/cation balance) and gaseous (O_2/CO_2) exchanges. PSB has ability of the solubilization of phosphorus with straight correlation to pH of the medium. Release of root exudates, for example, organic ligands can also change the concentration of P in the soil solution (Hinsinger, 2001). Organic acids formed by PSB solubilize insoluble phosphates by decreasing the pH, chelation of cations, and challenging with phosphate for adsorption sites in the soil (Nahas, 1996). Inorganic acids, for example, hydrochloric acid can also solubilize phosphate other than they are less effective compared to organic acids at the same pH (Kim et al., 1998). In addition, the microorganisms concerned in P solubilization can augment plant growth by enhancing the availability of erstwhile trace element such as iron (Fe), zinc (Zn), etc. (Ngoc et al., 2006), synthesize enzymes that can modulate plant hormone level, may limit the accessible iron via siderophore production, and can also kill the pathogen with antibiotic (Akhtar and Siddiqui, 2009). The PSB has the capability to solubilize soil P and applied phosphates that result in higher crop yields (Gull et al., 2004). The PSB strains show inorganic P-solubilizing capabilities ranging between 25 and 42 μg P/mL and organic P mineralizing capabilities between 8 and 18 μg P/mL (Tao et al., 2008). The PSB in conjunction with single super phosphate and rock phosphate decrease the P dose by 25% and 50%, respectively (Sundara et al., 2002). *Pseudomonas putida*, *P. fluorescens*, and *P. fluorescens* released 51%, 29%, and 62% P, respectively; with highest cost of 0.74 mg P/50 mL from Fe_2O_3 (Ghaderi et al., 2008). Several isolates, that is, *P. striata* and *Bacillus polymyxa* solubilized 156 and 116 mg P/L, respectively (Rodríguez and Fraga, 1999). One of the *P. fluorescens* isolates solubilized 100 mg P/L containing $Ca_3(PO_4)_2$ or 92 and 51 mg P/L containing $AlPO_4$ and $FePO_4$, respectively (Henri et al., 2008).

11.3.3 FUNGAL BIOFERTILIZERS

Arbuscular mycorrhizal fungi are also a type of biofertilizer, which are possibly the most copious fungi in agricultural soil (Khan, 2006; Marin, 2006). The fungal inocula progress crop yield because of improved availability or uptake or absorption of nutrients, stimulation of plant growth by hormone accomplishment or antibiosis, and by decomposition of organic deposits (Wani and Lee, 2002). Some of the selected fungal species which are commonly used as biofertilizers are mentioned below.

11.3.3.1 MYCORRHIZAL FUNGI USED AS BIOFERTILIZERS

Mycorrhizae form mutualistic symbiotic relationships with the plant roots of more than 80% of the land plants including numerous significant crops and forest tree species (Gentili and Jumpponen, 2006; Rinaldi et al., 2008). Presently, seven types of readily available fungi are of mycorrhiza: arbutoid mycorrhiza, ectomycorrhiza, endomycorrhiza or arbuscular mycorrhiza (AM), ect-endomycorrhiza, ericoid mycorrhiza, monotropoid mycorrhiza, and orchidoid mycorrhiza (Das et al., 2007; Gentili and Jumpponen, 2006; Raina et al., 2000; Tao et al., 2008; Zhu et al., 2008). The two dominant types of mycorrhizae presently reported are ectomycorrhizae (ECM) and AM, which can advance water and nutrient uptake and supply protection from pathogens but only a few families of plants are capable to outline functional associations with both AM and ECM fungi (Siddiqui and Pichtel, 2008). However, AM fungi are most generally established in the rhizosphere roots of a broad range of herbaceous and woody plants (Rinaldi et al., 2008). ECM fungi appearance mutualistic symbioses with several tree species (Anderson and Cairney, 2007). Large number of ECM fungi do not appear to penetrate the plant living cells in the roots but only surround them of particular plant roots (Das et al., 2007; Gupta et al., 2000; Raina, 2000). ECM fungi arise naturally in association with many forest trees, for example, pine, spruce, larch, hemlock, willow, poplar, oak birch, and eucalyptus (Raja, 2006; Rinaldi et al., 2008). Large number of ECM fungi that are associated with the forest trees are basidiomycetes such as *Amanita* sp., *Lactarius* sp., *Pisolithus* sp., and *Rhizopogon* sp. and numerous of these are suitable for eating (Buyck et al., 2008; Rinaldi et al., 2008). Some ascomycetes also form mycorrhizae, for example, *Cenococcum* sp., *Elaphomyces* sp., and *Tuber* sp. (Das et al., 2007; Rinaldi et al., 2008). The significance of ECM fungi to trees is in their capability to augment the tree growth due to improved nutrient attainment (Gentili and Jumpponen, 2006). ECM fungi involved in the growth and expansion of trees because the roots colonized with ectomycorrhiza are capable to absorb and accumulate nitrogen, phosphorus, potassium, and calcium more rapidly and over a longer phase than nonmycorrhizal roots. ECM fungi help to split the complex minerals and organic substances in the soil and transport nutrients to the tree. ECM fungi also emerge to augment the tolerance of trees to drought, high soil temperatures, soil toxins, and edges of soil pH. ECM fungi can also defend roots of trees from several pathogens (Dahm, 2006). The most frequently widespread ECM product is inoculum of *Pisolithus tinctorius* (Gentili and Jumpponen, 2006; Schwartz et al., 2006). The ECM *P. tinctorius* has a large host choice and their inoculum can be formed and applied as vegetative mycelium during a peat vermiculite carrier to the plant. These fungus inocula are applied to nursery or forestry plantations (Gentili and Jumpponen, 2006). *Piriformospora indica* (Hymenomycetes, Basidiomycota) is another ECM fungus that is used as a biofertilizer. This taxon can encourage plant growth and biomass production and assist plant tolerance to herbivory, heat, salt, disease, drought, and increased below- and above-ground biomass (Tejesvi et al., 2007).

Endomycorrhizae fungi form mutually symbiotic relationship between fungi and plant root (Ipsilantis and Sylvia, 2007). The plant roots supply substances for the fungi and the fungi shift nutrients and water to the plant roots (Adholeya et al., 2005; Chen, 2006). Endomycorrhizal fungi are intercellular and interred in the cells by breaking the root cortical cells and form structures called arbuscular vesicles and well known as vesicular AM, but in a few cases, no vesicles are formed, and they are commonly known as AM (Gupta et al., 2000). The agriculturally produced crop plants that structured endomycorrhizae of the vesicular-AM type are currently called AM fungi. AM fungi belong to nine genera: *Acaulospora*, *Archaeospora*, *Enterophospora*, *Gerdemannia*, *Geosiphon*, *Gigaspora*, *Glomus*, *Paraglomus*, and *Scutellospora* (Das et al., 2007). AM fungi are a widespread collection of fungi and are found as of the arctic to tropics. AM fungi are present in most agricultural and natural ecosystems of the world. These fungi play a very important role in the plant growth, health, and their productivity (Douds et al., 2005; Marin, 2006). AM fungi also help plants to absorb nutrients, especially the less available mineral nutrients, for example, copper, molybdenum, phosphorus, and zinc from the soil (Yeasmin et al., 2007). AM fungi augment seedling tolerance to drought, high temperatures, lethal heavy metals, high or low pH, and still extreme soil acidity to the several plants (Chen, 2006; Kannaiyan, 2002). AM fungi can also influence plant growth indirectly by recovering the soil structure, providing antagonist possessions against pathogens and distorted water relationships (Smith and Zhu, 2001). AM fungi also have the capacity to reduce the severity of soilborne pathogens and augment resistance in roots against root rot disease of several plants (Akhtar and Siddiqui, 2008a, 2008b; Chen, 2006). The above findings showed the competition for colonization sites or nutrients in the similar root tissues and production of fungistatic amalgams (Johansson et al., 2004; Marin, 2006). AM fungi encompass to have benefits to the several host plants including increasing herbivore forbearance, increasing pollination, increasing soil immovability, and heavy metal tolerance (Hart and Trevors, 2005). From almost two decades, AM fungi are used as biofertilizers, as they have been produced for use in agriculture, horticulture, landscape restoration, and soil remediation by the farmers and researchers (Hart and Trevors, 2005). There are the reports on mass production of AM fungi, which has been achieved with several species, such as *Acaulospora laevis*, *Glomus clarum*, *Glomus etunicatum*, *Glomus intraradices*, *Glomus mosseae*, *Gigaspora ramisporophora*, and *Gigaspora rosea* (Adholeya et al., 2005; Akhtar and Siddiqui, 2008b; Schwartz et al., 2006; Wu et al., 2005). Effective supervision of AM fungi occupies increasing populations of propagules such as spores, colonized root fragments, and hyphae which is commonly using by the host plants and also by adoption of soil management performance (Kapoor et al., 2008; Smith and Zhu, 2001; Tiwari et al., 2004).

11.3.3.2 OTHER FUNGI USED AS BIOFERTILIZERS

Penicillium species have been also used as a fungal biofertilizers which improve plant growth. These biofertilizers are PSMs that progress phosphorus absorption in plants and encourage plant growth (Pradhan and Sukla, 2005; Wakelin et al., 2004). Isolate of *Penicillium bilaiae* has been formulated as a commercial product named Jumpstart® and was released to promote the market as a wettable powder in 1999 (Burton and Knight, 2005). *P. bilaiae* is also applied to increase dry matter, phosphorus (P) uptake, and seed yield in canola (*Brassica napus*) (Burton and Knight, 2005; Grant et al., 2002). *Penicillium radicum* and *Penicillium italicum* are placed under the phosphate-solubilizing taxa (El-Azouni, 2008; Wakelin et al., 2004). A strain of *P. radicum*, isolated from the rhizosphere of wheat roots, has revealed a good assure in plant growth encouragement (Whitelaw et al., 1999), whereas *P. italicum* isolated from the rhizosphere soil was investigated for its capacity to solubilize tricalcium phosphate and could encourage the growth of soybean (El-Azouni, 2008).

Several species of *Aspergillus* have been accounted to be involved in the solubilization of inorganic phosphates, that is, *Aspergillus flavus*, *Aspergillus niger*, and *Aspergillus terreus* (Akintokun et al., 2007). These fungi are capable to solubilize of inorganic phosphate during the production of acids, for example, citric, gluconic, glycolic, oxalic acids, and succinic acid (Barroso et al., 2006). An isolate of *Aspergillus fumigatus* isolated from compost has been accounted to be potassium-releasing fungus (Lian et al., 2008). The product of *Chaetomium* species used as fungal biofertilizers, for example, Ketomium® which is formulated from *Chaetomium globosum* and *Chaetomium cupreum* is not only used as a mycofungicide but also used as plant growth stimulant because tomato, corn, rice, pepper, citrus, durian, birds of paradise, and carnation applied with Ketomium® have a greater plant growth and high yields than nontreated plants (Soytong et al., 2001).

Trichoderma species reduce the occurrence of disease and also inhibit pathogen growth when used as mycofungicides; however, they also augment the growth and yield of plants (Harman et al., 2004; Vinale et al., 2008). These mycofungicides are also augmenting the survival of seedlings, plant height, leaf area, and dry weight. Isolate of *Trichoderma* species advance mineral uptake, discharge minerals from soil and organic matter, enhance plant hormone invention, persuade systematic resistance mechanisms, and induce root organization in hydroponics (Yedidia et al., 1999). Due to the above capabilities, *Trichoderma* species are now best known as PGPF (Herrera-Estrella and Chet, 2004; Hyakumachi and Kubota, 2004) or commonly known as plant growth-increasing agent (biofertilization) (Benitez et al., 2004). *Trichoderma* species have successfully used as biofungicides and biofertilizers in the greenhouse and field plant for the successful production against the biotic stresses (Harman et al., 2004; Vinale et al., 2008). There are several *Trichoderma* products as fungal biofertilizers obtainable in the market. Their applications are however associated to their ability to manage plant diseases and encourage plant growth and development (Harman et al., 2004; Vinale et al., 2006). *Trichoderma* also has diverse applications and significant sources of antibiotics, enzymes, decomposers, and plant-growth promoters (Daniel and Filho, 2007).

11.4 PLANT-GROWTH-PROMOTING RHIZOBACTERIA

PGPR is a group of rhizosphere bacteria (rhizobacteria) that exerts a useful outcome on plant growth and development (Schroth and Hacock, 1981). PGPR belongs to numerous genera, for example, *Agrobacterium*, *Alcaligenes*, *Arthrobacter*, *Actinoplanes*, *Azotobacter*, *Bacillus*, *Pseudomonas* sp., *Rhizobium*, *Bradyrhizobium*, *Erwinia*, *Enterobacter*, *Amorpho sporangium*, *Cellulomonas*, *Flavobacterium*, *Streptomyces*, and *Xanthomonas* (Weller, 1988). PGPR augmented recently as a consequence of the abundant studies covering a wider choice of plant species and because of the advances through in bacterial taxonomy and the advancement in the understanding of the diverse mechanisms of deed of PGPR. In all thriving plant–microbe interactions, the fitness to colonize plant habitats is significant. Single bacterial cells can connect to root surfaces and, after cell division and proliferation, form dense aggregates usually referred to as macrocolonies or biofilms. Steps of colonization comprise attraction, recognition, adherence, invasion (just endophytes and pathogens), colonization and growth, and numerous strategies to found interactions (Nihorimbere et al., 2011). Plant roots begin crosstalk with soil microbes via producing signals that are recognized by the microbes, which in turn generate signals that start colonization (Berg, 2009). PGPR arrived at root surfaces by active motility facilitated by flagella and are guided by chemotactic responses. This involves that PGPR competence highly depends either on their abilities to obtain advantage of a specific environment or on their abilities to adapt to changing conditions or plant species (Nihorimbere et al., 2011).

11.5 METHODS OF APPLICATION OF BIOFERTILLIZER INOCULANTS FOR AGRICULTURE

It is important to be attentive of the method(s) to be used for the appliance of the ready inoculants (bacteria-carrier mixture) to the crops. There are several factors to be considered including the type of plants or seeds to be biofertilized, accessibility of the biofertilizers carrier, season, and age of the crop to be used. It must be make sure that suitable and efficient strain(s) of organisms are used. Biofertilizers should not be used in concern of strong doses of plant protection chemicals and other chemicals should not be combined with the biofertilizers (Balasubramanian et al., 2013; Son et al., 2014). The environmental situation of agricultural soil must be taken into deliberation: high soil temperature or low soil moisture, acidity or alkalinity in soil, poor accessibility of phosphorous and molybdenum, occurrence of elevated native population, or presence of bacteriophages. Consecutively, to obtain high-quality results from the use of biofertilizers, farmers must guarantee that they use the right scheme of application and apply the fertilizer at the correct time (Table 11.2).

11.6 CHARACTERISTICS OF SUITABLE CARRIERS FOR BIOFERTILIZERS

Expansion of suitable inoculants has to do with procurement of a suitable carrier substrate. This is significant since it is the carrier that will host and in that way decide the growth of the organism and probably sustain

TABLE 11.2 Methods of Applying Biofertilizers in different Crops and their Descriptions (Adapted from Lawal and Babalola 2014).

Methods	Different Crops	Application Method and their Description	References
Seed treatment	Pulses, oilseeds, and fodder	Keep the seeds required for sowing one acre in a heap on a clean cemented floor or gunny bag. It is applied at the rate of crops 100 g per 5 kg of seeds. Slurry is prepared with water and the biofertilizer is mixed with water at ratio 1:2	Singh et al., 2013a
Seedling treatment	Tomatoes, potato, and onion	Get the seedlings required for one acre and make small bundles of seedlings. Dip the root portion of these seedlings in this suspension for 15-30 minutes and transplant immidiately.	Andrade et al., 2013
Set treatment	Sugarcane, banana, and grapes	For set treatment, the ratio of bio-fertilizer to water is approximately 1:50. The pieces of materials to be planted are immersed in biofertilizers mixture for 30 min. The sets are dried and planted in the field.	Shen et al., 2013
Soil treatment	Maize and wheat	Soil treatment basically has been done before planting or sowing of the crops. The possible carriers are compost, farmyard manure, rice husks, and lignite at the rate of 1 kg per 25 kg of carrier. Irrigation follows immediately after the application.	Schoebitz et al., 2014
Spraying/Irrigation	Recommended for standing citrus plants, vines, mango, peach orchards	Biofertilizers firstly mixed with water and other micronutrients properly in a tank. The mixtures applies and it reaches plants like guava and custard apple through irrigation or spraying.	Singh et al., 2013b

the inoculants bacteria such that they can grow and proliferate suitably. A suitable carrier should contain a significant level of organic matter and elevated level of nitrogen and must be affordable and nontoxic. Deployment of inoculants in a carrier grants long-standing storage and makes the handling easier and effective (Schoebitz et al., 2014). The preparation of seed carrier occupies the milling of the material to powder to sizes which differs between 8 and 41 μm. Good carriers must (1) maintain good moisture capacity, (2) not be toxic to the inoculants bacterial strain, (3) be easy to process and devoid of lump forming substances, (4) be easy to sterilize by autoclaving and irradiation, (5) be available in adequate quantity, (6) be low-cost, (7) have good quality adhesion to seeds, (8) have good pH buffering capability, and (9) not be toxic to the plants. Separately from these mentioned criteria that resolve the suitability of a material as a carrier, it must be worried that it is also significant that the carrier must be capable to carry the survival of the bacteria even ahead of the seeds are sown or before the seedlings are transplanted in the field. This becomes important because in the case of seed coating, the seeds are not always sown immediately. Besides, the survival of inoculant bacteria when the seeds are stored must be supported by the carrier materials. Desirably, the carrier material must be able to support the biofertilizers when in the soil. This is necessary because the biofertilizers compete with the native soil microorganisms for nutrient, and they will have to cope and stay alive in the soil in spite of the protozoa in the soil as well. Therefore, the carriers must offer a sufficient microporous structures that will ensure the survival of the inoculants bacteria (Singh et al., 2013a). Some of the identified carriers now include: various clays, animal manure such as poultry manure, composted plant materials, or other complex organic matrices. Some users opined that animal manure possibly contains pathogens and antibiotic which can lead to serious soil degradation and phytotoxicity if uncomposted animal manure is applied to soil (Babalola and Glick, 2012b; Yousefzadeh et al., 2013).

Liquid formulations use liquid materials as carrier, which is typically water, oil, or some solvents in appearance of suspension, concentrates, or emulsions. Most popular liquid inoculant formulations (Chandra et al., 2011) contain particular organism's broth 10–40%, suspender ingredient 1–3%, dispersant 1–5%, surfactant 3–8%, and transporter liquid (oil and/or water) 35–65% by weight. Viscosity is adjusted at equal to the setting time of the particles, which is attained by the use of colloidal clays, polysaccharide gums, starch, cellulose, or synthetic polymers.

11.7 STERILIZATION OF THE CARRIER MATERIALS FOR BIOFERTILIZERS

Sterilization of carrier material is essential to keep elevated number of inoculant bacteria on carrier for long storage period to devoid of other microorganisms that may hinder the survival of the inoculants and also to prolong the shelf life of the inoculants. Essentially, there are two main ways of sterilizing the carrier materials, gamma irradiation and autoclaving, but the more adopted or recommended way is the gamma-irradiation procedure. This is because this method does not alter the chemical or physical properties of the carrier materials required (Babalola and Oladele, 2011; Minaxi and Saxena, 2011). Most of the times, the irradiation is done by packing the carrier material inside the polyethylene bag and gamma irradiate at 50 kGy (5 Mrads). Gamma rays effect the sterilization by crashing with the atoms of nutrients such as protein, carbohydrates, lipids, and nucleic acid. The rays ionize the atoms and thus damage them. From the above-listed materials, the most susceptible to ionizing radiation from the gamma rays is the nucleic acid and it is only 1% of the total composition of the cell. Furthermore, during the radiation treatment, the DNA strand breaks and the base is subsequently damaged. The break of the DNA strands leads to break in the flow of genetic information and this destroys replication process, and subsequently death of the cell results (Li et al., 2012). Besides, the sterilization by autoclaving is carried out by putting the carriers in polyethylene bags and autoclaved at 121°C for 60 min. However, it should be known that during autoclaving, some carriers undergo changes and their physicochemical properties may change and produce toxic substances to some bacteria strains (Li and Yu, 2011).

11.8 RHIZOSPHERE COMPETENCE OF BIOFERTILIZERS

Bacteria inoculation is often carried out when coating the seeds or when they are placed very close to the plant via a carrier. The inoculated bacteria should have the ability to establish themselves in the vicinity of the rhizoids at such a number that will be enough to have beneficial influence on the plants. Expectedly, inoculants bacteria should not only live in the vicinity of the rhizoids, but they should be able to maximally use nutrients produced from the root, multiply, and subsequently colonize the entire root area (Son et al., 2014). In summary, biofertilizers function as soil microbes, and thereby convert ambient nitrogen into forms that the plants can use (nitrate and ammonia), enlarges soil porosity by gluing soil particles together, defend plants against pathogens by outcompeting pathogens for food irradiation, and worthy of note is the fact that saprophytic fungi in the soil break leaf litter down into usable nutrients. It is important to note that at present, biofertilization is responsible for approximately 65% of the nitrogen supply to crops all over the world. The biofertilizers bacteria form a host-specific relationship with legumes. This relationship begins by initiation of root or stem nodules as a result of the presence of the bacterium. Lipooligosaccharide information activates molecules that are produced by the bacterium and plays an important role in this process. The bacteria percolate the cortex, stimulate formation of root nodules, increase and eventually break into bacteroids, which elicit the nitrogenized enzyme production. In the root nodules, the plant provides a low oxygen concentration, which promotes bacterial nitrogenase to change nitrogen in the atmosphere into ammonia. As a result of this, the plant supplies the bacteria with needed carbon source for multiplication and existence (Beneduzi et al., 2013).

Agricultural soil is said to be healthy if among other things, it contains sufficient strains of microorganisms that can terminate, prevent, or hinder bacteria, fungi, and species of nematodes that can cause root rots. Moreover, organisms like mycorrhizal fungi produce compounds that are antibiotic or bactericidal to many plant pathogens (Babalola, 2010). For the decomposition of the toxic materials, through the procedure of cooxidation, bacteria and fungi require organic materials to feed on with the toxic compounds (Asensio et al., 2013).

11.8.1 BIOFERTILIZERS AS STRENGTHENERS OF AGRICULTURAL PLANTS

Plant strengtheners refer to "plant-resistance improvers." It is now commonly referred to as agents, and it maintains plant health or which guide crop plants against nonparasitic adverse conditions. The association of plant with microorganisms explains important roles for plant wellbeing and health. Microbes are able to influence plants' health by improving nutrient uptake and hormonal stimulation. Different methods or ways are involved in the minimization of activities of plant pathogens, and this will influence and affect plant growth (Manivasagan et al., 2013). Although bacterial genera *Azospirillum* and *Rhizobium* are known to be good for plant growth improvement, other microorganisms like *Bacillus*, *Pseudomonas*, *Serratia*, *Stenotrophomonas*, *Streptomyces*, *Ampelomyces*, and *Coniothyrium* are yet to be fully explored. Developments in recent agricultural practices aim at developing harvest yields and directed toward minimizing preharvest and postharvest losses occasioned by devastating abiotic and biotic agents (Prashanth and Tharanathan, 2007). These developments have potential of reducing the effects of pests and diseases by about 20–40%. Recent pest management strategies in crop plants involve the use of classical and molecular marker-based resistance breeding, genetic manipulation of plant tolerance, and the use of chemicals as pesticides or strengtheners of plant wellbeing (Gomes and Silva, 2007).

11.8.2 BIOFERTILIZERS AS PHYTOSTIMULATORS FOR AGRICULTURAL PLANTS

Phytostimulators promotes crops growth. PGPRs include *Azospirillum*, which is a popular genus among the PGPR that exhibit positive effects on plants growth. A considerable volume of carbon gets below ground through the activities of plants' roots. Invariably, plants release exudates which serve as nutrients (carbohydrates, proteins, and other nutrients) for microbes around the roots area (Ramos et al., 2011). By this, crops get the right types of microbes around its roots. Eventually, these "well-fed" microbes will produce enzymes and growth hormones and protect the plants against pathogens. It is estimated that on an average, a gram of healthy soil should have or contain 100 million organisms (Drogue et al., 2012). Meanwhile, in the proximity of crops roots, there can be up to a trillion organisms per gram of soil and they live symbiotically. Many strains of *A. brasilense* and *A. lipoferum* have been explored in recent times as crop inoculants to maximize yield. The results obtained when *Azospirillum* was used as a phytostimulatory PGPR has elicited comprehensive studies on the biology of these bacteria. Other benefits from the use of *Azospirillum* are nitrogen fixation, deamination of the ethylene precursor-1-aminocyclopropane-1-carboxylate, and production of nitric oxide properties and phytohormones, chiefly IAA (Aimey et al., 2013). Moreover, these benefits have the ability to promote or enhance good rooting system which enhances root hair density and invariably, this will lead to better water mineral uptake by crops.

11.8.3 PLANT GROWTH PROMOTING BIOFERTILIZERS AS CROP HEALTH IMPROVERS

For about 60 years, PGPR have been confirmed to prompt development of various host plants, and they also benefit from the root exudates. The PGPR are classified into different groups according to their actions on crops. First, the phytostimulating rhizobacteria that promotes crop growth directly by providing nutrients and phytohormones; second, mycorrhiza and root nodule symbiosis which assist rhizobacteria and this positively affect functioning of plant and microorganisms in the symbiotic relationship; and third, the biocontrol rhizobacteria that defend plants and crops from pathogens via exudates from antimicrobial agents or by promoting plant resistance (Manivasagan et al., 2013). Due to their potential use as biofertilizers and biopesticides, their *modus operandi* has been largely studied in model bacteria such as *Azospirillum* spp. and *Pseudomonas* spp. The genotype of the host plant determines PGPR densities both in terms of the number, size, and composition. Furthermore, plant-growth-promoting effects of these bacteria have been shown to rely both on host-plant genes and bacterial strain (Son et al., 2014).

11.9 MICROBIAL BIOPESTICIDES

Biopesticides are certain types of pesticides obtained from such natural materials as animals, plants, bacteria, and certain minerals. In commercial terms, biopesticides comprise microorganisms that manage pests (microbial pesticides), naturally occurring substances that control pests (biochemical pesticides), and pesticidal materials produced by plants containing added genetic material (plant incorporated protectants). Biopesticides are engaged in agricultural use for the purposes of insect control, disease control, weed control, nematode control, and plant physiology and efficiency. Biopesticides are usually inherently less toxic than conventional pesticides. They provide growers with valuable tools by delivering solutions that are highly effective in managing pests, without creating negative collisions on the environment. They generally affect only the target pest and closely related organisms, in contrast to the broad range conventional pesticides that may affect organisms as different as birds, insects, and mammals. Overall, the biopesticides have very partial toxicity to birds, fish, bees, and other wildlife thus helping in maintaining beneficial insect populations.

The most commonly used biopesticides are living organisms, which are pathogenic for the pest of interest. These comprise biofungicides (*Trichoderma*), bioherbicides (*Phytopthora*), and bioinsecticides (*Bacillus thuringiensis* [*Bt*]). The probable benefits to agriculture and public health programs through the use of biopesticides are extensive. The interest in biopesticides is based on the advantages associated with such products which are (1) inherently least harmful and less environmental load, (2) designed to affect only one specific pest or, in some cases, a few goal organisms, (3) often effective in very small quantities and often decompose quickly, thereby resulting in lower exposures and mostly avoiding the pollution problems, and (4) when used as a component of integrated pest management programs, biopesticides can contribute greatly.

Microbial pesticides contain a microorganism (bacterium, fungus, virus, protozoan, or alga) as the dynamic component. Microbial pesticides can control many different kinds of pests, although each split active ingredient is relatively specific for its target pest(s). The most widely known microbial pesticides are the ranges of the bacterium *Bt*, which can control certain insects in cabbage, potatoes, and erstwhile crops. *Bt* produces a protein that is harmful to specific insect pests.

11.9.1 BACILLUS THURINGIENSIS

Bt is the most frequently used biopesticide worldwide. *Bt* is a Gram-positive spore-forming bacterium that manufactures crystalline proteins called deltaendotoxins during its stationary phase of growth (Schnepf et al., 1998). The crystal is discharged to the environment after lysis of the cell wall at the end of sporulation, and it can report for 20–30% of the dry weight of the sporulated cells (Schnepf et al., 1998). This bacterium is distributed worldwide. The soil has been described as its major habitat; however, it has also been isolated from foliage, water, storage grains, dead insects, etc. Isolation of the strains from the dead insects has been the main source for commercially used varieties, which contain *kurstaki*, isolated from *A. kuehniella*; *israelensis*, isolated from mosquitoes; and *tenebrionis*, isolated from *Tenebrio monitor* larvae (Joung and Cote, 2002; Iriarte and Caballero, 2001).

The Cry proteins comprise at least 50 subgroups with more than 200 components (Bravo et al., 2007). The members belong to a three-domain family, and the larger group of *Cry* proteins is spherical molecules with three structural domains connected by single linkers. The protoxins are characteristic of this family and have two diverse lengths. The C-terminal extension found in the long protoxins is necessary for toxicity and is believed to play a task in the formation of the crystal within the bacterium (de Maagd et al., 2001). Their mode of action involves several events that must be completed several hours after ingestion in order to lead to insect death. After ingestion, the crystals are solubilized by the alkaline conditions in the insect midgut and are subsequently proteolytically converted into a toxic-core fragment (Hofte and Whiteley, 1989). During proteolytic activation, peptides from the N terminus and C terminus are sliced from the full protein. Activated toxin binds to receptors located on the apical microvillus membranes of epithelial midgut cells. For *Cry*1A toxins, at least four different binding sites have been explained in diverse lepidopteran insects: a cadherin-like protein, a glycosylphosphatidyl-inositol (GPI)-anchored aminopeptidase-N, a GPI-anchored alkaline phosphatase, and a 270-kDa glycoconjugate (Agrawal et al., 2002; Valaitis et al., 2001). After binding, toxin adopts a conformation allowing its insertion into the cell membrane. Subsequently, oligomerization transpires, and this oligomer forms a pore or ion channel induced by an increase in cationic permeability inside the functional receptors contained on the brush-border membranes (Lorence et al., 1995). This allows the free flux of ions and liquids, causing disruption of membrane transport and cell lysis, leading to insect death. During the intoxication process, in lepidopterans as in coleopterans, many histopathological changes have been explained, including swelling and disruption of the microvilli, vacuolization of the cytoplasm, hypertrophy of epithelial cells, and necrosis of the nuclei (Lacey and Federici, 1979; Mathavan et al., 1989). The diversity of *Bt* is established in the almost 70 serotypes and the 92 subspecies described to date (Galan-Wong et al., 2006). The biological activity of *Bt* strains or their products toward different target organisms has been the focus of patent coverage for many years. Numerous patents belong to companies engaged in commercial endeavors, whereas others remain as a part of the basic research domain.

11.9.2 BACILLUS THURINGIENSIS-BASED FORMULATIONS

Bt-based biopesticide invention depends on high-quality and high-efficiency formulation processes. Formulations must be safe and effective products, must be easy to use, and should have a long-shelf life. The active ingredient in industrial formulations is the sporecrystal complex, which is more effective to use and cheaper to obtain than the crystals alone, which are commonly used in experimental tests. The spore–crystal complex must be carried by suitable percipients that can function to defend the spore–crystal complex or to increase palatability to insects. Formulation developed for killing fire ants that included a purified and activated Cry toxin from a novel strain of *Bt* is a good example, along with an attractant consisting of a biodegradable, environmentally sound glycoprotein (Bulla and Candas, 2002).

Many perishable compounds are often thought about to be used as inert carriers, counting on the sort of formulation needed. The bioinsecticide have to exhibit constancy in storage, requiring improvement of its biological and physical properties. The utilization of additives is critical so as to cut back evaporation and avoid formulation loss and to produce an additional extended coverage on and elevated adherence to foliage, enhanced dispersion, and a protracted residual impact. An excellent kind of ingredients are used to arrange formulations as well as liquid or solid carriers, surfactants, coadjuvants, liquidity agents, adherents, dispersants, stabilizers, moisturizers, attractants, and protecting agents among others (Morales-Ramos, 1996). A noteworthy and up-to-date example of those reasonably inert ingredients is the super absorbent starch-graft polymer that combined with a *Bt* strain among the several different pesticides that constitutes a unique formulation that would be applied in Associate in nursing agriculture atmosphere (Savich et al., 2008).

11.9.3 ENTOMOPATHOGENIC FUNGI

Use of entomopathogenic fungi as BCAs for several insect species has increased the global attention during the last few decades. Different strains of entomopathogenic fungus, which stranded arise on *Beauveria bassiana* (Balsamo) Vaillemin, *Paecilomyces fumosoroseus* (Wize) Brown and Smith, *Metarhizium anisopliae*, and *Verticillium lecanii* (Zimm.). Viegas have been utilized to control various insect pests (Alter and Vandenberg, 2000; Avery et al., 2004; Babu et al., 2001; Sharma, 2004). Production of adequate quantities of good quality inoculums is a necessary component of the biocontrol program. The production of entomopathogen can be taken up by two traditions either a relatively small quantity of the inoculums for laboratory experimentation and field-testing through the development of mycopesticide or by the development of a basic production system for large-scale production by following the labor-exhaustive and economically viable methods for relatively small-sized markets.

11.9.4 BEAUVERIA BASSIANA

B. bassiana is a fungus that grows naturally in soils throughout the world and acts as a parasite on diverse arthropod species, causing white muscardine disease. *B. bassiana* thus belongs to the entomopathogenic fungi. It is being used as a biological insecticide to manage a number of pests like termites, thrips, whiteflies, aphids, and completely different beetles. Its use within the management of bedbugs (Barbarin et al., 2012) and malaria-transmitting mosquitos is beneath investigation (Donald and McNeil, 2005). The insect malady caused by the flora may be a muscardine that has been referred to as white muscar dine malady. Once the microscopic spores of the flora acquire contact with the body of Associate in nursing insect host, they germinate, penetrate the cuticle, and grow within, killing the insect in a matter of days. Afterward, a white mildew emerges from the corpse and produces new spores. A typical isolate of *B. bassiana* will attack a broad variety of insects; numerous isolates dissent in their host vary. The factors liable for host condition don't seem to be renowned. *B. bassiana* parasitizing the Colorado Leptinotarsa decamping eat has been according to be, in turn, the host of a mycoparasitic flora *Syspastospora parasitica* (Posada, 2004). This organism conjointly attacks connected insect-pathogenic species of the Clavicipitaceae. *B. bassiana* may be used as a biological pesticide to regulate variety of pests like termites, whiteflies, and lots of different insects. It is frequently used for the management of different protozoal infection transmittal mosquitoes (Donald and McNeil, 2005). As Associate in nursing pesticide, the spores of *B. bassiana* sprayed as area unit on affected crops as the Associate in nursing blended suspension or wet table powder or applied to dipterans nets as dipterans management agent. As a species, *B. bassiana* parasitizes a really big selection of invertebrate hosts. However, completely different strains differ in their host ranges, some having rather slim ranges, like strain Bba 5653 that's terribly virulent to the larvae of the diamondback rattlesnake lepidopterous insect and kills solely few different styles of caterpillars. Some strains do have a good host vary and may so be thoughtabout nonselective biological pesticides.

11.9.5 FUNGUS

V. lecanii (formerly called *Cephalosporum lecanii*) was initial delineate in 1861 and may be a cosmopolitan flora found on insects. It's a standard microorganism of scale insects in tropical and climatic zone climates. *V. lecanii* is thought as a "white-halo" flora due to the white mycelial growth on the perimeters of infected scale insects. The conidia (spores) of *V. lecanii* area unit slimed and fasten to the cuticle of insects. The flora infects the insects by manufacturing hyphae from germinating spores that penetrates the insect's integument; the flora then destroys the inner contents and therefore the insect dies. The flora ultimately grows out through the cuticle and sporulates on the surface of the body. Infected insects seem as white to Xanthus soft particle. However, the unhealthy insect are seen sometimes within 7 days. However, because of environmental conditions, there is also a few substantial lag time from infection to death of insects. *V. lecanii* works best at temperatures of 15–25°C and ratio of 85:90. The flora wants high humidness for a minimum of 10–12 h. This maybe a haul as several plant-infective fungi (e.g., *Botrytis*) favor the same environmental conditions. The spore area units of *V. lecanii* are broken by ultraviolet radiation. In greenhouses, heating pipes might cut back the effectiveness of the flora, as a result of this creates a microclimate wherever the air is drier and humidness is lower. Additionally, *V. lecanii* is usually not helpful in interior scapes due the low humidness conditions in these environments (Cloyd, 1999).

The flora plant structure of *V. lecanii* produces a cyclodepsi peptide poison, referred as bassianolide, which has been shown to kill silkworm. The flora produces different insecticidal toxins like dipicolinic acid. The activity of *V. lecanii* spore depends on the strain of the flora. *V. lecanii* strains with little spores infect aphids, whereas flora strains with the massive spores infect whiteflies. Bound strains of *V. lecanii* have cojointly been infective on rust fungi. Flora virulence varies with the strategy of conidial production. Less virulent conidia area unit is obtained from sourced media as copared to jolted liquid or solid media. Developed product from the conidial production will last up to 1 year. These product area units are simple to wet and dilute for the spraying purpose. Also, the flora will stick with the surface of leaves and host insects. Studies have shown that combining entomopathogenic fungi with Associate in nursing pesticide might enhance its performance because the fungi produce wounds that make it easier for the pesticide to enter the insect (Cloyd, 1999). *V. lecanii* has been commercially accessible in Europe for management of aphids (Vertalec) and whiteflies (Mycotol). Vertalec has been used to manage inexperienced peach plant louse, *Myzus persicae*, on chrysanthemums in the greenhouse in the England. One other application further provided for the management for 3 months. Additionally, the melon plant louse, *Aphis gossypii*, has been inhibited with applications of *V. lecanii*. Infected aphids function as an extra supply of the matter and spores that will simply disperse among greenhouses.

V. lecanii is compatible with most parasitic and predatory arthropods. The flora of *V. leconii* will kill immature Encarsia formosa, a land parasitizing Trialeurodes vaporariorum however, these flora has no effects on adults (Cloyd, 1999).

11.9.6 METARHIZIUM

M. anisopliae, once called fungus genus *anisopliae* (basionym), may be a flora that grows naturally in soils throughout the planet and causes malady in numerous insects by acting as a parasitoid. Ilya I. Mechnikov named it once the insect species; it absolutely was originally isolated from the beetle *Anisoplia austriaca*. It is a mitosporic flora with asexual reproduction that was once classified within the syntactic category Hyphomycetes of the shape phylum Deuteromycota (also typically referred to as fungi Imperfecti). These practices has been recognized from the past with several very specific isolates and they were allotted on selection standing mode (Driver et al., 2000); however, they have currently been allotted as new *Metarhizium* species (Bischoff et al., 2009) like *M. anisopliae*, *M. majus*, and *M. acridum* (which was *M. anisopliae* var. *acridum* and enclosed the isolates used for locust control). *Metarhizium taii* was placed in *M. anisopliae* var. *anisopliae* (Huang et al., 2005); however, these have currently been delineated as a word of *Metarhizium guizhouense*. The commercially vital isolate M.a. 43 (a.k.a. F52, Met52, etc.) that infects animal order and alternative insect orders has currently been allotted to *Metarhizium brunneum* (Reddy et al., 2014). The malady caused by the flora is typically referred to as inexperienced muscardine malady due to the inexperienced color of its spores. Once these mitotic (asexual) spores (called conidia) of the flora get in contact with the body of an insect host, they germinate and also the hyphae that emerge penetrate the cuticle. The flora then develops within the body eventually killing the insect once in a number of days; this fatal impact is incredibly probably motor-assisted by the assembly of insecticidal cyclic peptides (destruxins). The cuticle of the remains typically becomes red. If the close humidness is high enough, a white mold then grows on the remains that presently turn inexperienced as spores are created. Most insects living close to the soil have evolved natural defenses alongside entomopathogenic fungi like *M. anisopliae*. This flora is thus barred in an organic process battle to beat these defenses that has crystal rectifier to an outsized range of isolates (or strains) which are custom-made to bound teams of insects (Freimoser et al., 2003). *M. anisopliae* and its connected species are used as biological pesticides to manage variety of pests like termites, thrips, etc., and its use within the management of malaria-transmitting mosquitoes is under investigation (Cloyd, 1999; McNeil and Donald, 2005). *M. anisopliae* doesn't seem to infect humans or alternative animals and is taken into account safe as an insect powder. The microscopic spores are usually sprayed on affected areas. A potential technique for protozoal infection management is to coat dipterous insect nets or cotton sheets hooked up to the wall with them.

11.9.7 BACULOVIRUSES

Baculoviruses are target-specific viruses which might infect and destroy variety of the vital plant pests. They are significantly effective against the lepidopterous pests of cotton, rice, and several vegetables. Their large-scale production poses difficulties; thus, their use has been restricted to tiny areas. Within the past, the appliance of baculoviruses for the protection of different agricultural annual crops, fruit orchards, and forests has not matched their potential. The amont of registered pesticides supported by baculovirus that acts slowly further will increase steady. At present, it exceeds 50 formulations, a number of them being similar baculovirus preparations distributed under completely different trade name in the several countries. NPVs and GVs are frequently used as pesticides; however, the cluster supported nucleo-polyhedrosis viruses is far larger. The primary infective agent insect powder Elcar™ was introduced by Sandoz Inc. in 1975 (Ignoffo and Couch, 1981). Elcar™ was a preparation of genus *Heliothis zea* NPV that is comparatively broad varies baculovirus and infects several species belongings to genera *Helicoverpa* and *Heliothis*. HzSNPV used for the management of *H. zea*, however, HzSNPV conjointly used for control of pests of the genera offensive i.e. soybean, sorghum, maize, tomato, and beans. In 1982, Sandoz determined to discontinue the assembly. The resistance to several chemical pesticides as well as pyrethroids revived the interest in HzSNPV and also the same virus was registered under the name GemStar™. HzSNPV may be an alternative product for biocontrol of *Helicoverpa armigera* (Mettenmeyer, 2002). Countries with massive areas of such crops resembling cotton, pigeonpea, tomato, pepper, and maize, for example, Asian nation and China introduced special programs for the reduction of this persecutor by biological suggests that in central Asian nation, *H. armigera* within the past was sometimes removed by shaking pigeonpea plants till caterpillars fell from the plants onto cotton sheets. This method is currently to get caterpillars that devored virus-infected seeds. Baculovirus preparations obtained during this process are utilized by farmers to organize a bioinsecticide spray applied on pigeonpea fields. Another baculovirus, HaSNPV is a sort of clone of HzSNPV. It had been registered in China as a chemical in 1993 (Zhang et al., 1995). These have been used as a big scale biopesticide production and have been extensively used on cotton fields. Moths of caterpillar contentment to *Spodoptera* genus are of main concern for the agricultural trade in several countries of the world. Two important industrial preparations supported *Spodoptera* NPV are accessible within the USA and Europe. These are SPOD-X™ containing *Spodoptera exigua* NPV to regulate insects on vegetable crops and Spodopterin™ containing *Spodoptera littolaris* NPV that is employed to shield cotton, corn, and tomatoes. Upto 20,000 ha of maize annually are controlled with *Spodoptera frugiperda* NPV in Brazil (Moscardi, 1999). Several alternative species belonging to the Noctuidae family is economically vital pests of sugarcane, legume, rice, and other cereals. *Autographa californica* and *Anagrapha falcifera* NPVs were registered within the USA and were field-tested at a restricted scale. These two NPVs have comparatively broad host spectrum and without doubt is used on a range of crops overrun with pests happiness to variety of genera, together with *Spodoptera* and *Helicoverpa*. The well-known success of using baculovirus as a biopesticide is that the case of *Anticarsia gemmatalis* nucleopolyhedrovirus accustomed management the velvet been caterpillar in soybean (Moscardi, 1999). This program was enforced in Brazil within the early eighties, and came up to over 2000,000 ha of soybean treated annually with the virus. Recently, this range borne down in the main attributable to new rising pests within the soybean advanced, though utilization of above-mentioned virus in Brazil is the most spectacular example of the bioregulation with microorganism chemical worldwide. The virus remains obtained by in vivo production in the main by infection of larvae in soybean farms. The requirement of virus production has increased staggeringly for defense of 4 million hectares of soybean annually.

11.9.8 MYCOFUNGICIDES

Mycofungicides are microbial antagonists which will suppress plant diseases and organisms that suppress pathogens are also commonly mentioned as biological management agents (BCA) (Alabouvette et al., 2006; Pal and Gardener, 2006). Varied flora species are used as BCAs and will offer effective activity agaist the varied harmful microorganisms. Examples are *Trichoderma harzianum*, a species with biocontrol potential against *Botrytis cineria*, *Fusarium*, fungus, and *Rhizoctonia* (Khetan, 2001); *Ampelomyces quisqualis*, a hyperparasite of mildew (Liang et al., 2007; Viterbo et al., 2007); *C. globosum* and *C. cupreum*, covering biocontrol activity against illness disease caused by *Fusarium*, *Phytophthora*, and fungus (Soytong et al., 2001); *Gliocladium virens*, effective biocontrol of soilborne pathogens (Viterbo et al., 2007); *Coniothyrium minitans*, a mycoparasite of fungus (Whipps et al., 2008); and *Fusarium oxysporum* (nonpathogenic species), having biocontrol potential against *F. oxysporum* (Fravel et al., 2003). An efficient BCA ought to be genetically stable, effective at low concentrations, straightforward to mass manufacture in culture on cheap media, and be effective against a good variety of pathogens (Irtwange, 2006; Wraight et al., 2001). The flora BCA ought to conjointly occur in Associate simply distributed kind, be nontoxic to humans, have resistance to pesticides, be compatible with alternative treatments, and be nonpathogenic against the host plant (Fravel, 2005; Irtwange, 2006). Many flora, taxa are reportable to be antagonist against the plant pathogens and are with the success developed as mycofungicides or biological management merchandise, for example, *A. quisqualis*, genus *A. niger*, fungus oleophila, *C. cupreum*, *C. globosum*, *C. minitans*, *Cryptococcus albidus*, *G. virens*, *G. catenulatum*, *F. oxysporum*, *Phlebiosis gigantean*, fungus oligandrum, *Rhodotorula glutinis*, *T. harzianum*, *Trichoderma polysporum*, *Trichoderma viride* (Boyetchko et al., 1999; Butt, 2000; Butt et al., 1999; Ezziyyani et al., 2007; Fravel, 2005; Ghisalberti, 2002; Hofstein and Chapple, 1999; Khetan, 2001; Soytong et al., 2001).

11.9.9 TRICHODERMA

Trichoderma spp. are free-living fungi that are very common in soil and root ecosystems. They are extremely interactive in root, soil, and foliar agricultural environments. They manufacture or discharge a mixture of compounds that encourage localized or systemic resistance responses in plants (Kodsueb et al., 2008; Thormann and Rice, 2007; Vinale et al., 2008). *Trichoderma* are simply isolated from soil, decaying wood, and different other alternative organic material (Howell, 2003; Zeilinger and Omann, 2007). There are several reports on the utilization of *Trichoderma* species as biological agents against plant pathogens (Harman et al., 2004; Zeilinger and Omann, 2007). *Trichoderma* species are used as BCAs against a good variety of harmful fungi, for example, *Rhizoctonia* spp., fungus spp., *Botrytis cinerea*, and *Fusarium* spp. *Phytophthora palmivora*, *Phytophthora parasitica*, and totally different species is used, for example, *T. harzianum*, *T. viride*, *Trichoderma virens* (Benitez et al., 2004; Sunantapongsuk et al., 2006; Zeilinger and Omann, 2007). Among them, *T. harzianum* is reportable to be most generally used as an efficient BCA (Abdel-Fattah et al., 2007; Szekeres et al., 2004). *T. harzianum* strain T-22 was made by protoplast fusion between *T. harzianum* T-95 and T-12, and this strain was developed as granular named RootSield® and powder named PlantShield® by Biworks, Geneva, NY. *T. harzianum* T-22 has effectiveness against a broad range of plant pathogenic fungi including *B. cinerea*, *Fusarium*, *Pythium*, *Rhizoctonia* in several cereal and pulse crops like corn, soybean, potato, tomato, beans, cotton, peanut, and varied trees (Paulitz and Belanger, 2001). One of the *T. harzianum* strains T-39 is marketed as TRICHODEX, 20P by Makhteshim Ltd. for the management of pink rot and stem rot of tomato caused by fungus genus *erythroseptica* (Etebarian et al., 2000) and management of blight disease caused by *B. cinerea* (Paulitz and Belanger, 2001). The biocontrol mechanism in *Trichoderma* may be a combination of different mechanisms (Benitez et al., 2004; Zeilinger and Omann, 2007). The most common mechanism is mycoparasitism and antibiosis (Howell, 2003; Vinale et al., 2008). Mycoparasitism depends on the popularity, binding, and catalyst disruption of the host flora cell membrane (Woo and Lorito, 2007). *Trichoderma* species have been very successfully used as mycofungicides because they are fast growing, have high reproductive capacity, inhibit a broad spectrum of fungal diseases, have a diversity of control mechanisms, are excellent competitors in the rhizosphere, have a capacity to modify the rhizosphere, are tolerant or resistance to soil fungicides, have the ability to survive under unfavorable conditions, are efficient in utilizing soil nutrients, have strong aggressiveness against phytopathogenic fungi, and also promote plant growth (Benitez et al., 2004; Vinale et al., 2006). Their ability to colonize and grow in association with plant roots is known as rhizosphere competence. The taxonomy of *Trichoderma* species is very complex and has been the subject of many recent taxonomic studies (Samuels, 2006; Woo et al., 2006). They also have a high level of genetic diversity (Harman, 2006; Harman et al., 2004). Thus, it is likely that only a few of the species available have been utilized as mycofungicide. However, *Trichoderma* species are the most common fungal biocontrol control agents and are commercially formulated as biofungicides, biofertilizers, and soil amendments (Harman, 2006; Vinale et al., 2006).

11.9.10 AMPELOMYCES

One mycoparasitic anamorphic ascomycete namely *A. quisqualis* reduces the growth and kills powdery mildews. It will have an effect on the infective agent through antibiosis and mutuality (Kiss, 2003; Viterbo et al., 2007). The plant life *A. quisqualis* was the primary organism reported to be a hyperparasite of mildew, and it will be simply found related to mildew colonies (Paulitz and Belanger, 2001). Hyphae of *Ampelomyces* penetrate the hyphae of powdery mildews and grow internally then kill all the parasitized cells (Kiss, 2003). *A. quisqualis* isolate M-10 has been developed as AQ10 biofungicide, developed by Ecogen, Inc., USA. This mycofungicide contains conidia of *A. quisqualis* and developed as water-dispersible granules for the management of mildew of carrot, cucumber, and mango (Kiss, 2003; Shishkoff and McGrath, 2002; Viterbo et al., 2007).

11.9.11 CHAETOMIUM

Chaetomium species are usually originated in soil and organic compost (Soytong et al., 2001). The genus *Chaetomium* was first recognized in 1817 by Gustav Kunze (Soytong and Quimio, 1989). The application of *Chaetomim* as a BCA to manage plant pathogens first started in about 1954 when Martin Tviet and M. B. Moor found *C. globosum* and *C. cochliodes* occurring on oat seeds and that these taxa provided some control of *Helminthosporium victoriae. Chaetomium* species have been accounted to be potential antagonists of various plant pathogens, especially soilborne and seedborne pathogens (Aggarwal et al., 2004; Dhingra et al., 2003; Park et al., 2005). Many species of *Chaetomium* with potential to be BCAs restrain the growth of bacteria and fungi through competition (for substrate and nutrients), mycoparasitism, antibiosis, or diverse combinations of these (Marwah et al., 2007; Zhang and Yang, 2007). *C. globosum* and *C. cupreum* in particular have been extensively studied and successfully used to control root-rot disease of citrus, black pepper, strawberry and have been shown to reduce damping off disease of sugar beet (Soytong et al., 2001; Tomilova and Shternshis, 2006). These taxa have been formulated in the form of powder and pellets as Ketomium®, a broad spectrum mycofungicide. Ketomium® has been also registered as a biological biofertilizers for degrading organic matter and for inducing plant immunity and stimulating plant growth (Soytong et al., 2001). The mycofungicide Ketomium® which comprises a *Chaetomium* spore suspension has been evaluated for its effect on Siberian isolates of the phytopathogenic fungi *B. cinerea*, *Didymella applanata*, *F. oxysporum*, and *Rhizoctonia solani.* It was found that Ketomium-mycofungicide was most efficient in suppressing raspberry spur blight caused by *D. applanata* and could also reduce potato disease caused by *R. solani*, increasing potato yield (Shternshis et al., 2005). After 2 years in storage, this mycofungicide was still capable of inhibiting the growth of phytopathogens but at higher doses (Tomilova and Shternshis, 2006). Other species of *Chaetomium* which can act as biological control mediator include *C. globosum* isolate CgA-1 which can reduce soybean stem canker disease caused by *Diaporthe phaseolorum* f. sp. *meridionalis* (Dhingra et al., 2003) and *C. cochliodes* CTh05 and VTh01 which has activity against *F. oxysporum* f. sp. *lycopersici* causing tomato wilt, whereas isolate CTh05 showed activity against *P. parasitica* causing citrus root rot (Phonkerd et al., 2008). *Chaetomium* species are reported as a broad spectrum mycofungicide that is not only used for protection but also for curative effect as well. Moreover, a new strain of *C. cupreum* RY202 has preliminary proved to be antagonistic against *Rigidoporus microporus* which causes white root disease of rubber trees variety RRIM600. This promising strain is being investigated as a potential BCA against *R. microporus*.

11.9.12 GLIOCLADIUM

Gliocladium species are frequent soil saprobes and several species have been reported to be parasites of many plant pathogens (Viterbo et al., 2007), such as *G. catenulatum* parasities *Sporidesmium sclerotiorum* and *Fusarium* spp. It destroys the fungal host through the hyphal contact and forms pseudoappressoria (Punja and Utkhede, 2004; Viterbo et al., 2007). *G. catenulatum* strain JI446 has also been used as a wettable powder named Primastop® by Kemira Agro Oy, Finland. These manufactured goods can be applied to soils, roots, and foliage to reduce the incidence of damping-off disease caused by two pathogens, *Pythium ultimum* and *R. solani* in the greenhouse (Punja and Utkhede, 2004). *G. virens* has been used as a biological control means against a wide range of soilborne pathogens, such as *Pythium* and *Rhizoctonia* under greenhouse and field conditions (Viterbo et al., 2007). *G. virens* isolate GL-21 was formulated as one of the alginate prills named GlioGard® by W. R. Grace Co. and a granular formulation with the trade name SoilGard® produced by the Thermo Triology Corp., Columbia, MD. SoilGard® was developed for greenhouse application (Paulitz and Belanger, 2001; Punja and Utkhede, 2004). *G. virens* produces antibiotic metabolites such as gliotoxin which have antibacterial, antifungal, antiviral, and antitumor activities. New molecular evidence indicates that *G. virens* is more closely related to *Trichoderma* than those *G. virens*. This sustains suggestions that this taxon should be referred to as *T. virens* (Hebbar and Lumsden, 1999; Punja and Utkhede, 2004; Paulitz and Belanger, 2001).

11.10 CONCLUSION

Agriculture is the most valuable sector in the world and is more dependent on fertile soils and a stable climate than some other trades. At the same time, it has a huge influence on the ecological balance, water and soil quality, and on the preservation of biological diversity. Since the last century, agricultural techniques and economic framework situation worldwide have undergone such a radical transformation that agriculture has become a major source of environmental pollution. The research about ecologically compatible techniques in agriculture and environmental sciences can take necessary benefit from the use of beneficial microorganisms as plant–microbe interactions fulfill important ecosystem functions. Plant diseases are main causes of yield losses and ecosystem instability worldwide. Novel biotechnological methods for crop protection are based on the use of beneficial microorganisms applied as biofertilizers and/or biocontrol representative; this approach represents an important utensil for plant disease control and could lead to a substantial reduction of chemical fertilizer utilize, which is a significant resource of environmental pollution. In short, from the examples and references cited above, it manifests that useful microorganisms of agricultural importance represent an alternative and ecological strategy for disease management to reduce the use of chemicals in agriculture and to improve cultivar performance. At the same time, their application is a highly efficient way to resolve environmental problems, for example, through bioremediation and bioengineering. For the future development of biotechnology in this field, the contribution of a combination of scientific disciplines is of primary importance to promote sustainable practices in plant production system, as well as in conservation and ecosystem restoration. Further, relevance of microbial symbiotic signals and their altered derivatives for the remodeling of the plant developmental or defensive occupations may symbolize a promising field for agricultural biotechnology. The prospects for a future development of agricultural microbiology may absorb the creation of new multipartite endo- and ectosymbiotic communities supported by the extended genetic and molecular (metagenomic) analyses.

KEYWORDS

- **sustainable agriculture**
- **microbial diversity**
- **beneficial microorganisms**
- **antimicrobials**
- **phytohormones**

REFERENCES

Abdel-Fattah, M. G.; Shabana, M. Y.; Ismail, E. A.; Rashad, M. Y. *Trichoderma harzianum*: A Biocontrol Agent against *Bipolaris oryzae*. *Mycopathologia* **2007,** *164*, 81–89.

Adholeya, A.; Tiwari, P.; Singh, R. Largescale Inoculum Production of Arbuscular Mycorrhizal Fungi on Root Organs and Inoculation Strategies. In: *Soil Biology, In Vitro Culture of Mycorrhizae*; Declerck, S., Strullu, D.-G., Fortin, A., Eds.; Springer-Verlag: Berlin-Heidelberg, 2005; Vol 4, pp 315–338.

Aggarwal, R.; Tewari, A. K.; Srivastava, K. D.; Singh, D. V. Role of Antibiosis in the Biological Control of Spot Blotch (*Cochliobolus sativus*) of Wheat by *Chaetomium globosum*. *Mycopathologia* **2004,** *157*, 369–377.

Agrawal, N.; Malhotra, P.; Bhatnagar, R. K. Interation of Gene-Cloned and Insect Cell-Expressed Aminopeptidase N of *Spodoptera litura* with Insecticidal Crystal Protein *Cry*1C. *Appl. Environ. Microbiol.* **2002,** *68* (9), 4583–4592.

Aimey, V.; Hunte, K.; Whitehall, P.; Sanford, B.; Trotman, M.; Delgado, A.; Lefrancois, T.; Shaw, J.; Hernandez, J. Prevalence of and Examination of Exposure Factors for *Salmonella* on Commercial Egg-Laying Farms in Barbados. *Prevent. Vet. Med.* **2013,** *110*, 489–496.

Akhtar, M. S.; Siddiqui, Z. A. Effect of Phosphate Solubilizing Microorganisms and *Rizobium* sp. on the Growth, Nodulation, Yield and Root-Rot Disease Complex of Chickpea under Field Condition. *Afr. J. Biotechnol.* **2009,** *8* (15), 3489–3496.

Akhtar, M. S.; Siddiqui, Z. A. Arbuscular Mycorrhizal Fungi as Potential Bioprotectants against Plant Pathogens. In: *Mycorrhizae: Sustainable Agriculture and Forestry*; Siddiqui, Z. A. Akhtar, M. S.; Futai, K., Eds.; Springer, 2008a; pp 61–97.

Akhtar, M. S.; Siddiqui, Z. A. Biocontrol of a Root-Rot Disease Complex of Chickpea by *Glomus intraradices*, *Rhizobium* sp. and *Pseudomonas straita*. *Crop Prot.* **2008b,** *27*, 410–417.

Akintokun, A. K.; Akande, G. A.; Akintokun, P. O.; Popoola, T. O. S.; Babalola, A. O. Solubilization on Insoluble Phosphate by Organic Acid-Producing Fungi Isolated from Nigerian Soil. *Int. J. Soil Sci.* **2007,** *2*, 301–307.

Alabouvette, C.; Olivain, C.; Steinberg, C. Biological Control of Plant Diseases: The European Situation. *Eur. J. Plant Pathol.* **2006,** *114*, 329–341.

Alam, S.; Khalil, S.; Ayub, N.; Rashid, M. In Vitro Solubilization of Inorganic Phosphate by Phosphate Solubilizing Microorganism (PSM) from Maize Rhizosphere. *Int. J. Agric. Biol.* **2002,** *4*, 454–458.

Alter, J. A.; Vandenberg, J. J. D. Factors that Influencing the Infectivity of Isolates of *Paecilomyces fumosoroseus* against Diamond Back Moth. *J. Invertebr Pathol.* **2000,** *78*, 31–36.

Altieri, M. A. Linking Ecologists and Traditional Farmers in the Search for Sustainable Agriculture. *Front. Ecol. Environ.* **2004,** *2*, 35–42.

Anderson, C. I.; Cairney, W. G. J. Ectomycorrhizal Fungi: Exploring the Mycelial Frontier. *FEMS Microb. Rev.* **2007,** *31*, 388–406.

Andrade, M. M. M.; Stamford, N. P.; Santos, C. E. R. S.; Freitas, A. D. S.; Sousa, C. A.; Lira-Junior, M. A. Effects of Biofertilizer with Diazotrophic Bacteria and Mycorrhizal Fungi in Soil Attribute, Cowpea Nodulation Yield and Nutrient Uptake in Field Conditions. *Sci. Hortic.* **2013,** *162*, 374–379.

Annesini, M. C.; Piemonte, V.; Tomei, M. C.; Daugulis, A. J. Analysis of the Performance and Criteria for Rational Design of a Sequencing Batch Reactor for Xenobiotic Removal. *J. Chem. Eng.* **2014,** *235*, 167–175.

Asensio, V.; Rodriguez-Ruiz, A.; Garmendia, L.; Andre, J.; Kille, P.; Morgan, A. J.; Soto, M.; Marigomez, I. Towards an Integrative Soil Health Assessment Strategy: A Three Tier (Integrative Biomarker Response) Approach with *Eisenia fetida* Applied to Soils Subjected to Chronic Metal Pollution. *J. Sci. Total Environ.* **2013,** *442*, 344–365.

Avery, P. B.; Faulla, J.; Simmands, M. S. J. Effect of Different Photoperiods on the Infectivity and Colonization of *Paecilomyces fumosoroseus*. *J. Insect Sci.* **2004,** *4*, 38.

Babalola, O. O. Beneficial Bacteria of Agricultural Importance. *Biotech. Lett.* **2010,** *32*, 1559–1570.

Babalola, O. O.; Glick, B. R. Indigenous African Agriculture and Plant Associated Microbes: Current Practice and Future Transgenic Prospects. *Sci. Res. Essays* **2012a,** *7*, 2431–2439.

Babalola, O. O.; Glick, B. R. The Use of Microbial Inoculants in African Agriculture: Current Practice and Future Prospects. *J. Food Agricul. Environ.* **2012b,** *10*, 540–549.

Babalola, O. O.; Oladele, O. I. Biotechnology in Agriculture: Implications for Agricultural Extension and Advisory Services. *J. Food Agric. Environ.* **2011,** *9*, 486–491.

Babu, V.; Murugan, S.; Thangaraja, P. Laboratory Studies on the Efficacy of Neem and the Entomopathogenic Fungus *Beauveria bassiana* on Spodoptera Litura. *Entomology* **2001,** *56*, 56–63.

Bakulin, M. K.; Grudtsyna, A. S.; Pletneva, A. Biological Fixation of Nitrogen and Growth of Bacteria of the Genus *Azotobacter* in Liquid Media in the Presence of *Perfluoro carbons*. *Appl. Biochem. Microbiol.* **2007,** *4*, 399–402.

Balasubramanian, D.; Arunachalam, K.; Arunachalam, A.; Das, A. K. Water Hyacinth [*Eichhornia crassipes* (Mart.) Solms.] Engineered Soil Nutrient Availability in a Low-Land Rain-Fed Rice Farming System of North-East India. *J. Ecol.* **2013,** *58*, 3–12.

Banerjee, S.; Palit, R.; Sengupta, C.; Standing, D. Stress Induced Phosphate Solubilization by *Arthrobacter* sp. and *Bacillus* sp. Isolated from Tomato Rhizosphere. *Aust. J. Crop Sci.* **2010,** *4* (6), 378–383.

Banik, S.; Dey, B. K. Available Phosphate Content of an Alluvial Soil as Influenced by Inoculation of Some Isolated Phosphate Solubilizing Microorganisms. *Plant Soil* **1982,** *69*, 353–364.

Barbarin, A. M.; Jenkins, N. E.; Rajotte, E. G.; Thomas, M. B. A Preliminary Evaluation of the Potential of *Beauveria bassiana* for Bed Bug Control. *J. Invert. Pathol.* **2012,** *111*, 82–85.

Barea, J. M.; Pozo, M. J.; Azcon, R.; Azcon-Aguilar, C. Microbial Interactions in the Rhizosphere. In: *Molecular Microbial Ecology of the Rhizosphere*; de Bruijn, F. J. Ed.; Wiley Blackwell: Hoboken, NJ, 2013; vol 1, pp 29–44.

Barroso, C. B.; Pereira, G. T.; Nahas, E. Solubilization of $CAHPO_4$ and $ALPO_4$ by *Aspergillus niger* in Culture Media with Different Carbon and Nitrogen Sources. *Braz. J. Microbiol.* **2006,** *37*, 434–438.

Baudoin, E.; Benizri, E.; Guckert, A. Impact of Growth Stages on Bacterial Community Structure along Maize Roots by Metabolic and Genetic Finger Printing. *Appl. Soil Ecol.* **2002,** *19*, 135–145.

Beneduzi, A.; Moreira, F.; Costa, P. B.; Vargas, L. K.; Lisboa, B. B.; Favreto, R.; Baldani, J. I.; Passaglia, L. M. P. Diversity and Plant Growth Promoting Evaluation Abilities of Bacteria Isolated from Sugarcane Cultivated in the South of Brazil. *Appl. Soil Ecol.* **2013,** *63*, 94–104.

Benitez, T.; Rincon, M. A.; Limon, M. C.; Codon, C. A. Biocontrol Mechanisms of *Trichoderma* Strains. *Int. Microbiol.* **2004,** *7*, 249–260.

Berg, G. Plant–Microbe Interactions Promoting Plant Growth and Health: Perspectives for Controlled Use of Microorganisms in Agriculture. *Appl. Microbiol Biotechnol.* **2009,** *84*, 11–18.

Bischoff, J. F.; Rehner, S. A.; Humber, R. A. A Multilocus Phylogeny of the *Metarhizium anisopliae* Lineage. *Mycologia* **2009,** *101* (4), 512–530.

Boyetchko, S.; Pedersen, E.; Punja, Z.; Reddy, M. Formulations of Biopesticides. In: *Methods in Biotechnology: Biopesticides: Use and Delivery*; Frinklin, R. H., Julius, J. M.; Humana Press: Totowa, NJ, 1999; Vol 5, pp 487–508.

Bravo, A.; Gill, S. S.; Soberon, M. Mode of Action of *Bacillus thuringiensis Cry* and *Cyt* Toxins and their Potential for Insect Control. *Toxicon* **2007,** *49*, 423–435.

Bulla, L. A.; Candas, M. Patent No. WO0140476, 2002.

Burdman, S.; Jurkevitch, E.; Okon, Y. Recent Advance in the Use of Plant Growth Promoting Rhizobacteria (PGPR) in Agriculture. In: *Microbial Interaction in Agriculture Forestry*, Subba Rao, N. S., Dommergues, Y. R., Eds., 2000; Vol II, pp 229–250.

Burton, E. M.; Knight, S. D. Survival of *Penicillium bilaiae* Inoculated on Canola Seed Treated with Vitavax RS and Extender. *Biol. Fert. Soils* **2005,** *42*, 54–59.

Butt, T. M.; Copping, L. G. Fungal Biological Control Agents. *Pesticide Outlook* **2000,** 186–191.

Butt, T. M.; Harris, J. G.; Powell, K. A. Microbial Biopesticides: The European Scene. In: *Methods in Biotechnology: Biopesticides Use and Delivery*; Frinklin, R. H., Julius, J. M., Eds.; Humana Press Inc.: Totowa, NJ, 1999; Vol 5, pp 23–44.

Buyck, B.; Hofstetter, V.; Eberhardt, U.; Verbeken, A.; Kauff, F. Walking the Thin Line between *Russula* and *Lactarius*: The Dilemma of *Russula* Subject. *Orchricompactae. Funal Div.* **2008,** *28*, 15–40.

Chandra, K.; Greep, S.; Ravindranath, P.; Srivathsa, R. S. H. *Liquid Biofertilizers*. Ministry of Agriculture Department of Agriculture & Co-operation, Government of India, 2011.

Chang, C. H.; Yang, S. S. Thermotolerant Phosphate-Solubilizing Microbes for Multi-functional Biofertilizer Preparation. *Biores. Technol.* **2009,** *100*, 1648–1658.

Chen, J. H. The Combined Use of Chemical and Organic Fertilizers and/or Biofertilizer for Crop Growth and Soil Fertility. In: International Workshop on Sustained Management of the Soil–Rhizosphere System for Efficient Crop Production and Fertilizer Use, 16–20 October 2006, Land Development Department: Bangkok 10900 Thailand, 2006; pp 1–11.

Chen, Y. P.; Rekha, P. D.; Arunshen, A. B.; Lai, W. A.; Young, C. C. Phosphate Solubilizing Bacteria from Subtropical Soil and their Tri-calcium Phosphate Solubilizing Abilities. *Appl. Soil Ecol.* **2006,** *34*, 33–41.

Cloyd, R. A. The Entomopathogenic Fungus *Metarhizium anisopliae. Midwest Biol. Control News* **1999,** *VI* (7).

Dahm, H. Role of Mycorrhizae in the Forestry. In: *Handbook of Microbial Biofertilizers*; Rai, M. K., Eds.; Food Products Press, Binghamton, NY, USA, 2006; pp 241–270.

Daniel, J. F. S.; Filho, R. E. Peptaibols of *Trichoderma. Nat. Prod. Rep.* **2007,** *24*, 1128–1141.

Das, A.; Prasad, R.; Srivastava, A.; Giang, H. P.; Bhatnagar, K.; Varma, A. Fungal Siderophores: Structure, Functions and Regulation. In: *Soil Biology: Microbial Siderophores*; Varma, A., Chincholkar, S. B., Eds.; Springer-Verlag: Berlin-Heidelberg, 2007; Vol 12, pp 1–42.

de Maagd, R. A.; Bravo, A.; Crickmore, N. How *Bacillus thuringiensis* Has Evolved Specific Toxins to Colonize the Insect World. *Trends Genet.* **2001,** *17*, 193–199.

Deubel, A.; Gransee, G.; Merbach, W. Transformation of Organic Rhizodeposits by Rhizoplane Bacteria and its Influence on the Availability of Tertiary Calcium Phosphate. *J. Plant Nutr. Soil Sci.* **2000,** *163*, 387–392.

Dhingra, O. D.; Mizubuti, E. S. G.; Santana, F. M. *Chaetomium globosum* for Reducing Primary Inoculum of *Diaporthe phaseolorum* f. sp. *meridionalis* in Soil-Surface Soybean Stable in Field Condition. *Biol. Control.* **2003,** *26*, 302–310.

Dobbelaere, S.; Vanderleyden, J.; Okon, Y. Plant Growth-Promoting Effects Diazotrophs in the Rhizosphere. *Crit. Rev. Plant Sci.* **2003,** *22*, 107–149.

Donald, G.; McNeil, J. Fungus Fatal to Mosquito May aid Global War on Malaria, *The New York Times*, 2005.

Douds, J. D. D.; Nagahashi, G.; Pfeffer, P. E.; Kayser, W. M.; Reider, C. On-Farm Production and Utilization of Arbuscular Mycorrhizal Fungus Inoculum. *Can. J. Plant Sci.* **2005,** *85*, 15–21.

Driver, F.; Milner, R. J.; Trueman, W. H. A. A Taxonomic Revision of *Metarhizium* Based on Sequence Analysis of Ribosomal DNA. *Mycol. Res.* **2000,** *104* (2), 135–151.

Drogue, B.; Dore, H.; Borland, S.; Wisniewski-Dye, F.; Prigent-Combaret, C. Which Specificity in Cooperation between Phytostimulating Rhizobacteria and Plants? *Resour. Microbiol.* **2012,** *163*, 500–510.

Duponnois, R.; Kisa, M.; Plenchette, C. Phosphate Solubilizing Potential of the Nemato Fungus *Arthrobotrys oligospora. J. Plant Nutr. Soil Sci.* **2006,** *169*, 280–282.

Dutton, V. M.; Evans, C. S. Oxalate Production by Fungi: Its Role in Pathogenicity and Ecology in the Soil Environment. *Can. J. Microbiol.* **1996,** *42*, 881–895.

El-Azouni, I. M. Effect of Phosphate Solubilising Fungi on Growth and Nutrient Uptake of Soybean (*Glycine max* L.) Plants. *J. Appl. Sci. Res.* **2008,** *4*, 592–598.

El-Komy, H. M. A. Co-immobilization of *A. lipoferum* and *B. megaterium* for Plant Nutrition. *Food Technol. Biotechnol.* **2005,** *43* (1), 19–27.

Etebarian, H. R.; Scott, E. S.; Wicks, T. J. *Trichoderma harzianum* T39 and *T. virens* DAR74290 as Potential Biological Control Agents for *Phytophthora erythroseptica. Eur. J. Plant Pathol.* **2000,** *106*, 329–337.

Ezziyyani, M.; Requena, M. E.; Egea-Gilabert, C.; Candel, M. E. Biological Control of *Phytophthora* Root Rot of Pepper Using *Trichoderma harzianum* and *Streptomyces rochei* in Combination. *J. Phytopathol.* **2007,** *155*, 342–349.

Fravel, D.; Olivain, C.; Alabouvette, C. *Fusarium oxysporum* and its Biocontrol. *New Phytol.* **2003,** *157*, 493–502.

Fravel, R. D. Commercialization and Implementation of Biocontrol. *Ann. Rev Phytopathol.* **2005,** *43*, 337–359.

Freimoser, F. M.; Screen, S.; Bagga, S.; Hu, G.; St. Leger, R. J. EST Analysis of Two Subspecies of *Metarhizium anisopliae* Reveals a Plethora of Secreted Proteins with Potential Activity in Insect Hosts. *Microbiology* **2003,** *149*, 239–247.

Galan-Wong, L. J.; Pereyra-Alferez, B.; Luna-Olvera, H. A. Bacterias entomopatogenas. In: *Biotecnologia financiera aplicada a bioplaguicidas*; Garcia-Gutierrez C., Medrano-Roldan H., Eds.; Mexico, 2006; pp 43–63.

Gentili, F.; Jumpponen, A. Potential and Possible Uses of Bacterial and Fungal Biofertilizers. In: *Handbook of Microbial Biofertilizers*; Rai, M. K., Eds.; Food Products Press, Binghamton, NY, USA, 2006, 1–28.

Ghaderi, A.; Aliasgharzad, N.; Oustan, S.; Olsson, P. A. Efficiency of Three *Pseudomonas* Isolates in Releasing Phosphate from an Artificial Variable-Charge Mineral (Iron III Hydroxide). *Soil Environ.* **2008,** *27*, 71–76.

Ghisalberti, E. L. Anti-infective Agents Produced by the Hyphomycetes Genera *Trichoderma* and *Gliocladium*. *Curr. Med. Chem. Anti-Infect. Agents* **2002,** *1*, 343–374.

Gomes, C. D. S. F.; Silva, J. B. P. Minerals and Clay Minerals in Medical Geology. *App. Clay Sci.* **2007,** *36*, 4–21.

Gothwal, R. K.; Nigam, V. K.; Mohan, M. K.; Sasmal, D.; Ghosh, P. Screening of Nitrogen Fixers from Rhizospheric Bacterial Isolates Associated with Important Desert Plants. *Appl. Ecol. Environ. Res.* **2007,** *62*, 101–109.

Grant, C. A.; Bailey, L. D.; Harapiak, J. T.; Flore, N. A. Effect of Phosphate Source, Rate and Cadmium Content and Use of *Penicillium bilaii* on Phosphorus, Zinc and Cadmium Concentration in Durum Wheat Grain. *J. Sci. Food Agric.* **2002,** *82*, 301–308.

Gull, M.; Hafeez, F. E.; Saleem, M.; Malik, K. A. Phosphorus Uptake and Growth Promotion of Chickpea by Co-inoculation of Mineral Phosphate Solubilizing Bacteria and a Mixed Rhizobial Culture. *Austr. J. Exp. Agric.* **2004,** *44*, 623–628.

Guo, J.; Dong, R.; Clemens, J.; Wang, W. Thermal Modelling of the Completely Stirred Anaerobic Reactor Treating Pig Manure at Low Range of Mesophilic Conditions. *J. Environ. Manage.* **2013,** *127*, 18–22.

Gupta, A. K. The Complete Technology Book on Biofertilizers and Organic Farming. National Institute of Industrial Research Press: India, 2004.

Gupta, V.; Satyanarayana, T.; Garg, S. General Aspects of Mycorrhiza. In: *Mycorrhizal Biology*; Mukerji, K. G., Singh, J., Chamola, B. P., Eds.; Kluwer Academic/Plenum Publishers: New York, 2000; pp 27–44.

Prashanth, H. K.V.; Tharanathan, R.N. Chitin/chitosan: Modifications and their unlimited application potential- An overview. *Trends Food Sci. Technol.* **2007,** *18*, 117–131.

Harman, G. E. Overview of Mechanisms and Uses of *Trichoderma* spp. *Phytopathology* **2006,** *96*, 190–194.

Harman, G. E.; Howell, C. R.; Viterbo, A.; Chet, I.; Lorito, M. *Trichoderma* Species Opportunistic, Avirulent Plant Symbionts. *Nat. Rev. Microbiol.* **2004,** *2*, 43–56.

Hart, M. M.; Trevors, J. T. Microbe Management: Application of Mycorrhyzal Fungi in Sustainable Agriculture. *Front. Ecol. Environ.* **2005,** *3*, 533–539.

Hebbar, P. K.; Lumsden, R. D. Biological Control of Seedling Diseases. In: *Methods in Biotechnology: Biopesticides: Use and Delivery*; Frinklin, R. H., Julius J. M., Eds.; Humana Press: Totowa, NJ, 1999; Vol 5, pp 103–116.

Henri, F.; Laurette, N. N.; Annette, A.; John, Q.; Wolfgang, M.; Francois-Xavier, E.; Dieudonne, E. Solubilization of Inorganic Phosphates and Plant Growth Promotion by Strains of *Pseudomonas fluorescens* Isolated from Acidic Soils of Cameroon. *Afr. J. Microbiol. Res.* **2008,** *2*, 171–178.

Hermosa, R.; Botella, L.; Alonso-Ramirez, A.; Arbona, V.; Gomez-Cadenas, A.; Monte, E.; Nicolsa, C. Biotechnological Applications of the Gene Transfer from the Beneficial Fungus *Trichoderma harzianum* spp. to Plants. *Plant Signal. Behav.* **2011,** *6*, 1235–1236.

Herrera-Estrella A.; Chet I. The biological control agent Trichoderma – from fundamentals to applications. p. 147–156. In: "Fungal Biotechnology in Agricultural, Food and Environmental Applications" (D.K. Arora, M. Dekker, eds.). Vol. 21. CRC Press, New York, USA, 2004; pp 700.

Hilda, R.; Fraga, R. Phosphate Solubilising Bacteria and their Role in Plant Growth Promotion. *Biotechnol. Adv.* **2000,** *17*, 319–359.

Hinsinger, P. Bioavailability of Soil Inorganic P in the Rhizosphere as Affected by Root Induced Chemical Changes: A Review. *Plant Soil* **2001,** *237*, 173–195.

Hofstein, R.; Chapple, A. C. Commercial Development of Biofungicide. In: *Methods in Biotechnology: Biopesticides Use and Delivery*; Frinklin, R. H., Julius J. M., Eds.; Humana Press Inc.: Totowa, NJ, 1999; Vol 5, pp 77–102.

Hofte, H.; Whiteley, H. R. Insecticidal Crystal Proteins of *Bacillus thuringiensis*. *Microbiol. Rev.* **1989,** *53*, 242–255.

Howell, R. C. Mechanisms Employed by *Trichoderma* Species in the Biological Control of Plant Diseases: The History and Evolution of Current Concepts. *Plant Dis*. **2003,** *87*, 4–10.

Huang, B.; Li, C.; Humber, R. A.; Hodge, K. T.; Fan, M.; Li, Z. Molecular Evidence for the Taxonomic Status of *Metarhizium taii* and its Teleomorph, *Cordyceps taii* (Hypocreales, Clavicipitaceae). *Mycotaxon* **2005,** *94*, 137–147.

Hyakumachi, M.; M. Kubota. Fungi as Plant Growth Promoter and Disease Suppressor. In: *Fungal Biotechnology in Agriculture, Food and Environmental Applications*, Arora, D. K., Ed.; Dekker: New York, 2004; pp. 101–110.

Ignoffo, C. M.; Couch, T. L. The Nucleopolyhedrosis Virus of *Heliothis* Species as a Microbial Pesticide. In: *Microbial Control of Pests and Plant Diseases*; Burges, H. D., Ed., Academic Press: London, 1981; pp. 329–362.

Igual, J. M.; Valverde, A.; Cervantes, E.; Velazquez, E. Phosphate-Solubilizing Bacteria Asinoculants for Agriculture: Use of Updated Molecular Techniques in their Study. *Agronomie* **2001,** *21*, 561–568.

Ipsilantis, I.; Sylvia, D. M. A Bundance of Fungi and Bacteria in a Nutrient-Impacted Florida Wetland. *Appl. Soil Ecol.* **2007,** *35*, 272–280.

Iriarte, J.; Caballero, P. Biologíay Ecología de *Bacillus thuringiensis*. In: *Bioinsecticidas: Fundamentosy Aplicaciones de Bacillus thuringiensis en el Control Integrado de Plagas*; Caballero, P., Ferré, J., Eds.; Phytoma-Espana, 2001; pp 15–44.

Irtwange, V. S. Application of Biological Control Agents in Pre- and Postharvest Operations. *Agric. Eng. Int.: CIGR J. Invited Overview* **2006,** *3*, 1–12.

Jeeva, S. Studies on the Effect of *Azospirillum* on the Growth and Development of Banana cv. Poovan (AAB). M.Sc. (Hort.) Thesis, TNAU: Coimbatore, Tamil Nadu, India, 1987.

Jin, M.; Wang, Y.; Huang, M.; Lu, Z.; Wang, Y. Sulphation Can Enhance the Antioxidant Activity of Polysaccharides Produced by *Enterobacter cloacae* Z0206. *Carbohydr. Polym.* **2014,** *99*, 624–629.

Johansson, F. J.; Paul, R. L.; Finlay, D. R. Microbial Interactions in the Mycorrhizosphere and their Significance for Sustainable Agriculture. *FEMS Microbiol. Ecol.* **2004,** *48*, 1–13.

Joung, K. J.; Cote, J. C. A Review of the Enviromental Impacts of the Microbial Insecticida *Bacillus thuringiensis*. *Technical Bulletin No. 29*, Horticultural Research and Development Centre: Canada, 2002.

Kannaiyan, S. Biofertilizers for Sustainable Crop Production. In: *Biotechnology of Biofertilizers*; Kannaiyan, S.; Kluwer Academic Publishers, Narosa, New Delhi, 2002; pp 9–50.

Kapoor, R.; Sharma, D.; Bhatnagar, A. K. Arbuscular Mycorrhizae in Micropropagation Systems and their Potential Applications. *Sci. Hortic.* **2008,** *116*, 227–239.

Khan, A. G. Mycorrhizoremediation—An Enhanced Form of Phytoremediation. *J. Zhejiang Univ. Sci. B* **2006,** *7*, 503–514.

Khan, M. S.; Zaidi, A.; Wani, P. A. Role of Phosphate-Solubilizing Microorganisms in Sustainable Agriculture—A Review. *Agron. Sustain. Dev.* **2007,** *27*, 29–43.

Khetan, S. K. *Microbial Pest Control*. Marcel Dekker, Inc.: New York, Basel, 2001; p 300.

Khiari, L.; Parent, L. E. Phosphorus Transformations in Acid Light-Textured Soils Treated with Dry Swine Manure. *Can. J. Soil Sci.* **2005,** *85*, 75–87.

Kim, K. Y.; Jordan, D.; Mc Donald, G. A. Effect of Phosphate-Solubilizing Bacteria and Vesicular-Arbuscular Mycorrhizae on Tomato Growth and Soil Microbial Activity. *Biol. Fertil. Soils* **1998,** *26*, 79–87.

Kiss, L. A Review of Fungal Antagonists of Powdery Mildews and their Potential as Biocontrol Agents. *Pest Manage. Sci.* **2003,** *59*, 475–483.

Kodsueb, R.; McKenzie, E. H. C.; Lumyong, S.; Hyde, K. D. Diversity of Saprobic Fungi on Magnoliaceae. *Fungal Div.* **2008,** *30*, 37–53.

Kouvoutsakis, G.; Mitsi, C.; Tarantilis, P. A.; Polissiou, M. G.; Pappas, C. S. Geographical Differentiation of Dried Lentil Seed (*Lens culinaris*) Samples Using Diffuse Reflectance Fourier Transform Infrared Spectroscopy (DRIFTS) and Discriminant Analysis. *Food Chem.* **2014,** *145*, 1011–1014.

Ku, N. S.; Kim, Y. C.; Kim, M. H.; Song, J. E.; Oh, D. H.; Ahn, J. Y. Risk Factors for 28-day Mortality in Elderly Patients with Extended-Spectrum â-Lactamase (ESBL)-Producing *Escherichia coli* and *Klebsiella pneumoniae* bacteremia. *Arch. Gerontol. Geriatr.* **2014,** *58*, 105–109.

Kumari, S.; Sarkar, P. K. In Vitro Model Study for Biofilm Formation by *Bacillus cereus* in Dairy Chilling Tanks and Optimization of Clean-in-Place (CIP) Regimes Using Response Surface Methodology. *Food Control* **2014,** *36*, 153–158.

Lacey, L. A.; Federici, B. A. Pathogenesis and Midgut Histopathology of *Bacillus thuringiensis* in *Simulium vittatum* (Dipetra: Simulidae). *J. Invertebr. Pathol.* **1979,** *33*, 171–182.

Lawal, T. E.; Babalola, O. O. Relevance of Biofertilizers to Agriculture. *J Hum Ecol*, **2014,** *47*(1), 35-43.

Li, L.; Ma, J.; Li, Y.; Wang, Z.; Gao, T.; Wang. Q. Screening and Partial Characterization of *Bacillus* with Potential Applications in Biocontrol of Cucumber Fusarium Wilt. *Crop Prot.* **2012,** *35*, 29–35.

Li, W. W.; Yu, H. Q. From Wastewater to Bioenergy and Biochemicals via Two-stage Bioconversion Processes: A Future Paradigm. *Biotechnol. Adv.* **2011,** *29*, 972–982.

Lian, B.; Wang, B.; Pan, M.; Liu, C.; Teng, H. H. Microbial release of potassium from Kbearing minerals by thermophlic fungus *Aspergillus fumigatus. Geochim. Cosmochim. Acta* **2008,** *72*, 87–98.

Liang, C.; Yang, J.; Kovacs, G. M.; Szentivanyi, O.; Li, B.; Xu, X. M.; Kiss, L. Genetic Diversity of *Ampelomyces mycoparasites* Isolated from Different Powdery Mildew Species in China Inferred from Analyses of rDNA ITS Sequences. *Fungal Div.* **2007,** *24*, 225–240.

Lorence, A.; Darszon, A.; Diaz, C.; Lievano, A.; Quintero, R.; Bravo, A. Endotoxins Induce Cation Channels in *Spodoptera frugiperda* Brush Border Membranes in Suspension and in Planar Lipid Bilayers. *FEBS Lett.* **1995,** *360*, 217–222.

Lubambo, A. F.; Benelli, E. M.; Klein, J. J.; Schreiner, W. H.; Silveira, E.; de Camargo, P. C. Tuning Protein GlnBHs Surface Interaction with Silicon: FTIR-ATR, AFM and XPS Study. *Biointerfaces* **2013,** *102*, 348–353.

Lugtenberg, B. Life of Microbes in the Rhizosphere. In: *Principles of Plant–Microbe Interactions*; Lugtenberg, B., Ed.; Springer International Publishing: Switzerland, Heidelberg, 2015; pp 7–15.

Manivasagan, P.; Venkatesan, J.; Sivakumar, K.; Kim, S. K. Marine Actinobacterial Metabolites: Current Status and Future Perspectives. *Microb. Res.* **2013,** *168*, 311–332.

Marin, M. Arbuscular Mycorrhizal Inoculation in Nursery Practice. In: *Handbook of Microbial Biofertilizers*; Rai, M. K., Eds.; Food Products Press, Binghamton, NY, 2006; pp 289–324.

Marwah, R. G.; Fatope, M. O.; Deadman, M. L.; Al-Maqbali, Y. M.; Husband, J. Musanahol: A New Aureonitol-Related Metabolite from a *Chaetomium* sp. *Tetrahedron* **2007,** *63*, 8174–8180.

Mathavan, S.; Sudha, P. M.; Pechimithu, S. M. Effect on *Bacillus thuringiensis* on the Midgut Cells of *Bombix mori* Larvae: A Histopahological and Histochemical Study. *J. Invertebr. Pathol.* **1989,** *53*, 217–227.

McNeil, J.; Donald, G. Fungus Fatal to Mosquito May Aid Global War on Malaria. *New York Times* **2005,** *104*, 135–151.

Mettenmeyer, A. Viral Insecticides Hold Promise for Bio-control. *Farm. Ahead* **2002,** *124*, 50–51.

Minaxi, H.; Saxena, J. Efficacy of Rhizobacterial Strains Encapsulated in Nontoxic Biodegradable Gel Matrices to Promote Growth and Yield of Wheat Plants. *Appl. Soil Ecol.* **2011,** *48*, 301–308.

Mohammadi, K.; Ghalavand, A.; Aghaalikhani, M.; Sohrabi, Y.; Heidari, G. R. Impressibility of Chickpea Seed Quality from Different Systems of Increasing Soil Fertility. *Electron. J. Crop Prod.* **2010,** *3* (1), 103–119.

Morales-Ramos, L. H. In: Formulacion de Bioinsecticidas Avances recientes en la Biotecnlogía en *Bacillus thuringiensis*; Galán-Wong, L. J., Rodriguez-Padilla, C., Luna-Olvera, H. A., Eds.; Universidad Autónoma de Nuevo Leon, 1996; pp 157–177.

Moscardi, F. Assessment of the Application of Baculoviruses for Control of Lepidoptera. *Ann. Rev. Entomol.* **1999,** *44*, 257–289.

Nahas, E. Factors Determining Rock Phosphate Solubilization by Microorganism Isolated from Soil. *World J. Microb. Biotechnol.* **1996,** *12*, 18–23.

Nawrocka, J.; Maolepsza, U. Diversity in Plant Systemic Resistance Induced by Trichoderma. *Biol. Control.* **2013,** *67*, 149–156.

Ngoc, S. T. T.; Ngoc, D. C.; Giang, T. T. Effect of Bradyrhizobia and Phosphate Solubilizing Bacteria Application on Soybean in Rotational System in the Mekong Delta. *J. Omonrice.* **2006,** *14*, 48–57.

Nihorimbere, V.; Ongena, M.; Smargiassi, M.; Thonart, P. Beneficial Effect of the Rhizosphere Microbial Community for Plant Growth and Health. *Biotechnol. Agron. Soc. Environ.* **2011,** *15* (2), 327–337.

Okon, Y. Response of Cereal and Forage Grasses to Inoculation with N_2-Fixing Bacteria. In: *Advances in the Nitrogen Fixation Research*; Veeger C., Newton W. E., Ed.; Nijoff/Junk: The Hague, 1984; pp 30–39.

Oldroyd, G. E. D.; Dixon, R. Biotechnological Solutions to the Nitrogen Problem. *Curr. Opin. Biotechnol.* **2014,** *26*, 19–24.

Pal, K.; Gardener, B. M. Biological Control of Plant Pathogens. *Plant Health Instruct.* **2006.** doi:10.1094/PHI-A-2006-1117-02. APSnet: 1–25.

Park, J. H.; Choi, G. J.; Jang, S. K.; Lim, K. H.; Kim, T. H.; Cho, Y. K.; Kim, J. C. Antifungal Activity against Plant Pathogenic Fungi of Chaetoviridins Isolated from *Chaetomium globosum*. *FEMS Microbiol. Lett.* **2005,** *252*, 309–313.

Paulitz, T. C.; Belanger, R. R. Biological Control in Greenhouse System. *Ann. Rev. Phytopathol.* **2001,** *39*, 103–133.

Phonkerd, N.; Kanokmedhakul, S.; Kanokmedhakul, K.; Soytong, S.; Prabpai, S.; Kongsearee, P. Bis-Spiro-Azaphilones and Azaphilones from the Fungi *Chaetomium cochliodes* VTh01 and *C. cochliodes* CTh05. *Tetrahedron* **2008,** *64*, 9636–9645.

Ponmurugan, P.; Gopi, G. Distribution Pattern and Screening of Phosphate Solubilizing Bacteria Isolated from Different Food and Forage Crops. *J. Agron.* **2006,** *5* (4), 600–604.

Posada, F. *Syspastospora parasitica*, a Mycoparasite of the Fungus *Beauveria bassiana* Attacking the Colorado Potato Beetle *Leptinotarsa decemlineata*: A Tritrophic Association. *J. Insect Sci.* **2004,** *4*, 24.

Pradhan, N.; Sukla, L. B. Solubilization of Inorganic Phosphates by Fungi Isolated from Agriculture Soil. *Afr. J. Biotechnol.* **2005,** *5*, 850–854.

Punja, Z. K.; Utkhede, R. S. Biological Control of Fungal Diseases on Vegetable Crops with Fungi and Yeasts. In: *Fungal Biotechnology in Agricultural, Food, and Environmental Applications*; Arora, D. K., Ed.; CRC Press: New York Basel, 2004; pp 157–171.

Qin, H.; Brookes, P. C.; Xu, J. *Cucurbita* spp. and *Cucumis sativus* Enhance the Dissipation of Polychlorinated Biphenyl Congeners by Stimulating Soil Microbial Community Development. *Environ. Pollut.* **2014,** *184*, 306–312.

Raaijmakers, J. M.; Lugtenberg, B. J. J. Perspectives for Rhizosphere Research. In: *Molecular Microbial Ecology of the Rhizosphere*; de Bruijn, F. J., Ed.; Wiley Blackwell: Hoboken, NJ, 2013; vol 2, pp 1227–1232.

Raaijmakers, J. M.; Paulitz, T. C.; Steinberg, C.; Alabouvette, C.; Moenne-Loccoz, Y. The Rhizosphere: a Playground and Battlefield for Soilborne Pathogens and Beneficial Microorganisms. *Plant Soil* **2009,** *321*, 341–361.

Raina, S.; Chamola, B. P.; Mukerji, K. G. Evolution of Mycorrhiza. In: *Mycorrhizal Biology* Mukerji, K. G., Singh, J., Chamola, B. P., Eds. Kluwer Academic/Plenum Plublishers: New York, 2000; pp 1–25.

Raja, P. Status of Endomycorrhizal (AMF) Biofertilizer in the Global Market. In: *Handbook of Microbial Biofertilizers*; Rai, M. K., Ed.; Food Products Press, Binghamton, NY, 2006; pp 395–416.

Ramos, H. J. O.; Yates, M. G.; Pedrosa, F. O.; Souza, E. M. Strategies for Bacterial Tagging and Gene Expression in Plant–Host Colonization Studies. *Soil Biol. Biochem.* **2011,** *43*, 1626–1638.

Reddy, G. V. P.; Zhao, Z.; Humber, R. A. Laboratory and Field Efficacy of Entomopathogenic Fungi for the Management of the Sweet Potato Weevil, *Cylas formicarius* (Coleoptera: Brentidae). *J. Invert. Pathol.* **2014,** *122*, 10–15.

Rinaldi, A. C.; Comandini, O.; Kuyper, T. W. Ectomycorrhizal Fungal Diversity: Separating the Wheat from the Chaff. *Fungal Div.* **2008,** *33*, 1–45.

Rodríguez, H.; Fraga, R. Phosphate Solubilising Bacteria and their Role in Plant Growth Promotion. *Biotechnol. Adv.* **1999,** *17*, 319–339.

Sagoe, C.; Ando, T.; Kouno, K.; Nagaoka, T. Relative Importance of Protons and Solution Calcium Concentration in Phosphate Rock Dissolution by Organic Acids. *Soil Sci. Plant Nutr.* **1998,** *44*, 617–625.

Samuels, G. J. Trichoderma: Systematics, the Sexual State, and Ecology. *Phytopathology* **2006,** *96*, 195–206.

Savich, M. H.; Olson, G. S.; Clark, E. W.; Doane, W. M.; Doane, S. W. US20087425595, 2008.

Schnepf, E. N.; Crickmore, J.; Van Rie, J.; Lereclus, D.; Baum, J.; Feitelson, J.; Zeigler, D. R.; Dean, D. H. *Bacillus thuringiensis* and its Pesticidal Crystal Proteins. *Microbiol. Mol. Biol. Rev.* **1998,** *62* (3), 775–806.

Schoebitz, M.; Mengual, C.; Roldan, A. Combined Effects of Clay Immobilized *Azospirillum brasilense* and *Pantoea dispersa* and Organic Olive Residue on Plant Performance and Soil Properties in the Revegetation of a Semiarid Area. *Sci. Total Environ.* **2014,** *466*, 67–73.

Schroth, M. N.; Hancock, J. G. Selected Topics in Biological Control. *Ann. Rev. Microbiol.* **1981,** *35*, 453–476.

Schwartz, M. W.; Hoeksema, J. D.; Gehring, C. A.; Johnson, N. C.; Klironomos, J. N.; Abbott, L. K.; Pringle, A. The Promise and the Potential Consequences of the Global Transport of Mycorrhizal Inoculum. *Ecol. Lett.* **2006,** *9*, 501–515.

Serelis, J.; Papaparaskevas, J.; Stathi, A.; Sawides, A. L.; Karagouni, A. D.; Tsakris, A. Granulomatous Infection of the Hand and Wrist Due to *Azospirillum* spp. *Diagn. Microbiol. Infect. Dis.* **2013,** *76*, 513–515.

Shanmugam, P. M.; Veeraputhran, R. Effect of Organic Manure, Biofertilizers, Inorganic Nitrogen and Zinc on Growth and Yield of Rabi Rice. *Madras Agric. J.* **2000,** *2*, 87–90.

Sharma, K. Bionatural Mangement of Pests in Organic Farming. *Agrobios Newsl.* **2004,** *2*, 296–325.

Shen, Z.; Zhong, S.; Wang, Y.; Wang, B.; Mei, X.; Li, R. Induced Soil Microbial Suppression of Banana Fusarium Wilt Disease Using Compost and Biofertilizers to Improve Yield and Quality. *Eur. J. Soil Biol.* **2013,** *57*, 1–8.

Shishkoff, N.; McGrath, M. T. AQ10 Biofungicide Combined with Chemical Fungicides or AddQ Spray Adjuvant for Control of Cucurbit Powdery Mildew in Detached Leaf Culture. *Plant Dis.* **2002,** *86*, 915–918.

Shternshis, M.; Tomilova, O.; Shpatova, T.; Soytong, K. Evaluation of Ketomium Mycofungicide on Siberian Isolates of Phytopathogenic Fungi. *J. Agric. Technol.* **2005,** *1*, 247–253.

Siddiqui, Z. A.; Pichtel, J. Mycorrhixae: An Overview. In: *Mycorrhizae: Sustainable Agriculture and Forestry*; Siddiqui, Z. A., Akhtar, M. S., Futai, K., Eds.; Springer, Netherlands, Germany, 2008; pp 1–36.

Simmons, C. W.; Claypool, J. T.; Marshall, M. N.; Jabusch, L. K.; Reddy, A. P.; Simmons, B. A.; Singer, S. W.; Stapleton, J. J.; Vander Gheynst, J. S. Characterization of Bacterial Communities in Solarized Soil Amended with Lignocellulosic Organic Matter. *Appl. Soil Ecol.* **2014,** *73*, 97–104.

Singh, R.; Soni, S. K.; Patel, R. P.; Kalra, A. Technology for Improving Essential Oil Yield of *Ocimum basilicum* L. (Sweet Basil) by Application of Bioinoculant Colonized Seeds under Organic Field Conditions. *Ind. Crops Prod.* **2013b,** *45*, 335–342.

Singh, Y. V.; Singh, K. K.; Sharma, S. K. Influence of Crop Nutrition on Grain Yield, Seed Quality and Water Productivity under Two Rice Cultivation Systems. *Rice Sci.* **2013a,** *20*, 129–138.

Smith, S. E.; Zhu, Y. G. Application of Arbuscular Mycorrhizal Fungi: Potentials and Challenges. In: *Bio-Exploitation of Filamentous Fungi*; Stephen, B. P., Hyde, K. D., Eds.; *Fungal Div. Res. Ser.* **2001,** *6*, 291–308.

Son, J. S.; Sumayo, M.; Hwang, Y. J.; Kim, B. S.; Ghim, S. Y. Screening of Plant Growth-Promoting Rhizobacteria as Elicitor of Systemic Resistance against Gray Leaf Spot Disease in Pepper. *Appl. Soil Ecol.* **2014,** *73*, 1–8.

Soytong, K.; Quimio, T. H. A Taxonomic Study on the Philippines Species of *Chaetomium*. *Philipp. Agric.* **1989,** *72*, 59–72.

Soytong, K.; Kanokmadhakul, S.; Kukongviriyapa, V.; Isobe, M. Application of *Chaetomium* Species (Ketomium®) as a New Broad Spectrum Biological Fungicide for Plant Disease Control: A Review Article. *Fungal Div.* **2001,** *7*, 1–15.

Subbarao, N. S. Phosphate Solubilizing Microorganism. In: *Biofertilizer in Agriculture and Forestry*. Regional Biofert. Dev. Centre: Hissar, India, 1988; pp. 133–142.

Sunantapongsuk, V.; Nakapraves, P.; Piriyaprin, S.; Manoch, L. Protease Production and Phosphate Solubilization from Potential Biological Control Agants *Trichoderma viride* and *Azomonas agilis* from Vetiver Rhizosphere. In: International Workshop on Sustained

Managament of Soil–Rhizosphere System for Efficient Crop Production and Fertilizer Use, Land Development Department: Bangkok, Thailand, 2006; pp 1–4.

Sundara, B.; Natarajan, V.; Hari, K. Influence of Phosphorus Solubilizing Bacteria on the Changes in Soil Available Phosphorus and Sugarcane Yields. *Field Crops Res.* **2002,** *77*, 43–49.

Szekeres, A.; Kredics, L.; Antal, Z.; Kevei, F.; Manczinger, L. Isolation and Characterization of Protease Overproducing Mutants of *Trichoderma harzianum*. *Microbiol. Lett.* **2004,** *233*, 215–222.

Tao, G.; Tian, S.; Cai, M.; Xie, G. Phosphate Solubilizing and Mineralizing Abilities of Bacteria Isolated from Soils. *Pedosphere* **2008,** *18*, 515–523.

Taylor, M. S.; Stahl-Timmins, W.; Redshaw, C. H.; Osborne, N. J. Toxic Alkaloids in *Lyngbya majuscula* and Related Tropical Marine Cyanobacteria. *Harmf. Algae* **2014,** *31*, 1–8.

Tejesvi, M. V.; Kini, K. R.; Prakash, H. S.; Subbiah, V.; Shetty, H. S. Genetic Diversity and Antifungal Activity of Species of Pestalotiopsis Isolated as Endophytes from Medicinal Plants. *Fungal Div.* **2007,** *24*, 37–54.

Thormann, M. N.; Rice, A. V. Fungal from Peatlands. *Fungal Div.* **2007,** *24*, 241–299.

Tiwari, P.; Adholeya, A.; Prakash, A. Commercialization of Arbuscular Mycorrhizal Biofertilizers. In: *Fungal Biotechnology in Agricultural, Food, and Environmental Applications*; Arora, D. K., Ed.; CRC Press: New York Basel, 2004; pp 195–203.

Tomilova, O. G.; Shternshis, M. V. The Effect of a Preparation from *Chaetomium* Fungi on the Growth of Phytopathogenic Fungi. *Appl. Biochem. Microbiol.* **2006,** *42*, 76–80.

Valaitis, A. P.; Jenkins, J. L.; Lee, M. K.; Dean, D. H.; Garner, K. J. Isolation and Partial Characterization of Gypsy Moth BTR-270 an Anionic Brush Border Membrane Glycoconjugate that Binds *Bacillus thuringiensis Cry*1A Toxins with High Affinity. *Arch. Ins. Biochem. Physiol.* **2001,** *46*, 186–200.

Vandamme, P.; Goris, J.; Chen, W. M.; de Vos, P.; Willems, A. *Burkholderia tuberum* sp. nov. and *Burkholderia phymatum* sp. nov., Nodulate the Roots of Tropical Legumes. *Syst. Appl. Microb.* **2002,** *25*, 507–512.

Vazquez, P.; Holguin, G.; Puente, M.; Cortes, A. E.; Bashan, Y. Phosphate Solubilizing Microorganisms Associated with the Rhizosphere of Mangroves in a Semi Arid Coastal Lagoon. *Biol. Fertil. Soils* **2000,** *30*, 460–468.

Vinale, F.; Marra, R.; Scala, F.; Ghisalbert, E. L.; Lorito, M.; Sivasithamparam, K. Major Secondary Metabolotes Produced by Two Commercial *Trichoderma* Strains Active against Different Phytopathogens. *Lett. Appl. Microbiol.* **2006,** *43*, 143–148.

Vinale, F.; Sivasithamparam, K.; Ghisalberti, E. L.; Marra, R.; Woo, S. L.; Lorito, M. Trichoderma Plant Pathogen Interactions. *Soil Biol. Biochem.* **2008,** *40*, 1–10.

Viterbo, A.; Inbar, J.; Hadar, Y.; Chet, I. Plant Disease Biocontrol and Induced Resistance via Fungal Mycoparasites. In: *Environmental and Microbial Relationships. The Mycota IV*, 2nd ed.; Kubicek, C. P., Druzhinina, I. S.; Springer-Verlag: Berlin-Heidelberg, 2007; pp 127–146.

Vos, C.; Schouteden, N.; van Tuinen, D.; Chatagnier, O.; Elsen, A.; De Waele, D. Mycorrhiza-Induced Resistance against the Root-Knot Nematode *Meloidogyne incognita* Involves Priming of Defense Gene Responses in Tomato. *Soil Biol. Biochem.* **2013,** *60*, 45–54.

Wakelin, S. A.; Werren, P. R.; Ryder, H. M. Phosphate Solubilization by *Penicillium* spp. Closely Associated with Wheat Root. *Biol. Fert. Soils* **2004,** *40*, 36–43.

Wani, S. P. Inoculation with Associative Nitrogen-Fixing Bacteria: Role in Cereal Grain Production Improvement. *Indian J. Microbiol.* **1990,** *30*, 363–393.

Wani, S. P.; Lee, K. K. Biofertilizers for Sustaining Cereal Crops Production. In: *Biotechnology of Biofertilizers*; Kannaiyan, S., Ed.; Narosa Publishing House, 2002; pp 50–64.

Weller, D. M. Biological Control of Soil Borne Plant Pathogens in the Rhizosphere with Bacteria. *Ann. Rev. Phytopathol.* **1988,** *26*, 379–407.

Whipps, J. M.; Sreenivasaprasad, S.; Muthumeenakshi, S.; Rogers, C. W.; Challen, M. P. Use of *Coniothyrium minitans* as a Biocontrol Agent and Some Molecular Aspects of Sclerotial Mycoparasitism. *Eur. J. Plant Pathol.* **2008,** *121*, 323–330.

Whitelaw, M. A. Growth Promotion of Plants Inoculated with Phosphate Solubilizing Fungi. *Adv. Agron.* **2000,** *69*, 99–151.

Whitelaw, M. A.; Hardena, T. J.; Helyar, K. R. Phosphate Solubilisation in Solution Culture by the Soil Fungus *Penicillium radicum*. *Soil Biol. Biochem.* **1999,** *3*, 655–665.

Woo, S. L.; Scala, F.; Ruocco, M.; Lorito, M. The Molecular Biology of the Interactions between *Trichoderma* spp., Phytopathogenic Fungi, and Plants. *Phytopathology* **2006,** *96*, 181–185.

Woo, L. S.; Lorito, M. Exploiting the Interactions between Fungal Antagonists, Pathogens and the Plant for Biocontrol. In: *Novel Biotechnologies for Biocontrol Agent Enhancement and Management*; Vurro, M. and Gressel, J., Eds., Springer: Germany, 2007; pp 107–130.

Wraight, S. P.; Jackson, M. A.; de Kock, S. L. Production, Stabilization and Formulation of Fungi Biocontrol Agents. In: *Fungi as Biocontrol Agents Progress, problem and Potential*;

Wu, S. C.; Cao, Z. H.; Li, Z. G.; Cheung, C.; Wong, M. H. Effects of Biofertilizer Containing N-fixer, P and K Solubilizers and AM Fungi on Maize Growth: A Greenhouse Trial. *Geoderma* **2005,** *125*, 155–166.

Yeasmin, T.; Zaman, P.; Rahman, A.; Absar, N.; Khanum, N. S. Arbuscular Mycorrhizal Fungus Inoculum Production in Rice Plants. *Afr. J. Agricul. Res.* **2007,** *2*, 463–467.

Yedidia, I.; Benhamou, N.; Chet, I. Induction of Defense Responses in Cucumber Plants (*Cucumis sativus* L.) by the Biocontrol Agent *Trichoderma harzianum*. *Appl. Environ. Microbiol.* **1999,** *65*, 1061–1070.

Yousefzadeh, S.; Modarres-Sanavy, S. A. M.; Sefidkon, F.; Asgarzadeh, A.; Ghalavand, A.; Sadat-Asilan, K. Effects of Azocompost and Urea on the Herbage Yield and Contents and Compositions of Essential Oils from Two Genotypes of Dragonhead (*Dracocephalum moldavica* L.) in Two Regions of Iran. *Food Chem.* **2013,** *138*, 1407–1413.

Zeilinger, S.; Omann, M. Trichoderma Biocontrol: Signal Tranduction Pathways Involved in Host Sensing and Mycoparasitism. *Gene Regul. Systems Biol.* **2007,** *1*, 227–234.

Zhang, G. Y.; Sun, X. L.; Zhang, Z. X.; Zhang, Z. F.; Wan, F. F. Production and Effectiveness of the New Formulation of Helicoverpa Virus Pesticide-Emulsifiable Suspension. *Virol. Sin.* **1995,** *10*, 242–247.

Zhang, H. Y.; Yang, Q. Expressed Sequence Tags-Based Identification of Genes in the Biocontrol Agent *Chaetomium cupreum*. *Appl. Microb. Biotechnol.* **2007,** *74*, 650–658.

Zhu, G. S.; Yu, Z. N.; Gui, Y.; Liu, Z. Y. A Novel Technique for Isolating Orchid Mycorrhizal Fungi. *Fungal Div.* **2008,** *33*, 123–137.

Zolla, G.; Bakker, M. G.; Badri, D. V.; Chaparro, J. M.; Sheflin, A. M.; Manter, D. K.; Vivanco, J. Understanding Root–Microbiome Interactions. In: de Bruijn, F. J., Ed.; *Mol. Microb. Ecol. Rhizosphere* **2013,** *2*, 745–754.

PART V
Oxidative Stress

CHAPTER 12

OXIDATIVE STRESS IN PLANTS: OVERVIEW ON REACTIVE OXYGEN SPECIES FORMATION IN CHLOROPLASTS, CONSEQUENCES, AND EXPERIMENTAL APPROACHES UNDER ABIOTIC STRESSES

DEEPAK KUMAR YADAV[1*], ANKUSH PRASAD[1,2], and ABHISHEK MANI TRIPATHI[3]

[1]*Department of Biophysics, Centre of the Region Haná for Biotechnological and Agricultural Research, Faculty of Science, Palacký University, Olomouc, Czech Republic (#Former address)*

[2]*Biomedical Engineering Research Center, Tohoku Institute of Technology, Sendai, Japan*

[3]*Global Change Research Institute, Academy of Sciences of the Czech Republic, Brno, Czech Republic*

[*]*Corresponding author. E-mail: deep_mns@yahoo.co.in*

CONTENTS

ABSTRACT

Plants are exposed to different types and variable degree of oxidative stress during their growth and development. Reactive oxygen species (ROS) are known to be formed during the abiotic stress conditions which is one of the inevitable processes. Effect of ROS formation in plants are known predominately in the context of the damage to the organic biomolecules such as lipids, proteins and nucleic acids. However, ROS can also regulates the biological processes that are involved in either acclimation or disruption under abiotic stresses. In plants, ROS are produced in several cellular compartments (mitochondria, chloroplast, etc.). Formation of ROS in chloroplasts is mainly associated with electron transport chain and/or by photosensitization of chlorophyll molecules. Detection of ROS in chloroplasts had always been a challenge since limitations exists with short half-life time, lower concentration, and unspecific detection probes for ROS. The current chapter is aimed to provide a summarized information on the different abiotic stresses (high light, temperature fluctuations, UV irradiation, drought etc.) and associated mechanisms in the formation of ROS in chloroplasts.

12.1 OXIDATIVE STRESS IN PLANT

Oxidative stress is a consequence reflecting an imbalance between the production of reactive oxygen species (ROS) and detoxification of either ROS or its reactive intermediates (Halliwell and Gutteridge, 2007). The imbalance in the redox state has been known to be associated with damage of biological components such as lipids, proteins, and nucleic acids which in turn leads to the disturbance in the physiological state of the living system. Biotic and abiotic stress factors are known to be involved in generation of oxidative stress in plants. The biotic stress factors include infection by viruses, bacteria, and fungi which are responsible for several plant diseases and are associated with boundless loss of agricultural crop over the past decades (Atkinson and Urwin, 2012; Rejeb et al., 2014). The abiotic stress factors that includes physical and chemical stress factors involves UV-irradiation, high light conditions, drought and temperature fluctuations, etc. (Aroca et al., 2012; Atkinson and Urwin, 2012; Lichtenthaler et al., 1983; Rozema et al., 1997; Wang et al., 2009). During the last few decades, the rise in the temperature around the world has become a global concern. This chapter highlights and evaluates the effects of various abiotic stress factors and its involvement in oxidative stress in plants (Figure 12.1).

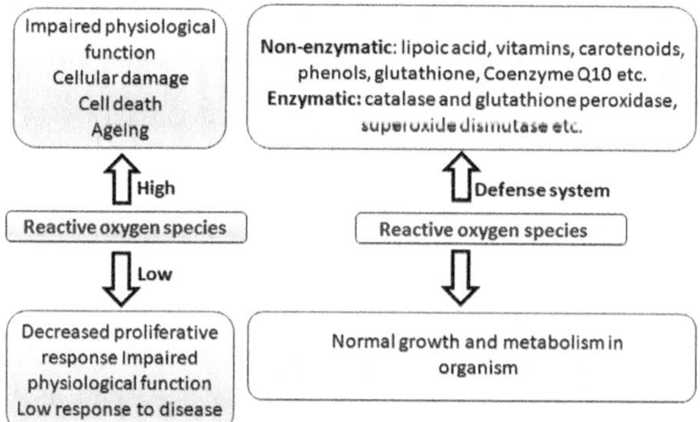

FIGURE 12.1 *The schematic representation showing the consequence resulting from ROS production in cells.* Conditions during homeostasis where balance between antioxidant and ROS production is maintained, the normal growth and metabolism is achieved. On the contrary, the cell function is known to be impaired during higher or lower production of ROS.

12.1.1 CONSEQUENCE OF OXIDATIVE STRESS

The ROS generation is one of the unavoidable conditions in plants. The level of ROS is enhanced during the exposure of plants to oxidative stress conditions. Oxidative stress damages the organic molecules, especially the lipids, proteins, and to some extent the nucleic acid as well. Damage to lipids can cause biological problem in cells, such as destruction of membranes composed of lipids which regulate the fluidity and permeability (Sharma et al., 2012; Stark, 2005). Proteins damage may inhibit the biological processes like protein–protein interaction, protein–nucleic acid interaction which are involved in many regulatory metabolic pathway and diseases (Berlett and Stadtman, 1997; Cabiscol et al., 2000; Cooke et al., 2003; Uttara et al., 2009). Damage of nucleic acid is one of the major targets of UV radiation and specially, UVB and UVC are known to affect DNA by causing lethal effects such as formation of cyclobutyl pyrimidines dimers, inter/intra cross links, 8-oxo-guanines, and double strand breaks (Cooke et al., 2003; Galloway et al., 1994). The inhibition or damage of these biological process is due to formation of highly reactive free radicals in plants under abiotic stresses conditions (Das and Roychoudhury, 2014, Gill and Tuteja, 2010; Sharma et al., 2012).

Molecular mechanism of the excessive formation of ROS in plant cells have been studied extensively in the last decades. Beside plant system, production of ROS have also been proposed in other biological studies such as pathologic conditions (human health and diseases) (Alfadda and Sallam,

2012; Castro and Freeman, 2001), ischemia–reperfusion injury (Braunersreuther and Jaquet, 2012; Verma et al., 2002), cancerous cell cytotoxicity (Liou and Storz, 2010; Schumacker, 2006; Waris and Ahsan, 2006), intrinsic cell death and homeostasis (Circu and Aw, 2010; Ray et al., 2012; Wu and Bratton, 2013), and cell signaling pathways (Apel and Hirt, 2004; Thannickal and Fanburg, 2000, Tripathy and Oelmüller, 2012). Thus, the study of ROS formation and elimination in biological system is required to determine precise level and localization of ROS in living cells. Several research findings suggest that the combination of different methods can be used for precise and appropriate ROS detection, as well as for the determination of their intracellular localization in cells under oxidative stress condition. A general overview for formation of ROS by different sources and consequently biomolecules damage in plants has been presented in Figure 12.2.

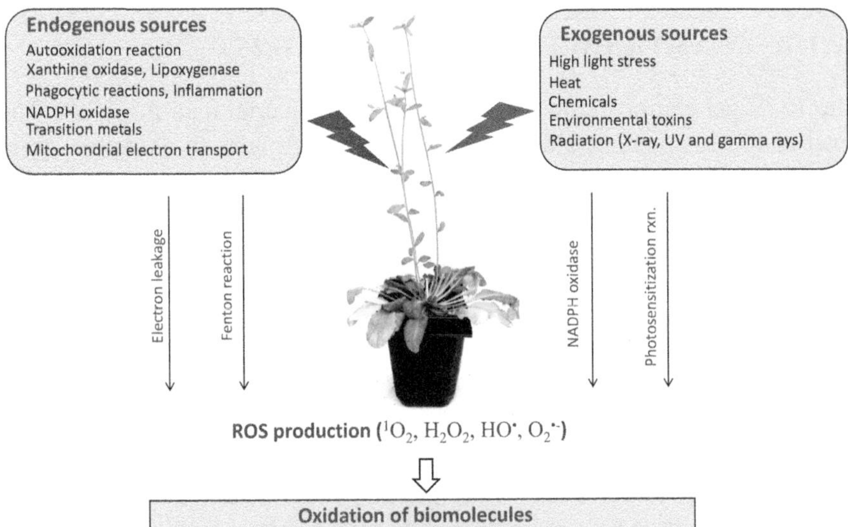

FIGURE 12.2 Schematic representation of formation of ROS by exogenous and endogenous sources and consequently oxidation of biomolecules in plants.

12.1.2 REACTIVE OXYGEN SPECIES IN PLANTS

Formation and elimination of ROS are crucial processes in plant system, as optimal level of ROS is required for diverse cellular responses in normal cells. ROS predominantly regulates the redox reaction or modification of critical biomolecules such as amino acids of regulatory proteins (Apel and Hirt, 2004; Breusengem and Dat, 2006). Formation of ROS in plant cell also regulates the signaling cascades that results into functional changes and normal cellular homeostasis (Foyer and Noctor, 2013; Foyer et al., 2012; Mittler et al., 2004; Suzuki et al., 2012). However, the increase in formation of ROS in different organelles of plants cell disturbs normal cellular homeostasis and leads to the damage of biological molecules (protein, lipids, and nucleic acids) which may result in functional impairment and in extremes conditions cause cell death (Breusengem and Dat, 2006; Petrov et al., 2015).

These ROS can be formed in different cell organelles such as chloroplasts, mitochondria, peroxisomes, endoplasmic reticulum, and plasma membranes (Table 12.1). Plant cell possess two organelles, chloroplast, and mitochondria for bioenergy process, that is, electron transport chain in photosynthesis and respiration (Møller, 2001; Renger and Holzwarth, 2005; Taiz and Zeiger, 2006). Chloroplast is the site for the conversion of light energy into chemical energy by process called photosynthesis. Photosynthesis is accomplished by the light dependent (absorption of light, electron-transport

TABLE 12.1 Formation of ROS in Different Organelles in Plant Cell.

Organelles	ROS	Comments
Chloroplast	$^1O_2$, H_2O_2, $HO^•$, $O_2^{•-}$	Occurs in PSI and PSII during electron transport chain and by chlorophyll pigments (Asada, 2006; Mattila et al., 2015; Pospíšil, 2012; Sharma et al., 2012)
Mitochondria	H_2O_2, $HO^•$, $O_2^{•-}$	Formed by complex I, II, and III during electron transport chain (Das and Roychoudhury 2014; Møller, 2001; Murphy, 2009a; Noctor et al., 2007; Turrens, 2003)
Microbodies	H_2O_2, $O_2^{•-}$	Microbodies organelles like peroxisomes, glyoxisome, etc.; ROS formed by xanthine oxidase and cytochrome *b* (Dat et al., 2000; Foyer and Noctor, 2003; Karuppanapandian et al., 2011; Ślesak et al., 2007)
Endoplasmic reticulum	H_2O_2 and $O_2^{•-}$	During redox cycling of certain quinones; by NAD(P)H oxidase and Cyt P_{450} (Bartosz, 1997; Das and Roychoudhury 2014; Mittler, 2002; Ślesak et al., 2007)
Plasma membranes	H_2O_2 and $O_2^{•-}$	Electron transporting oxidoreductase, NADPH oxidase (Apel and Hirt, 2004; Heyno et al., 2011; Karuppanapandian et al., 2011; Sharma et al., 2012)

chain) and light independent (assimilation of carbon) processes (Taiz and Zeiger, 2006). Thus, light-dependent pathway is one of the main sources for the formation of ROS in plants. The malfunctioning of these processes under environmental stress conditions enhances the ROS formation in chloroplast (Asada, 2006; Pospíšil, 2009; Vass, 2012) that leads to the damage of photosystem repair mechanism (Murata et al., 2007, 2012) and consequently could reduce the photosynthetic efficiency and the yield of plant biomass or bioenergy.

ROS is the form of molecular oxygen and is formed either by the energy transfer or the electron transfer. ROS formed by the energy transfer is also called Type II reaction (Halliwell and Gutteridge, 2007). Usually, ROS consists of one or two oxygen atoms and have either one or two unpaired or in some cases, no unpaired electron. Based on the electronic configuration, it can be divided in two groups, namely radical ROS [superoxide anion radical ($O_2^{\bullet -}$), hydroxyl radical ($HO^{\bullet}$)] and non-radical ROS [singlet oxygen ($^1O_2$) and hydrogen peroxide (H_2O_2)]. Since molecular oxygen is a non-reactive molecule, to convert it into ROS either excess energy or electron is required. ROS are highly reactive in nature and half-life time ranges from nanoseconds to minutes (Halliwell and Gutteridge, 2007).

12.1.2.1 SINGLET OXYGEN

Singlet oxygen is an energetically excited form of molecular oxygen with half-life in microsecond (Halliwell and Gutteridge, 2007). It is highly reactive and is formed by energy transfer from a triplet excited photosensitizers to molecular oxygen. In plants, chlorophyll pigments act as a photosensitizer which absorbs the light and results into excited triplet chlorophyll molecules. The excited triplet chlorophylls transfer excess energy to molecular oxygen by triplet–singlet state energy transfer (Mehrdad et al., 2002; Pospíšil, 2012). Two forms of $^1O_2$ can be formed based on the energy level either high-energy excited state or low-energy excited state. High-energy state $^1O_2$ is unstable and converts into low energy state by dissipating energy (Halliwell and Gutteridge, 2007; Klan and Wirz, 2009). Singlet oxygen emits energy around 1270 nm (Macpherson et al., 1993; Telfer, 2014; Tomo et al., 2012), whereas two $^1O_2$ molecules interact to form dimol and emit energy at 634 and 703–708 nm (Cifra and Pospíšil, 2014; Devaraj and Inaba, 1997; Lengfelder et al., 1983).

12.1.2.2 SUPEROXIDE ANION RADICALS, HYDROGEN PEROXIDE, AND HYDROXYL RADICAL

Superoxide anion radicals, H_2O_2, and $HO^•$ are reduced forms of molecular oxygen with half-life in microseconds, minutes, and nanoseconds, respectively (Halliwell and Gutteridge, 2007). These are also reactive molecules and formed by transfer of electron from donor molecules to molecular oxygen. In plants, usually $O_2^{•-}$ is formed by one electron reduction of molecular oxygen by the molecules having lower reduction potential than molecular oxygen (Asada, 2006; Pospíšil, 2012). Hydrogen peroxide is formed by either one electron reduction of $O_2^{•-}$ or two electrons reduction of molecular oxygen (Halliwell and Gutteridge, 2007). Similarly, $HO^•$ is also formed by the one electron reduction of H_2O_2 in presence of metal ion known as Fenton reaction (Gutteridge, 1984; Halliwell and Gutteridge, 2007; Kehrer, 2000). Despite above reaction mechanisms, it has been also reported that these ROS can also be formed by the oxidation of H_2O in certain conditions (Pospíšil, 2012). Detailed mechanisms for the formation of ROS in chloroplasts have been discussed further in Section 12.1.4.

12.1.3 ABIOTIC STRESSES

Plants are subjected to several unfavourable conditions and climatic changes throughout the year. Extreme temperatures (high and low temperatures), pollution (toxicity), herbivory, infections, drought, nutrient deficiency, light, etc. are among few listed abiotic stress generators. Plant bears the ability to respond to these fluctuations and prevents itself by mode of different acclimation and adaptation thereby maintaining its growth and development. Abiotic environmental factors (light, temperature, water, salinity, air, etc.) are the parameters and assets that determine the plant growth. Abiotic stress such as high light and UV-irradiation in photosynthetic process are among the widely studied stresses.

12.1.3.1 HIGH LIGHT

Sunlight is the dominating energy source for the plant's life on earth, whereas photosynthetically active radiation (wavelength 400–700 nm) is one of the main factors which helps to the plant growth, development, and energy production by photosynthesis. All these processes directly or indirectly depend on the light intensity, which changes along the elevation, latitude, leaf arrangement, and light quantity received by plants in a particular region and is also affected by the length of the day. Plants have different preferences for light strength based on the requirements, such as in some plants, flowering depends on length of daylight (Johansson and Staiger, 2015; Sawa and Kay, 2011) and some are not affected. Similarly, some plants like corn, cucurbits, and legumes need more intense light, whereas others like asparagus, carrot, lettuce spinach, etc. preferably grow under low intense light.

Excess of light exposure or high light stress induces several responses such as light induce adaptation in the photosynthetic apparatus, changes in the ultra-structure of the chloroplast, limit the photosynthetic activity (Lichtenthaler and Babani, 2004), as well as the ability of light to regulate plant growth independently (photomorphogenesis) (Nemhauser and Chory, 2002). In addition, the de-epoxidation to zeaxanthin from violaxanthin, photoinhibition of the photosynthetic pigment–protein complexes, and increase in heat emission are also involved in response to high light stress (Lichtenthaler and Burkart, 1999). ROS generation has been widely studied in photosynthetic plants *in vivo* and *ex-vivo*, leaves, and to the microscopic level including chloroplast, thylakoids membranes, and photosystems. ROS have been measured in plants under high light stress and have been described in details in Section 12.1.4.1.

12.1.3.2 TEMPERATURE

Temperature regulates the growth and productivity of plants, depending on whether it is a tropical or temperate region plants experiencing temperature variation. Plant function depends on varied (narrow to extreme) range of temperatures. Plant can survive on a temperature range between 0°C and 50°C and its growth, maximum yields, depends on optimal day and night temperature range which varies among plants species. A favourable soil temperature affects the seed germination (Finch-Savage and Phelps, 1993), root development (Kaspar and Bland, 1992), water and nutrient absorption by roots (Bassirirad, 2000; Lv et al., 2012), bacterial growth and development (Demoling et al., 2007; Pietikäinen et al., 2005), and biological matter decay (Vanhala et al., 2008). Due to increase of temperature or heat stress condition, the rate of photosynthesis and respiration along with overall enzymatic activity are also affected. Heat stress can inhibit plant enzymes (Chaitanya et al., 2002) and other proteins (Kotak et al., 2007) by denaturation and burdens huge quantity of water loss (due to water transpiration and evaporation), reduce the plants yield (Wahid et al., 2007), fades the color of flowers and also shortens the life span of the plants (Åkerfelt et al., 2010). Heat stress for the plants and soils are the significant problems across the world including developing and developed countries. Damage or inhibition of above mentioned biochemical processes in plants are linked to the excessive formation of ROS under heat stress.

12.1.3.3 UV IRRADIATION

With the increasing penetration of UV components into the atmosphere, the effects of UV radiation on plants have been widely studied. UV radiation contributes to only about 8–9% of total solar radiation (Carbonell-Bejerano et al., 2014), however, can induce varied damaging effects on the plants specifically targeting nucleic acid causing mutation. The long-wavelength UVA (320–400 nm) with photosynthetically active radiation plays an essential role in plant growth and sensitivity (Krizek, 2004). The next UV radiation range is biologically effective UVB (280–320 nm) induces ameliorating damage in plant cells (Krizek, 2004). Ballaré et al. (1996) reported UVB stress in plants, affects the physiological and morphological development, for example, inhibits the internodes growth and alters the leaves structure. The third range being UVC (100–280 nm) is biologically very active and has more energy than other types of UV radiations.

Since proteins bear aromatic amino acid residues (phenylalanine, tryptophan, and tyrosine) and due to strong absorption of these amino acids residues at about 280 nm, proteins could be the target for UV associated damage. In addition, direct excitation of tryptophan has been reported under the effect of small dose of UV-radiation (Kehoe et al., 2008). The loss of photosynthetic activity under UV stress is associated with damage of proteins and photosynthetic pigments. The UV radiation not only have severe deleterious effects on nucleic acids and proteins degradation but also to the membranes composed of lipids thereby increasing the overall permeability of the cellular system contributing to the membrane disassembly (Murphy, 1983). ROS have been known to be produced via type I and type II photosensitization reaction. The photons are absorbed by pigments in the photosynthetic organisms and lead to the formation of excited state of pigments referred to as Sen*. The excited Sen* undergoes reactions (type I and type II) and ultimately results in the chemical alteration of the substrate and finally leading to the formation of ROS.

12.1.3.4 DROUGHT

Drought can be defined as shortage of precipitation over an extended period, or in a meteorological term "a period without significant soil moisture" commonly a season or more, consequently leading to water scarcity causing adverse impacts on flora and fauna. There are three types of drought namely meteorological (extended period of below average precipitation) (Mishra and Cherkauer, 2010), agricultural (lack of ample moisture in the soil) (Manivannan et al., 2008), and hydrological (shortage in reservoirs) drought (Mishra and Singh, 2010). Water stress adversely affects the plant physiology, growth, biomass production (Osakabe et al., 2014), reduction in rate of cell division and growth, root multiplication, shoot elongation, leaf area, water use efficiency, disturbed stomatal oscillations (Li et al., 2009), and subsequently the productivity (Jaleel et al., 2009). The plants have developed molecular, physiological (osmotic process, antioxidant activities, and growth regulators) and morphological (drought escape, dehydration avoidance, and dehydration tolerance) adaptation (Farooq et al., 2011, 2012; Reddy et al., 2004; Zlatev and Lidon, 2012) to combat the water stress during drought condition. In response to water deficit, closure of stomata is stimulated by inducing changes in the turgor pressure in the guard cells through ion- and water-transport across membranes (Sirichandra et al., 2009). Besides the different adaptions evolved to overcome the stress, during extremes of drought, ROS has been known to be produced predominantly in the thylakoids (De Carvalho, 2008). The photorespiration under drought condition has been known to contribute by 70% to the oxidative stress (Hasanuzzaman et al., 2013).

12.1.3.5 SALINITY

Salinity is referred as the amount of salt dissolved in water and measured either in grams of salt per kilogram of water or in parts per thousand. Salinity stress in plants can be primary (natural) or secondary (due to human activities), the conditions where excessive salt in the soil influence the growth and development of the plant by accumulating in the root zone that has a detrimental effect on total yield (Al-Karaki, 2000; Ruiz-Lozano et al., 2012). The salts are toxic to the plants when present in excess in soil predominantly because of improper drainage and high temperature.

The high temperature leads to evaporation of water while accumulation of salt ions increases over the time. Several biochemical and physiological stress response mechanism for salt tolerance has been demonstrated in plants including ion-homeostasis, ion-transport, activation of antioxidant compounds and synthesis, hormone modulation, etc. (Gupta and Huang, 2014; Hasegawa et al., 2000; Sairam and Tyagi, 2004). Other biological activities like respiration, enzymatic activity, protein synthesis (Feng et al., 2002), nutrient imbalance, diminished transport to shoot, reduction of leaf expansion (Colla et al., 2006) are affected and ROS generation are known to be associated with the salt stress. It has been reported that production of ROS are enhanced during the salinity stress which leads to the oxidative damage in plants (Ahmad, 2010; Ahmad and Prasad, 2012; Apel and Hirt, 2004; Mahajan and Tuteja, 2005).

12.1.3.6 METAL TOXICITY

Metals are a natural part of the ecosystems found in soil, rock, air, water, and organisms; few of them (Cu, Mn, and Zn) serve as micronutrients which are important for the plants growth and development in trace amount. Some metals such as Cu and Zn are micronutrients at low concentration and however toxic at higher concentration, while other heavy metals are themselves toxic in nature even at very low concentration. Heavy metals are known to bear severe detrimental and toxic effects on the normal growth and functioning of the cellular components. The primary influence of heavy metals stress in plants occurs mainly via the imbalance in the cellular ionic homeostasis (Emamverdian et al., 2015). Metal toxicity is accountable for various visual symptoms in plants such as color change of leaf may results in interveinal foliar chlorosis symptom and affects root growth (Minnich et al., 1987; Zhu and Alva, 1993), necrotic brown spotting on stems, leaves, and petioles (Wu, 1994), and molecular level decreases the Fe uptake and transport (Fontes and Cox, 1998a, 1998b). Detoxification mechanisms have been evolved in plants which involves formation of heavy metal chelator phytochelatins (synthesized from glutathione) to combat with the severe effect of ionic imbalance (Ernst, 2006; Yadav, 2010).

12.1.4 FORMATION OF REACTIVE OXYGEN SPECIES IN CHLOROPLASTS UNDER ABIOTIC STRESSES

12.1.4.1 HIGH LIGHT STRESS

Photosynthesis is the light-dependent biological pathway, and inhibition of photosynthetic process due to high light stress is called photoinhibition (Aro et al., 1993; Tyystjärvi, 2013). Chloroplast is the site for photosynthesis and composed of two pigment–protein complex, known as photosystem I and II. These pigment–protein complexes absorb the light and emit the step-wise electron from donor side and cofactors accept the electron at acceptor side of PSI and PSII (Rappaport and Diner, 2008). Based on different molecular mechanisms, photoinhibition occurs on either donor or acceptor side of PSII in plants under high-light stress. Thus, in chloroplasts, ROS are formed either by excessive absorption of light energy or by the electron transfer pathway.

Absorption of light energy by chlorophyll molecules leads to sequential electron transfer from P680 to plastoquinone molecules at acceptor side resulting into oxidized chlorophyll and reduced plastoquinone (Cardona et al., 2012; Grundmeier and Dau, 2012). Oxidized chlorophyll takes stepwise electron from water mediated by water-splitting manganese complex, called as water oxidation by PSII and oxygen, protons are released (Brudvig, 2008; Dau and Haumann, 2008). Under high light stress, excess absorption of high light causes over-reduction of PQ-pool at the acceptor side results in the formation of triplet chlorophyll via charge recombination (Pospíšil, 2012; Tyystjärvi, 2008; Vass, 2012; Vass and Aro, 2007). Excited triplet chlorophyll molecule transfers energy to molecular oxygen and results into singlet oxygen formation in PSII (Hideg et al., 1994; Krieger-Liszkay, 2005; Krieger-Liszkay et al., 2008; Vass, 2012). This process occurs due to the acceptor side photoinhibition of PSII. However, under donor side photoinhibition, due to absence or limited supply of electron to excited chlorophyll molecules, it may provide enough time for highly reactive oxidized molecules to react with lipids and proteins in PSII membranes and can form $^1O_2$ via type II or Russell type of mechanism (Howard and Ingold, 1968; Miyamoto et al., 2003; Russell, 1957; Yadav and Pospíšil, 2012a). Furthermore, formation of triplet chlorophyll and $^1O_2$ has also been reported in PSII antenna complex by Type II mechanism (Rinalducci et al., 2004; Santabarbara et al., 2001, 2002; Zolla and Rinalducci, 2002).

Absorption of light by photosynthetic tissues leads to electron transfer chain from PSII donor side to PSI acceptor, and this provides the feasibility to form ROS by electron transfer under high light stress. Superoxide anion radical, H_2O_2, and HO$^•$, are formed by the involvement of many components of thylakoid membranes under impaired electron transport chain. These ROS are formed both either at the acceptor or donor side of PSII (Pospíšil, 2009, 2012; Vass, 2012) and likely acceptor side of PSI (Asada, 2006) under high light stress conditions by electron leakage during electron transport chain. The different components of thylakoid membranes involved in the formation of ROS are summarized in Table 12.2.

12.1.4.2 HEAT STRESS

Heat stress has been shown to affect electron transport processes in PSII, carbon assimilation by Rubisco and ATP synthesis by ATP synthase (Ahmad and Wani, 2014; Tóth et al., 2007). Among these processes, electron transport both on the electron donor and acceptor side of PSII is one of the most heat-sensitive processes in photosynthesis. Due to heat stress in PSII, limitation of electron transport on the both electron donor and acceptor side has been known to be associated with the formation of ROS. Under the condition of mild heat stress, ROS was shown to either directly damage PSII proteins or lipids or inhibit the synthesis of PSII proteins (Yamashita et al., 2008). In particular, H_2O_2 and HO$^•$ are known to be formed on the electron donor side of PSII. Using electron paramagnetic resonance (EPR) spin-trapping spectroscopy, it was demonstrated that the exposure of PSII membranes to heat stress results in HO$^•$ generation, as observed by the formation of EMPO–OH adduct EPR signal. The authors proposed that HO$^•$ is formed by one electron reduction of H_2O_2 by manganese from the water-splitting complex through the metal-catalyzed Fenton reaction (Pospíšil et al. 2007; Yadav and Pospíšil, 2012b; Yamashita et al., 2008). Apart from electron donor side of PSII, evidence has proposed that $^1O_2$ is also formed on the electron acceptor side of PSII (Pospíšil et al., 2007; Yamashita et al., 2008).

12.1.4.3 UV IRRADIATION STRESS

Photosynthetic organisms, in addition to visible photosynthetic active radiation (400–700 nm), are also exposed to ultraviolet components which penetrates into the atmosphere. The UV components are composed of UVA (320–400 nm), UVB (280–320 nm), and UVC (< 280 nm) (Krizek, 2004). The higher degree of UV penetration is known to influence the growth and productivity in crops (Nawkar et al., 2013). It is known that different UV

TABLE 12.2 Summary of the Involved Components of Thylakoid Membranes in ROS Formation under High Light Stress.

ROS	Component of Thylakoids Membranes	Comments
$O_2^{\bullet-}$	Pheophytine (Ananyev et al., 1994; Pospíšil et al., 2004), tightly (Cleland and Grace, 1999), or loosely (Yadav et al., 2014; Zhang et al., 2003) or free plastosemiquinone (Mubarakshina and Ivanov, 2010) Cytochrome b_{559} (Pospíšil et al., 2006), oxidized tyrosine *Z* (Pospíšil, 2012), ferredoxin (Asada, 2006), and phylloquinone A1 (Kozuleva et al., 2014)	Generally formed by one electron reduction of molecular oxygen or by oxidation of H_2O_2. Acceptor and donor side of PSII as well as component of PSI.
H_2O_2	Spontaneous dismutation of $O_2^{\bullet-}$ (Klimov et al., 1993), heme and non-heme iron (Pospíšil et al., 2004; Tiwari and Pospíšil, 2009), plastoquinol (Mubarakshina and Ivanov, 2010), ferredoxin (Jakob and Heber, 1996), and FeS centers (Terashima et al., 1998)	Generally formed by reduction of $O_2^{\bullet-}$ or by oxidation of water. Acceptor and donor side of PSII and acceptor side of PSI.
$HO^{\bullet}$	Free metals (Fe^{2+} or Mn^{2+}) (Pospíšil et al., 2004), non-heme iron (Pospíšil et al., 2004), water splitting manganese complex (Arató et al., 2004), and ferredoxin (Šnyrychová et al., 2006)	Reduction of H_2O_2 by free or bound metals. Acceptor and donor side of PSII and acceptor side of PSI.
$^1O_2$	Triplet chlorophyll (Krieger-Liszkay, 2005; Pospíšil, 2012)	In PSII membranes

Note: For details reactions mechanisms for ROS formation in photosynthesis refer recent reviews Asada (2006), Krieger-Liszkay et al. (2008), Vass (2012), and Pospíšil (2009, 2012).

components bring about different effects on plants growth and productivity. UVB and UVC are known to stimulate cellular damage to a high extent, whereas high UVC can lead to programmed cell death (Gill et al., 2015; Nawkar et al., 2013; Stapleton, 1992). It has been known that activity of NADPH oxidase is enhanced under UV irradiation thereby leading to enhancement in the production of $O_2^{\bullet-}$ which leads to cell damage (Hideg and Vass, 1996; Kalbina and Strid, 2006). Superoxide anion radical, H_2O_2, and $HO^{\bullet}$ are dominant form of the ROS formed in chloroplast under UV stress, whereas $^1O_2$ is less favourable (Barta et al., 2004; Hideg and Vass, 1996; Hideg et al., 2002). The mechanism of the $^1O_2$ formation in UV stress is unknown and suggested that it is a different mechanism than the light-induced $^1O_2$ production in PSII (Vass, 2012). In chloroplast of *Arabidopsis thaliana*, it has been recently reported that ROS are produced during the early response of programmed cell death induced by UVC irradiation (Gao et al., 2008; Nawkar et al., 2013).

12.1.5 BIOMOLECULES DAMAGE BY REACTIVE OXYGEN SPECIES IN PLANTS

Damage of biomolecules (proteins, lipids, and nucleic acid) is the primary function of ROS in oxidative stress condition in plants. Biomolecules-damage reaction mechanisms depend on the type of ROS generated. Singlet oxygen damages the molecules either due to addition reaction on carbon–carbon double bond by *ene* reaction or cycloaddition with aliphatic or aromatic to form hydroperoxide and dioxetane or endoperoxide, respectively (Mattila et al., 2015). It oxidizes unsaturated fatty acids into lipid hydro-peroxide and some specific amino acids into their adduct form such as histidine (Rehman et al., 2013; Telfer et al., 1994; Triantaphylidès et al., 2008). The presence of $^1O_2$ scavengers have also been reported in plants to protect the damage (Krieger-Liszkay et al., 2008; Pospíšil, 2012; Rastogi et al., 2014; Triantaphylidès and Havaux, 2009; Yadav and Pospíšil, 2010). Studies showed that the reaction of $^1O_2$ with other biomolecules leads to the formation of other type of radicals. Likewise, $^1O_2$ interact with lipids and plastoquinol to form lipid radicals (Triantaphylidès and Havaux, 2009) and H_2O_2 in thylakoids (Khorobrykh et al., 2015), respectively.

Superoxide anion radical is either dismutase by interacting with another $O_2^{\bullet-}$ or by enzymatic reaction of superoxide dismutase (SOD) (Asada, 1996; Halliwell and Gutteridge, 2007; Navari-Izzo et al., 1999). In a presence of proton donor, it gets protonated which oxidizes organic molecules by several ways including protonation, reduction or nucleophilic addition, etc. (Mattila et al., 2015). Hydrogen peroxide is the main source of HO$^•$ formation by the reduction process in presence of either heavy metals (Fenton reaction) or $O_2^{•-}$ (Haber–Weiss mechanism) (Kehrer, 2000; Liochev, 1999). Hydrogen peroxide also interacts specially with thiol-group containing compound (Winterbourn and Metodiewa, 1999) and cysteine residue as the target site in proteins (D'Autréaux and Toledano, 2007). Peroxidases enzymes are having catalytic activity against H_2O_2, interactions results into the formation of water or oxygen (Asada, 2006; Mattila et al., 2015). Furthermore, HO$^•$ is one of the most reactive ROS and interacts with organic molecules by several ways, through abstraction of hydrogen atom (Gutteridge, 1995; Pogozelski and Tullius, 1998), addition of HO$^•$ to double bond aliphatic and aromatic compound (hydroxylation) (Montgomery et al., 1999). Hydroxyl radical initiates the cascade of lipid peroxidation by abstracting the hydrogen atom, resulting in the formation of alkyl radical. In the propagation step, the radical molecules form additional lipid radicals by interacting with molecular oxygen and another lipid molecule. Lipid radicals formed by the action of HO$^•$ (alkyl, peroxyl, and alkoxyl radicals) convert into non radical species by reacting with each other known as termination step of lipid peroxidation (Miyamoto et al., 2006, 2007; Prasad and Pospíšil, 2015).

12.2 DETECTION OF ROS IN PLANT—AN OVERVIEW ON METHODS

In this section, the commonly used techniques for ROS measurement in plants are described, for more details refer to recent review Mattila et al. (2015). In general, there is a good agreement between different approaches to measure ROS.

12.2.1 ELECTRON PARAMAGNETIC RESONANCE SPECTROSCOPY

Electron paramagnetic resonance (EPR) also known as electron spin resonance (ESR), was first measured by Yevgeny Zavoisky in 1944. This technique is used for the detection of radical molecules having unpaired electron or in paramagnetic nature. In theory, EPR is usually comparable or is an analogue technique to nuclear magnetic resonance (NMR). Mostly, EPR is used to study the nature of radical species in chemical, physical processes, and its formation during stress condition in cells, whereas NMR is used for structural study. Commercially available EPR has four major components, klystrons (source of microwave radiation), magnets (generates electromagnetic field), sample cavity (holder located between the magnets), and detectors (detects resulting microwave radiation). To measure resonance spectra, a suitable EPR setup is needed with a proper range of microwave frequency, magnet strength, and waveguide. Based on the microwave frequency used, a set of different EPR has been classified such as L,S,C,X,P,K,Q,U,V,E,W, F,D,J, etc. Most commonly used EPR is X-band (~10 GHz), then others important range are L,S,Q, W-band (source: https://en.wikipedia.org/wiki/Electron_paramagnetic_resonance).

12.2.1.1 EPR SPIN-TRAPPING SPECTROSCOPY

EPR spin-trapping is based on same principal and radial species are measured in presence of additional molecule called spin-trap. Spin-traps are stable compounds, reacts with radical species to increase the radical stability (Khan et al., 2003; Samuni et al., 1986; Venkataraman et al., 2004). Due to its interaction with radicals, it results in the formation of a stable radical-spin-trap adduct, that gives a spectra during measurements. In EPR spectra, the spectral shape, number of lines, and its splitting gives information about the interacting nuclei with electron (Buettner, 1987; Murphy, 2009b). In plants or its tissue, measurement of ROS is very difficult due to short half-life and high reactivity. Furthermore, plants also possess many other components, which may hinder the specific quantification and localization of ROS itself. Thus, EPR spin-trapping could be a better solution for the detection of ROS in plant tissue. The pros of using EPR spin-trapping is the formation of radical-spin-trap adduct can be stable up to hours, which makes the measurement easy and feasible. The cons are that some spin-traps are not very specific and needs to be used according to the measuring conditions. Improved and new spin traps have been developed and reported in different studies in literature, which enhanced the use of spin-trapping for ROS detection in plants. For detailed reactions on mechanisms of ROS detection by EPR in plants, refer to recent reviews by Bačić and Mojović (2005), Steffen-Heins and Steffen (2015), and Mattila et al. (2015).

12.2.2 MICROSCOPIC IMAGING

Several molecular probes such as singlet oxygen sensor green (SOSG), 3-[N-(β-diethylaminoethyl)-N-dansyl]-aminomethyl-2,2,5,5-tetramethyl-

2,5-dihydro-1H-pyrrole (DanePy), 3,3'-diaminobenzidine (DAB), amplex red (AR), amplex ultra red (AUR), etc. have been used for microscopic imaging of ROS in *in-vivo, ex-vivo,* and *in-vitro* plant system. The major challenges faced using the molecular probes are its specificity and sensitivity toward ROS and have always been a concern of researchers. The specificity and selectivity has always been questioned with molecular probes such as SOSG for the detection of $^1O_2$ (Flors et al., 2006; Gollmer et al., 2011). The sensitivity with molecular probes is generally within the range of micromole concentration and thus ROS detection at the level of unicellular organism and the level of cellular components have always been a challenge. Also, the toxicity caused by the exogenous addition of molecular probes cannot be ruled out completely. In the next sections, a description on commonly used molecular probes for various ROS has been summarized.

12.2.2.1 SINGLET OXYGEN SENSOR GREEN AND 3-[N-(β-DIETHYLAMINOETHYL)-N-DANSYL]-AMINOMETHYL-2,2,5,5-TETRAMETHYL-2,5-DIHYDRO-1H-PYRROLE

The singlet oxygen sensor green (SOSG) is commonly available and widely used fluorescent sensor for detection of $^1O_2$ and is composed of fluorescein and anthracene moiety. The SOSG florescence has been widely used for measurement of the formation of $^1O_2$ in different living system (Flors et al., 2006; Gollmer et al., 2011; Rác et al., 2015; Sinha et al., 2011). The SOSG fluorescence was measured with excitation at 488 nm and subsequent emission measured at 505–525 nm (Flors et al., 2006). Besides, the recent wide application of SOSG in different living sample such as unicellular green algae, cyanobacteria, and at level of leaves, its applicability in studies with visible light which coincides with the absorption and emission range of the probe itself has always been questioned (Kim et al., 2013). In addition to this, the photodecomposition of SOSG and production of $^1O_2$ in itself is known to be a drawback of the probe (Kim et al., 2013). DanePy, a dansyl-based ROS fluorescent probe, has been used similar to SOSG. DanePy has been used in chloroplasts for the *in-vivo* detection of $^1O_2$ and its localization in the sub-cellular structure (Hideg et al., 2002). DanePy has also been tested for its applicability in leaf sample (Hideg et al., 2001). It has been found that penetration of DanePy in dicotyledonous plant such as spinach can be achieved with ease and that with a pinhole at the edge of the leaf, the diffusion of DanePy to different sub-cellular components of the leaf can be achieved and fluorescence imaging can be measured subsequently (Hideg et al., 2002).

12.2.2.2 3,3'-DIAMINOBENZIDINE AND NITROBLUE TETRAZOLIUM

3,3'-diaminobenzidine (DAB) for the detection of H_2O_2 and nitroblue tetrazolium (NBT) for the detection of $O_2^{\bullet -}$ are among the most widely used test in the plant cells and tissue. It has been used as a universal test from the past few decades because of the procedure being very simple. DAB reacts with H_2O_2 forming a brown precipitate (Litwin, 1979; Liu et al., 2014) and thus localization on tissues is rather simple. DAB precipitate is stable and being insensitive to light, it has been widely applied for experiments to study light stress on plant samples. NBT, as similar to DAB, has also been applied for histochemical studies of $O_2^{\bullet -}$ formation in roots, leaves, and extracted components of plants (Liszkay et al., 2004). Its application however is limited because of its low solubility in aqueous medium.

12.2.2.3 AMPLEX RED AND AMPLEX ULTRA RED

Fluorescent compounds such as amplex red (AR) (10-acetyl-3,7-dihydroxyphenoxazine) and amplex ultra-red (AUR) (modified 10-acetyl-3,7-dihydroxyphenoxazine) have been used in the recent studies for the detection of H_2O_2 from different components of the cell system (Serrano et al., 2009; Šnyrychová et al., 2009; Votyakova and Reynolds, 2004; Yadav and Pospíšil, 2012b; Zhou et al., 1997). It can be used with highest efficacy among the other fluorescent probes because of the high permeability inside the cell as compared to other fluorescent probes, however, the toxicity caused by the chemical and the photosensitivity of the probes itself limits its wide application in photosynthetic research (Šnyrychová et al., 2009).

12.2.3 ELECTROCHEMICAL MEASUREMENTS

Mediator and non-mediator based electrode have been used in the past for the electrochemical measurement of oxygen consumption or generation of ROS in varied living system. Among the modified electrode, the mediator, horseradish peroxidase (HRP) based electrode is among the most commonly and widely used (Ahammad, 2013). In HRP modified electrode, the enzyme HRP is converted to its oxidized form followed by its reduction at the surface of the carbon electrode by the transfer of the electron through the mediator (Ahammad, 2013; Prasad et al., 2015).

To study the photosynthetic activity under the effect of benzoquinone, electrochemical measurements using amperometry have been employed at the level of protoplast during the last decade (Yasukawa et al., 1999). Direct detection of H_2O_2 using electrochemical methods was very recently demonstrated using catalytic amperometry in PSII membrane under high light (Prasad et al., 2015). The real time simultaneous measurement of oxygen consumption and H_2O_2 production reflected by changes in reduction current were demonstrated using platinum microelectrode and osmium-horseradish peroxidase (Os-HRP) modified carbon electrode. Exogenous addition of SOD enhanced the production of H_2O_2 which was immediately suppressed by exogenous addition of catalase. The authors claimed that catalytic amperometric method could be potentially applied for precise measurement of H_2O_2 in localized structures of plants and tissues with a detection limit in the concentration range of micromoles down to nanomoles (Prasad et al., 2015).

12.2.4 LOW-LEVEL CHEMILUMINESCENCE

Low-level chemiluminescence, also referred as ultraweak photon emission, has been used as an indirect method for detection of ROS in plant species. Low-level chemiluminescence is known to originate by the formation of electronically excited species during the oxidative radical reaction (Pospíšil et al., 2014; Shen and van Wijk, 2005). ROS are known to be involved in the initiation of oxidative radical reaction either by lipid peroxidation or protein oxidation (Cadenas et al., 1980; Fedorova et al., 2007; Miyamoto et al., 2007; Pospíšil et al., 2014). Several experimental evidences have been reported, where addition of different exogenous ROS scavenger has shown considerable suppression in the low-level chemiluminescence (Prasad and Pospíšil, 2011a, 2011b) and hence the involvement of ROS in the oxidative reactions has been claimed. The fluctuation in the low-level chemiluminescence and its kinetics has been widely applied to study the effect of heat stress, chemical stress, light stress including UV during the recent past also for the *in-vivo* study (Hao et al., 2004; Pospíšil et al., 2014). The imaging of low-level chemiluminescence does serve as a non-invasive method to imaging oxidative stress in plant system. Two-dimensional imaging of low-level chemiluminescence has also been studied and is known to provide spatial and temporal information on the distribution of oxidative stress, thereby reflecting the physiological state of an organism (Kobayashi et al., 2009).

KEYWORDS

- **abiotic stresses**
- **chloroplasts**
- **oxidative stress**
- **reactive oxygen species**

REFERENCES

Ahammad, A. J. S. Hydrogen Peroxide Biosensors Based on Horseradish Peroxidase and Hemoglobin. *Biosens. Bioelectron.* **2013,** *S9*, 001.

Ahmad, P.; Prasad, M. N. V. *Abiotic Stress Responses in Plants: Metabolism, Productivity and Sustainability*. Springer: New York, 2012.

Ahmad, P.; Wani, M. R. *Physiological Mechanisms and Adaptation Strategies in Plants under Changing Environment*; Springer: New York, 2014, p 2.

Ahmad, P. Growth and Antioxidant Responses in Mustard (*Brassica juncea* L.) Plants Subjected to Combined Effect of Gibberellic Acid and Salinity. *Arch. Agron. Soil Sci.* **2010,** *56*, 575–588.

Åkerfelt, M.; Morimoto, R. I.; Sistonen, L. Heat Shock Factors: Integrators of Cell Stress, Development and Lifespan. *Nat. Rev. Mol. Cell Biol.* **2010,** *11*, 545–555.

Alfadda, A. A.; Sallam, R. M. Reactive Oxygen Species in Health and Disease. *J. Biomed. Biotechnol.* **2012,** Article ID 936486, 14 pp.

Al-Karaki, G. N. Growth of Mycorrhizal Tomato and Mineral Acquisition under Salt Stress. *Mycorrhiza* **2000,** *10*, 51–54.

Ananyev, G. M.; Renger, G.; Wacker, U.; Klimov, V. The Photoproduction of Superoxide Radicals and the Superoxide-Dismutase Activity of Photosystem II. The Possible Involvement of Cytochrome b_{559}. *Photosynth. Res.* **1994,** *41*, 327–338.

Apel, K.; Hirt, H. Reactive Oxygen Species: Metabolism, Oxidative Stress and Signal Transduction. *Annu. Rev. Plant Biol.* **2004,** *55*, 373–399.

Arató, A.; Bondarava, N.; Krieger-Liszkay, A. Production of Reactive Oxygen Species in Chloride- and Calcium-Depleted Photosystem II and their Involvement in Photoinhibition. *Biochim. Biopys. Acta* **2004,** *1608*, 171–180.

Aro, E. M.; Virgin, I.; Andersson, B. Photoinhibition of Photosystem II. Inactivation, Protein Damage and Turnover. *Biochim. Biophys. Acta* **1993,** *1143*, 113–134.

Aroca, R.; Porcel, R.; Ruiz-Lozano, J. M. Regulation of Root Water Uptake under Abiotic Stress Conditions. *J. Ex. Bot.* **2012,** *63*, 43–57.

Asada, K. Production and Scavenging of Reactive Oxygen Species in Chloroplasts and their Functions. *Plant Physiol.* **2006,** *141*, 391–396.

Asada, K. Radical Production and Scavenging in the Chloroplasts. In: *Photosynthesis and the Environment*; Baker, N. R., Eds.; Kluwer Academic Publishers: Dordrecht, the Netherlands, 1996; pp 123–150.

Atkinson, N. J.; Urwin, P. E. The Interaction of Plant Biotic and Abiotic Stresses: From Genes to the Field. *J. Exp. Bot.* **2012,** *63* (10), 3523–543.

Bačić, G.; Mojović, M. EPR Spin Trapping of Oxygen Radicals in Plants: A Methodological Overview. *Ann. N.Y. Acad. Sci.* **2005,** *1048*, 230–243.

Ballaré, C. L.; Scopel, A. L.; Stapleton, A. E.; Yanovsky, M. J. Solar Ultraviolet-B Radiation Affects Seedling Emergence, DNA Integrity, Plant Morphology, Growth Rate, and Attractiveness to Herbivore Insects in *Datura ferox*. *Plant Physiol.* **1996,** *112*, 161–170.

Barta, C.; Kalái, T.; Hideg, K.; Vass, I.; Hideg, É. Differences in the ROS-Generating Efficacy of Various Ultraviolet Wavelengths in Detached Spinach Leaves. *Funct. Plant Biol.* **2004,** *31*, 23–28.

Bartosz, G. Oxidative Stress in Plants. *Acta Physiol. Plant* **1997,** *19*, 47–64.

Bassirirad, H. Kinetics of Nutrient Uptake by Roots: Responses to Global Change. *New Phytol.* **2000,** *147*, 155–169.

Berlett, B. S.; Stadtman, E. R. Protein Oxidation in Aging, Disease, and Oxidative Stress. *J. Biol. Chem.* **1997,** *272*, 20213–20316.

Braunersreuther, V.; Jaquet, V. Reactive Oxygen Species in Myocardial Reperfusion Injury: From Physiopathology to Therapeutic Approaches. *Curr. Pharm. Biotechnol.* **2012,** *13*, 97–114.

Breusengem, F. V.; Dat, J. F. Reactive Oxygen Species in Plant Cell Death. *Plant Physiol.* **2006,** *141*, 384–390.

Brudvig, G. W. Water Oxidation Chemistry of Photosystem II. *Philos. Trans. R. Soc. B* **2008,** *363*, 1211–1219.

Buettner, G. R. Spin Trapping: ESR Parameters of Spin Adducts. *Free Radic. Biol. Med.* **1987,** *3*, 259–303.

Cabiscol, E.; Tamarit, J.; Ros, J. Oxidative Stress in Bacteria and Protein Damage by Reactive Oxygen Species. *Int. Microbiol.* **2000,** *3*, 3–8.

Cadenas, E.; Arad, I. D.; Boveris, A.; Fisher, A. B.; Chance, B. Partial Spectral-Analysis of the Hydroperoxide-Induced Chemi-luminescence of the Perfused Lung. *FEBS Lett.* **1980,** *111*, 413–418.

Carbonell-Bejerano, P.; Diago, M. P.; Martínez-Abaigar, J.; Martínez-Zapater, J. M.; Tardáguila, J.; Núñez-Olivera, E. Solar Ultraviolet Radiation Is Necessary to Enhance Grapevine Fruit Ripening Transcriptional and Phenolic Responses. *BMC Plant Biol.* **2014,** *14* (1), 183.

Cardona, T.; Sedoud, A.; Cox, N.; Rutherford, A. W. Charge Separation in Photosystem II: A Comparative and Evolutionary Overview. *Biochim. Biophys. Acta* **2012,** *1817*, 26–43.

Castro, L.; Freeman, B. A. Reactive Oxygen Species in Human Health and Disease. *Nutrition* **2001,** *17*, 161–165.

Chaitanya, K. V.; Sundar, D.; Masilamani, S.; Reddy, A. R. Variation in Heat Stress-Induced Antioxidant Enzyme Activities among Three Mulberry Cultivars. *Plant Growth Regul.* **2002,** *36*, 175–180.

Cifra, M.; Pospíšil, P. Ultra-Weak Photon Emission from Biological Samples: Definition, Mechanisms, Properties, Detection and Applications. *J. Photochem. Photobiol. B* **2014,** *139*, 2–10.

Circu, M. L.; Aw, T. Y. Reactive Oxygen Species, Cellular Redox System and Apoptosis. *Free Radic. Biol. Med.* **2010,** *48*, 749–763.

Cleland, R. E.; Grace, S. C. Voltametric Detection of Superoxide Production by Photosystem II. *FEBS Lett.* **1999,** *457*, 348–352.

Colla, G.; Roupahel, Y.; Cardarelli, M.; Rea, E. Effect of Salinity on Yield, Fruit Quality, Leaf Gas Exchange, and Mineral Composition of Grafted Watermelon Plants. *HortScience* **2006,** *41*, 622–627.

Cooke, M. S.; Evans, M. D.; Dizdaroglu, M.; Lunec, J. Oxidative DNA Damage: Mechanisms, Mutation, and Disease. *FASEB J.* **2003,** *17*, 1195–1214.

Das, K.; Roychoudhury, A. Reactive Oxygen Species (ROS) and Response of Antioxidants as ROS-Scavengers during Environmental Stress in Plants. *Front. Eviron. Sci.* **2014,** *2*, Article 53.

D'Autréaux, B.; Toledano, M. B. ROS as Signalling Molecules: Mechanisms that Generate Specificity in ROS Homeostasis. *Nat. Rev. Mol. Cell Biol.* **2007,** *8*, 813–824.

Dat, J.; Vandenabeele, S.; Vranová, E.; Van Montagu, M.; Inzé, D.; Breusegem, F. Dual Action of the Active Oxygen Species during Plant Stress Responses. *CMLS Cell. Mol. Life. Sci.* **2000,** *57*, 779–795.

Dau, H.; Haumann, M. The Manganese Complex of Photosystem II in its Reaction Cycle-Basic Framework and Possible Realization at the Atomic Level. *Coord. Chem. Rev.* **2008,** *252*, 273–295.

Demoling, F.; Figueroa, D.; Bååth, E. Comparison of Factors Limiting Bacterial Growth in Different Soils. *Soil Biol. Biochem.* **2007,** *39*, 2485–2495.

Devaraj, B.; Inaba, H. Biophotons: Ultraweak Light Emissions from Living Systems. *Curr. Opin. Solid State Matter Sci.* **1997,** *2*, 188–193.

De Carvalho, M. H. C. Drought Stress and Reactive Oxygen Species. *Plant Signal. Behav.* **2008,** *3*, 156–165.

Emamverdian, A.; Ding, Y.; Mokhberdoran, F.; Xie, Y. Heavy Metal Stress and Some Mechanisms of Plant Defense Response. *Sci. World J.* **2015,** *5*, Article ID 756120.

Ernst, W. H. O. Evolution of Metal Tolerance in Higher Plants. *For. Snow Landsc. Res.* **2006,** *80*, 251–274.

Farooq, M.; Bramley, H.; Palta, J. A.; Siddique, K. H. Heat Stress in Wheat during Reproductive and Grain-Filling Phases. *CRC Crit. Rev. Plant Sci.* **2011,** *30*, 491–507.

Farooq, M.; Hussain, M.; Wahid, A.; Siddique, K. H. M. Drought Stress in Plants: An Overview. In: *Plant Responses to Drought Stress*; Aroca, R., Eds.; Springer-Verlag: Berlin-Heidelberg, 2012.

Fedorova, G. F.; Trofimov, A. V.; Vasil'ev, R. F.; Veprintsev, T. L. Peroxy-Radical-Mediated Chemiluminescence: Mechanistic Diversity and Fundamentals for Antioxidant Assay. *Arkivoc.* **2007,** *163*, 215.

Feng, G.; Zhang, F.; Li, X.; Tian, C.; Tang, C.; Rengel, Z. Improved Tolerance of Maize Plants to Salt Stress by Arbuscular Mycorrhiza Is Related to Higher Accumulation of Soluble Sugars in Roots. *Mycorrhiza* **2002,** *12*, 185–190.

Finch-Savage, W. E.; Phelps, K. Onion (*Allium cepa* L.) Seedling Emergence Patterns can be Explained by the Influence of Soil Temperature and Water Potential on Seed Germination. *J. Exp. Bot.* **1993,** *44*, 407–414.

Flors, C.; Fryer, M. J.; Waring, J.; Reeder, B.; Bechtold, U.; Mullineaux, P. M.; Nonell, S.; Wilson, M. T.; Baker, N. R. Imaging the Production of Singlet Oxygen *In Vivo* Using a New Fluorescent Sensor, Singlet Oxygen Sensor Green. *J. Exp. Bot.* **2006,** *57*, 1725–1734.

Fontes, R. L. F.; Cox, F. R. Iron Deficiency and Zinc Toxicity in Soybean Grown in Nutrient Solution with Different Levels of Sulfur. *J. Plant Nutr*. **1998a,** *21*, 1715–1722.

Fontes, R. L. F.; Cox, F. R. Zinc Toxicity in Soybean Grown at High Iron Concentration in Nutrient Solution. *J. Plant Nutr.* **1998b,** *21*, 1723–1730.

Foyer, C. H.; Neukermans, J.; Queval, G.; Noctor, G.; Harbinson, J. Photosynthetic Control of Electron Transport and the Regulation of Gene Expression. *J. Exp. Bot.* **2012,** *63*, 1637–1661.

Foyer, C. H.; Noctor, G. Redox Sensing and Signalling Associated with Reactive Oxygen in Chloroplasts, Peroxisomes and Mitochondria. *Physiol. Plant.* **2003,** *19*, 355–364.

Foyer, C. H.; Noctor, G. Redox Signaling in Plants. *Antioxid. Redox Signal.* **2013,** *18*, 2087–2090.

Galloway, A. M.; Liuzzi, M.; Paterson, M. C. Metabolic Processing of Cyclobutyl Pyrimidine Dimers and (6–4) Photoproducts in UV-Treated Human Cells. Evidence for Distinct Excision-Repair Pathways. *J. Biol. Chem.* **1994,** *269*, 974–980.

Gao, C.; Xing, D.; Li, L.; Zhang, L. Implication of Reactive Oxygen Species and Mitochondrial Dysfunction in the Early Stages of Plant Programmed Cell Death Induced by Ultraviolet-C Overexposure. *Planta* **2008,** *227*, 755–767.

Gill, S. S.; Anjum, N. A.; Gill, R.; Jha, M.; Tuteja, N. DNA Damage and Repair in Plants under Ultraviolet and Ionizing Radiations. *Sci. World J.* **2015,** *250158*, 1–11.

Gill, S. S.; Tuteja, N. Reactive Oxygen Species and Antioxidant Machinery in Abiotic Stress Tolerance in Crop Plants. *Plant Physiol. Biochem.* **2010,** *48*, 909–930.

Gollmer, A.; Arnbjerg, J.; Blaikie, F. H.; Pedersen, B. W.; Breitenbach, T.; Daasbjerg, K.; Glasius, M.; Ogilby, P. R. Singlet Oxygen Sensor Green®: Photochemical Behavior in Solution and in a Mammalian Cell. *Photochem. Photobiol.* **2011,** *87*, 671–679.

Grundmeier, A.; Dau, H. Structural Models of the Manganese Complex of Photosystem II and Mechanistic Implications. *Biochim. Biophys. Acta* **2012,** *1817*, 88–105.

Gupta, B.; Huang, B Mechanism of Salinity Tolerance in Plants: Physiological, Biochemical, and Molecular Characterization. *Int. J. Genomics* **2014,** Article ID: 701596, 18.

Gutteridge, J. M. C. Lipid Peroxidation Initiated by Superoxide-Dependent Hydroxyl Radicals Using Complexed Iron and Hydrogen Peroxide. *FEBS Lett.* **1984,** *172*, 245–249.

Gutteridge, J. M. C. Lipid Peroxidation and Antioxidants as Biomarkers of Tissue Damage. *Clin. Chem.* **1995,** *41*, 1819–1828.

Halliwell, B.; Gutteridge, J. M. C. *Free Radicals in Biology and Medicine*, fourth ed. Oxford University Press: Oxford, 2007.

Hao, O. Y.; Stamatas, G.; Saliou, C.; Kollias, N. A Chemiluminescence Study of UVA-Induced Oxidative Stress in Human Skin In Vivo. *J. Invest. Dermatol.* **2004,** *122*, 1020–1029.

Hasanuzzaman, M.; Nahar, K.; Gill, S. S.; Fujita, M. Drought Stress Responses in Plants, Oxidative Stress, and Antioxidant Defense. In: *Climate Change and Plant Abiotic Stress Tolerance*; Tuteja, N., Gill, S. S., Eds.; Wiley: Weinheim, Germany, 2013.

Hasegawa, P. M.; Bressan, R. A.; Zhu, J. K.; Bohnert, H. J. Plant Cellular and Molecular Responses to High Salinity. *Annu. Rev. Plant Biol.* **2000,** *51*, 463–499.

Heyno, E.; Mary, V.; Schopfer, P.; Krieger-Liszkay, A. Oxygen Activation at the Plasma Membrane: Relation between Superoxide and Hydroxyl Radical Production by Isolated Membranes. *Planta* **2011,** *234*, 35–45.

Hideg, É.; Vass, I. UV-B Induced Free Radical Production in Plant Leaves and Isolated Thylakoid Membranes. *Plant Sci.* **1996,** *115*, 251–260.

Hideg É, Barta, C.; Kalái, T.; Vass, I.; Hideg, K.; Asada, K. Detection of Singlet Oxygen and Superoxide with Fluorescent Sensors in Leaves under Stress by Photoinhibition or UV Radiation. *Plant Cell Physiol.* **2002,** *43*, 1154–1164.

Hideg, E.; Oqawa, K.; Kalai, T.; Hideg, K. Singlet Oxygen Imaging in *Arabidopsis thaliana* Leaves under Photoinhibition by Excess Photosynthetically Active Radiation. *Physiol. Plant.* **2001,** *112*, 10–14.

Hideg, É.; Spetea, C.; Vass, I. Singlet Oxygen and Free Radical Production during Acceptor and Donor Side Induced Photoinhibition: Studies with Spin Trapping EPR Spectroscopy. *Biochim. Biophys. Acta* **1994,** *1186*, 143–152.

Howard, J. A.; Ingold, K. U. Self-Reaction of Sec-Butylperoxy Radicals Confirmation of Russell Mechanism. *J. Am. Chem. Soc.* **1968,** *90*, 1056–1058.

Jakob, B.; Heber, U. Photoproduction and Detoxification of Hydroxyl Radicals in Chloroplasts and Leaves and Relation to Photoinactivation of Photosystems I and II. *Plant Cell Physiol.* **1996,** *37*, 629–635.

Jaleel, C. A.; Manivannan, P.; Wahid, A.; Farooq, M.; Somasundaram, R.; Panneerselvam, R. Drought Stress in Plants: A Review on Morphological Characteristics and Pigments Composition. *Int. J. Agric. Biol.* **2009,** *11*, 100–105.

Johansson, M.; Staiger, D. Time to Flower: Interplay between Photoperiod and the Circadian Clock. *J. Exp. Bot.* **2015,** *66*, 719–730.

Kalbina, I.; Strid, A. The Role of NADPH Oxidase and MAP Kinase Phosphatase in UV-B-Dependent Gene Expression in *Arabidopsis*. *Plant Cell Environ.* **2006,** *29*, 1783–1793.

Karuppanapandian, T.; Moon, J.-C.; Kim, C.; Manoharan, K.; Kim, W. Reactive Oxygen Species in Plants: Their Generation, Signal Transduction, and Scavenging Mechanisms. *Aust. J. Crop Sci.* **2011,** *5*, 709–725.

Kaspar, T. C.; Bland, W. L. Soil Temperature and Root Growth. *Soil Sci.* **1992,** *154*, 290–299.

Kehoe, J. J.; Remondetto, G. E.; Subirade, M.; Morris, E. R.; Brodkorb, A. Tryptophan-Mediated Denaturation of β-Lactoglobulin A by UV Irradiation. *J. Agric. Food Chem.* **2008,** *56*, 4720–4725.

Kehrer, J. P. The Haber–Weiss Reaction and Mechanisms of Toxicity. *Toxicology* **2000,** *149* (1), 43–50.

Khan, N.; Wilmot, C. M.; Rosen, G. M.; Demidenko, E.; Sun, J.; Joseph, J.; Julia O'Hara, Kalyanaraman, B.; Swartz, H. M. Spin Traps: In Vitro Toxicity and Stability of Radical Adducts. *Free Radic. Biol. Med.* **2003,** *34,* 1473–1481.

Khorobrykh, S. A.; Karonen, M.; Tyystjärvi, E. Experimental Evidence Suggesting that H_2O_2 Is Produced within the Thylakoid Membrane in a Reaction between Plastoquinol and Singlet Oxygen. *FEBS Lett.* **2015,** *589,* 779–786.

Kim, S.; Fujitsuka, M.; Majima, T. Photochemistry of Singlet Oxygen Sensor Green. *J. Phys. Chem. B* **2013,** *117,* 13985–13992.

Klan, P.; Wirz, J. Photochemistry of Organic Compounds: From Concepts to Practice. Wiley-Blackwell: Chichester West, Sussex, UK, 2009.

Klimov, V.; Ananyev, G.; Zastryzhnava, O.; Wydrzynski, T.; Renger, G. Photoproduction of Hydrogen Peroxide in Photosystem II Membrane Fragments: A Comparison of Four Signals. *Photosynth. Res.* **1993,** *38,* 409–416.

Kobayashi, M.; Kikuchi, D.; Okamura, H. Imaging of Ultraweak Spontaneous Photon Emission from Human Body Displaying Diurnal Rhythm. *PLoS ONE* **2009,** *4,* e6256.

Kotak, S.; Larkindale, J.; Lee, U.; von Koskull-Döring, P.; Vierling, E.; Scharf, K. D. Complexity of the Heat Stress Response in Plants. *Curr. Opin. Plant Biol.* **2007,** *10,* 310–316.

Kozuleva, M. A.; Petrova, A. A.; Mamedov, M. D.; Semynov, A. Y.; Ivanov, B. N. O_2 Reduction by Photosystem I involves Phylloquinone under Steady-State Illumination. *FEBS Lett.* **2014,** *588,* 4364–4368.

Krieger-Liszkay, A. Singlet Oxygen Production in Photosynthesis. *J. Exp. Bot.* **2005,** *56,* 337–346.

Krieger-Liszkay, A.; Fufezan, C.; Trebst, A. Singlet Oxygen Production in Photosystem II and Related Protection Mechanism. *Photosynth. Res.* **2008,** *98,* 551–564.

Krizek, D. T. Influence of PAR and UV-A in Determining Plant Sensitivity and Photomorphogenic Response to UV-B Radiation. *Photochem. Photobiol.* **2004,** *79,* 307–315.

Lengfelder, E.; Cadenas, E.; Sies, H. Effect of DABCO (1,4-diazabicyclo[2,2,2]-octane) on Singlet Oxygen Monomol (1270 nm) and Dimol (634 and 703 nm) Emission. *FEBS Lett.* **1983,** *164* (2), 366–370.

Li, Y.; Ye, W.; Wang, M.; Yan, X. Climate Change and Drought: A Risk Assessment of Crop-Yield Impacts. *Climate Res.* **2009,** *39,* 31–46.

Lichtenthaler, H. K.; Babani, F. Light Adaptation and Senescence of the Photosynthetic Apparatus. Changes in Pigment Composition, Chlorophyll Fluorescence Parameters and Photosynthetic Activity. In: *Chlorophyll a Fluorescence: A Signature of Photosynthesis*; Papageorgiou, G. C., Ed.; Springer: Dordrecht, 2004; Vol 19, pp 713–736.

Lichtenthaler, H. K.; Burgstahler, R.; Buschmann, C.; Meier, D.; Prenzel, U.; Schönthal, A. Effect of High Light and High Light Stress on Composition, Function and Structure of the Photosynthetic Apparatus. In: *Effects of Stress on Photosynthesis*; Marcelle, R. Eds.; Nijhoff, The Hague, 1983, pp 353–70.

Lichtenthaler, H. K.; Burkart, S. Photosynthesis and High Light Stress. *Bulg. J. Plant Physiol.* **1999,** *25,* 3–16.

Liochev, S. I. The Mechanism of "Fenton-Like" Reactions and their Importance for Biological Systems. A Biologist's View. In: *Metals in Biological Systems*; Sigel, A., Sigel, H., Eds.; Marcel Dekker, Inc.: New York, 1999; Vol 36, pp 1–39.

Liou, G.-Y.; Storz, P. Reactive Oxygen Species in Cancer. *Free Radic. Res.* **2010,** *44*. doi:10.3109/10715761003667554.

Liszkay, A.; van der Zalm, E.; Schopfer, P. Production of Reactive Oxygen Intermediates ($O_2^{\cdot-}$, H_2O_2, and $^{\cdot}OH$) by Maize Roots and their Role in Wall Loosening and Elongation Growth. *Plant Physiol.* **2004,** *136*, 3114–3123.

Litwin, J. A. Histochemistry and Cytochemistry of 3,3′-Diaminobenzidine. A Review. *Folia Histochem. Cytochem.* **1979,** *17*, 3–28.

Liu, Y. H.; Offler, C. E.; Yong-Ling, R. Simple, Rapid, and Reliable Protocol to Localize Hydrogen Peroxide in Large Plant Organs by DAB-Mediated Tissue Printing. *Front. Plant Sci.* **2014,** *5*.

Lv, G.; Hu, W.; Kang, Y.; Liu, B.; Li, L.; Song, J. Root Water Uptake Model Considering Soil Temperature. *J. Hydrol. Eng.* **2012,** *18*, 394–400.

Macpherson, A. N.; Telfer, A.; Barber, J.; Truscott, T. G. Direct Detection of Singlet Oxygen from Isolated Photosystem II Reaction Centres. *Biochim. Biophys. Acta* **1993,** *1143*, 301–309.

Mahajan, S.; Tuteja, N. Cold, Salinity and Drought Stresses: An Overview. *Archiv. Biochem. Biophys.* **2005,** *444* (2), 139–158.

Manivannan, P.; Jaleel, C. A.; Somasundaram, R.; Panneerselvam, R. Osmoregulation and Antioxidant Metabolism in Drought-Stressed *Helianthus annuus* under Triadimefon Drenching. *C. R. Biol.* **2008,** *331*, 418–425.

Mattila, H.; Khorobrykh, S.; Havurinne, V.; Tyystjärvi, E. Reactive Oxygen Species: Reactions and Detection from Photosynthetic Tissues. *J. Photochem. Photobiol. B* **2015,** *152*, 176–214.

Mehrdad, Z.; Noll, A.; Grabner, E.-W.; Schmidt, R. Sensitization of Singlet Oxygen via Encounter Complexes and via Exciplexes of π–π* Triplet Excited Sensitizers and Oxygen. *Photochem. Photobiol. Sci.* **2002,** *1*, 263–269.

Minnich, M. M.; McBride, M. B.; Chaney, R. L. Copper Activity in Soil Solution. II. Relation to Copper Accumulation in Young Snapbeans. *Soil Sci. Soc. AM. J.* **1987,** *51*, 573–578.

Mishra, A. K.; Singh, V. P. A Review of Drought Concepts. *J. Hydrol.* **2010,** *391*, 202–216.

Mishra, V.; Cherkauer, K. A. Retrospective Droughts in the Crop-Growing Season: Implications to Corn and Soybean Yield in the Midwestern United States. *Agric. Forest. Meteorol.* **2010,** *150*, 1030–1045.

Mittler, R. Oxidative Stress, Antioxidants and Stress Tolerance. *Trends Plant Sci.* **2002,** *7*, 405–410.

Mittler, R.; Vanderauwera, S.; Gollery, M.; Breusengem, F. V. Reactive Oxygen Gene Network of Plants. *Trends Plant Sci.* **2004,** *9*, 490–498.

Miyamoto, S.; Martinez, G. R.; Rettori, D.; Augusto, O.; Medeiros, M. H. G.; Mascio, P. D. Linoleic Acid Hydroperoxide Reacts with Hypochlorous Acid, Generating Peroxyl Radical Intermediates and Singlet Molecular Oxygen. *Proc. Natl Acad. Sci. U.S.A.* **2006,** *103*, 293–298.

Miyamoto, S.; Martinez, G. R.; Medeiros, M. H. G.; Mascio, P. D. Singlet Molecular Oxygen Generated from Lipid Hydroperoxide by the Russell Mechanism: Studies Using ^{18}O-Labeled Linoleic Acid Hydroperoxide and Monomol Light Emission Measurements. *J. Am. Chem. Soc.* **2003,** *125*, 6172–6179.

Miyamoto, S.; Ronsein, G. E.; Prado, F. M.; Uemi, M.; Correa, T. C.; Toma, I. N.; Bertolucci, A.; Oliveira, M. C. B.; Motta, F. D.; Medeiros, M. H. G.; Mascio, P. D. Biological Hydroperoxides and Singlet Molecular Oxygen Generation. *IUBMB Life* **2007,** *59*, 322–331.

Montgomery, J.; Ste-Marie, L.; Boismenu, D.; Vachon, L. Hydroxylation of Aromatic Compounds as Indices of Hydroxyl Radical Production: A Cautionary Note Revisited. *Free Radic. Biol. Med.* **1999,** *19*, 927–933.

Møller, I. M. Plant Mitochondria and Oxidative Stress: Electron Transport, NADPH Turnover, and Metabolism of Reactive Oxygen Species. *Annu. Rev. Plant Physiol. Plant Mol. Biol.* **2001,** *52*, 561–591.

Mubarakshina, M. M.; Ivanov, B. N. The Production and Scavenging of Reactive Oxygen Species in the Plastoquinone Pool of Chloroplast Thylakoid Membranes. *Physiol. Plant.* **2010,** *140*, 103–110.

Murata, N.; Allakhverdiev, S. I.; Nishiyama, Y. The Mechanism of Photoinhibition In Vivo: Re-evaluation of the Roles of Catalase, α-Tocopherol, Non-photochemical Quenching, and Electron Transport. *Biochim. Biophys. Acta* **2012,** *1817*, 1127–1133.

Murata, N.; Takahashi, S.; Nishiyama, Y.; Allakhverdiev, S. I. Photoinhibition of Photosystem II under Environmental Stress. *Biochim. Biophys. Acta* **2007,** *1767*, 414–421.

Murphy, D. M. EPR (Electron Paramagnetic Resonance) Spectroscopy of Polycrystalline Oxide Systems. In: *Metal Oxide Catalysis*; Jackson, S. D., Hargreaves, J. S. J., Eds.; Wiley-VCH Verlag: Weinheim, 2009b.

Murphy, M. P. How Mitochondria Produce Reactive Oxygen Species. *Biochem. J.* **2009a,** *417*, 1–13.

Murphy, T. M. Membranes as Targets of Ultraviolet Radiation. *Physiol. Plant.* **1983,** *58* (3), 381–388.

Navari-Izzo, F.; Pinzino, C.; Quartacci, M. F.; Sgherri, C. L. M. Superoxide and Hydroxyl Radical Generation, and Superoxide Dismutase in PSII Membrane Fragments from Wheat. *Free Radic. Res.* **1999,** *33*, 3–9.

Nawkar, G. M.; Maibam, P.; Park, J. H.; Sahi, V. P.; Lee, S. Y.; Kang, H. O. UV-Induced Cell Death in Plants. *Int. J. Mol. Sci.* **2013,** *14*, 1608–1628.

Nemhauser, J.; Chory, J. Photomorphogenesis. *The Arabidopsis Book*; 2002; pp 1–12. http://www.ncbi.nlm.nih.gov/pmc/articles/PMC3243328/pdf/tab.0054.pdf.

Noctor, G.; De Paepe, R.; Foyer, C. H. Mitochondrial Redox Biology and Homeostasis in Plants. *Trend. Plant Sci.* **2007,** *12*, 125–134.

Osakabe, Y.; Osakabe, K.; Shinozaki, K.; Tran, L. S. P. Response of Plants to Water Stress. *Front. Plant Sci.* **2014,** *5*, 86.

Petrov, V.; Hille, J.; Mueller-Roeber, B.; Gechev, T. S. ROS-Mediated Abiotic Stress-Induced Programmed Cell Death in Plants. *Front. Plant Sci.* **2015,** *6*, 69.

Pietikäinen, J.; Pettersson, M.; Bååth, E. Comparison of Temperature Effects on Soil Respiration and Bacterial and Fungal Growth Rates. *FEMS Microbiol. Ecol.* **2005,** *52*, 49–58.

Pospíšil, P. Molecular Mechanism of Production and Scavenging of Reactive Oxygen Species by Photosystem II. *Biochim. Biophys. Acta* **2012,** *1817*, 218–231.

Pospíšil, P. Production of Reactive Oxygen Species by Photosystem II. *Biochim. Biophys. Acta* **2009,** *1787*, 1151–1160.

Pospíšil, P.; Arató, A.; Krieger-Liszkay, A.; Rutherford, A. W. Hydroxyl Radical Generation by Photosystem II. *Biochemistry* **2004,** *43*, 6783–6792.

Pospíšil, P.; Prasad, A.; Rác, M. Role of Reactive Oxygen Species in Ultra-Weak Photon Emission. *J. Photochem. Photobiol. B* **2014,** *139*, 11–23.

Pospíšil, P.; Šnyrychová, E.; Kruk, J.; Strzalka, K.; Nauš, J. Evidence that Cytochrome b_{559} Is Involved in Superoxide Production in Photosystem II: Effect of Synthetic Short-chain Plastoquinones in a Cytochrome b_{559} Tobacco Mutant. *Biochem. J.* **2006,** *397*, 321–327.

Pospíšil, P.; Šnyrychová, I.; Nauš, J. Dark Production of Reactive Oxygen Species in Photosystem II Membrane Particle at Elevated Temperature: EPR Spin-Trapping Study. *Biochim. Biophys. Acta* **2007,** *1767*, 854–859.

Pogozelski, W. K.; Tullius, T. D. Oxidative Strand Scission of Nucleic Acids: Routes Initiated by Hydrogen Abstraction from the Sugar Moiety. *Chem. Rev.* **1998,** *98*, 1089–108.

Prasad, A.; Kumar, A.; Suzuki, M.; Kikuchi, H.; Sugai, T.; Kobayashi, M.; Pospíšil, P.; Tada, M.; Kasai, S. Detection of Hydrogen Peroxide in Photosystem II (PSII) Using Catalytic Amperometric Biosensors. *Front. Plant Sci.* **2015,** *6*, 682.

Prasad, A.; Pospíšil, P. Linoleic Acid-Induced Ultra-Weak Photon Emission from *Chlamydomonas reinhardtii* as a Tool for Monitoring of Lipid Peroxidation in the Cell Membranes. *PLoS ONE* **2011a,** *6*, e22345.

Prasad, A.; Pospíšil, P. Two-Dimensional Imaging of Spontaneous Ultra-Weak Photon Emission from the Human Skin: Role of Reactive Oxygen Species. *J. Biophotonics* **2011b,** *4*, 840–849.

Prasad, A.; Pospíšil, P. Photon Source within the Cell. In: *Fields of the Cell*; Fels, D., Cifra, M., Scholkmann, F., Eds.; Research Signpost: India, 2015.

Rác, M.; Sedlářová, M.; Pospíšil, P. The formation of electronically excited species in the human multiple myeloma cell suspension. *Sci. Rep.* **2015,** *8882*.

Rappaport, F.; Diner, B. A. Primary Photochemistry and Energetics Leading to the Oxidation of the $(Mn)_4Ca$ Cluster and to the Evolution of Molecular Oxygen in Photosystem II. *Coord. Chem. Rev.* **2008,** *252*, 259–272.

Rastogi, A.; Yadav, D. K.; Szymańska, R.; Kruk, J.; Sedlářová, M.; Pospíšil, P. Singlet Oxygen Scavenging Activity of Tocopherol and Plastochromanol in *Arabidopsis thaliana*: Relevance to Photooxidative Stress. *Plant Cell Environ.* **2014,** *37*, 392–401.

Ray, P. D.; Huang, B. W.; Tsuji, Y. Reactive Oxygen Species (ROS) Homeostasis and Redox Regulation in Cellular Signaling. *Cell Signal.* **2012,** *24*, 981–990.

Reddy, A. R.; Chaitanya, K. V.; Vivekanandan, M. Drought-Induced Responses of Photosynthesis and Antioxidant Metabolism in Higher Plants. *J. Plant Physiol.* **2004,** *161*, 1189–1202.

Rehman, A. U.; Cser, K.; Sass, L.; Vass, I. Characterization of singlet oxygen production and its involvement in photodamage of Photosystem II in the cyanobacterium *Synechocystis* PCC 6803 by Histidine-Mediated Chemical Trapping. *Biochim. Biophys. Acta* **2013,** *1827*, 689–698.

Rejeb, I.; Victoria, P.; Brigitte, M. M. Plant Responses to Simultaneous Biotic and Abiotic Stress: Molecular Mechanisms. *Plants* **2014,** *3* (4), 458–475.

Renger, G.; Holzwarth, A. R. Primary Electron Transfer. In: *Photosystem II: The Light-Driven Water: Plastoquinone Oxidoreductase*; Wydrzynski, T. J., Satoh, K., Eds.; Springer: Dordrecht, 2005; pp 139–175.

Rinalducci, S.; Pedersen, J. Z.; Zolla, L. Formation of Radicals from Singlet Oxygen Produced during Photoinhibition of Isolated Light Harvesting Proteins of Photosystem II. *Biochim. Biophys. Acta* **2004,** *1608*, 63–73.

Rozema, J.; van de Staaij, J.; Björn, L. O.; Caldwell, M. UV-B as an Environmental Factor in Plant Life: Stress and Regulation. *Trends Ecol. Evol.* **1997,** *12*, 22–28.

Ruiz-Lozano, J. M.; Porcel, R.; Azcón, C.; Aroca, R. Regulation by Arbuscular Mycorrhizae of the Integrated Physiological Response to Salinity in Plants: New Challenges in Physiological and Molecular Studies. *J. Exp. Bot.* **2012,** *63*, 4033–4044.

Russell, G. A. Deuterium-Isotope Effects in the Autooxidation of Aralkyl Hydrocorbons—Mechanism of Interaction of Peroxy Radicals. *J. Am. Chem. Soc.* **1957,** *79*, 3871–3877.

Sairam, R. K.; Tyagi, A. Physiology and Molecular Biology of Salinity Stress Tolerance in Plants. *Curr. Sci.* **2004,** *86*, 407–421.

Samuni, A.; Carmichael, A. J.; Russo, A.; Mitchell, J. B.; Riesz, P. On the Spin Trapping and ESR Detection of Oxygen-Derived Radicals Generated Inside Cells. *Proc. Natl. Acad. Sci. USA* **1986,** *83* (20), 7593–7597.

Santabarbara, S.; Cazzalini, I.; Rivadossi, A.; Garlaschi, F. M.; Zucchelli, G.; Jennings, R. C. Photoinhibition In Vivo and In Vitro Involves Weakly Coupled Chlorophyll Protein Complexes. *Photochem. Photobiol.* **2002,** *75*, 613–618.

Santabarbara, S.; Neverov, K. V.; Garlaschi, F. M.; Zucchelli, G.; Jennings, R. C. Involvement of Uncoupled Antenna Chlorophylls in Photoinhibition in Thylakoids. *FEBS Lett.* **2001,** *491*, 109–113.

Sawa, M.; Kay, S. A. GIGANTEA Directly Activates Flowering Locus T in *Arabidopsis thaliana*. *Proc. Natl. Acad. Sci. USA* **2011,** *108*, 11698–11703.

Schumacker, P. T. Reactive Oxygen Species in Cancer Cells: Live by the Sword, Die by the Sword. *Cancer Cell* **2006,** *10*, 175–176.

Serrano, J.; Jové, M. Boada, J.; Bellmunt, M. J.; Pamplona, R. Dietary Antioxidants Interfere with Amplex Red-Coupled-Fluorescence Assays. *Biochem. Biophys. Res. Commun.* **2009,** *388*, 443–449.

Sharma, P.; Jha, A. B.; Dubey, R. S.; Pessarakli, M. Reactive Oxygen Species, Oxidative Damage, and Antioxidative Defense Mechanism in Plants under Stressful Conditions. *J. Bot.* **2012,** Article ID 217037, 26 pp.

Shen, X.; van Wijk, R. Biophotonics: Optical Science and Engineering for the 21st Century. Springer Press, Springer-Verlag USA, 2005.

Sinha, R. K.; Komenda, J.; Knoppová, J.; Sedlářová, M.; Pospíšil, P. Small CAB-Like Proteins Prevent Formation of Singlet Oxygen in the Damaged Photosystem II Complex of the Cyanobacterium *Synechocystis* sp. PCC 6803. *Plant Cell Environ.* **2011,** *35*, 806–818.

Sirichandra, C.; Wasilewska, A.; Vlad, F.; Valon, C.; Leung, J. The Guard Cell as a Single-Cell Model towards Understanding Drought Tolerance and Abscisic Acid Action. *J. Exp. Bot.* **2009,** *60*, 1439–1463.

Ślesak, I.; Libik, M.; Karpinska, B.; Karpinski, S.; Miszalski, Z. The Role of Hydrogen Peroxide in Regulation of Plant Metabolism and Cellular Signalling in Response to Environmental Stresses. *Acta Biochim. Pol.* **2007,** *54*, 39–50.

Šnyrychová, I.; Ayaydin, F.; Hideg, E. Detecting Hydrogen Peroxide in Leaves In Vivo—A Comparison of Methods. *Physiol. Plant.* **2009,** *135*, 1–18.

Šnyrychová, I.; Pospíšil, P.; Nauš, J. Reaction Pathways Involved in the Production of Hydroxyl Radicals in Thylakoid Membrane: EPR Spin-Trapping Study. *Photochem. Photobiol. Sci.* **2006,** *5*, 472–476.

Stapleton, A. E. Ultraviolet Radiation and Plants: Burning Questions. *Plant Cell* **1992,** *4*, 1353–1358.

Stark, G. Functional Consequences of Oxidative Membrane Damage. *J. Membrane Biol.* **2005,** *205*, 1–16.

Steffen-Heins, B.; Steffen, A. EPR Spectroscopy and Its Use In Planta—A Promising Technique to Disentangle the Origin of Specific ROS. *Front. Environ. Sci.* **2015,** *3*, Article 15.

Suzuki, N.; Koussevitzky, S.; Mittler, R.; Miller, G. ROS and Redox Signalling in Response of Plants to Abiotic Stress. *Plant Cell Environ.* **2012,** *35*, 259–270.

Taiz, L.; Zeiger, E. *Plant Physiology*, fourth ed. Sinauer Associates: Sunderland MA, 2006.

Telfer, A. Singlet Oxygen Production by PSII under Light Stress: Mechanism, Detection and the Protective Role of β-Carotene. *Plant Cell Physiol.* **2014,** *55*, 1216–1223.

Telfer, A.; Bishop, S. M.; Phillips, D.; Barber, J. Isolated Photosynthetic Reaction Center of Photosystem II as a Sensitizer for the Formation of Singlet Oxygen. Detection and Quantum Yield Determination Using a Chemical Trap Technique. *J. Biol. Chem.* **1994,** *269*, 13244–13253.

Terashima, I.; Noguchi, K.; Itoh-Nemoto, T.; Park, Y. M.; Kubo, A.; Tanaka, K. The Cause of PSI Photoinhibition at Low Temperatures in Leaves of *Cucumis sativus*, a Chilling-Sensitive Plant. *Physiol. Plant.* **1998,** *103*, 296–303.

Thannickal, V. J.; Fanburg, B. L. Reactive Oxygen Species in Cell Signaling. *Am. J. Physiol. Lung Cell Mol. Physiol.* **2000,** *279*, 1005–1028.

Tiwari, A.; Pospíšil, P. Superoxide Oxidase and Reductase Activity of Cytochrome b_{559} in Photosystem II. *Biochim. Biophys. Acta* **2009,** *1787*, 985–994.

Tomo, T.; Kusakabe, H.; Nagao, R.; Ito, H.; Tanaka, A.; Akimoto, S.; Mimuro, M.; Okazaki, S. Luminescence of Singlet Oxygen in Photosystem II Complexes Isolated from Cyanobacterium *Synechocystis* sp. PCC6803 Containing Monovinyl or Divinyl Chlorophyll *a*. *Biochim. Biophys. Acta* **2012,** *1817*, 1299–1305.

Tóth, S. Z.; Schansker, G.; Garab, G.; Strasser, R. J. Photosynthetic Electron Transport Activity in Heat-Treated Barley Leaves: The Role of Internal Alternative Electron Donors to Photosystem II. *Biochim. Biophys. Acta* **2007,** *1767*, 295–305.

Triantaphylidès, C.; Havaux, M. Singlet Oxygen in Plants: Production, Detoxification and Signaling. *Trends Plant Sci.* **2009,** *14*, 219–228.

Triantaphylidès, C.; Krischke, M.; Hoeberichts, F. A.; Ksas, B.; Gresser, G.; Havaux, M.; Van Breusegem, F.; Mueller, M. J. Singlet Oxygen is the Major Reactive Oxygen Species Involved in Photooxidative Damage to Plants. *Plant Physiol.* **2008,** *148*, 960–968.

Tripathy, B. C.; Oelmüller, R. Reactive Oxygen Species Generation and Signaling in Plants. *Plant Signal. Behav.* **2012,** *7*, 1621–1633.

Turrens, J. F. Mitochondrial Formation of Reactive Oxygen Species. *J. Physiol.* **2003,** *552*, 335–344.

Tyystjärvi, E. Photoinhibition of Photosystem II and Photodamage of the Oxygen Evolving Manganese Cluster. *Coord. Chem. Rev.* **2008,** *252*, 361–376.

Tyystjärvi, E. Photoinhibition of Photosystem II. *Int. Rev. Cell Mol. Biol.* **2013,** *300*, 243–303.

Uttara, B.; Singh, A. V.; Zamboni, P.; Mahajan, R. T. Oxidative Stress and Neurodegenerative Diseases: A Review of Upstream and Downstream Antioxidant Therapeutic Options. *Curr. Neuropharmacol.* **2009,** *7*, 65–74.

Vanhala, P.; Karhu, K.; Tuomi, M.; Björklöf, K.; Fritze, H.; Liski, J. Temperature Sensitivity of Soil Organic Matter Decomposition in Southern and Northern Areas of the Boreal Forest Zone. *Soil Biol. Biochem.* **2008,** *40*, 1758–1764.

Vass, I. Molecular Mechanisms of Photodamage in the Photosystem II Complex. *Biochim. Biophys. Acta* **2012,** *1817*, 1127–1133.

Vass, I.; Aro, E.-M. Photoinhibition of Photosynthetic Electron Transport. In: *Primary Processes in Photosynthesis, Basic Principles and Apparatus*; Renger, G., Eds.; The Royal Society of Chemistry: Cambridge, 2007; pp 393–425.

Venkataraman, S.; Schafer, F. Q.; Buettner, G. R. Detection of Lipid Radicals Using EPR. *Antioxid. Redox Signal.* **2004,** *6*, 619–629.

Verma, S.; Fedak, P. W. M.; Weisel, R. D.; Butany, J.; Rao, V.; Maitland, A.; Li, R. K.; Dhillon, B.; Yau, T. M. Fundamentals of Reperfusion Injury for the Clinical Cardiologist. *Circulation* **2002,** *105*, 2332–2336.

Votyakova, T. V.; Reynolds, I. J. Detection of Hydrogen Peroxide with Amplex Red: Interference by NADH and Reduced Glutathione Auto-oxidation. *Arch. Biochem. Biophys.* **2004,** *431*, 138–144.

Wahid, A.; Gelani, S.; Ashraf, M.; Foolad, M. R. Heat Tolerance in Plants: An Overview. *Environ. Exp. Bot.* **2007,** 61, 199–223.

Wang, Y.; Zhilong, B.; Ying, Z.; Jian, H. Analysis of Temperature Modulation of Plant Defense against Biotrophic Microbes. *Mol. Plant Microbe Interact.* **2009,** *22*, 498–506.

Waris, G.; Ahsan, H. Reactive Oxygen Species: Role in the Development of Cancer and Various Chronic Conditions. *J. Carcinog.* **2006,** *5*, 14.

Winterbourn, C. C.; Metodiewa, D Reactivity of Biologically Important Thiol Compounds with Superoxide and Hydrogen Peroxide. *Free Radic. Biol. Med.* **1999,** *27*, 322–328.

Wu, C. C.; Bratton, S. B. Regulation of the Intrinsic Apoptosis Pathway by Reactive Oxygen Species. *Antioxid. Redox Signal.* **2013,** *19*, 546–558.

Wu, S. Effect of Manganese Excess on the Soybean Plant Cultivated under Various Growth Conditions. *J. Plant Nutr.* **1994,** *17*, 993–1003.

Yadav, D. K.; Pospíšil, P. Evidence on the Formation of Singlet Oxygen in the Donor Side Photoinhibition of Photosystem II: EPR Spin-Trapping Study. *PLoS ONE* **2012a,** *7*, e45883.

Yadav, D. K.; Pospíšil, P. Role of Chloride Ion in Hydroxyl Radical Production in PSII under Heat Stress: Electron Paramagnetic Resonance Spin-Trapping Study. *J. Bioenerg. Biomembr.* **2012b,** *44*, 365–372.

Yadav, D. K.; Pospíšil, P. Singlet Oxygen Scavenging Activity of Plastoquinol in Photosystem II of Higher Plants: Electron Paramagnetic Resonance Spin-Trapping Study. *Biochim. Biophys. Acta* **2010,** *1797*, 1807–1811.

Yadav, D. K.; Prasad, A.; Kruk, J.; Pospíšil, P. Evidence for the Involvement of Loosely Bound Plastosemiquinones in Superoxide Anion Radical Production in Photosystem II. *PLoS ONE* **2014,** *9*, e115466.

Yadav, S. K. Heavy Metals Toxicity in Plants: An Overview on the Role of Glutathione and Phytochelatins in Heavy Metal Stress Tolerance of Plants. *S. Afr. J. Bot.* **2010,** *76*, 167–179.

Yamashita, A.; Nijo, N.; Pospíšil, P.; Morita, N.; Takenaka, D.; Aminaka, R.; Yamamoto, Y. Quality Control of Photosystem II: Reactive Oxygen Species Are Responsible for the Damage to Photosystem II under Moderate Heat Stress. *J Biol. Chem.* **2008,** *283*, 28380–28391.

Yasukawa, T.; Uchida, I.; Matsue, T. Microamperometric Measurements of Photosynthetic Activity in a Single Algal Protoplast. *Biophys. J.* **1999,** *76*, 1129–1135.

Zhang, S.; Weng, J.; Tu, T.; Yao, S.; Xu, C. Study on the Photo-generation of Superoxide Radicals in Photosystem II with EPR Spin Trapping Techniques. *Photosynth. Res.* **2003,** *75*, 41–48.

Zhou, M.; Diwu, Z.; Panchuk-Voloshina, N.; Haugland, R. P. A Stable Nonfluorescent Derivative of Resorfin for the Fluorometric Determination of Trace Hydrogen Peroxide: Applications in Detecting the Activity of Phagocyte NADPH Oxidase and Other Oxidases. *Anal. Biochem.* **1997,** *253*, 162–168.

Zhu, B.; Alva, A. K. Effect of pH on Growth and Uptake of Copper by Swingle Citrumelo Seedlings. *J. Plant Nutr.* **1993,** *16*, 1837–1845.

Zlatev, Z.; Lidon, F. C. An Overview on Drought Induced Changes in Plant Growth, Water Relations and Photosynthesis. *Emirates J. Food Agric.* **2012,** *24*, 57–72.

Zolla, L.; Rinalducci, S. Involvement of Active Oxygen Species in Degradation of Light Harvesting Proteins under Light Stresses. *Biochemistry* **2002,** *42*, 14391–14402.

PART VI

Plant Disease Diagnostics and Management

CHAPTER 13

BIOTECHNOLOGICAL APPROACHES FOR PLANT DISEASE DIAGNOSIS AND MANAGEMENT

RAVI RANJAN KUMAR[1*], GANESH PATIL[2], KUMARI RAJANI[3], SHAILESH YADAV[4], NIMMY M. S.[5], and VINOD KUMAR[1]

[1]*Department of Molecular Biology and Genetic Engineering, Bihar Agricultural University, Sabour 813210, Bihar, India*

[2]*Vidya Pratisthan's College of Agriculture Biotechnology, Vidyanagari, Baramati 413133, Maharashtra, India*

[3]*Department of Seed Science and Technology, Bihar Agricultural University, Sabour 813210, Bihar, India*

[4]*International Rice Research Institute (IRRI), ICRISAT, Hyderabad 502324, Telangana, India*

[5]*National Research Centre on Plant Biotechnology, New Delhi 110012, India*

**Corresponding author. E-mail: ravi1709@gmail.com*

CONTENTS

ABSTRACT

Plant pathogens cause heavy yield loss in cultivated crops and plants which leads to decremented production and productivity. In order to enhance the production and productivity, there is much necessity of improving crop varieties against various biotic stresses. The first step of improvement of crop/plant species is the proper diagnosis of the disease and identification of its causative agent. Better diagnosis and identification of microorganism would pave a way to formulate the strategy for its control and management. At present, there are several methods available for disease diagnosis, each having their own potential and shortcomings. Traditionally, available detection and diagnostic techniques for plant pathogenic microorganisms have been morphological, microscopic and biochemical characterization. Serological techniques like ELISA has sped up the accuracy of disease diagnosis. In the present era, the disease diagnosis is dominated by molecular methods comprising nucleic acid detection, molecular markers and serology-based diagnostic applications comprising antigens and antibodies etc. Besides, recent advancements and applications of genomics and proteomics offer new age solutions. In this chapter, the method of disease diagnosis and the methods involved in improvement of the crops/plants through conventional as well as molecular strategies like tissue culture, molecular breeding, transgenic, RNAi are discussed in details.

13.1 INTRODUCTION

Plant pathogenic microorganisms such as fungi, bacteria, phytoplasmas, viruses, and viroids cause harmful and economically important diseases in a very broad range of plant species worldwide. Damage is often sufficient to cause significant yield losses in cultivated plants. The two major effects of pathogens on agriculture are decremented production and, in a less direct way, the need of implementation of extravagant management, control procedures, and strategies. In addition, efficient registered products for the chemical control of bacteria are lacking, and there is no chemical control available for viruses. Consequently, prevention is essential to avoid the dissemination of the pathogens (Alvarèz, 2004; De Boer et al., 2007; Martin et al., 2000; López et al., 2003). The prevention measures demand efficient pathogen detection methods with high sensitivity, specificity, and reliability, because many phytopathogenic microorganisms can remain latent in subclinical infections, and/or in below detectable numbers (Grey and Steck, 2001;

Ordax et al., 2006). Accurate detection of phytopathogenic organisms is crucial for virtually all aspects of plant pathology, from basic research on the biology of pathogens to the control of the diseases they cause. Moreover, the need for rapid techniques of high accuracy is the need of the hour (López et al., 2009). In this review, we discuss implications of recent biotechnological interventions for detection of plant pathogens, disease diagnosis of agriculture crops. Emphasis is on the recent advances in the field of molecular detection methods, genomics, and proteomics-based applications to plant disease diagnosis of field crops.

Standard protocols for detection of plant pathogens are based on isolation and further identification are time-consuming and not always sensitive and specific enough. Consequently, they are not suited for routine analysis of a large number of samples. Other confines are the low reproducibility of identification by phenotypic traits, frequent lack of phylogenetic significance and false negatives due to stressed or injured microorganisms, or those in the viable but nonculturable state, which escape from isolation. There are currently many methods, which have been used to detect and/or characterize specific viral, viroids, or graft-transmissible virus-like associated pathogens in plant material. The most frequently employed are biological indexing and serological tests. Biological assays were developed first and are still in widespread use, because they are simple; and require minimal knowledge of the pathogen. Furthermore, biological indexing is still the only method of choice to detect uncharacterized but graft transmissible agents. Its sensitivity is considered to be very high due to the viral multiplication in the host plant used as indicator.

Traditionally, available detection and diagnostic techniques for plant pathogenic microorganisms have been microscopic examination, isolation, biochemical characterization. However, the pathologists have turned their attention to new methods of detection such as enzyme-linked immunosorbent assay (ELISA) using polyclonal and/or monoclonal antibodies and bioassays and pathogenicity tests. For viruses and viroids, nucleic acid-based molecular polymerase chain reaction (PCR), electrophoresis, electron microscopy, and ELISA-based techniques have been the choice.

Accuracy of plant pathogen detection has greatly improved due to the development of serological techniques, especially ELISA (Clark and Adams, 1977). Applying ELISA to detection of viruses has revolutionized face of disease diagnosis, making the accurate analysis of large number of samples feasible, simpler, and with both low cost and high sensitivity. Polyclonal antibodies used earlier times represented problems of specificity. However, the availability of specific monoclonal and recombinant antibodies solved this problem (Terrada et al., 2000). Nevertheless, one of the major circumscriptions of this technique was its low sensitivity in plant reproductive stage because the titer of some viral pathogens is detected less in reproductive stage as compare plants vegetative periods.

Currently, plant disease diagnosis space is dominated by molecular diagnosis methods comprising nucleic acid detection and serology-based diagnostic applications comprising antigens and antibodies. Besides, recent advancements and applications of genomics and proteomics offer new age solutions, which are briefly discussed in this review. These -omics-based solutions have opened up new possibilities for the identification of new disease diagnostic targets in the form of up- or downregulation of genes at molecular level expression. Besides, expression of their pathogen/disease specific components could be identified as possible indicators of disease conditions. All these applications are briefly discussed in the chapter below.

13.2 DIAGNOSIS OF PLANT DISEASES

There are different approaches to diagnose plant disease problems. A plant disease is defined as anything that adversely affects plant health. The definition usually includes a persistent irritation resulting in damage to plant health. In precisely, definition also includes only those (living) things that replicate themselves and spread to adjacent plants. The living things include biological agents such as, bacteria, fungi, viruses, nematodes, etc. Plants damaged by macroscopic organisms, such as birds rodents and deer usually are not considered to be diseased.

The condition of a plant health is diagnosed by their symptoms and signs. The physical characteristics of disease expressed by the plant are called symptoms. It can include cankers, rots, chlorosis, necrosis, wilt, galls, and reduced growth, whereas signs are the physical evidence of the pathogen causing the disease which may include mycelia, bacterial slime, fungal spores, nematodes or insects presence, the presence of insect holes accompanied by sawdust or frass.

The diagnostic approaches differ from lab to lab. The simplest way to diagnose plant problems are by making a personal, onsite inspection. Subtle influences of the site, plant environment, and possible management practices can also be seen for better diagnosis.

13.2.1 CONVENTIONAL METHODS OF DISEASE DIAGNOSIS

Several methods have been reported till date to identify plant pathogens and each method is having its own advantages and limitations. In traditional or conventional method, identification of pathogens include visual appearance of symptoms which may be small necrotic or chlorotic spots called local lesions develop at the site of infection. Typical leaf symptoms of disease include chlorotic or necrotic lesions; mosaic patterns; yellowing; vein banding vein clearing; rolling; and curling leaf, stripes, or streaks. Typical symptoms are possible only after major damage has already been done to the crop and treatments have limited or of no use.

13.2.1.1 SYMPTOMOLOGY

Symptomology can be defined as the study of signs and symptoms of disease-affected plant based on its morphological appearances in comparison with the healthy plants. Two types of symptoms namely *external symptoms* and *internal symptoms* are observed. Plant growth, signs of yellowness, lack of pigmentation, stunted growth, abnormal physiological growth, wrinkling of leaves are the initial symptoms to identify diseased plants. After initial observation of symptoms, diseased plant specimen undergoes several laboratory tests for diagnosis purpose.

13.2.1.2 DIAGNOSING PLANT DISEASE PROBLEM

Diagnosis is the process of gathering information about a plant's health problem and determining the cause. Once the cause of disease is determined, it is then possible to find the solution. Sometimes, the primary cause of a problem is hidden by more obvious due to insufficient information. The success in diagnosing plant problems depends on the information available about the host plant, their plant problems in general, and the quality of information obtained from the farmers.

13.2.1.2.1 Basic Steps in Reaching a Diagnosis

13.2.1.2.1.1 Identification of Diseased Plant

The identification of infected plants is majorly based on the morphological and physiological responses of the plants. The plant identification skill should be better in order to diagnose a problem. Several literatures on plant pests and diseases are published which may help to differentiate between healthy and diseased plants. Several new diseases are arising due to environmental pressure which also needs to be identified by keen observation of symptoms.

13.2.1.2.1.2 Distribution Pattern of Plant Disease

The pattern of distribution of disease should be observed by disease distribution in the plants and their different plant parts, information on plant species/cultivars, the site where the plant is growing (orchard, garden, field, green house, etc.). The previous crop history of the site must be noted for disease diagnosis.

13.2.1.2.1.3 Symptoms and Signs of Diseased Plant

For a hypothetical diagnosis of diseases in plants, the symptoms and signs of the disease should be recorded carefully. Symptoms are the internal or external alterations of a plant in response to a disease-causing agent, for example, blight, canker, wilt, lesion, leaf spot, rot, gall, necrosis, witches broom, mosaic, chlorosis, etc., while sign is referred as the pathogen that causes the disease and produces a characteristic growth or structures on the diseased plant. Sclerotia, mushrooms, conks, mold, mildew, etc. are the examples of sign.

13.2.1.2.1.4 Tentative Diagnosis of Disease

On the basis of the knowledge of the plant and information from literatures, tentative diagnosis can be formulated. It will help for examination of the plant and aid in gathering relevant information.

13.2.1.2.1.5 Reconfirmation the Diagnosis

Once disease diagnosis is done, it is necessary to reconfirm it with experts. A successful plant disease diagnosis also depends on a combination of various factors: (1) the knowledge of the plant and its basic cultural requirements, and (2) recognition of the potential problems that might affect the plant health.

13.2.2 BIOTECHNOLOGICAL APPROACHES TO PLANT DISEASE DIAGNOSIS

In comparison with conventional diagnosis method, biotechnological techniques offer the promise of rapid and economical detection of plant diseases.

In fact, emergence and advancement of biotechnology have revolutionized the face of plant pathology. The following section gives overview of various biotechnological techniques in plant disease pathogenesis.

13.2.2.1 GEL ELECTROPHORESIS METHOD

Electrophoresis is the movement of molecules through a fluid or gel under the action of an electrical current. Traditionally, this technique is being used to separate and characterize biomolecules according to their size in conventional biochemistry and molecular biology labs. However, in the recent time this application has been extended to the field of plant pathology. Thus, extraction and electrophoretic separation of pathogen specific proteins from infected plant tissues has emerged as one of the most powerful techniques. Here by electrophoresis, we mainly refer to sodium dodecyl sulfate–polyacrylamide gel electrophoresis method (SDS–PAGE). By this method, disease-related entities, such as peptides, protein, antigens, and antibodies are electrophoresed, studied, and analyzed. Thus, it has emerged as one of the very promising and convenient methods for studying disease related components. In this method, cellular proteinacious mixture is obtained in well resolved and fixed form. This method is widely used for detection of plant pathogenic viruses. Structural components of viruses, such as surface glycoproteins, envelope, capsid, and core can be resolved and studied.

This resolved gel proteins can be further subjected to electrotransfer to nitrocellulose membrane by blotting and method called Western blotting. Thereby, the presence of pathogen-specific proteins (antigens) could be confirmed by reacting with known pathogen specific antibodies available in the market. This is an example of mechanism called as protein–protein interaction, giving strong evidence of disease and pathogen associated with it.

Two-dimensional gel electrophoresis is one of the recent variants of SDS–PAGE. Although laborious, it is one of the interesting and very informative methods for separating proteins. This method can be coupled in tandem with mass spectrometry to identify disease/pathogenesis-related proteins leading to novel biomarker discovery.

The diagnostic protocol by SDS–PAGE method consists of following steps:

- obtaining a diseased plant sample from an infected plant and healthy plant as control;
- extracting the proteins from the sample;

- sample preparation as per standard SDS–PAGE protocols;
- apply the prepared samples to PAGE, and electrophoresing the resultant protein samples;
- stain and destain the as per standard protocol; and
- the resultant electrophoretic migration pattern of disease infected and noninfected plant can be compared and studied further.

13.2.2.2 IMMUNOCHEMICAL METHODS

Immunochemical methods are used to track down various antigens and antibodies implicated in particular plant diseases. Immunochemical techniques provide extremely specific and sensitive method for studying structural and functional relationships of antigens derived from plant pathogenic microorganism or viruses. Antibodies specific for the topographical features of many type of biomolecules, including proteins and carbohydrates, polysaccharides, toxins, enzymes, and nucleic acids can be produced in immunized animals.

Plant virologists for decades have used serological techniques for the rapid identification and taxonomic classification of plant viruses and regarded as extremely sensitive and reliable methods for diagnosing virus infections. Various viral, bacterial, and fungal proteins comprise various immunoreactive determinants or epitopes, and therefore serve as antigens. Similar analyses have been made for bacterial and lesser extent with fungal diseases.

These series of immunoreactive substances involved in plant diseases are detected with the help of immunochemical techniques such as ELISA, immunoprecipitation and radioimmunoassay methods.

13.2.2.3 MOLECULAR DETECTION METHODS

Nucleic-acid-based detection methods are sensitive, specific, and allow molecular level identification of microorganisms. In plant pathology, molecular detection methods are the most frequently used techniques for detection of pathogen by its nucleic acid content. PCR involving known pathogen specific set of primers is widely used method for this purpose. Amplification of pathogen-specific genes could be achieved using thermal cycler. Detection of amplified DNA bands in agarose gel electrophoresis is the indication of pathogen specific nucleic acid. Molecular hybridization tests are the second frequently used methods for this purpose. Multiple variants of PCR have emerged and being increasingly used for plant-disease diagnosis purpose.

Compared to traditional methods, PCR offers several advantages, because organisms do not need to be cultured before their detection. It affords high sensitivity, enabling a single target molecule to be detected in a complex mixture, and it is also rapid and versatile. In fact, the different variants of PCR have increased the accuracy of detection and diagnosis and opened new insights into our knowledge of the ecology and population dynamics of many pathogens, providing a valuable tool for basic and applied studies in plant pathology (López et al., 2009). Detection of DNA provides evidence for the presence/absence of targets; rRNA is an indicator of cell activity or viability, and mRNA signals specific activity and expression of certain metabolic processes (Chandler and Jarrell, 2005).

However, nucleic-acid-extraction protocols are usually necessary to obtain a successful result when processing plant samples by molecular methods. Although in the recent times, colony PCR have emerged as an alternative and is being increasingly used without necessity of nucleic acid extraction procedures. In addition, knowledge of nucleotide sequence of pathogen is must and also primer design plays a crucial step for PCR-based diagnostic step.

Molecular approaches developed over the last 10 years to detect many fungi, bacteria, viruses, viroids, spiroplasma, and phytoplasmas in plant or environmental samples (Alvarez, 2004; Louws et al., 1999). In fact, molecular diagnosis represents the most frequently used method for all plant-type plant pathogens. According to Bonants et al. (2005) and López et al. (2009), various molecular methods emerged in plant disease diagnosis can be grouped as follows (Table 13.1).

TABLE 13.1 RNA and DNA-Based Molecular Methods for Disease Diagnosis.

RNA Level	DNA Level
Reverse transcriptase PCR (RT-PCR)	Fluorescence in situ hybridization (FISH)
Nucleic acid sequence-based amplification (NASBA)	Southern hybridization
AmpliDet RNA	Conventional PCR and its variants such as nested PCR, cooperative PCR, multiplex PCR, real-time PCR

Technological advances in PCR-based methods enable fast, accurate detection, characterization, and quantification of plant pathogens and are now being applied to solve practical problems. For example, the use of molecular techniques in bacterial taxonomy allows different taxa of etiologically significant sublevel species differentiation of bacteria (De Boer et al., 2007). Therefore, molecular diagnostics can provide the degree of discrimination needed to detect and monitor plant diseases, which is not always obtained by other types of analysis.

An interesting application of PCR-based detection method is PCR-based virus indexing method. This method serves as benchmark for confirming presence or absence of viruses in plant diseases. Virus indexing method is applicable in case of Banana bunchy top virus (BBTV) disease (Selvarajan et al., 2011; Shelake et al., 2013). BBTV is a serious cause of concern for yield losses in Banana crop. In order to ensure the supply of disease-free planting material, government agencies have made virus indexing as mandatory step before release of planting material for nurseries involved in producing and supplying tissue-culture-based plantlets (Selvarajan et al., 2011). This example demonstrate changing scenario in plant disease diagnostics.

13.2.2.4 GENOMICS METHODS

Nowadays, genomics-based applications are increasingly being used in plant disease diagnosis. An example of a computational genomics pipeline was used to compare sequenced genomes of *Xanthomonas* spp. and to rapidly identify unique regions for development of highly specific diagnostic markers (Lang et al., 2010). A suite of diagnostic primers was selected to monitor diverse *loci* and to distinguish the rice bacterial blight (BB) and bacterial leaf streak pathogens, *Xanthomonas oryzae* pv. *oryzae* and *X. oryzae* pv. *oryzicola*, respectively. A subset of these primers was combined into a multiplex PCR set that accurately distinguished the two rice pathogens in a survey of a geographically diverse collection of *X. oryzae* pv. *oryzae*, *X. oryzae* pv. *oryzicola*, other xanthomonads, and several genera of plant-pathogenic and plant- or seed-associated bacteria. This computational approach for identification of unique *loci* through whole-genome comparisons is a powerful tool that can be applied to other plant pathogens to expedite development of diagnostic primers. This example highlights the emerging trend in plant disease diagnosis.

13.2.2.5 PROTEOMICS APPROACH

Understanding the proteome, the structure and function of each protein and the complexities of protein–protein interactions is critical for developing the effective diagnostic techniques and disease treatments in the future. Proteomics is highly useful in identification of candidate biomarkers (proteins in body fluids that are of value for diagnosis), identification of the bacterial antigens that are targeted by the immune response, and identification of possible markers of infectious diseases. A number of techniques such as 2D-PAGE, mass spectrometry, and protein microarray are widely used for identification of proteins produced during a particular disease, which help in plant disease diagnosis.

González-Fernández et al. (2010) have reviewed various reports in the field of proteomics of agriculturally important plant pathogenic fungi. The relevance of proteomics in plant fungal pathogens research is very well illustrated by the pioneer work on the *Cladosporium fulvum*–tomato interaction carried out by the Pierre de Wit research group back in 1985 (De Wit et al., 1986) that allowed the characterization of the first avirulence gene product (*Avr9*) after purification from tomato apoplastic fluids by preparative PAGE followed by reverse-phase HPLC and EDMAN N-terminal sequencing (Schottens-Toma and De Wit, 1988). Later on, a number of avirulence gene product effectors have been discovered, mainly by genomic approaches (Ellis et al., 2009). Curiously, this pioneer work followed the typical proteomics strategy even before Murashige and Skoog's medium (MS)-based powerful techniques were developed. Another good example is the tomato *F. oxysporum* pathosystem, in which the first effector of root invading fungi was identified and sequenced, in this case by MS, the Six1, corresponding to a 12-kDa cysteine-rich protein (Rep et al., 2004). Other further protein effectors have been characterized in different fungi (Rep, 2005).

Zorn et al. (2005) reported systematic shotgun proteomics analysis at different stages of development of powdery mildew in the host Barley to gain further understanding of the biology during infection of fungus *Blumeria graminis* (Bindschedler et al., 2009). In another report, proteomic approach was applied to study Pea (*Pisum sativum*) responses to powdery mildew (*Erysiphe pisi*) for the identification of proteins implicated in powdery mildew resistance (Curto et al., 2006). Murad et al. (2008) reported proteomic analysis of *Metarhizium anisopliae* secretion in the presence of the insect pest *Callosobruchus maculates*. These are few interesting applications of proteomics in plant-disease diagnosis.

13.2.2.6 DNA-BASED DIAGNOSTIC KITS

DNA diagnostic kits are based on the ability of single-stranded nucleic acids to bind to other single-stranded nucleic acids that are complementary in sequence. The tool used in DNA diagnostic kits is the PCR. There are three steps involved in PCR. The DNA is first unwound, and its strands separated by high temperatures. As the temperature is lowered, short, single-stranded DNA sequences called primers are free to bind to the DNA strands at regions of homology, allowing the *Taq* polymerase enzyme to make a new copy of the molecule. This cycle of denaturation–annealing–elongation is repeated 30–40 times, yielding millions of identical copies of the segment (Fig. 13.1).

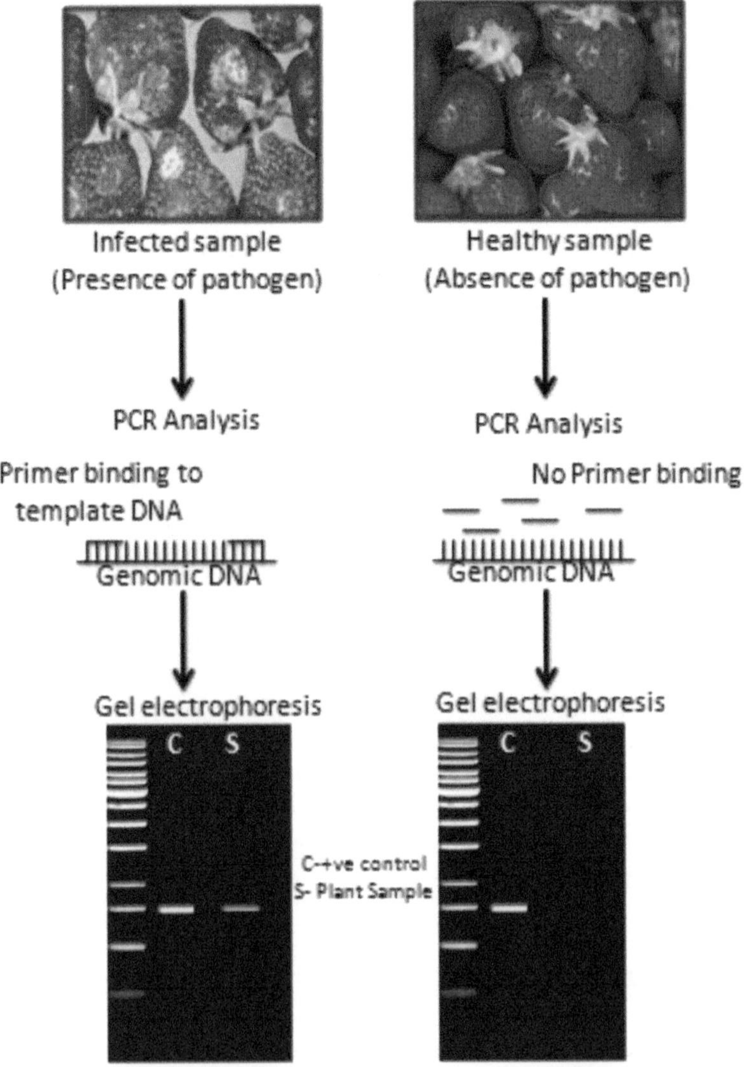

FIGURE 13.1 PCR-based diagnostic method for pathogen detection.

13.2.2.7 PROTEIN-BASED DIAGNOSTIC KITS

The first step in a defense-response reaction is the recognition of an invader by a host's immune system. This recognition is due to the ability of specific host proteins, called antibodies, to recognize and bind proteins that are unique to a pathogen (antigens) and to trigger an immune reaction.

Protein-based diagnostic kits for plant diseases contain an antibody (the primary antibody) that can either recognize a protein from either the pathogen or the diseased plant. Because the antibody–antigen complex cannot be seen by the naked eye, diagnostic kits also contain a secondary antibody, which is joined to an enzyme. This enzyme will catalyze a chemical reaction that will result in a color change only when the primary antibody is bound to the antigen. Therefore, if a color change occurs in the kit's reaction mixture, then it is the indication of presence of plant pathogen (Fig. 13.2).

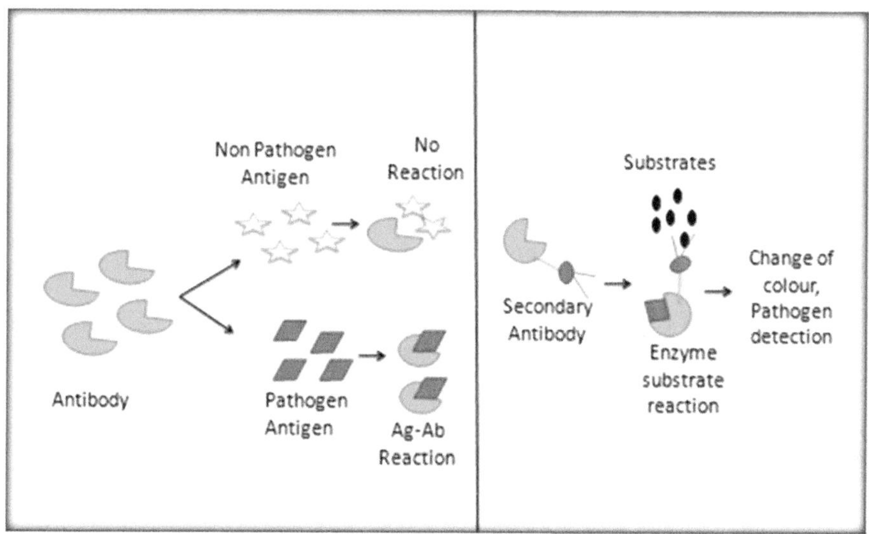

FIGURE 13.2 Antigen–Antibody interaction for pathogen detection.

The ELISA method makes use of this detection system and forms the basis of some protein-based diagnostic kits. ELISA kits are convenient to use because test takes only a few minutes to perform and does not require sophisticated laboratory equipment or training. There are already numerous ELISA test kits available on the market. Some of them detect diseases of root crops (e.g., cassava, beet, potato), ornamentals (e.g., lilies, orchids), fruits (e.g., banana, apple, grapes), grains (e.g., wheat, rice), and vegetables.

ELISA techniques can detect ratoon stunting disease of sugarcane, tomato mosaic virus, papaya ringspot virus, banana bract mosaic virus, BBTV, watermelon mosaic virus, and rice tungro virus. In addition to these applications, virus indexing can also be performed using ELISA kits.

One of the first ELISA kits developed to diagnose plant disease was made by the International Potato Center (Priou, 2001). It can detect the presence of all races, biovars, and serotypes of *Ralstonia solanacearum*, the pathogen that causes bacterial wilt or brown rot in potato. They also developed a kit that samples for the presence of any member of sweet potato family viruses.

13.2.2.8 MICROARRAY TECHNOLOGY

Since the development of microarray technology for gene expression studies (Schena et al., 1995), new approaches are extending their application to the detection of pathogens. Microarrays are generally composed of thousands of specific probes spotted onto a solid surface (usually nylon or glass). Each probe is complementary to a specific DNA sequence (genes, ITS, ribosomal DNA) and hybridization with the labeled complementary sequence provides a signal that can be detected and analyzed. Although there is great potential for microarray technology in the diagnosis of plant diseases, the practical development of this application is still in progress. For example, following the methodology utilized for genetic analysis (Brown and Botstein, 1999), large numbers of DNA probes used in two-dimensional arrays have allowed thousands of hybridization reactions to be analyzed at the same time (Hadidi et al., 2004). Until now, the microarray technology focuses its use in multiplex format of similar or very different pathogens, taking advantage of the number of probes that can be employed in one chip (Bonants et al., 2002, 2005).

With the availability of genomic sequences of pathogens and the rapid development of microarray technology, as well as a renewed emphasis on detection and characterization of quarantine pathogens, there is a rush in the European Union to set up this technology and apply it to detection.

The probes can be prepared in at least three basic formats: (1) PCR fragments arrayed on nylon membranes, hybridized against cDNA samples radioactively labeled, called macroarrays (Richmond et al., 1999); (2) PCR products spotted onto glass slides and DNA labeled with fluorescent dyes (Richmond et al., 1999); and (3) oligonucleotides of different length (from 18 to 70 bp) arrayed and hybridized with the same type of labeled DNA material (Lockhart et al., 1996; Loy et al., 2002). For bacterial detection, the material spotted until now is almost universally oligonucleotides targeting the 16S–23S rDNA genes (Crocetti et al., 2000; Franke-Whittle et al., 2005). The microarrays are analyzed either by scanning or by a direct imaging system.

The potential of microarray technology in the detection and diagnosis of plant diseases is very high, due to the multiplex capabilities of the system. DNA microarrays are also of great use for simultaneous pathogen detection. This is important, as plants are often infected with several pathogens, some of which may act together to cause a disease complex. Microarrays consist of pathogen-specific DNA sequences immobilized onto a solid surface. Sample DNA is amplified by PCR, labeled with fluorescent dyes, and then hybridized to the array (Fig. 13.3; modified from Alberts et al., 2002).

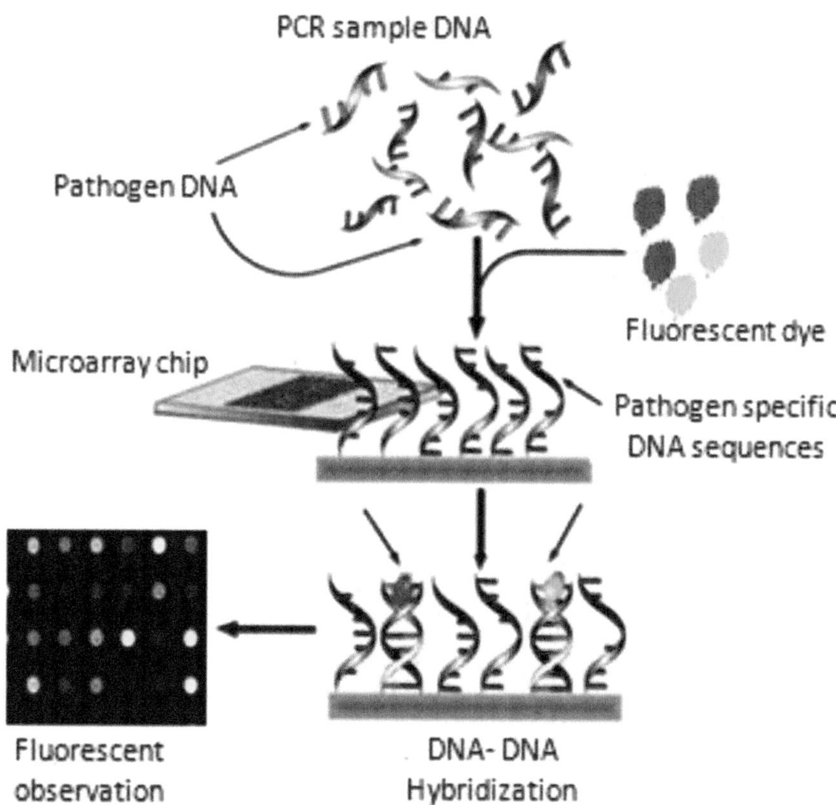

FIGURE 13.3 Overview of DNA microarray.

13.3 MANAGEMENT OF PLANT DISEASES THROUGH BIOTECHNOLOGICAL APPROACHES

Since the beginning of agriculture, generations of farmers have been evolving practices for combating the various plagues suffered by our crops. Following discovery of the causes of plant diseases in the early 19th century, our growing understanding of the interactions of pathogen and host has enabled us to develop a wide array of measures for the control of specific plant diseases. Managing plant disease deals largely with prevention of infection in plant populations rather than with cure or therapy of diseased individuals. Therefore, it is imperative that action be taken in advance of infection. Essentials for sound management planning include a basic knowledge of the host plant, pathogen life cycle, and environmental factors such as temperature, moisture, and light intensity that influence pathogen disease dynamics. The principles of plant diseases management was first articulated by H. H. Whetzel in 1929 and modified by various authors over the years have been widely adopted. The rationale or justification for disease management and experimental research is found in past disease experience rather than immediate crises. Traditional methods for preventing, curing, or reducing the severity of disease are directed at the inducing agent following one or more basic principles as follows:

1. **Avoiding the pathogenic agent**—prevent disease by selecting a time of the year or a site where there is no inoculum or where the environment is not favorable for infection.
2. **Exclusion of the pathogen from an area**—prevent the introduction of inoculum.
3. **Eradication of an established pathogen**—eliminate, destroy, or inactivate the inoculum.
4. **Protection of the plant by placing a barrier**—prevent infection by means of a toxicant or some other barrier to infection.
5. **Curing infected plants**—cure plants that are already infected.
6. **Improving host resistance**—utilize cultivars that are resistant to or tolerant of infection.

While these principles are as valid today as they were in 1929, in the context of modern concepts of plant disease management, they have some critical shortcomings. First of all, these principles are stated in absolute terms (e.g., "exclude," "prevent," and "eliminate") that imply a goal of zero disease. Plant disease "control" in this sense is not practical and, in most cases, is not even possible. Indeed, we need not eliminate a disease; we merely need to reduce its progress and keep disease development below an acceptable level. Instead of plant disease control, we need to think in terms of plant disease management.

A second shortcoming is that the traditional principles of plant disease control do not take into consideration the dynamics of plant disease, that is, the changes in the incidence and severity of disease in time and space. Furthermore, considering that different diseases differ in their dynamics, they do not indicate the relative effectiveness of the various tactics for the control of a particular disease. They also fail to show how the different disease control measures interact in their effects on disease dynamics. We need some means of assessing quantitatively the effects of various control measures, singly and in combination, on the progress of disease.

Finally, the traditional principles of plant disease control tend to emphasize tactics without fitting them into an adequate overall strategy.

13.3.1 MOLECULAR APPROACHES FOR PLANT DISEASE MANAGEMENT

Despite the systematic and continuous efforts through conventional disease management methods like biological and chemical means, substantial success has not been achieved due to high genotype × environment (G×E) interactions on the expression of important quantitative traits leading to slow gain in genetic improvement, besides severe losses caused by susceptibility to several biotic stresses. These issues require an immediate attention, and, overall, a paradigm shift is needed in the management strategies to strengthen our traditional crop improvement programs. One way is to utilize genomics and molecular tools in selection of desirable genotypes or growing of transgenic crops. The use of transgenic crops is especially required for those traits that are not easy to improve genetically through conventional approaches because of the lack of satisfactory sources of desirable gene(s) in crossable gene pools. However, the ongoing debate on biosafety and ethical issues involving use of transgenic crops for commercial cultivation suggests that molecular marker-aided conventional methods may be the main short-term option for controlling plant diseases.

The use of tissue culture and genetic engineering for controlling plant diseases has been reviewed by Fuchs and Gonsalves (1996), while the role of biotechnology in controlling plant disease has been discussed by Mandahar and Khurana (1998). Plant biotechnology impinges or helps plant pathology to management of disease in many ways:

1. To obtain pathogen-free mother plants through tissue culture
2. Transgenic technology/genetic engineering
3. RNA interference (RNAi) technology
4. Molecular breeding techniques.

13.3.1.1 TISSUE CULTURE

As an emerging technology, the plant tissue culture has a great impact on both agriculture and industry, through providing disease-free plants needed to meet the ever-increasing world demand. It has made significant contributions to the advancement of agricultural sciences in recent times, and today they constitute an indispensable tool in modern agriculture. Almost all tissue culture techniques are used in management of plant diseases by producing disease-free plants. The widely used tissue culture techniques are meristem/shoot-tip culture and protoplast fusion whose importance to plant pathology is briefly described.

13.3.1.1.1 Meristem or Shoot Tip Culture

Most of the horticultural and forest crops are infected by systemic disease caused by fungi, viruses, bacteria, mycoplasma, and nematode. Although plant infected with bacteria and fungi may respond to treatments with bactericidal and fungicidal compounds, there is no commercially available treatment to cure virus infected plants. It is possible to produce disease-free plants through tissue culture. Apical meristems in the infected plants are generally either free or carry a very low concentration of the viruses. The following possibilities have been suggested to explain the mechanisms underlying the resistance of meristems to viruses (Lizarraga et al., 1986).

1. **High metabolic activity:** Viruses replicate by taking over the host metabolic pathways. Due to the high metabolic activity in these cells, viruses are unable to take over control of the host biosynthetic machinery.
2. **Lack of vascular system:** Viruses spread rapidly through the vascular system. Phloem-restricted viruses cannot invade the meristematic tissues due to the absence of cell differentiation. In this meristematic region, viruses which infect nonvascular tissues spread from cell to cell through the plasmodesmata. This is a slow process

which makes it relatively difficult for viruses to completely infect the rapidly dividing cells.

3. **High auxin concentration:** Plant's meristematic tissues have a higher auxin concentration than tissues from the other plant regions. These auxins have been reported to inhibit the replication of viruses.

13.3.1.1.2 Culture Medium for Meristem Culture

The nutrients, growth regulators, and nature of the medium highly influence the development of virus-free plants from meristem-tip cultures. Maximum success is achieved from MS medium which promoted healthy, green shoot development compared to other nutrient media. The main reason for the suitability of medium for meristem-tip culture could be the presence of high levels of K^+ and NH_4^+ ions. There is no critical assessment on the role of various vitamins or amino acids but sucrose or glucose is the most commonly used carbon source in the medium, at the range of 2–4%, to raise virus-free plants from meristem-tip cultures.

Large meristem-tip explants, measuring 500 µm or more in length, may give rise to plants even in the basal medium, but generally the presence of an auxin or a cytokinin or both plays a major role in the development of excised apical meristem. In angiosperms, the meristematic dome in the shoot-tip does not synthesize auxin on its own, but it is supplied by the second pair of youngest leaf primordia. Therefore, for development of excised meristem in culture, without the leaf primordia, requires the supply of exogenous auxin. The plants requiring only auxin must have a high endogenous cytokinin level in their meristems. Among auxins, the use of 2,4-D should be avoided which promotes only callusing. Napthalene acetic acid (NAA) and Indole Acetic Acid (IAA) are widely used auxins, and NAA being preferred due to better stability. The gibberellic acid (GA3) is also known to promote better growth and differentiation and suppresses callusing from meristem explants. Both liquid and semisolid (agar) media have been tried for meristem-tip culture, but agar medium is generally preferred.

13.3.1.1.3 Factors Affecting Virus Eradication

Factors such as culture medium, explant size, and incubation conditions affecting plant regeneration from meristem-tip cultures have pronounced effect on virus eradication. Besides, thermotherapy or chemotherapy and physiological stage of the explants also affect virus elimination by shoot-tip culture. The success in obtaining complete plants can be considerably improved by the choice of the culture medium. The major features of the culture medium to be considered are its nutrients, growth regulators, and physical mature.

1. The size of meristem tip is an important factor governing regeneration capacity of meristems and to obtain virus-free plants. For example, in cassava, meristems exceeding 0.2-mm size regenerated to plantlets, but those less than 0.2-mm size developed either gallus or callus with roots. In general, the larger the meristem, the greater is the number of regenerated plants, but the number of virus-free plant is inversely proportional to the size of meristem cultured.
2. For meristem-tip cultures, light incubation has generally proved better than dark incubation. The optimum light intensity for initiating tip cultures of potato is 100 lx, which should be increased to 200 lx after 4 weeks. The cultures are generally stored under standard culture room temperatures (25 ± 2°C).
3. Meristem tips should preferably be taken from actively growing buds. Tips taken from terminal buds gave better results than those from axillary buds.

13.3.1.1.4 Production of Virus-Free Plant through Meristem Culture

Crop plants have a greater potential for improved yield and quality when they are free from harmful diseases. Stocks of the vegetatively propagated crops like potato, sugarcane, cassava, sweet potato, beet, strawberry, blueberry, banana, and certain ornamental plants are multiplied continuously for many years have ample chances for infection with one or more viruses/viroids and show a decline in growth and yield. The production and distribution of virus-free propagating materials has been proven to be highly successful in controlling virus diseases in many crops and promise to be of wider application in others (Sastry and Zitter, 2014).

Some of the successful stories of meristem-tip culture were reported in crops like cassava (Wasswa et al., 2010), potato (Awan et al., 2007; Faccioli, 2001), sweet potato (Mervat and Far, 2009), and sugarcane (Mishra et al., 2010). Throughout the world, meristem-tip culture technique is being used for production of virus and virus-like diseases-free planting materials primarily for vegetative propagated plants. In majority of the countries, the responsibility of production and distribution of the virus-free plant material lies on the state and central governments (Table 13.2).

TABLE 13.2 Production of Virus-Free Plants through Meristem Culture.

Crop	Virus	Reference(s)
Alstromeria sp.	*Alstroemeria mosaic virus* (AlMV)	Chiari and Bridgen (2002)
Chrysanthemum morifolium	mixed infection by CMV and *Tomato aspermy virus* (TAV)	Kumar et al. (2009)
I. hawkerii	*Tomato spotted wilt virus* (TSWV)	Milošević et al. (2011)
Dianthus gratianopolotanus	*Carnation latent virus* (CLV), potyviruses	Fraga et al. (2004)
Apple	Apple mosaic virus	Hansen and Lane (1985), Bhardwaj et al. (1998)
Banana	Cucumber mosaic virus	Helliot et al. (2004), Kenganal et al. (2008)
	Banana streak virus	
Blueberry	Blueberry scorch carlavirus	Postman (1997)
Cassava	Cassava brown streak virus	Ng et al. (1992), Wasswa et al. (2010)
Chilli	Chilli leaf curl virus (CLCV)	Meena et al. (2014)
Dahlia	Dahlia mosaic virus	Sediva et al. (2006)
Grapevine	Grapevine viruses	Milkus et al. (2000)
Potato	Potato virus Y	Nascimento et al. (2003), Awan et al. (2007), Al-Taleb et al. (2011)
	Potato viruses	
Strawberry	Strawberry mottle virus Strawberry viruses	Cieslinska (2003), Biswas et al. (2007)
Sugarcane	Sugarcane mosaic virus	Waterworth and Kahn (1978), Ramgareeb et al. (2010), Mishra et al. (2010)
	Sugarcane yellow leaf virus	
Sweet potato	Sweet potato feathery mottle virus	Gichuki et al. (2005), Mervat and Far (2009)

13.3.1.2 PROTOPLAST FUSION TECHNIQUE

Plant protoplasts provide a unique single cell system to underpin several aspects of modern biotechnology. Protoplasts are the cells whose cell walls are removed and cytoplasmic membrane is the outermost layer in such cells. The specific lytic enzymes are being used to remove cell wall. Protoplast fusion is a physical phenomenon, during fusion two or more protoplasts come in contact and adhere with one another either spontaneously or in presence of fusion-inducing agents. By protoplast fusion, it is possible to transfer some useful genes such as disease resistance, nitrogen fixation, rapid growth rate, protein quality, drought resistance, herbicide resistance, heat and cold resistance from one species to another. Protoplast fusion may be used to produce interspecific or even intergeneric hybrids. Protoplast fusion becomes an important tool of gene manipulation because it breakdown the barriers to genetic exchange imposed by conventional mating systems. It has been used to combine genes from different organisms to create strains with desired properties (Tomar and Dantu, 2010). The basic steps involved in production of somatic hybrids by protoplast isolation, fusion, and regeneration is discussed in this section.

13.3.1.2.1 Protoplast Isolation

The first and foremost step in the isolation of protoplast is the removal of cell wall either by mechanical rupture or enzyme digestion. The application of protop lasts to many areas of biochemical, morphological, physiological, and genetical studies demands large-scale production of purified viable protoplasts. Enzymatic method is preferred as it provides better protoplast yield with low tissue damage, whereas mechanical method causes maximum tissue chopping with lower protoplast yields. Both of these methods are described below:

13.3.1.2.1.1 Mechanical Method

Klercker in 1892 pioneered the isolation of protoplasts by mechanical methods. In this method, the cells were kept in suitable plasmolyticum, for example, CPW containing 13% w/v mannitol. Once the plasmolysis is complete, while remaining in the osmoticum, the leaf lamina would be cut with a sharp-edged knife. In this process, some of the plasmolyzed cells were cut only through the cell wall, releasing intact protoplasts while some of the protoplasts may be damaged inside many cells. Protoplasts that were trapped in a cell and only the corner had been cut off could be encouraged to come out by reducing the osmolarity slightly to force the protoplasts swell to force their way out of the cut surface. The released protoplasts then have to be separated from damaged ones and cell debris. Mechanical procedures, involving slicing of plasmolyzed tissues, are now rarely employed for protoplast isolation but are useful with large cells and when limited numbers of protoplasts are required. This approach has been used successfully to isolate protoplasts of the giant marine alga, *Valonia utricularis*, for patch clamp analyses of their electrical properties, including physiological changes of the plasma membrane induced by exposure of isolated protoplasts to enzymes normally used to digest cell walls (Binder et al., 2003).

13.3.1.2.1.2 Enzymatic Method

In 1960, E. C. Cocking demonstrated the possibility of enzymatic isolation of a large number of protoplasts from roots of tomato seedlings. This method involves leaf sterilization followed by peeling of the lower epidermis to release cells which are plasmolyzed and added to enzyme mixture followed by harvest of protoplast. Either of the procedures for enzymatic isolation can be used: sequential enzymatic hydrolysis or mixed enzymatic hydrolysis. Major advancement in protoplast isolation was attained with the discovery of several commercially available enzymes, namely; cellulase, hemicellulase, β-glucuronidase, chitinase, pectinase/macerozyme, etc. (Lalithakumari, 1996). The commercial preparation Novozyrn 234 has been widely used to produce high yields of protoplasts from several fungi.

13.3.1.2.2 Protoplast Fusion

Protoplast fusion can be broadly classified into two categories:

Spontaneous fusion: Protoplast during isolation often fuses spontaneously and this phenomenon is called spontaneous fusion. During the enzyme treatment, protoplasts from adjoining cells fuse through their plasmodesmata to form multinucleate protoplasts. The occurrence of multinucleate fusion bodies is more frequent when the protoplasts are prepared from actively dividing callus cells or suspension cultures. Since the somatic hybridization or cybridization requires fusion of protoplasts of different origin, the spontaneous fusion has no value.

Induced fusion: Fusion of freely isolated protoplasts from different sources with the help of fusion-inducing chemicals agents is known as induced fusion. Normally, isolated protoplast do not fuse with each other because the surface of isolated protoplast carries negative charges (−10 mV to −30 mV) around the plasma membrane (outer surface) which repels each other. Therefore, fusion of protoplast needs chemicals which reduce the electronegativity of the isolated protoplast and allow them to fuse with each other. Induced fusion may be performed either in the presence of suitable chemical agents (fusogen) like, $NaNO_3$, high Ca^{2+}, polyethylene glycol (PEG), or electric stimulus.

1. **Fusion by sodium nitrate ($NaNO_3$):** It was first demonstrated by Kuster in 1909 that the hypotonic solution of $NaNO_3$ induces fusion of isolated protoplast forming heterokaryon (hybrid). This method was fully described by Evans and Cocking (1975); however, this method has a limitation of generating few number of hybrids, especially when highly vacuolated mesophyll protoplasts are involved.
2. **High pH and Ca^{2+} treatment:** This technique lead to the development of intra- and interspecific hybrids (Keller and Melcher, 1973). The isolated protoplasts from two plant species are incubated in 0.4 M mannitol solution containing high Ca^{2+} (50 mM $CaCl_2 \cdot 2H_2O$) with highly alkaline pH of 10.5 at 37°C for about 30 min. Aggregation of protoplasts takes place at once and fusion occurs within 10 min.
3. **Polyethylene glycol treatment:** PEG is the most popularly known fusogen as it has an ability of forming high frequency, binucleate heterokaryons with low cytotoxicity. With PEG, the aggregation occurred mostly between two and three protoplasts unlike Ca^{2+} induced fusion which involves large clump formation. The freshly isolated protoplasts from two selected parents are mixed in appropriate proportions and treated with 15–45% PEG (1500–6000 MW) solution for 15–30 min followed by gradual washing of the protoplasts to remove PEG. Protoplast fusion occurs during washing. The washing medium may be alkaline (pH 9–10) and contain a high Ca^{2+} ion concentration (50 mM). This combined approach of PEG and Ca^{2+} is much more efficient than the either of the treatment alone. PEG is negatively charged and may bind to cation like Ca^{2+}, which in turn, may bind to the negatively charged molecules present in plasmalemma; they can also bind to cationic molecules of plasma membrane. During the washing process, PEG molecules may pull out the plasmalemma components bound to them. This would disturb plasmalemma organization and may lead to the fusion of protoplasts located close to each other. The technique is nonselective, thus, induce fusion between any two or more protoplasts.

4. **Electrofusion:** The chemical fusion of plant protoplast has many disadvantages: (1) the fusogen are toxic to some cell systems, (2) it produces random, multiple cell aggregates, and (3) must be removed before culture. Compared to this, electrofusion is rapid, simple, synchronous, and more easily controlled. Moreover, the somatic hybrids produced by this method show much higher fertility than those produced by PEG-induced fusion.

Zimmerman and Scheurich (1981) demonstrated that batches of protoplasts could be fused by electric fields by devising a protocol which is now widely used. This protocol involves a two-step process. First, the protoplasts are introduced into a small fusion chamber containing parallel wires or plates which serve as electrodes. Second, a low voltage and rapidly oscillating AC field is applied, which causes protoplasts to become aligned into chains of cells between electrodes. This creates complete cell-to-cell contact within a few minutes. Once alignment is complete, the fusion is induced by application of a brief spell of high-voltage DC pulses (0.125–1 kV/cm). A high-voltage DC pulse induces a reversible breakdown of the plasma membrane at the site of cell contact, leading to fusion and consequent membrane reorganization. The entire process can be completed within 15 min.

13.3.1.2.3 Selection of Fusion Products

The somatic hybridization by electrofusion of protoplasts allows one-to-one fusion of desired pairs of protoplasts and, therefore, it is easy to know the fate of fusion products. However, protoplast suspension recovered after chemical treatments (fusogen) consists of the following cell types:

1. unfused protoplasts of the two species/strains,
2. products of fusion between two or more protoplasts of the same species (homokaryons), and
3. "hybrid" protoplasts produced by fusion between one (or more) protoplasts of each of the two species (heterokaryons).

The heterokaryons which are the potential source of future hybrids constitute of a very small (0.5–10%) proportion of the mixture. Therefore, an effective strategy has to be employed for their identification and isolation. Various protocols have been proposed and practiced for the effective selection of hybrids, including morphological basis, complementation of biochemical and genetic traits of the fusing partners, and manual or electronic sorting of heterokaryons/hybrid cells.

13.3.1.2.3.1 Morphophysiological Basis

The whole mixtures of the protoplasts are cultured after fusion treatment and the resulting calli or regenerants are screened for their hybrid characteristics. Occasionally, the hybrid calli outgrow the parental cell colonies and are identified by their intermediate morphology, that is, green with purple-colored cells. However, the process is labor intensive and requires glass-house facilities. It is limited to certain combinations showing differences in their regeneration potential under specific culture conditions.

13.3.1.2.3.2 Complementation

In this case, complementation or genetic or metabolic deficiencies of the two fusion partners are utilized to select the hybrid component. When protoplasts of two parents (one parent bearing cytoplasmic albino trait and the other parent bearing green trait) each parent carrying a nonallelic genetic or metabolic defect are fused, it reconstitutes a viable hybrid cell of wild type in which both defects are mutually abolished by complementation, and the hybrid cells are able to grow on minimal medium nonpermissive to the growth of the parental cells bearing green trait. Later, the calli of hybrid nature could be easily distinguished from the parental type tissue (albino trait) by their green color. The complementation selection can also be applied to dominant characters, such as dominant resistance to antibiotics, herbicides, or amino acid analogs.

13.3.1.2.3.3 Isolation of Heterokaryons or hybrid cells

The manual or electronic isolation of heterokaryons or hybrid cells is the most reliable method. Manual isolation requires that the two parental type protoplasts have distinct morphological markers and are easily distinguishable. For example, green vacuolated, mesophyll protoplasts from one parent and richly cytoplasmic, nongreen protoplasts from cultured cells of another parent. The dual fluorescence method also helps easy identification of fusion products. In this case, the protoplast labeled green by treatment with fluorescein diacetate (1–20 mg/L) are fused with protoplasts emitting a red fluorescence, either from chlorophyll autofluorescence or from exogenously applied rhodamine isothiocyanate (10–20 mg/L). The labeling can be achieved by adding the compound into the enzyme mixture. This can be applied even for morphologically indistinguishable protoplasts from two parents. The diagrammatic representation of protoplast isolation, fusion, and culture is provided in Figure 13.4.

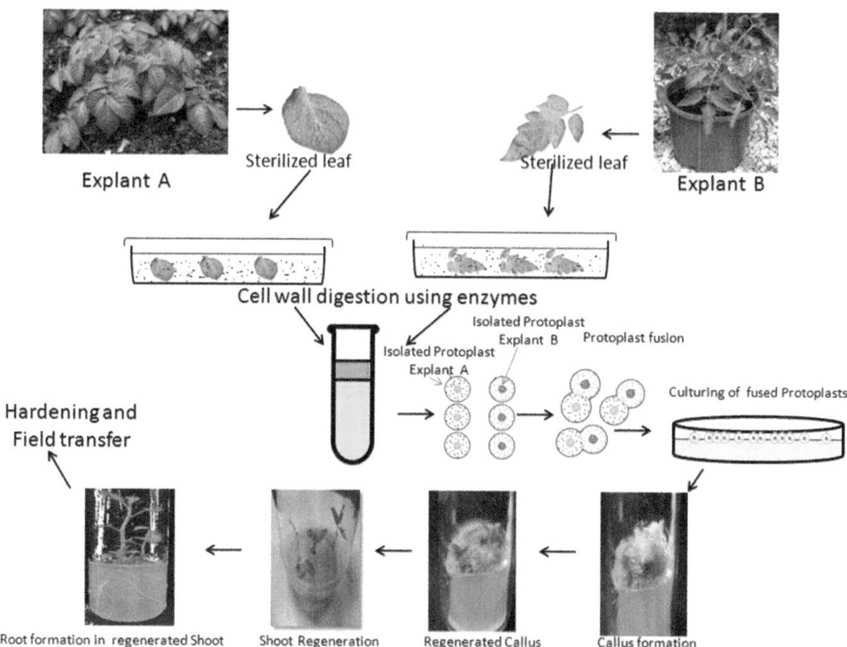

FIGURE 13.4 Schematic representation of protoplast isolation, fusion, and culture technique.

13.3.1.2.4 Verification and Characterization of Somatic Hybrids

No system is foolproof, and they have their own advantages and disadvantages. Therefore, even after selecting the desired hybrids/cybrids following protoplast fusion, it is required to carry out one or more tests to compare the parent protoplast lines with the putative hybrids. Some of the techniques that can be tried are as follows:

1. **Morphology:** Somatic hybrids in most of the cases show characters intermediate between the two parents, such as shape of leaves, pigmentation of corolla, plant height, root morphology, and other vegetative and floral characters. The method is not much accurate as tissue culture conditions may also alter some morphological

characters or the hybrid may show entirely new traits not shown by any of the parents.

2. **Isozyme analysis:** Multiple molecular forms of same enzyme which catalyses similar or identical reactions are known as isozymes. Electrophoresis is performed to study banding pattern as a check for hybridity. If the two parents exhibit different band patterns for a specific isozyme, the putative hybrid can be easily verified. The isozymes commonly used for hybrid identification include, acid phosphatase, esterase, peroxidase.
3. **Cytological analysis:** Chromosome counting of the hybrid is an easier and reliable method to ensure hybridity as it also provides the information of ploidy level. Cytologically, the chromosome count of the hybrid should be sum of number of chromosomes from both the parents. Besides number of chromosomes, the size and structure of chromosomes can also be monitored. However, the approach is not applicable to all species, particularly where fusion involves closely related species or where the chromosomes are very small. Moreover, sometimes the somaclonal variations may also give rise to different chromosome number.
4. **Molecular analysis:** Specific restriction pattern of nuclear, mitochondrial, and chloroplast DNA characterizes the plastomes of hybrids and cybrids. Molecular markers such as RFLP, RAPD, and ISSR can be employed to detect variation and similarity in banding pattern of fused protoplasts to verify hybrid and cybrid.

13.3.1.2.5 Disease Resistance Plant through Protoplast Fusion

Applications of somatic hybridization in crop improvement are constantly evolving, and original experiments generally targeted gene transfer from wild accessions to cultivated selections that were either difficult or impossible to accomplish by conventional methods (Grosser and Gmitter, 2011). Plant somatic hybridization via protoplast fusion has become an important tool for ploidy manipulation in plant improvement particularly disease resistance, allowing researchers to combine somatic cells from different cultivars, species, or genera, resulting in novel allotetraploid and autotetraploid genetic combinations. The successful transfer of disease resistance through protoplast fusion is reported in potato, tomato, citrus, and *Brassica* plant species (Davey et al., 2005). The examples of the application of protoplast fusion to transfer disease resistant traits are given in Table 13.3.

TABLE 13.3 Transfer of Disease Resistance Traits in Crop Species through Protoplast Fusion Technology.

Crop Species	Trait Transferred	Reference(s)
Brassica		
B. napus (+) *Sinapsis arvensis*	Enhanced resistance to Blackleg (*Leptosphaeria maculans*)	Hu et al. (2002)
Citrus		
C. limonia (+) *C. sunki* cv. *Tanaka*	Tolerance to citrus blight, Tristeza virus, and phytophthora	Costa et al. (2003)
C. reticulata cv. *Blanco* (+) *C. paradise*	Production of mixoploid plants tolerant to citrus exocortis virus (CEV)	Liu and Deng (2002)
C. reticulata cv. *Blanco* (+) *C. volkameriana*	Tolerance to citrus blight, tristeza virus, and phytophthora	Costa et al. (2003)
C. reticulata cv. *Blanco* (+) *Poncirus trifoliate*	Resistance to CEV	Guo et al. (2002)
C. sinensis cv. *Rohde Red* (+) *C. volkameriana*	Tolerance to citrus blight, Tristeza virus, and phytophthora	Costa et al. (2003)
C. sinensis cv. *Ruby Blood* (+) *C. volkameriana*	Tolerance to citrus blight, Tristeza virus, and phytophthora	Costa et al. (2003)
Potato		
S. tuberosum (+) *S. brevidens*	Resistance to Potato leaf roll virus (PLRV), late blight	Helgeson et al. (1998)
S. melongena (+) *S. aethiopicum*	Resistance to bacterial wilt (*Ralstonia solanacearum*)	Collonnier et al. (2001)
S. melongena (+) *S. sisymbrifolium*	Resistance to bacterial and fungal wilts	Collonnier et al. (2003)
S. tuberosum (+) *S. etuberosum*	Resistance to potato virus Y	Gavrilenko et al. (2003)
S. tuberosum (+) *S. nigrum*	Resistance to potato blight (*Phytophthora infestans*)	Szczerbakowa et al. (2003)
S. tuberosum (+) *S. stenotomum*	Resistance to bacterial wilt (*R. solanacearum*)	Fock et al. (2001)
Tomato		
Lycopersicon esculentui (+) *L. hirsutum*	Resistance to *Fusarium oxysporium*	Hardy (2011)

13.3.1.3 GENETIC ENGINEERING/TRANSGENIC TECHNOLOGY

Plant diseases cause diverse problems ranging from total crop loss to loss of product quality. Solutions implemented to control the damage caused by disease include cultural practice, treatment with pesticides, and disease resistance. Disease resistance is the panacea, the ultimate universal answer, since this is the only method which does not require input by the grower. However, resistance is often not available, or is not durable, primarily because pathogen populations adapt to overcome resistance. Disease resistance also incurs metabolic costs leading to reduced yield. Conventional plant-breeding programs provide a cost-effective and morally uncontroversial strategy for introducing disease resistance against many plant pathogens, although sources for disease resistance have not been identified for many pathosystems. While breeding techniques are time consuming and have disadvantage of introducing undesired traits by linkage, genetic engineering is relatively fast and allows transference of individual traits into crops in a calculated manner. In addition, genetic engineering may allow for rapid introduction of desirable traits from one species or organisms into crops and the precise manipulation of temporal or tissue specific expression of the trait of interest.

Nowadays, it has become routine to transfer genes from one organism to another, it is possible to introduce genes conferring disease resistance into crop plants. Such gene transfers could be accomplished by two methods, direct methods and vector-mediated methods. Gene gun or Biolistic method and *Agrobacterium*-mediated method are the best examples of the direct method and vector-mediated method, respectively.

13.3.1.3.1 Agrobacterium-Mediated Gene Transfer

Agrobacterium-mediated transformation is the most commonly used method for plant genetic engineering. The pathogenic soil bacteria *Agrobacterium tumefaciens* (*A. tumefaciens*) that causes crown gall disease has the ability to introduce part of its plasmid DNA (called transfer DNA or T-DNA) into the nuclear genome of infected plant cells. *A. tumefaciens* has the exceptional ability to transfer a particular DNA segment (T-DNA) of the tumor-inducing (Ti) plasmid into the nucleus of infected cells where it is then stably integrated into the host genome and transcribed, causing the crown gall disease. T-DNA contains two types of genes: the oncogenic genes, encoding for enzymes involved in the synthesis of auxins and cytokinins and responsible for tumor formation; and the genes encoding for the synthesis of opines (Fig. 13.5A). These compounds, produced by condensation between amino acids and sugars, are synthesized and excreted by the crown gall cells and consumed by *A. tumefaciens* as carbon and nitrogen sources. Outside the T-DNA are located the genes for the opine catabolism, the genes involved in the process of T-DNA transfer from the bacterium to the plant cell and the genes involved in bacterium–bacterium plasmid conjugative transfer (de la Riva et al., 1998).

Virulent strains of *A. tumefaciens* and *A. rhizogenes*, when interacting with susceptible dicotyledonous plant cells, induce diseases known as crown gall and hairy roots, respectively. These strains contain a large megaplasmid (more than 200 kb) which plays a key role in tumor induction, and for this reason, it was named Ti plasmid, or Ri in the case of *A. rhizogenes*. Ti plasmids are classified according to the opines, which are produced and excreted by the tumors they induce. During infection the T-DNA, a mobile segment of Ti or Ri plasmid, is transferred to the plant cell nucleus and integrated into the plant chromosome. The T-DNA fragment is flanked by 25-bp direct repeats, which act as a *cis* element signal for the transfer apparatus. The process of T-DNA transfer is mediated by the cooperative action of proteins encoded by genes determined in the Ti plasmid virulence region (*vir* genes) and in the bacterial chromosome. The Ti plasmid also contains the genes for opine catabolism produced by the crown gall cells, and regions for conjugative transfer and for its own integrity and stability. The 30-kb virulence (*vir*) region is a regulon organized in six operons that are essential for the T-DNA transfer (*virA*, *virB*, *virD*, and *virG*) or for the increasing of transfer efficiency (*virC* and *virE*) (Jeon et al., 1998). Different chromosomal-determined genetic elements have shown their functional role in the attachment of *A. tumefaciens* to the plant cell and bacterial colonization: the loci *chvA* and *chvB*, involved in the synthesis and excretion of the β-1,2 glucan (Cangelosi et al., 1989); the *chvE* required for the sugar enhancement of *vir* genes induction and bacterial chemotaxis (Ankenbauer and Nester, 1990); the *cel* locus, responsible for the synthesis of cellulose fibrils; the *pscA* (*exoC*) locus, playing its role in the synthesis of both cyclic glucan and acid succinoglycan; and the *att* locus, which is involved in the cell surface proteins. The initial results of the studies on T-DNA transfer process to plant cells demonstrate three important facts for the practical use of this process in plants transformation. First, the tumor formation is a transformation process of plant cells resulted from transfer and integration of T-DNA and the subsequent expression of T-DNA genes. Second, the T-DNA genes are transcribed only in plant cells and do not play any role during the transfer process. Third, any foreign DNA placed between the T-DNA borders can be transferred to plant cells, no matter where it comes from. These well-established facts, allowed the construction of the first vector and bacterial strain systems for plant transformation (Torisky et al., 1997).

The discovery that the *vir* genes do not need to be in the same plasmid with a T-DNA region to lead its transfer and insertion into the plant genome led to the construction of a system for plant transformation where the T-DNA region and the vir region are on separate plasmids. A co-integrative vector produced by integration of recombinant intermediate vector (IV containing the DNA inserts) into a disarmed pTi. Transformed gene is initially cloned in *E. coli* for easy in cloning procedure. A suitably modified *E. coli* plasmid is used to initiate cloning of gene. The subsequent gene transfer into plants is obtained by co-integrative vectors. Co-integration of the two plasmids is achieved with in *Agrobacterium* by homologous recombination. A binary vector consists of a pair of plasmids of which one contain *vir* region and other contains disarmed T-DNA sequence with right and left border sequences (Fig. 13.5B). The plasmid containing disarmed T-DNA are called micro-Ti or mini-Ti for, that is, Bin 19.

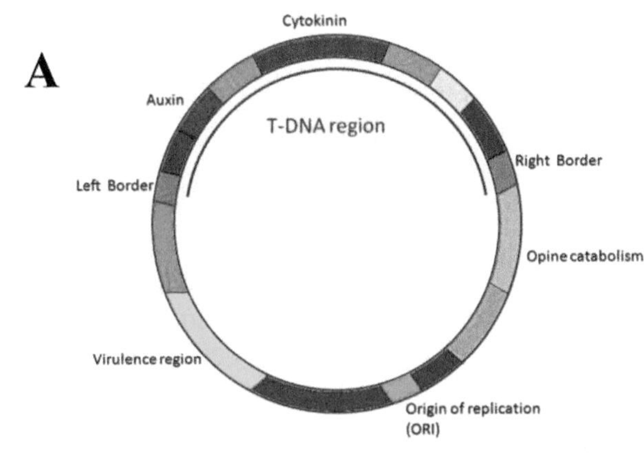

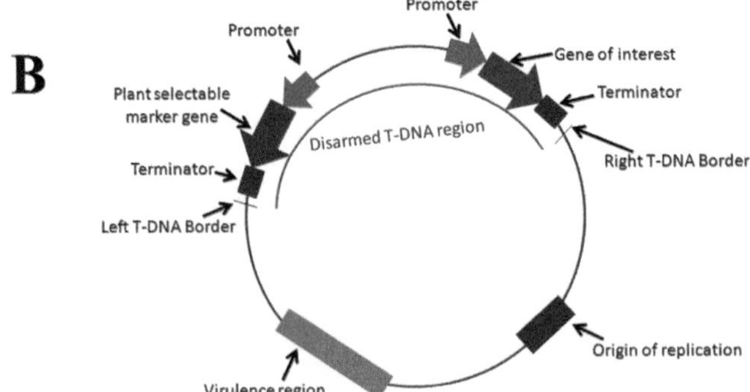

FIGURE 13.5 General structure of Ti and co-integrated plasmid: (A) Ti plasmid and (B) co-integrated plasmid.

13.3.1.3.2 Disease Resistant Plants through Transgenic Technology

Significant progress has been made over the past three decades to develop genetically engineered plants with resistance to biotic and abiotic stress factors. Genes from bacteria such as *Bacillus thuringiensis* (*Bt*) have been deployed on a commercial scale for pest management through transgenic crops. In addition to *Bt* genes, genes encoding for protease inhibitors, plant lectins, secondary plant metabolites, and vegetative insecticidal proteins have also been used to develop transgenic plants for crop protection. Insect-resistant cotton and maize, herbicide-resistant soybean, and tomato with a long shelf-life have been deployed on a commercial scale in several countries, and transgenic crops are now grown on over 100 million hectares. In contrast to herbicide- or insect-resistant transgenic plants, which have been grown extensively worldwide for more than 10 years, the development of transgenic plants with enhanced resistance to fungal and bacterial pathogens has received only limited success. Much of the limitation toward successful implementation in transgenic strategies for increasing plant tolerance toward pathogens stems from generally achieving low levels of resistance that are below the threshold desired by producers, or high levels of resistance against only a specific pathogen or even a single strain. This generally observed low levels of resistance coupled with the negative perception of GM plants has resulted in a relatively small number of transgenic lines being brought to late-stage field testing and even fewer that have been successfully brought to market (Wally and Punja, 2010). A list of transgenic plants developed against various diseases is enlisted in Table 13.4.

13.3.1.4 RNA INTERFERENCE TECHNOLOGY

Despite substantial advances in plant disease management strategies, our global food supply is still threatened by a multitude of pathogens and pests. This changed scenario warrants us to respond more efficiently and effectively to this problem. The situation demands judicious blending of conventional, unconventional, and frontier technologies. In this sense, RNAi technology or antisense RNA technology has emerged as one of the most potential and promising strategies for enhancing the building of resistance in plants to combat various fungal, bacterial, viral, and nematode diseases causing huge losses in important agricultural crops. The nature of this

TABLE 13.4 Disease Resistance Plants Developed through Transgenic Technology.

Crop	Disease Resistance	Source of Gene	Reference(s)
Bacterial resistance			
Tomato	Bacterial spot	*R* gene from pepper	Horvath et al. (2012)
Rice	Bacterial blight and bacterial streak	Engineered *E* gene	Hummel et al. (2012)
Potato	Multibacterial resistance	*PRR* from Arabidopsis	Lacombe et al. (2010)
Banana	*Xanthomonas* wilt	Novel gene from pepper	Tripathi et al. (2010)
Potato	Late blight	*R* gene from wild relatives	Bradeen et al. (2009), Foster et al. (2009), Halterman et al. (2008)
Rice	Bacterial Streak	*R* gene from maize	Zhao et al. (2005)
Fungal resistance			
Tobacco	*Sclerotinia sclerotiorum* and *Botrytis cinerea*	*Chi1* gene from *Rhizopus oligosporus*	Terakawa et al. (1997)
Grapevine	powdery mildew by *Uncinula necator*	*RCC2* gene from rice	Yamamoto et al. (2000)
Potato	Multifungal resistance	β-1,3-Glucanase and chitinase gene from pea	Chang et al. (2002)
Rice	Sheath blight	*MOD1* gene from maize; *RCH10* gene from rice	Kim et al. (2003)
Wheat	*Fusarium* head blight	α-1-purothionin from wheat; tlp-1 and β-1,3-glucanase from barley	Mackintosh et al. (2007)
Indian mustard	*Alternaria brassicae*	Gulcanase gene from tomato	Mondal et al. (2007)
Tobacco, peanut	Multifungal resistance	*BjD* gene from mustard	Anuradha et al. (2008)
Rice	*Rhizoctonia solani*	*chi11* from rice; glucfrom tobacco	Sridevi et al. (2008)
Carrot	Foliar fungal pathogens	Chitinase and glucanase gene from wheat; *POC1* from rice	Wally et al. (2009)

TABLE 13.4 *(Continued)*

Crop	Disease Resistance	Source of Gene	Reference(s)
Apple	Apple scab fungus	Thionine gene from barley	Krens et al. (2011)
Finger millet	Leaf blast	*Chi11* gene from Rice	Ignacimuthu and Ceaser (2012)
Wheat	Powdery mildew	Over expression of *R* gene from wheat	Brunner et al. (2012)
Viral resistance			
Potato	Potato Virus Y	Pathogen derived resistance gene	Bravo-Almonacid et al. (2011)
Papaya	Ring spot virus	Pathogen derived resistance gene	Ferreira et al. (2002)
Tobacco	Tobacco Mosaic Virus	*p35* gene from baculovirus *Autographa californica*	Wang et al. (2008)
Potato	Potato Mosaic Virus	Coat protein gene of Potato mosaic virus	Malnoe et al. (1994)
Tobacco	Cucumber mosaic virus	Replicases from Cucumber mosaic virus (CMV)	Morroni et al. (2008)
Tobacco	Grapevine berry inner necrosis virus	Movement protein (P50) and partially functional deletion mutants (*DeltaA* and *DeltaC*) of the *Apple chlorotic leaf spot virus* (ACLSV)	Yoshikawa et al. (2006)

biological phenomenon has been evaluated in a number of host–pathogen systems and effectively used to silence the action of pathogen (Wani et al., 2010).

During the last decade, our knowledge repertoire of RNA-mediated functions has been greatly increased with the discovery of small noncoding RNAs which play a central part in a process called RNA silencing. Ironically, the very important phenomenon of co-suppression has recently been recognized as a manifestation of RNAi, an endogenous pathway for negative posttranscriptional regulation. RNAi has revolutionized the possibilities for creating custom "knock-downs" of gene activity. RNAi operates in both plants and animals and uses dsRNA as a trigger that targets homologous mRNAs for degradation or inhibiting its transcription or translation (Karthikeyan et al., 2013).

RNAi is a biological process where RNA molecule inhibits the expression of a particular gene by targeting and destructing of specific mRNA molecules. RNAi is also known as posttranscriptional gene silencing, co-suppression, and quelling. The discovery of RNAi was totally serendipity. The concept of RNAi for the first time came into the existence while the study of transcriptional inhibition by antisense RNA expressed in transgenic *Petunia* plant was conducted by Napoli et al. (1990). These plant scientists were trying to introduce additional copies of chalcone synthase gene responsible for darker pigmentation of flowers. The transgenic copy, intended to make more corresponding gene products. But instead of darker flowers, white or less pigmented flowers were observed indicating the suppressed/decreased expression of endogenous chalcone synthase gene (Napoli et al., 1990). This suggests downregulation of endogenous gene by the event posttranscriptional inhibition due to their mRNA degradation. Silencing of target genes by RNAi technology came into the limelight just after discovery of plant defense mechanism against virus, where it was believed that plant encode short, noncoding region of viral RNA sequences, which after infection recognize and degrades viral mRNA. These short and noncoding RNA sequences might be against viral DNA/RNA polymerase and other important genes necessary for viral infection and multiplication. On the theme of above concept, plant virologist introduced short nucleotides sequence into the viruses, and expression of target genes in the infected plants was found to be suppressed (Sharma et al., 2013; Wani et al., 2010). This most popular phenomenon is known as "virus-induced gene silencing" and brings the boom in the era of biotechnologists. Just a year later in 1998,

Craig Mello and Andrew Fire's performed works in the laboratory to study the effect of RNAi in *C. elegans*, and interestingly they found that dsRNA effectively silenced the target gene in comparison to antisense ssRNA (100 folds more potent). The term RNAi was coined by these two scientists for the first time and they were awarded Nobel Prize in the field of medicine in 2006 for this breakthrough (Fire et al., 1998). After this great discovery of dsRNA as an extremely potent trigger for gene silencing, it became very realistic to unravel the mechanism of RNAi action in various biological systems.

13.3.1.4.1 General Mechanism of RNA Interference

The RNAi pathway, ubiquitous to most of the eukaryotes, consists of short RNA molecule binds to specific target mRNA, forms a dsRNA hybrid, and inactivate the mRNA by preventing from producing a protein. Apart from their role in defense against viruses, protozoans, it also influences the development of organisms. During RNAi, the dsRNA formed in cells by DNA- or RNA-dependent synthesis of complementary strands, or introduced into cells by viral infection or artificial expression is processed to 20-bp double-stranded small interfering RNAs (siRNAs) containing 2-nucleotides 3′ overhangs (Filipowicz, 2005). RNAi operates by triggering the action of dsRNA intermediates, which are processed into RNA duplexes of 21–24 nucleotides by a ribonuclease III-like enzyme called Dicer. The siRNAs are then incorporated into an RNA-induced silencing complex (RISC), which mediates the degradation of mRNAs with sequences fully complementary to the siRNA. The siRNAs within RISC acts as a guide to target the degradation of complementary messenger RNAs (mRNAs). The host genome codifies for small RNAs called miRNAs that are responsible for endogenous gene silencing. The dsRNAs triggering gene silencing can originate from several sources such as expression of endogenous or transgenic antisense sequences, expression of inverted repeated sequences, or RNA synthesis during viral replication (Wani et al., 2010). In another recent pathway, occurring in the nucleus, siRNAs formed from repeat element transcripts and incorporated into the RNAi-induced transcriptional silencing complex may guide chromatin modification and silencing (Fig. 13.6).

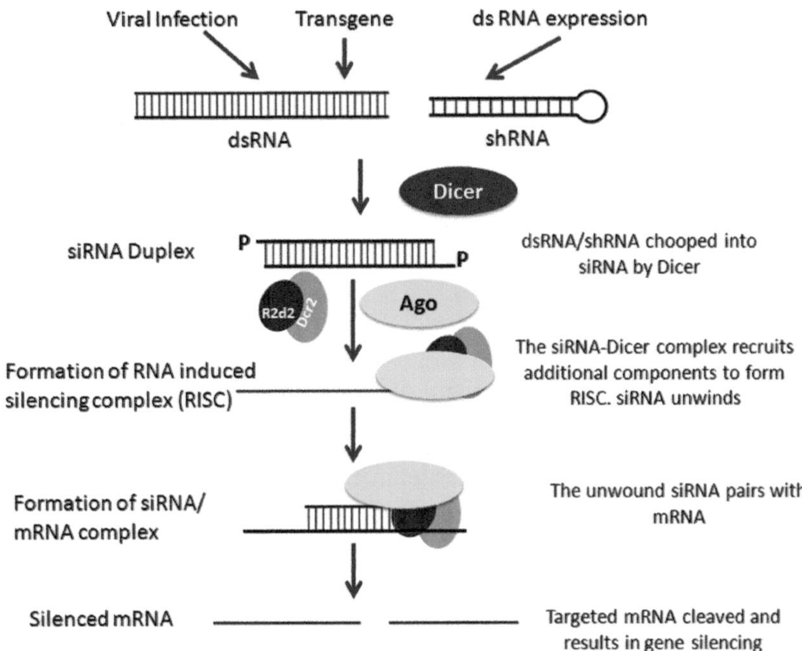

FIGURE 13.6 General mechanism of RNA interference.

13.3.1.4.2 RNAi in Plant Disease Management

RNA silencing has become a major focus of molecular biology and biomedical research around the world. To reduce the losses caused by plant pathogens, plant biologists have adopted numerous methods to engineer resistant plants. Among them, RNA silencing-based resistance has been a powerful tool that has been used to engineer resistant crops during the last two decades. RNA-mediated gene silencing is used as a reverse tool for gene targeting in fungi including *Fusarium graminearum* (Nakayashiki, 2005), *C. fulvum* (Hamada and Spanu, 1998), *Aspergillus nidulans* (Hammond and Keller, 2005), *Magnaporthae oryzae* (Chen et al., 2010; Kim et al., 2009), and *Venturia inaequalis* (Fitzgerald et al., 2004), whether it is suitable for large-scale mutagenesis in fungal pathogens remains to be tested. Hypermorphic mechanism of RNAi implies that this technique can also be applicable to all those plant pathogenic fungi, which are polyploid and polykaryotic in nature, and also offers a solution to the problem where frequent lack of multiple marker genes in fungi is experienced. Homology-based gene silencing induced by transgenes (co-suppression), antisense, or dsRNA has been demonstrated in many plant pathogenic fungi. The utilization of RNAi technology has also resulted in inducing immunity reaction against several viruses in different plant–virus systems (Table 13.5). The effectiveness of the technology in generating virus resistant plants was first reported to PVY in potato, harboring vectors for simultaneous expression of both sense and antisense transcripts of the helper-component *proteinase* (*HC-Pro*) gene (Waterhouse et al., 1998).

13.3.1.5 MOLECULAR BREEDING TECHNIQUES

The modern molecular techniques make it possible to use markers and probes to track the introgression of several *R*-genes into a single cultivar from various sources during a crossing program. Although conventional breeding has a significant impact on improving resistance cultivars, the time-consuming process of making crosses and backcrosses, and the selection of the desired resistant progeny make it difficult to react adequately to the evolution of new virulent pathogens. DNA markers serve as a new tool to detect the presence of allelic variation in the genes underlying the economic traits. DNA markers have enormous potential to improve the efficiency and precision of conventional plant breeding *via* marker-assisted selection (MAS) by reducing the reliance on laborious and fallible screening procedures. MAS is most useful for traits that are difficult to select, for example, disease resistance, salt tolerance, drought tolerance, heat tolerance, quality traits (aroma of *basmati* rice, flavor of vegetables). The approach involves selecting plants at early generation with a fixed, favorable genetic background at specific loci, conducting a single large-scale MAS, while maintaining as much as possible the allelic segregation in the population and the screening of large populations to achieve the objectives of the scheme. No selection is applied outside the target genomic regions to maintain as much as possible the Mendelian allelic segregation among the selected genotypes. After selection with DNA markers, the genetic diversity at unselected loci may allow breeders to generate new varieties and hybrids through conventional breeding in response to targets set in breeding program (Datta et al., 2011; Kumar et al., 2015).

13.3.1.5.1 Material Required for MAS

Molecular markers, a set of authentic lines carrying trait of interest and a population to validate the markers to be used, for example, F_2 or BC_1F_2 for

TABLE 13.5 Gene Silencing through RNA Interference Technology in Plant–Virus System.

Host	Virus	Gene Silenced	Reference(s)
N. benthamiana, *M. esculenta*	African cassava mosaic virus	*pds*, *su*, *cyp79d2*	Fofana et al. (2004)
Barley, wheat	Barley stripe mosaic virus	*Pds*	Holzberg et al. (2002), Scofield et al. (2005), Cakir et al. (2010)
Soybean	Bean pod mottle virus	*Actin*, *Pds*	Zhang and Ghabrial (2006), Zhang et al. (2009)
Barley, rice, maize	Brome mosaic virus	*pds*, *actin 1*, *rubisco activase*	Ding et al. (2006)
Arabidopsis	Cabbage leaf curl virus	*gfp*, *CH42*, *pds*	Turnage et al. (2002)
P. sativum	Pea early browning virus	*pspds*, *uni*, *kor*, *pds*	Constantin et al. (2004)
N. benthamiana, *S. tuberosum*	Potato virus X	*pds*, *gfp*	Ruiz et al. (1998), Faivre-Rampant et al. (2004)
N. benthamiana	Tomato bushy shunt virus	*Gfp*	Hou et al. (2003)
N. benthamiana, *Tomato*, *Solanum* sp., *chilli pepper*	Tobacco rattle virus	*Rar1*, *EDS1*, *NPR1/NIM1*, *pds*, *rbcS*	Ratcliff et al. (2001), Hileman et al. (2005)
Arabidopsis	Cabbage leaf curl virus	*CH42*, *pds*	Turnage et al. (2002)
N. benthamiana, *N. tobaccum*	Tomato yellow leaf curl virus	*Pcna*, *pds*, *su*, *gfp*	Kjemtrup et al. (1998)

each of the individual traits/genes (Datta et al., 2011). Following are the basic prerequisites for MAS:

- Evaluating molecular markers that are linked to the trait of interest
- Validation of markers in parents and breeding population
- Designing and validation of new markers in case of nonavailability of the markers
- Designing of selection scheme and breeding strategy
- Fix the minimum population to be assayed to capture all beneficial alleles
- Progeny testing for fixation of traits.

13.3.1.5.2 Limitations of MAS

- Cost factor
- Requirement of technical expertise
- Automated techniques for maximum benefit
- Per se, DNA markers are not affected by environment but traits may be affected by the environment and show G×E interactions. Therefore, while developing markers, phenotyping should be carried out in multiple environments and implications of G×E should be understood and markers should be used judiciously.
- DNA marker has to be validated for each of the breeding population.

13.3.1.5.3 Identification of Disease Resistance Genes through Molecular Markers

DNA-based markers have shown great promises in expediting plant-breeding methods. The identification of molecular markers closely linked with resistance genes would facilitate expeditious pyramiding of major genes into elite background, making it more cost effective. Once the resistance genes are tagged with molecular marker, the selection of resistant plant in the segregating generations becomes easy (Kumar et al., 2015). *R* genes have remarkable property of rapid diversification under selective pressure from the pathogens. Most plant species contain a large number of highly polymorphic disease resistance genes having common structural domains (Gururani et al., 2012). The DNA rearrangements have been advocated to play a crucial role in *R*-gene evolution allowing plants to generate novel resistance specificities to match the changing virulence pattern of the pathogen. The R genes pave a way for developing durable resistance in plants through molecular breeding approaches (Sharma et al., 2014). Sekhwal et al. (2015) provided comprehensive details on the R gene cloned and characterized in various plant species which would be helpful in generating useful genetic resources to create novel resistant cultivars (Table 13.6).

13.3.1.5.4 Marker-Assisted Backcross Breeding

A backcross breeding program is aimed at gene introgression from a "donor" line into the genomic background of a "recipient" line. The potential utilization of molecular markers in such programs has received considerable attention in the recent past. Markers can be used to assess the presence of the introgressed gene ("foreground selection") when direct phenotypic evaluation is not possible, or too expensive, or only possible late in the development. Markers can also be used to accelerate the return to the recipient parent genotype at other loci ("background selection"). It is assumed that the introgressed gene can be detected without ambiguity, and the theoretical study was restricted to background selection only. The use of molecular markers for background selection in backcross programs has been tested experimentally and proved to be very efficient. Introgressing the favorable allele of QTL by recurrent backcrossing can be a powerful mean to improve the economic value of a line, provided the expression of the gene is not reduced in the recipient genomic background. Yet, recent results show that for many traits of economic importance QTLs have rather small effects. In this case, the economic improvement resulting from the introgression of the favorable allele at a single QTL may not be competitive when compared with the improvement resulting from conventional breeding methods over the same duration. Marker-assisted introgression of superior QTL alleles can then compete with classical phenotypic selection only if several QTLs could be manipulated (Datta et al., 2011; Xu et al., 2013).

Marker-assisted backcross breeding (MABC) has been widely used to transfer or introgress genes from one elite line to another. To improve the hybrid rice currently widely grown in China, a series of MAS were performed. *Xa21*, *Xa7*, and *Xa23*, three wide-spectrum BB resistance genes, were introgressed to the restorers Minghui 63 and 9311 by MAS (Chen et al., 2000). Two genes, *Pi1* and *Pi2*, showing broad spectrum resistance to fungi blast, were introgressed into the maintainer Zhenshan 97. Three genes, *Bph14*, *Bph15*, and *Bph18*, highly resistant to brown planthop per (BPH), were introgressed to Zhenshan 97, Minghui 63 and 9311 to improve their

TABLE 13.6 Cloned and Characterized R Genes in Different Plant Species.

Plant	Disease	Resistance Genes	Reference
Arabidopsis	Downy mildew	*RPM1, RPS2, RPP8/HRT, RPP13, RPP1, RPP4, RPP5, RPS5, RPP27*	Sekhwal et al. (2015)
	Powdery mildew	*RPS4*	
	Fusarium wilt	*RFO1*	
	Bacterial wilt	*RRS1*	
Rice	BLAST	*Pib, Pi-ta, Pi36, Pia,* $Pi\text{-}K^{h}$*, Pi37, Pi54, Pi9, Piz-t/Pi2, Rpr1, Pid3, Pi-d2*	
	Bacterial blight	*Xa21, Xa3/Xa26, CEBiP, Xa10, Xa25, Xa27, Xa5, Xa13, Xa1*	
Wheat	Leaf rust	*Lr10, Lr1, Lr21, Lr34*	
	Stem rust	*Sr33, Sr35*	
	Cereal cyst	*Cre3, Cre1*	
	Powdery mildew	*Pm3b, Stpk-V (Pm21), Lr34*	
	Stripe rust	*Yr36, Lr34*	
Barley	Powdery mildew	*Mla6, Mla1, Mla13, Mlo*	
	Stem rust	*Rpg1*	
Tomato	Leaf mold	*Cf-2, Cf-4, Cf-5, Cf-9, Hcr9-4E*	
	Bacterial speck	*Fen, Pto, Pti1, Prf*	
	Root knot	*Mi*	
	Fusarium wilt	*I2*	
	Late blight	*Ph- 3*	
	Tomato spotted wilt	*Sw-5*	

TABLE 13.6 *(Continued)*

Plant	Disease	Resistance Genes	Reference
	Tobacco mosaic	*Tm-2*	
	Bacterial spot	*Bs4*	
	Potato cyst	*Hero*	
	Verticillium wilt	*Vel 2*	
Potato	PVX	*Rx*, *Rx2*	
	Late blight	*RB*, *R1*	
Maize	Rust	*Rp1-D*	
	Corn leaf blight	*Hm1*	
Chickpea	Fusarium wilt	*foc-0*, *foc-1*, *foc-2*, *foc-3*, *foc-4*, *foc-5*	Varshney et al. (2015)

BPH resistance. To improve the disease resistance for Basmati rice, Pusa 1460 was utilized as the donor for introgressing BB resistance genes *xa13* and *Xa21* into Pusa6B and PRR78, the two parental lines for aromatic hybrid rice Pusa RH10 (Basavaraj et al., 2010). Two BB-resistant rice cultivars, Improved Pusa Basmati-1 (Pusa 1460) and Improved Sambha Mahsuri, were developed (Joseph et al., 2004; Sundaram et al., 2009) and released for commercial cultivation. Traditional basmati varieties were also improved for BB resistance and plant height using MABC to transfer two BB resistance genes, *xa13* and *Xa21*, and semidwarfig gene, *sd-1*, into two traditional basmati varieties, Basmati 370 and Basmati 386 (Bhatia et al., 2011).

In wheat, a study was undertaken to assess the effect on improving FHB resistance and on possible unwanted side effects (linkage drag) of two resistance QTL, *Fhb1* and *Qfhs.ifa-5A*, from the spring wheat line CM-82036 when transferred by MABC into several European winter wheat lines (Salameh et al., 2011). In USA, 27 different disease and pest resistance genes and 20 alleles with beneficial effects on bread-making and pasta quality were incorporated into about 180 lines adapted to the primary US wheat production regions (Sorrells, 2007).

Efforts are being made to introgress resistance to different races independently as well as pyramiding of resistance to two races for Fusarium Wilt in some elite Chickpea varieties in India. ICRISAT (India) is pyramiding resistances for *Foc1* and *Foc*3 from WR 315 and 2 QTLs for Ascochyta blight resistance from ILC 3279 line into C 214 (Varshney et al., 2013).

13.3.1.5.5 Genomics-Assisted Breeding Approach

The advent of markers based on simple sequence repeats and single nucleotide polymorphisms and the availability of high-throughput genotyping platforms has further accelerated the generation of dense genetic linkage maps and the routine use of the markers for marker-assisted breeding in several crops (Collard and Mackill, 2008). However, despite the routine use of markers for genome-wide profiling and trait-specific MAS, breeding of crops with many traits of interest such as yield, improved nutritive value, and resistance to several biotic and abiotic stresses is still a challenge due to complex inheritance of these traits. Plant genomics has enormous potential to revolutionize crop improvement by providing extensive knowledge from the analysis of genomes which in turn can be used for rapid and efficient plant breeding toward crop improvement (Kumpatla et al., 2012).

The advent of NGS technologies has changed the dynamics and the pace of genomic research in cereals and pulses against biotic stresses because of their rapid, inexpensive and highly accurate sequencing capabilities. Unlike Sanger sequencing method which depends upon capillary electrophoresis, these NGS technologies are highly dependent on massive parallel sequencing, high resolution imaging, and complex algorithms to deconvolute the signal data to generate sequence data. NGS technologies offer a wide variety of applications such as whole genome de novo and resequencing, transcriptome sequencing (RNA-seq), miRNA sequencing, amplicon sequencing, targeted sequencing, chromatin immuno-precipitated DNA sequencing (ChIP-seq), methylome sequencing, etc. (Varshney et al., 2015).

To facilitate crop improvement, NGS and other accessory technologies can be used for whole genome sequencing, transcriptome sequencing, genome wide and candidate gene marker development, targeted enrichment and sequencing and other applications. These NGS technologies even hold promise for a methodological leap toward genotyping by sequencing and genetic mapping applications. Analysis of NGS data from genome wide association studies, transcriptomics and epigenomics in combination with data from proteomics, metabolomics, and other "omics" can provide an integrative systems biology approach to understand the regulation of complex traits.

13.4 CONCLUSIONS

With advancement in molecular biology, biotechnology, genomics and proteomics, researchers and farmers alike would be able to improve plant disease diagnosis and management effectively. Although, there are several techniques developed for disease diagnosis, each having their own potential and shortcomings. Efforts are already underway to produce better diagnostic kits to detect pathogens in crops important to developing countries. Advances in tissue culture and protoplast fusion technique are being evaluated to develop pathogen free as well as disease resistant varieties. RNAi technology is being explored to develop disease resistance varieties. Emphasis is being given on disease prevention than control. The Whole genome sequencing of both model and other crop species are expected to offer new perspectives into develop resistance against various biotic as well as abiotic stresses. Nevertheless, the collaborative and coordinated efforts made during the last decade, contributed to development of large-scale genomic resources in cereals, pulses, and horticultural crops. As a result, crops have become "genomic resource rich" crops which can be used to understand the genetics of several traits and as a result, approaches like MABC and MARS are being used in these crops. Genomic selection seems to be a potential approach to be used very soon in crop species. While genome sequence has become available in several plant species, molecular-breeding approaches will have major milestones to combat against major diseases and pests. Although conventional breeding based exclusively on phenotypic selection remains the mainstay for most breeding programs, adoption of molecular methods is increasing and in some cases are superseding conventional approaches.

KEYWORDS

- **biotechnology**
- **molecular breeding**
- **molecular diagnosis**
- **pathogen**
- **RNAi**
- **transgenic**

REFERENCES

Alberts, B.; Johnson, A.; Lewis, J.; Raff, M.; Roberts, K.; Walter, A. P. *Molecular Biology of the Cell*, 4th ed. Garland Science: New York, 2002.

Al-Taleb, M. M.; Hassawi, D. S.; Abu-Rommau, S. M. Production of Virus-Free Potato Plants Using Meristem Culture from Cultivars Grown under Jordanian Environment. *J. Agric. Environ. Sci.* **2011,** *11*, 467–472.

Alvarez, A. M. Integrated Approaches for Detection of Plant Pathogenic Bacteria and Diagnosis of Bacterial Diseases. *Annu. Rev. Phytopathol.* **2004,** *42*, 339–366.

Ankenbauer, R. G.; Nester, E. W. Sugar-Mediated Induction of *Agrobacterium tumefaciens* Virulence Genes: Structural Specificity and Activities of Monosaccharides. *J. Bacteriol.* **1990,** *172*, 6442–6446.

Anuradha, T. S.; Divya, K.; Jami, S. K.; Kirti, P. B. Transgenic Tobacco and Peanut Plants Expressing a Mustard Defensin Show Resistance to Fungal Pathogens. *Plant Cell Rep.* **2008,** *27*, 1777–1786.

Awan, A. R.; Mughal, S. M.; Iftikhar, Y.; Khan, H. Z. *In Vitro* Elimination of Potato Leaf Roll Polerovirus from Potato Varieties. *Eur. J. Sci. Res.* **2007,** *18*, 155–164.

Basavaraj, S. H.; Singh, V. K.; Singh, A.; Singh, A.; Singh, A.; Anand, D.; Yadav, S.; Ellur, R. K.; Singh, D.; Krishnan, S. G.; Nagarajan, M.; Mohapatra, T.; Prabhu, K. V.; Singh, A.

K. Marker-Assisted Improvement of Bacterial Blight Resistance in Parental lines of Pusa RH10, a Superfine Grain Aromatic Rice Hybrid. *Mol. Breed.* **2010,** *26*, 293–305.

Bhardwaj, S. V.; Rai, S. J.; Thakur, P. D.; Handa, A. Meristem Tip Culture and Heat Therapy for Production of Apple Mosaic Virus Free Plants in India. *Acta Hortic.* **1998,** *472*, 65–68.

Bhatia, D.; Sharma, R.; Vikal, Y.; Mangat, G. S.; Mahajan, R.; Sharma, N.; Singh, K. Marker-assisted Development of Bacterial Blight Resistant, Dwarf, and High Yielding Versions of Two Traditional Basmati Rice Cultivars. *Crop Sci.*, **2011**. *51* (2), 759–770.

Binder, K. A.; Wegner, L. H.; Heidecker, M.; Zimmermann, U. Gating of Cl-currents in Protoplasts from the Marine Alga *Valonia utricularis* Depends on the Transmembrane Cl-gradient and Is Affected by enzymatic cell wall degradation. *J. Membrane Biol.*, **2003**. *191* (3), 165–178.

Bindschedler, L. V.; Burgis, T. A.; Mills, D. J. S.; Ho, J. T. C.; Cramer, R.; Spanu, P. D. In Planta Proteomics and Proteogenomics of the Biotrophic Barley Fungal Pathogen *Blumeria graminis* f. sp. *Hordei. Mol. Cell. Proteomics* **2009,** *8* (10), 2368–2381.

Biswas, M. K.; Hossain, M.; Islam, R. Virus-free Plantlets Production of Strawberry through Meristem Culture. *World J. Agric. Sci.* **2007,** *3*, 757–763.

Bonants, P. J. M.; Schoen, C. D.; Szemes, M.; Speksnijder, A.; Klerks, M. M.; van den Boogert, P. H. J. F.; Waalwijk, C.; van der Wolf, J. M.; Zijlstra, C. From Single to Multiplex Detection of Plant Pathogens: pUMA, a New Concept of Multiplex Detection Using Microarrays. *Phytopathol. Pol.* **2005,** *35*, 29–47.

Bonants, P.; De Weerdt, M.; Van Beckhoven, J.; Hilhorst, R.; Chan, A.; Boender, P.; Zijlstra, C.; Schoen, C. Multiplex Detection of Plant Pathogens by Microarrays: An Innovative Tool for Plant Health Management. In: Abstracts Agricultural Biomarkers for Array Technology, Management Committee Meeting, Wadenswil, 2002, p 20.

Bradeen, J. M.; Iorizzo, M.; Mollov, D. S.; Raasch, J.; Kramer, L. C.; Millett, B. P.; Carputo, D. Higher Copy Numbers of the Potato RB Transgene Correspond to Enhanced Transcript and Late Blight Resistance Levels. *Mol. Plant-Microbe Interact.* **2009,** *22* (4), 437–446.

Bravo-Almonacid, F.; Rudoy, V.; Welin, B.; Segretin, M.; Bedogni, M.; Stolowicz, F.; Criscuolo, M.; et al. Field Testing, Gene Flow Assessment and Pre-commercial Studies on transgenic spp. (cv. Spunta) Selected for PVY Resistance in Argentina. *Transgenic Res.* **2012,** *5* (21), 967–982.

Brown, P. O.; Botstein, D. Exploring the New World of the Genome with DNA Microarrays. *Nat. Genet.* **1999,** *21*, 33–37.

Brunner, S.; Stirnweis, D.; Diaz Quijano, C.; Buesing, G.; Herren, G.; Parlange, F.; Keller, B. Transgenic Pm3 Multilines of Wheat Show Increased Powdery Mildew Resistance in the Field. *Plant Biotechnol. J.* **2012,** *10* (4), 398–409.

Cakir, C.; Tör, M. Factors Influencing Barley Stripe Mosaic Virus-Mediated Gene Silencing in Wheat. *Physiol. Mol. Plant Pathol.* **2010,** *74*, 246–253.

Cangelosi, G. A.; Martinetti, G.; Leigh, J. A.; Lee, C. C.; Theines, C.; Nester, E. W. Role of *Agrobacterium tumefaciens* chvA Protein in Export of b-1,2 Glucan. *J. Bacteriol.* **1989,** *171*, 1609–1615.

Chandler, D. P.; Jarrell, A. E. Taking Arrays from the Lab to the Field: Trying to Make Sense of the Unknown. *BioTechniques* **2005,** *38*, 591–600.

Chang, M. M.; Culley, D.; Choi, J. J.; Hadwiger, L. A. Agrobacterium-mediated co-transformation of a Pea β-1, 3-glucanase and Chitinase Genes in Potato (*Solanum tuberosum* L. cv Russet Burbank) Using a Single Selectable Marker. *Plant Sci.* **2002** *163* (1), 83–89.

Chen, L.; Shiotani, K.; Togashi, T.; Miki, D.; Aoyama, M.; Wong, H. L.; Kawasaki, T.; Shimamoto, K. Analysis of the Rac/Rop Small GTPase Family in Rice: Expression, Subcellular Localization and Role in Disease Resistance. *Plant Cell Physiol.* **2010,** *51*, 585–595.

Chen, S.; Lin, X. H.; Xu, C. G.; Zhang, Q. Improvement of Bacterial Blight Resistance of 'Minghui63', an Elite Restorer Line of Hybrid Rice, by Molecular Marker-Assisted Selection. *Crop Sci.* **2000,** *40*, 239–244.

Chiari, A.; Bridgen, M. P. Meristem Culture and Virus Eradication in Alstromeria. *Plant Cell, Tissue Organ Cult.* **2002,** *68*, 49–55.

Cieslinska, M. Elimination of Apple Chlorotic Leaf Spot Virus (ACLSV) from Pear by *in vitro* Thermotherapy and Chemotherapy. *Acta Hortic.* **2002,** *596*, 481–484.

Clark, M. F.; Adams, A. M. Characteristics of the Microplate Method of Enzyme-Linked Immunosorbent Assay for the Detection of Plant Viruses. *J. General Virol.* **1977,** *34*, 475–483.

Collard, B. C.; Mackill, D. J. Marker-assisted Selection: An Approach for Precision Plant Breeding in the Twenty-First Century. *Philos. Trans. R. Soc. Lond. B: Biol Sci.* **2008,** *363* (1), 557–572.

Collonnier, U.; Fock, I.; Daunay, M. C.; Servaes, A.; Vedel, F.; Sijak-Yakovlev, S. Somatic Hybrids Between *Solanum melongena* and *S. sisymbrifolium*, as a Useful Source of Resistance against Bacterial and Fungal Wilts. *Plant Sci.* **2003,** *164*, 849–861.

Collonnier, U.; Mulya, K.; Fock, I.; Mariska, I.; Servaes, A.; Vedel, F. Source of Resistance against *Ralstonia solanacearum* in Fertile Somatic Hybrids of Eggplant (*Solanum melongena* L.) with *Solanum aethiopicum* L. *Plant Sci.* **2001,** *160*, 30–13.

Constantin, G. D.; Krath, B. N.; MacFarlane, S. A.; Nicolaisen, M.; Johansen, I. E.; Lund, O. S. Virus-Induced Gene Silencing as a Tool for Functional Genomics in a Legume Species. *Plant J.* **2004,** *40*, 622–631.

Costa, M. A. P. D.; Mendes, B. M. J.; Mourao, F. A. A. Somatic Hybridization for Improvement of Citrus Rootstock: Production of Five New Combinations with Potential for Improved Disease Resistance. *Austr. J. Exp. Agric.* **2003,** *43*, 1151–1156.

Crocetti, G. R.; Hugenholtz, P.; Bond, P. L.; Schuler, A.; Keller, J.; Jenkins, D.; Blackall, L. L. Identification of Polyphosphate-Accumulating Organisms and Design of 16S rRNA-Directed Probes for their Detection and Quantitation. *Appl. Environ. Microbiol.* **2000,** *66* (3), 1175–1182.

Curto, M.; Camafeita, E.; Lopez, J. A.; Maldonado, A. N. A. M.; Rubiales, D.; Jorrín, J. V. A Proteomic Approach to Study Pea (*Pisum sativum*) Responses to Powdery Mildew (*Erysiphe pisi*). *Proteomics* **2006,** *6* (Suppl. 1), S163–S174.

Datta, D.; Gupta, S.; Chaturvedi, S. K.; Nadarajan, N. Molecular Markers in Crop Improvement. Indian Institute of Pulses Research: Kanpur, 2011, pp 1–54.

Davey, M. R.; Anthony, P.; Power, J. B.; Lowe, K. C. Plant Protoplasts: Status and Biotechnological Perspectives. *Biotechnol. Adv.* **2005** *23* (2), 131–171.

De Boer, S. H.; Elphinstone, J. G.; Saddler, G. Molecular Detection Strategies for Phytopathogenic Bacteria. In: *Biotechnology and Plant Disease Management*; Punja, Z. K., De Boer, S. H., Sanfançon, H., Eds.; CAB International: Oxfordshire, UK, 2007; pp 165–194.

de la Riva, G. A.; González-Cabrera, J.; Vázquez-Padrón, R.; Ayra-Pardo, C. *Agrobacterium tumefaciens*: A Natural Tool for Plant Transformation. *Electron. J. Biotechnol.* **1998,** *1* (3), 1–16.

De Wit, P. J. G. M.; Buurlage, M. B.; Hammond, K. E. The Occurrence of Host–Pathogen-and Interaction-specific Proteins in the Apoplast of *Cladosporium fulvum* (syn. *Fulvia fulva*) Infected Tomato Leaves. *Physiol. Mol. Plant Pathol.* **1986,** *29* (2), 159–172.

Ding, X. S.; Schneider, W. L.; Chaluvadi, S. R.; Rouf Mian, R. M.; Nelson, R. S. Characterization of a Brome Mosaic Virus Strain and Its Use as a Vector for Gene Silencing in Monocotyledonous Hosts. *Mol. Plant Microbe Interact.* **2006,** *19*, 1229–1239.

Ellis, J. G.; Rafiqi, M.; Gan, P.; Chakrabarti, A.; Dodds, P. N. Recent Progress in Discovery and Functional Analysis of Effector Proteins of Fungal and Oomycete Plant Pathogens. *Curr. Opin. Plant Biol.* **2009,** *12* (4), 399–405.

Evans, P. K.; Cocking, E. C. techniques of plant cell culture and somatic cell hybridization. *New Tech Biophys & Cell Biol.* 1975.

Faccioli, G. Control of Potato Viruses Using Meristem and Stem-Cutting Cultures, Thermotherapy and Chemotherapy. In: Virus and Virus-Like Disease of Potatoes and Production of Seed-Potatoes; Loebenstein, G., et al., Eds.; Kluwer Academic Publishers: Dordrecht, 2001; pp 365–390.

Faivre-Rampant, O.; Gilroy, E. M.; Hrubikova, K.; Hein, I.; Millam, S.; Loake, G. J.; Birch, P.; Taylor, M.; Lacomme, C. Potato Virus X-Induced Gene Silencing in Leaves and Tubers of Potato. *Plant Physiol.* **2002,** *134*, 1308–1316.

Ferreira, S. A.; Pitz, K. Y.; Manshardt, R.; Zee, F.; Fitch, M.; Gonsalves, D. Virus Coat Protein Transgenic Papaya Provides Practical Control of Papaya Ringspot Virus in Hawaii. *Plant Dis.* **2002,** *86* (2), 101–105.

Filipowicz, W. RNAi: The Nuts and Bolts of the RISC Machine. *Cell* **2005,** *122*, 17–20.

Fire, A.; Xu, S.; Montgomery, M. K.; Kostas, S. A.; Driver, S. E.; Mello, C. C. Potent and Specific Genetic Interference by Double-stranded RNA in *Caenorhabditis elegans*. *Nature* **1998,** *391*, 806–811.

Fitzgerald, A.; Van Kha, J. A.; Plummer, K. M. Simultaneous Silencing of Multiple Genes in the Apple Scab Fungus *Venturia inaequalis*, by Expression of RNA with Chimeric Inverted Repeats. *Fungal Genetics Biol.* **2004,** *41*, 963–971.

Fock, I.; Collonnier, C.; Luisetti, J.; Purwito, A.; Souvannavong, V.; Vedel, F. Use of *Solanum stenotomum* for Introduction of Resistance to Bacterial Wilt in Somatic Hybrids of potato. *Plant Physiol. Biochem.* **2001,** *39*, 899–908.

Fofana, I. B.; Sangare, A.; Collier, R.; Taylor, C.; Fauquet, C. M. A Geminivirus-Induced Gene Silencing System for Gene Function Validation in Cassava. *Plant Mol. Biol.* **2004,** *56*, 613–624.

Foster, S. J.; Park, T. H.; Pel, M.; Brigneti, G.; Sliwka, J.; Jagger, L.; Jones, J. D. Rpi-vnt1.1, a Tm-22 Homolog from *Solanum venturii*, Confers Resistance to Potato Late Blight. *Mol. Plant-Microbe Interact.* **2009,** *22* (5), 589–600.

Fraga, M.; Alonso, M.; Ellul, P.; Borja, M. Micropropagation of *Dianthus gratianopolitanus*. *Hortic. Sci.* **2004,** *39* (5), 1083–1087.

Franke-Whittle, I. H.; Klammer, S. H.; Insam, H. Design and Application of an Oligonucleotide Microarray for the Investigation of Compost Microbial Communities. *J. Microbiol. Methods* **2005,** *62*, 37–56.

Fuchs, M.; Gonsalves, D. Genetic Engineering. In: *Environmentally Safe Approaches to Crop Disease Control*; Rechcigl, N. A., Rechcigl, J. E., Eds.; CRC Press, Boca Raton, New York, 1996; pp 333–368.

Gavrilenko, T.; Thieme, R.; Heimbach, U.; Thieme, T. Fertile Somatic Hybrids of *Solanum tuberosum* (+) Dihaploid *Solanum tuberosum* and their Backcrossing Progenies: Relationships of Genome Dosage with Tuber Development and Resistance to Potato Virus Y. *Euphytica* **2003,** *131*, 323–332.

Gichuki, S.; La Bonte, D.; Burg, K.; Kapinga, R.; Simon, J. C. Assessment and Genetic Diversity, Farmer Participatory Breeding and Sustainable Conservation of Eastern Africa Sweet Potato Germplasm. *Annual Report April 2004–March 2005*, 2005.

González-Fernández, R.; Prats, E.; Jorrín-Novo, J. V. Proteomics of plant pathogenic fungi. *J. Biomed. Biotechnol.* **2010,** *36*, 1–36.

Grey, B. E.; Steck, T. R. The Viable but Nonculturable State of *Ralstonia solanacearum* may be Involved in Long-term Survival and Plant Infection. *Appl. Environ. Micro.* **2001**, *67*(9), 3866–3872.

Grosser, J. W.; Gmitter, Jr., F. G. Protoplast Fusion for Production of Tetraploids and Triploids: Applications for Scion and Rootstock Breeding in Citrus. *Plant Cell, Tissue, Organ Cult.* **2011,** *104*, 343–357.

Guo, W.; Cheng, Y.; Deng, X. Regeneration and Molecular Characterization of Intergeneric Somatic Hybrids between Citrus reticulata and Poncirus trifoliata. *Plant Cell Reports* **2002,** *20* (9), 829–834.

Gururani, M. A.; Venkatesh, J.; Upadhyaya, C. P.; Nookaraju, A.; Pandey, S. K.; Park, S. W. Plant Disease Resistance Genes: Current Status and Future Directions. *Physiol. Mol. Plant Pathol.* **2012,** *78*, 51–65.

Hadidi, A.; Czosnek, H.; Barba, M. DNA Microarrays and their Potential Applications for the Detection of Plant Viruses, Viroids, and Phytoplasmas. *J. Plant Pathol.* **2004,** *86*, 97–104.

Halterman, D.; Kramer, L.; Wielgus, S.; Jiang, J. *Plant Dis.* **2008,** *92*, 339–343.

Hamada, W.; Spanu, P. D. Co-Suppression of the Hydrophobin Gene *Hcf-1* is Correlated with Antisense RNA Biosynthesis in *Cladosporium fulvum*. *Mol. Gen. Genet.* **1998,** *259*, 630–638.

Hammond, T. M.; Keller, N. P. RNA Silencing in *Aspergillus nidulans* is Independent of RNA-Dependent RNA Polymerase. *Genetics* **2005,** *169*, 607–617.

Hansen, A. J.; Lane, W. D. Elimination of Apple Chlorotic Leaf Spot Virus from Apple Shoot Cultures by Ribavirin. *Plant Dis.* **1985,** *69*, 134–135.

Hardy, M. S. *Use of Somatic Hybridization for Production of a New Pure Tomato Cell Line Resistant to Some Microbial Diseases*. AGRIS, 2011.

Helgeson, J. P.; Pohlman, J. D.; Austin, S.; Haberlach, G. T.; Wielgus, S. M.; Ronis, D. Somatic Hybrids between *Solanum bulbocastanum* and Potato: A New Source of Resistance to Late Blight. *Theor Appl. Genet.* **1998,** *96*, 738–742.

Helliot, B.; Panis, B.; Hernandez, R.; Swennen, R.; Lepoivre, P.; Frison, E. Development of *In Vitro* Techniques for the Elimination of Cucumber Mosaic Virus from Banana (*Musa* spp.). In: *Banana Improvement. Cellular, Molecular Biology and Induced Mutations*. Mohan, J. S., Swennen, R., Eds.; Sci. Publishers, Inc.: Enfield, 2004; pp 183–191.

Hileman, L. C.; Drea, S.; Martino, G.; Litt, A.; Irish, V. F. Virus Induced Gene Silencing is an Effective Tool for Assaying Gene Function in the Basal Eudicot Species *Papaver somniferum* (Opium Poppy). *Plant J.* **2005,** *44*, 334–341.

Holzberg, S.; Brosio, P.; Gross, C.; Pogue, G. P. Barley Stripe Mosaic Virus-Induced Gene Silencing in a Monocot Plant. *Plant J.* **2002,** *30*, 315–327.

Horvath, D. M.; Stall, R. E.; Jones, J. B.; Pauly, M. H.; Vallad, G. E.; Dahlbeck, D.; Scott, J. W. Transgenic Resistance Confers Effective Field Level Control of Bacterial Spot Disease in Tomato. *PLoS ONE* **2012,** *7* (8), 420–436.

Hou, H.; Qiu, W. A Novel Co-delivery System Consisting of a Tomato Bushy Stunt Virus and a Defective Interfering RNA for Studying Gene Silencing. *J. Virol Methods* **2003,** *111*, 37–42.

Hu, Q.; Andersen, S. B.; Dixelius, C.; Hansen, L. N. Production of Fertile Intergeneric Somatic Hybrids Between *Brassica napus* and *Sinapsis arvensis* for the Enrichment of the Rapeseed Gene Pool. *Plant Cell Rep.* **2002,** *1*, 147–152.

Hummel, A. W.; Doyle, E. L.; Bogdanove, A. J. Addition of Transcription Activator-Like Effector Binding Sites to a Pathogen Strain-Specific Rice Bacterial Blight Resistance Gene Makes It Effective Against Additional Strains and against Bacterial Leaf Streak. *N. Phytol.* **2012,** *195* (4), 883–893.

Ignacimuthu, S.; Ceasar, S. A. Development of Transgenic Finger Millet (*Eleusine coracana* (L.) Gaertn.) Resistant to Leaf Blast Disease. *J. Biosci.* **2012,** *37*, 135–147.

Jeon, G.A.; Eum, J.S.; Sim, W.S. The Role of Inverted Repeat (IR) Sequence of the *virE* Gene Expression in *Agrobacterium tumefaciens* pTiA6. *Molecules Cells* **1998,** *8*, 49–53.

Joseph, M.; Gopalakrishnan, S.; Sharma, R. K.; Singh, V. P.; Singh, A. K.; Singh, N. K.; Mohapatra, T. Combining Bacterial Blight Resistance and Basmati Quality Characteristics by Phenotypic and Molecular Marker Assisted Selection in Rice. *Mol. Breed.* **2004,** *13*, 377–387.

Karthikeyan, A.; Deivamani, M.; Shobhana, V. G.; Sudha, M.; Anandhan, T. RNA Interference: Evolutions and Applications in Plant Disease Management. *Arch. Phytopathol. Plant Protect.* **2013**, *46* (12), 1430–1441.

Keller, W. A.; Melchers, G. The effect of high pH and calcium on tobacco leaf protoplast fusion. *Zeitschrift für Naturforschung C*, **1973,** *28* (11-12), 737–741.

Kenganal, M. Y.; Amudai, J. A.; Patil, F. S.; Kulikarni, U. G. Feasibility of Meristem Culture for Management of Banana Streak Virus (BSV) through Micropropagation. *Res. Crop* **2008,** *9*, 605–609.

Kim, J. A.; Cho, K.; Singh, R.; Jung, Y. H.; Jeong, S. H.; Kim, S. H.; Lee, J. E.; Cho, Y. S.; Agrawal, G. K.; Rakwal, R.; Tamogami, S.; Kersten, B.; Jeon, J. S.; An, G.; Jwa, N. S. Rice OsACDR1 (*Oryza sativa* Accelerated Cell Death and Resistance 1) is a Potential Positive Regulator of Fungal Disease Resistance. *Mol. Cells* **2009,** *30*, 431–439.

Kim, J. K.; Jang, I. C.; Wu, R.; Zuo, W. N.; Boston, R. S.; Lee, Y. H.; Ahn, I. P.; Nahm, B. H. Co-expression of a Modified Maize Ribosome-Inactivating Protein and a Rice Basic Chitinase Gene in Transgenic Rice Plants Confers Enhanced Resistance to Sheath Blight. *Transgenic Res.* **2003,** *12*, 475–484.

Kjemtrup, S.; Sampson, K. S.; Peele, C. G.; Nguyen, L. V.; Conkling, M. A. Gene Silencing from Plant DNA Carried by a Geminivirus. *Plant J.* **1998,** *14*, 91–100.

Krens, F. A.; Schaart, J. G.; Groenwold, R.; Walraven, A. E. J.; Hesselink, T.; Thissen, J. T. Performance and Long-Term Stability of the Barley Hordothionin Gene in Multiple Transgenic Apple Lines. *Transgenic Res.* **2011,** *20* (5), 1113–1123.

Kumar, R. R.; Kumar, M.; Nimmy, M. S.; Kumar, V.; Sinha, S.; Shamim, Md.; Dharamsheela. Diagnosis of Pulse Diseases and Biotechnological Approaches for their Management. In: *Diseases of Pulse Crops and their Management*; Biswas, S. K.; Kumar, S.; Chand, G., Eds.; Biotech Books: New Delhi, 2015; pp 519–542.

Kumar, S.; Khan, M. S.; Raj, S. K.; Sharma, A. K. Elimination of Mixed Infection of Cucumber Mosaic and Tomato Aspermy Virus from *Chrysanthemum morifolium* Ramaty *cv*. Pooja by Shoot Meristem Culture. *Sci. Hortic.* **2009,** *119* (2), 108–112.

Kumpatla, S. P.; Buyyarapu, R.; Abdurakhmonov, I. Y.; Mammadov, J. A. Genomics-Assisted Plant Breeding in the 21st Century: Technological Advances and Progress. *Plant Breed*, In: INTECH Open Access Publisher, 2012, ISBN: 978-953-307-932-5.

Lacombe, S.; Rougon-Cardoso, A.; Sherwood, E.; Peeters, N.; Dahlbeck, D.; Van Esse, H. P.; ... ; Jones, J. D. Interfamily Transfer of a Plant Pattern-recognition Receptor Confers Broad-spectrum Bacterial Resistance. *Nature Biotech.* **2010,** *28* (4), 365–369.

Lalithakumari, D. Protoplasts—A Biotechnological Tool for Plant Pathological Studies. *Indian Phytopathol.* **1996,** *49*, 199–212.

Lang, J. M.; Hamilton, J. P.; Diaz, M. G. Q.; Van Sluys, M. A.; Burgos, M. R. G.; Vera Cruz, C. M.; Buell, C. R.; Tisserat, N. A.; Leach, J. E. Genomics-Based Diagnostic Marker Development for *Xanthomonas oryzae* pv. *oryzae* and *X. oryzae* pv. *oryzicola*. *Plant Dis.* **2010,** *94*, 311–319.

Liu, J.; Deng, X. Regeneration and Analysis of Citrus Interspecific Mixoploid Hybrid Plants from Asymmetric Somatic Hybridization. *Euphytica* **2002,** *125* (1), 13–20.

Lizarraga, R.; Tovar, P.; Jayasinghe, U.; Dodds, J. Tissue Culture for Elimination of Pathogens. *Specialized Technology Document 3*; International Potato Center: Lima, Peru, 1986; p 22.

Lockhart, D. J.; Dong, H.; Byrne, M. C.; Follettie, M. T.; Gallo, M. V.; Chee, M. S.; Mittmann, M.; Wang, C.; Kobayashi, M.; Horton, H.; Brown, E. L. Expression Monitoring by Hybridisation to High Density Oligonucleotide Arrays. *Nat. Biotechnol.* **1996,** *14*, 1675–1680.

López, M. M.; Bertolini, E.; Olmos, A.; Caruso, P.; Gorris, M. T.; Llop, P.; Penyalver, R.; Cambra, M. Innovative Tools for Detection of Plant Pathogenic Viruses and Bacteria. *Int. Microbiol.* **2003,** *6*, 233–243.

López, M. M.; Llo, P.; Olmos, A.; Marco-Noales, E.; Cambra, M.; Bertolini, E. Are Molecular Tools Solving the Challenges Posed by Detection of Plant Pathogenic Bacteria and Viruses? *Curr. Issues Mol. Biol.* **2009,** *11*, 13–46.

Louws, F. J.; Rademaker, J. L. W.; Brujin, F. J. The three Ds of PCR-Based Genomic Analysis of Phytobacteria: Diversity, Detection, and Disease Diagnosis. *Annu. Rev. Phytopathol.* **1999,** *37*, 81–125.

Loy, A.; Lehner, A.; Lee, N.; Adamczyk, J.; Meier, H.; Ernst, J.; Schleifer, K. H.; Wagner, M. Oligonucleotide Microarray for 16S rRNA Gene Based Detection of All Recognized Lineages of Sulfate-Reducing Prokaryotes in the Environment. *Appl. Environ. Microbiol.* **2002,** *68*, 5064–5081.

Mackintosh, C. A.; Lewis, J.; Radmer, L. E.; Shin, S.; Heinen, S. J.; Smith, L. A.; Muehlbauer, G. J. Overexpression of Defense Response Genes in Transgenic Wheat Enhances Resistance to Fusarium Head Blight. *Plant Cell Rep.* **2007,** *26* (4), 479–488.

Malnoe, P.; Farinelli, L.; Collet, G. F.; Reust, W. Small-Scale Field Test with Transgenic Potato cv. Bintje to Test Resistance to Primary and Second Infections with *Potato virus Y*. *Plant Mol. Biol.* **1994,** *25*, 963–975.

Mandahar, C. L.; Paul Khurana, S. M. Role of Plant Biotechnology in Controlling Plant Diseases. In: *Pathological Problems of Economic Crop Plants and their Management*; Paul Khurana, S. M., Eds.; Scientific Publishers: Jodhpur, India, 1998, pp 637–647.

Martin, R. R.; James, D.; Levesque, C. A. Impacts of Molecular Diagnostic Technologies on Plant Disease Management. *Annu. Rev. Phytopathol.* **2000,** *38*, 207–239.

Meena, R. K.; Gour, K.; Patni, V. Production of Leaf Curl Virus—Free Chilli by Meristem Tip Culture. *Int. J. Pharm. Sci. Rev. Res.* **2014,** *25* (2), 67–71.

Mervat, M. M.; Far, E. Optimization of Growth Conditions during Sweet Potato Micro-propagation. *Afr. Potato Assoc. Conf. Proc.* **2007,** *7*, 204–211.

Milkus, B. N.; Avery, J. D.; Pinska, V. N. Elimination of Grapevine Viruses by Heat Treatment and Meristem Shoot Tip Culture. In: Proceedings of the 13th International Council of Virus and Virus-Like of Grapevine, Adelaide (Australia), 12–17 March 2000; pp 174.

Milošević, S.; Subotić, A.; Bulajić, A.; Đekić, I.; Jevremović, S.; Vučurović, A.; Krstić, B. Elimination of TSWV from *Impatiens hawkerii* Bull. and Regeneration of Virus-Free Plant. *Electron. J. Biotechnol.* **2011,** *14* (1), ISSN: 0717-3458.

Mishra, S.; Singh, D.; Tiwari, A. K.; Lal, M.; Rao, G. P. Elimination of Sugarcane Mosaic Virus and Sugarcane Streak Mosaic Virus by Tissue Culture. *Sugar Cane Int.* **2010,** *28* (3), 119–122.

Mondal, K. K.; Bhattacharya, R. C.; Koundal, K. R.; Chatterjee, S. C. Transgenic Indian Mustard (*Brassica juncea*) Expressing Tomato Glucanase Leads to Arrested Growth of *Alternaria brassicae*. *Plant Cell Rep.* **2007,** *26*, 247–252.

Morroni, M.; Thompson, J. R.; Tepfer, M. Twenty Years of Transgenic Plants Resistant to Cucumber Mosaic Virus. *Mol Plant Microbe Interact.* **2008,** *21*, 675–684.

Murad, A. M.; Noronha, E. F.; Miller R. N. Proteomic Analysis of *Metarhizium anisopliae* Secretion in the Presence of the Insect Pest *Callosobruchus maculatus*. *Microbiology* **2008,** *154* (12), 3766–3774.

Nakayashiki, H. RNA Silencing in Fungi: Mechanisms and Applications. *Fed. Eur. Biochem. Soc. Lett.* **2005,** *579*, 5950–5970.

Napoli, C.; Lemieux, C.; Jorgensen, R. Introduction of a Chimeric Chalcone Synthase Gene into Petunia Results in Reversible Co-suppression of Homologous Genes in Trans. *Plant Cell* **1990,** *2* (4), 279–289.

Nascimento, L. C.; Pio Ribeiro, G.; Willadino, L.; Andrade, G. P. Stock Indexing and Potato Virus Y Elimination from Potato Plants Cultivated In Vitro. *Sci. Agric.* **2003,** *60*, 525–530.

Ng, S. Y. C.; Thottappilly, G.; Rossel, H. W. Tissue Culture in Disease Elimination and Micropropagation. In: *Biotechnology: Enhancing Research on Tropical Crops in Africa*; Thottappilly, G., Monti, L. M., Mohan-Raj, D., Moore, A. W., Eds.; CTA/IITA Copublication, IITA: Ibadan, 1992; pp 171–182.

Ordax, M.; Marco-Noales, E.; López, M. M.; Biosca, E. G. Survival Strategy of *Erwinia amylovora* against Copper: Induction of the Viable but Nonculturable State. *Appl. Environ. Microbiol.* **2006,** *72*, 3482–3488.

Postman, J. D. Blueberry Scorch Carlavirus Eliminated from Infected Blueberry (*Vaccinium corymbosum*) by Heat Therapy and Apical Meristem Culture. *Plant Dis.* **1997,** *81*, 111.

Priou, S. Kit ELISA-NCM para la deteccion de Ralstonia solanacearum en papa: Instrucciones para el usuario. **2001**.

Ramgareeb, S.; Snyman, S. J.; Van Antwerpen, T.; Rutherford, R. S. Elimination of Virus and Rapid Propagation of Disease Free Sugarcane (*Saccharum* spp. Cultivar NCO 376) Using Apical Meristem Culture. *Plant Cell Tissue Org.* **2010,** *100*, 175–181.

Ratcliff, F.; Martin-Hernandez, A. M.; Baulcombe, D. C. Tobacco Rattle Virus as a Vector for Analysis of Gene Functions by Silencing. *Plant J.* **2001,** *25*, 237–245.

Rep, M. Small Proteins of Plant–Pathogenic Fungi Secreted during Host Colonization. *FEMS Microbiol. Lett.* **2005,** *253* (1), 19–27.

Rep, M.; van der Does, H. C.; Meijer, M. A Small, Cysteine-Rich Protein Secreted by *Fusarium oxysporum* during Colonization of Xylem Vessels Is Required for I-3-Mediated Resistance in Tomato. *Mol. Microbiol.* **2004,** *53* (5), 1373–1383.

Richmond, C. S.; Glasner, J. D.; Mau, R.; Jin, H.; Blattner, F. R. Genome-Wide Expression Profiling in *Escherichia coli* K-12. *Nucleic Acids Res.* **1999,** *27*, 3821–3835.

Ruiz, M. T.; Voinnet, O.; Baulcombe, D. C. Initiation and Maintenance of Virus-Induced Gene Silencing. *Plant Cell* **1998,** *10*, 937–946.

Salameh, A.; Buerstmayr, M.; Steiner, B.; Neumayer, A.; Lemmens, M.; Buerstmayr, H. Effects of Introgression of Two QTL for Fusarium Head Blight Resistance from Asian Spring Wheat by Marker-Assisted Backcrossing into European Winter Wheat on Fusarium Head Blight Resistance, Yield and Quality Traits. *Mol. Breed.* **2011,** *28*, 485–494.

Sastry, K. S.; Zitter, T. A. Management of Virus and Viroid Diseases of Crops in the Tropics. In: *Plant Virus and Viroid Diseases in the Tropics,* Springer: Netherlands; 2014, 149–480.

Schena, M.; Shalon, D.; Davis, R. W.; Brown, P. O. Quantitative Monitoring of Gene Expression Patterns with a Complementary DNA Microarray. *Science* **1995,** *270*, 467–470.

Schottens-Toma, I. M. J.; De Wit, P. J. G. M. Purification and Primary Structure of a Necrosis-Inducing Peptide from the Apoplastic Fluids of Tomato Infected with *Cladosporium fulvum* (syn. *Fulvia fulva*). *Physiol. Mol. Plant Pathol.* **1988,** *33* (1), 59–67.

Scofield, S. R.; Huang, L.; Brandt, A. S.; Gill, B. S. Development of a Virus-Induced Gene Silencing System for Hexaploid Wheat and Its Use in Functional Analysis of the lr21-Mediated Leaf Rust Resistance Pathway. *Plant Physiol.* **2005,** *138*, 2165–2173.

Sediva, J.; Novak, P.; Laxa, J.; Kanka, J. Micropropagation, Detection and Elimination of DMV in the Czech Collection of Dahlia. *Acta Hortic.* **2006,** *725*, 495–498.

Sekhwal, M. K.; Li, P.; Lam, I.; Wang, X.; Cloutier, S.; You, F. M. Disease Resistance Gene Analogs (RGAs) in Plants. *Int. J. Mol. Sci.* **2015,** *16* (8), 19248–19290.

Selvarajan, R.; Balasubramanian, V.; Sheeba, M. M.; Raj Mohan, R.; Mustaffa, M. M. Virus-Indexing Technology for Production of Quality Banana Planting Material: A Boon to the Tissue-Culture Industry and Banana Growers in India. *Acta Hortic.* **2011,** *897*, 463–469.

Sharma, R.; Bhardwaj, V.; Dalamu, D.; Kaushik, S. K.; Singh, B. P.; Sharma, S. et al.. Identification of Elite Potato Genotypes Possessing Multiple Disease Resistance Genes Through Molecular Approaches. *Scientia Horticulturae* **2014,** *179*, 204–211.

Sharma, T. R.; Das, A.; Thakur, S.; Jalali, B. Recent Understanding on Structure, Function and Evolution of Plant Disease Resistance Genes. *Proc. Indian Nat. Sci. Acad.* **2014,** *1*, 83–93.

Sharma, V. K.; Sanghera, G. S.; Kashyap, P. L.; Sharma, B. B.; Chandel, C. RNA interference: A novel tool for plant disease management. *African J. Biotech.*, **2013**. *12*(18).

Shelake, R. M.; Senthil, K. T.; Angappan, K. PCR Detection of Banana Bunchy Top Virus (BBTV) at Tissue Culture Level for the Production of Virus-Free Planting Materials. *Int. Res. J. Biol. Sci.* **2013,** *2* (6), 22–26.

Sorrells, M. E. Application of New Knowledge, Technologies, and Strategies to Wheat Improvement. *Euphytica* **2007,** *157*, 299–306.

Sridevi, G.; Parameswari, C.; Sabapathi, N.; Raghupathy, V.; Veluthambi, K. Combined Expression of Chitinase and β-1,3-Glucanase Genes in Indica Rice (*Oryza sativa* L.) Enhances Resistance against *Rhizoctonia solani*. *Plant Sci.* **2008,** *175*, 283–290.

Sundaram, R. M.; Vishnupriya, M. R.; Laham, G. S.; Shobha Rani, N.; Srinivas Rao, P.; Balachandaran, S. M.; Asho Reddy, G.; Sarma, N. P.; Shonti, R. V. Introduction of Bacterial Blight Resistance into Triguna, a High Yielding, Mid-Early Duration Rice Variety. *Biotechnol. J.* **2009,** *4*, 400–407.

Szczerbakowa, A.; Maciejewska, U.; Zimnoch-Guzowska, E.; Wielgat, B. Somatic Hybrids *Solanum nigrum* (+) *S. tuberosum*: Morphological Assessment and Verification of Hybridity. *Plant Cell Rep.* **2003,** *21*, 577–584.

Terakawa, T.; Takaya, N.; Horiuchi, H.; Koike, M.; Takagi, M. A Fungal Chitinase Gene from *Rhizopus oligosporus* Confers Antifungal Activity to Transgenic Tobacco. *Plant Cell Rep.* **1997,** *16*, 439–443.

Terrada, E.; Kerschbaumer, R. J.; Giunta, G.; Galeffi, P.; Himmler, G.; Cambra, M. Fully "Recombinant Enzyme-Linked Immunosorbent Assays" Using Genetically Engineered Single-Chain Antibody Fusion Proteins for Detection of *Citrus tristeza* Virus. *Phytopathology* **2000,** *90*, 1337–1344.

Tomar, U. K.; Dantu, P. Protoplast Culture and Somatic Hybridization. *Cellular and Biochemical Science Book*; I. K. International Pvt Ltd.: New Delhi, 2010; pp 876–891.

Torisky, R. S.; Kovacs, L.; Avdiushko, S.; Newman, J. D.; Hunt, A. G.; Collins, G. B. Development of a Binary Vector System for Plant Transformation Based on Supervirulent *Agrobacterium tumefaciens* Strain Chry5. *Plant Cell Rep.* **1997,** *17*, 102–108.

Tripathi, L.; Mwaka, H.; Tripathi, J. N.; Tushemereirwe, W. K. Expression of Sweet Pepper Hrap Gene in Banana Enhances Resistance to *Xanthomonas campestris* pv. *musacearum*. *Mol. Plant Pathol.* **2010,** *11* (6), 721–731.

Turnage, M. A.; Muangsan, N.; Peele, C. G.; Robertson, D. Geminivirus-Based Vectors for Gene Silencing in Arabidopsis. *Plant J.* **2002,** *30* (1), 107–114.

Varshney, R. K.; Kudapa, H.; Pazhamala, L.; Chitikineni, A.; Thudi, M.; Bohra, A.; Gaur, P. M.; Janila, P.; Fikre, A.; Kimurto, P.; Ellis, N. Translational Genomics in Agriculture: Some Examples in Grain Legumes. *Crit. Rev. Plant Sci.* **2015,** *34* (1–3), 169–194.

Varshney, R. K.; Mohan, S. M.; Gaur, P. M.; Chamarthi, S. K.; Singh, V. K.; Srinivasan, S.; Swapna, N.; Sharma, M.; Singh, S.; Kaur, L.; Pande, S. Marker-Assisted Backcrossing to Introgress Resistance to Fusarium Wilt (FW) *Race 1* and Ascochyta Blight (AB) in C 214, an Elite Cultivar of Chickpea. *Plant Gene* **2013,** *7* (1), 1–11.

Wally, O. S.; Punja, Z. K. Carrot (*Daucus carota* L.). *Agrobacterium Protocols: Volume 2*, **2015,** 59–66.

Wally, O.; Jayaraj, J.; Punja, Z. Comparative Resistance to Foliar Fungal Pathogens in Transgenic Carrot Plants Expressing Genes Encoding for Chitinase, b-1,3-Glucanase and Peroxidise. *Eur. J. Plant Pathol.* **2009,** *123*, 331–342.

Wally, O.; Punja, Z. K. Genetic Engineering for Increasing Fungal and Bacterial Disease Resistance in Crop Plants. *GM Crops* **2010,** *1*, 199–206.

Wang, L.; Webster, D. E.; Campbell, A. E.; Dry, L. B.; Wesselingh, S. L.; Coppel, R. L. Immunogenicity of *Plasmodium yoelii* Merozoite Surface Protein 4/5 Produced Intransgenic Plants. *Int. J. Parasitol.* **2008,** *38*, 103–110.

Wani, S. H.; Sanghera, G. S.; Singh, N. B. Biotechnology and Plant Disease Control-Role of RNA Interference. *Am. J. Plant Sci.* **2010,** *1* (2), 55.

Wasswa, P.; Alicai, T.; Mukasa, S. B. Optimisation of In Vitro Techniques for Cassava Brown Streak Virus Elimination from Infected Cassava Clones. *Afr. Crop Sci. J.* **2010,** *18*, 235–241.

Waterhouse, P. M.; Graham, M. W.; Wang, M. B. Virus Resistance and Gene Silencing in Plants can be Induced by Simultaneous Expression of Sense and Antisense RNA. *Proc. Nat. Acad. Sci. USA* **1998,** *95*, 13959–13964.

Waterworth, P.; Kahn, R. P. Thermotherapy and Aseptic Bud Culture of Sugarcane to Facilitate the Exchange of Germplasm and Passage through Quarantine. *Plant Dis Rep*. **1978,** *62*, 72–776.

Xu, Y.; Xie, C.; Wan, J.; He, Z.; Prasanna, B. M. Marker-Assisted Selection in Cereals: Platforms, Strategies and Examples. *Cereal Genomics II*; Springer: Netherlands, 2013; pp 375–411.

Yamamoto, T.; Iketani, H.; Ieki, H.; Nishizawa, Y.; Notsuka, K.; Hibi, T.; Hayashi, T.; Matsuta, N. Transgenic Grapevine Plants Expressing a Rice Chitinase with Enhanced Resistance to Fungal Pathogens. *Plant Cell Rep*. **2000,** *19*, 639–646.

Yoshikawa, N.; Saitou, Y.; Kitajima, A.; Chida, T.; Sasaki, N.; Isogai, M. Interference of Long Distance Movement of Grapevine Berry Inner Necrosis Virus in Transgenic Plants Expressing a Defective Movement Protein of Apple Chlorotic Leaf Spot Virus. *Phytopathology* **2006,** *96*, 378–385.

Zhang, C.; Ghabrial, S. A. Development of Bean Pod Mottle Virus-Based Vectors for Stable Protein Expression and Sequence-Specific Virus-Induced Gene Silencing in Soybean. *Virology* **2006,** *344*, 401–411.

Zhang, C.; Yang, C.; Whitham, S. A.; Hill, J. H. Development and Use of an Efficient DNA-based Viral Gene Silencing Vector for Soybean. *Mol. Plant-Microbe Interact.* **2009,** *22*, 123–131.

Zhao, B.; Lin, X.; Poland, J.; Trick, H.; Leach, J.; Hulbert, S. A Maize Resistance Gene Functions against Bacterial Streak Disease in Rice. *Proc. Nat. Acad. Sci.* **2005,** *102* (43), 15383–15388.

Zimmerman, U.; Scheurich, P. High Frequency Fusion of Plant Protoplasts by Electric Fields. *Planta* **1981,** *151*, 26–32.

Zorn, H.; Peters, T.; Nimtz, M.; Berger, R. G. The Secretome of *Pleurotus sapidus*, *Proteomics*, **2005,** *5* (18), 4832–4838.

INDEX

R

S

T

U

V

Printed and bound by CPI Group (UK) Ltd, Croydon, CR0 4YY

23/10/2024

01777704-0018